妊娠和哺乳母猪

The Gestating and Lactating Sow

［加］Chantal Farmer 主编

李新建 殷跃帮 李 平 主译

中国农业大学出版社
·北京·

内 容 简 介

本书主要介绍了母猪营养、生理及管理的最新知识,如后备母猪最佳繁殖性能的调节、妊娠和哺乳母猪饲喂最新技术策略以及妊娠期母猪群体饲喂系统管理技术,人与动物相互作用对母猪福利和性能的影响及最新管理策略。本书涵盖的技术知识将帮助畜牧科学家、营养学家、兽医和养猪生产者学习关于影响母猪生产的相关和当前技术的最新信息。

图书在版编目(CIP)数据

妊娠和哺乳母猪 /(加)尚塔尔·法莫(Chantal Farmer)主编;李新建,殷跃帮,李平主译. —北京:中国农业大学出版社,2018.2(2021.1 重印)
书名原文:The Gestating and Lactating Sow
ISBN 978-7-5655-1797-6

Ⅰ.①妊… Ⅱ.①尚…②李…③殷…④李… Ⅲ.①母猪-饲养管理 Ⅳ.①S828.9

中国版本图书馆 CIP 数据核字(2017)第 330441 号

书　　名	妊娠和哺乳母猪	
作　　者	Chantal Farmer 主编　　李新建　殷跃帮　李平　主译	
策划编辑	梁爱荣　林孝栋	责任编辑　梁爱荣
封面设计	郑　川	
出版发行	中国农业大学出版社	
社　　址	北京市海淀区圆明园西路 2 号	邮政编码　100193
电　　话	发行部 010-62818525,8625	读者服务部 010-62732336
	编辑部 010-62732617,2618	出 版 部 010-62733440
网　　址	http://www.caupress.cn	E-mail　cbsszs@cau.edu.cn
经　　销	新华书店	
印　　刷	涿州市星河印刷有限公司	
版　　次	2018 年 4 月第 1 版　2021 年 1 月第 2 次印刷	
规　　格	787×1 092　16 开本　26 印张　495 千字	
定　　价	158.00 元	

图书如有质量问题本社发行部负责调换

中文简体版本翻译自 Chantal Farmer 主编的"*The Gestating and Lactating Sow*"。
Translation from the English language edition:
The original English language work has been published by Wageningen Academic Publishers.
Copyright © 2015 Wageningen Academic Publishers, The Netherlands.
All rights reserved.

Translation rights was arranged with the permission of the proprietor.
中文简体版本由 Wageningen Academic Publishers 授权中国农业大学出版社有限公司专有权利在中国大陆出版发行。

No part of this book may be reproduced or transmitted in any form or by any means, electronic or mechanical, including photocopying, recording or any information storage and retrieval system, without permission, in writing, from the proprietor.

本书任何部分之文字及图片，如未获得出版者之书面同意不得以任何方式抄袭、节录或翻译。

著作权合同登记图字：01-2018-3255

译者名单

主　译　李新建（河南农业大学）

　　　　　殷跃帮（Erasmus University Rotterdam）

　　　　　李　平（广东省农业科学院动物科学研究所）

参　译　徐　磊（西北农林科技大学）

　　　　　乔瑞敏（河南农业大学）

　　　　　李平华（南京农业大学）

　　　　　于太永（西北农林科技大学）

序

中国是世界上养猪历史最长、数量最多、消费量最大的国家。猪肉占我国人均肉类消费的60%以上，2017年我国出栏商品猪6.88亿头，占世界总量的54.4%。近年来，我国养猪生产水平显著提升，生产效率显著增强。但与新时代养猪产业转型升级、与欧美先进国家的养猪生产水平比较，仍有不少差距，尤其是在动物福利、母猪精准营养、母猪生产环境等新技术领域相对落后。现实生产中，营养和管理不善，导致母猪生产性能退化，尤其是妊娠母猪及哺乳母猪管理的不科学不精准，造成母猪使用年限变短、疾病困扰、淘汰率增加，严重影响了养猪生产水平及生产效益。因此，妊娠与哺乳母猪的精准营养与科学管理是进一步提升我国养猪生产效益的重要理论与实践需求。

欧美的一些养猪先进水平国家从20世纪80年代开始就关注母猪的营养、健康与福利，在妊娠与哺乳母猪的各阶段对母猪的精准营养与科学管理的关键技术均取得了很大进展，由河南农业大学李新建博士等翻译的《妊娠和哺乳母猪》一书，详细介绍了妊娠与哺乳母猪的营养、生理、行为及管理的最新知识，对我国现代养猪科研工作者及养猪生产一线的管理及从业人员具有很好的参考价值。希望本书的出版能对我国养猪企业生产水平与生产效益的提高、对我国新时代养猪产业的转型升级提供有力借鉴。

是为序。

中国畜牧兽医学会理事长：
中国科学院院士：
国家畜禽遗传资源委员会副主任：
江西农业大学教授：

致我生命中的男人……

——致我的爸爸 Jacques,在我研究生几年期间,给予我无微不至和坚定的支持。没有他,就不能成就现在的我。

——致我亲爱的丈夫 Roger,他鼓励我,并在完成此书的每一步献出他的热忱。没有他,我甚至可能还未开始写这本书。

——致我的儿子 Vincent,他相信自己母亲的梦想,并代表世间万物的延续,正如本书中的母猪为其后代的哺乳一样。

我爱你们,并且感谢你们……

目　　录

1. 调节后备母猪体况以获得最佳繁殖性能 ································· 1
　D. W. Rozeboom
　1.1　引言 ··· 1
　1.2　后备母猪在首次配种前的身体状况和发情活动 ······················· 3
　1.3　后备母猪在首次配种后体况和繁殖表现 ····························· 4
　1.4　结论 ··· 8
　参考文献 ··· 10

2. 妊娠早期的饲养管理以获得最佳繁殖性能 ······························· 14
　P. Langendijk
　2.1　引言 ··· 14
　2.2　早期妊娠问题：妊娠确立和胚胎死亡率 ······························ 15
　2.3　配种前营养对妊娠前期黄体功能的影响 ······························ 17
　2.4　配种后营养、胚胎损失和妊娠维持 ·································· 17
　2.5　激素干预妊娠维持的作用 ·· 24
　2.6　结论 ··· 25
　参考文献 ··· 25

3. 妊娠母猪的群养 ··· 32
　H. A. M. Spoolder and H. M. Vermeer
　3.1　引言 ··· 33
　3.2　群养系统的母猪行为表现 ·· 34
　3.3　群养系统类型 ·· 43
　3.4　结论 ··· 49
　参考文献 ··· 50

4. 乳腺发育 ··· 56
　C. Farmer and W. L. Hurley
　4.1　引言 ··· 56

4.2	乳腺个体发育	57
4.3	乳腺发育的控制	60
4.4	排乳的作用	62
4.5	营养对乳腺发育的影响	64
4.6	乳腺退化	66
4.7	影响乳腺发育的管理策略	68
4.8	结论	69
参考文献		70

5. 妊娠期高纤维饲喂 ... 76
M. C. Meunier-Salaün and J. E. Bolhuis

5.1	引言	76
5.2	日粮纤维:定义、来源和特征	77
5.3	日粮纤维和饱腹感:新陈代谢和生理影响	79
5.4	日粮纤维对行为活动的影响	80
5.5	日粮纤维对生产性能的影响	86
5.6	结论	90
致谢		91
参考文献		91

6. 母猪氨基酸和能量供给 ... 98
N. L. Trottier, L. J. Johnston and C. F. M. de Lange

6.1	引言	99
6.2	总体原则	99
6.3	能量供给	100
6.4	氨基酸营养	107
6.5	饲养管理	116
6.6	结论	118
参考文献		119

7. 围产期母猪饲养管理 ... 124
P. K. Theil

7.1	引言	124
7.2	围产期的重要性	125
7.3	现代高产母猪围产期饲养方法	134
7.4	最新进展	137
7.5	未来展望	141
参考文献		142

8. 初乳和常乳生产 · 147
H. Quesnel, C. Farmer and P. K. Theil

    8.1   引言 · 147
    8.2   初乳、过渡乳和常乳的定义 · 148
    8.3   初乳生产 · 149
    8.4   产奶量 · 152
    8.5   结论 · 158
    参考文献 · 159

9. 母猪初乳和常乳的组成 · 166
W. L. Hurley

    9.1   引言 · 166
    9.2   研究方法 · 169
    9.3   物理化学性质 · 170
    9.4   水和总固形物 · 170
    9.5   碳水化合物 · 173
    9.6   脂肪 · 174
    9.7   蛋白质 · 177
    9.8   能量 · 181
    9.9   矿物质 · 181
    9.10 维生素 · 183
    9.11 细胞 · 186
    9.12 生物活性成分 · 187
    9.13 生理状态的影响 · 189
    9.14 结论 · 190
    参考文献 · 190

10. 母猪分娩期间和泌乳早期的圈舍、管理和环境 · 202
O. A. T. Peltoniemi and C. Oliviero

    10.1 引言 · 202
    10.2 分娩生理机能 · 203
    10.3 行为和活动 · 204
    10.4 成功分娩 · 204
    10.5 分娩环境 · 205
    10.6 产程以及其对母猪繁殖能力的影响 · 207
    10.7 泌乳早期的环境影响 · 209
    10.8 分娩时的体况、脂肪代谢和肠道功能 · 210

10.9　分娩预测及管理技术 ………………………………………………… 215
　　10.10　结论 ………………………………………………………………… 217
　　参考文献 …………………………………………………………………… 217

11. 仔猪死亡：原因和预防 ……………………………………………………… 223
　　S. A. Edwards and E. M. Baxter
　　11.1　引言 …………………………………………………………………… 223
　　11.2　死亡原因 ……………………………………………………………… 224
　　11.3　预防死亡 ……………………………………………………………… 233
　　11.4　结论 …………………………………………………………………… 236
　　参考文献 …………………………………………………………………… 237

12. 母猪与人类相互作用对母猪内分泌、行为以及繁殖性能的影响 ………… 246
　　A. Prunier and C. Tallet
　　12.1　引言 …………………………………………………………………… 246
　　12.2　哺乳仔猪的处理 ……………………………………………………… 247
　　12.3　后备母猪和经产母猪的处理 ………………………………………… 249
　　12.4　人与动物相互作用对繁殖性能的影响 ……………………………… 252
　　12.5　更好地了解动物的表达信号，改善人与动物之间的相互作用 …… 254
　　12.6　结论 …………………………………………………………………… 255
　　参考文献 …………………………………………………………………… 256

13. 哺乳行为 ………………………………………………………………………… 261
　　M. Špinka and G. Illmann
　　13.1　引言 …………………………………………………………………… 261
　　13.2　分娩后哺乳 …………………………………………………………… 262
　　13.3　正常哺乳期间的哺乳行为 …………………………………………… 264
　　13.4　哺乳过程中哺乳/吮乳行为的功能 …………………………………… 270
　　13.5　结论 …………………………………………………………………… 274
　　参考文献 …………………………………………………………………… 274

14. 乳腺血流和营养摄取 …………………………………………………………… 280
　　C. Farmer, N. L. Trottier and J. Y. Dourmad
　　14.1　引言 …………………………………………………………………… 280
　　14.2　血流 …………………………………………………………………… 281
　　14.3　营养摄入 ……………………………………………………………… 284
　　14.4　结论 …………………………………………………………………… 290
　　参考文献 …………………………………………………………………… 291

15. 仔猪肠道和肠道相关淋巴组织发育:母体环境的作用 …………………… 295
I. Le Huërou-Luron and S. Ferret-Bernard

- 15.1　引言 ………………………………………………………………………… 295
- 15.2　肠道和肠相关淋巴组织的发育 ……………………………………………… 296
- 15.3　母体环境对调节肠和肠相关淋巴组织发育的重要性 ……………………… 303
- 15.4　结论 ………………………………………………………………………… 307
- 参考文献 …………………………………………………………………………… 307

16. 各种来源的脂肪饲喂母猪:对母猪和仔猪的免疫状态和性能的影响 ……… 315
V. Bontempo and X. R. Jiang

- 16.1　引言 ………………………………………………………………………… 315
- 16.2　母猪饲料中脂肪和油的使用 ………………………………………………… 316
- 16.3　ω-脂肪酸（n-3 与 n-6 FA） …………………………………………… 316
- 16.4　共轭亚油酸 ………………………………………………………………… 323
- 16.5　结论 ………………………………………………………………………… 327
- 参考文献 …………………………………………………………………………… 328

17. 优化哺乳期和断奶母猪繁殖生理及性能的最佳技术 ………………………… 334
N. M. Soede and B. Kemp

- 17.1　引言 ………………………………………………………………………… 334
- 17.2　繁殖生理 …………………………………………………………………… 335
- 17.3　哺乳期:在断奶时优化卵泡发育 …………………………………………… 337
- 17.4　断奶至发情间隔:支持卵泡发育 …………………………………………… 345
- 17.5　发情和授精:确保最大受精 ………………………………………………… 348
- 17.6　结论 ………………………………………………………………………… 352
- 参考文献 …………………………………………………………………………… 354

18. 母猪健康 ………………………………………………………………………… 363
R. M. Friendship and T. L. O'Sullivan

- 18.1　引言 ………………………………………………………………………… 363
- 18.2　母猪群体的疾病管理影响整个畜群 ………………………………………… 364
- 18.3　母猪健康管理 ……………………………………………………………… 365
- 18.4　母猪疾病概述 ……………………………………………………………… 366
- 18.5　影响母猪健康的传染病 …………………………………………………… 367
- 18.6　结论 ………………………………………………………………………… 372
- 参考文献 …………………………………………………………………………… 373

19. 母猪利用年限 ………………………………………………… 374
J. A. Calderón Díaz, M. T. Nikkilä and K. Stalder
19.1 引言 …………………………………………………………… 374
19.2 用于评估母猪利用年限的常用指标 …………………………… 375
19.3 母猪利用年限的经济重要性：对母猪群的影响 ……………… 377
19.4 为什么母猪通常离开繁殖群？ ………………………………… 379
19.5 提高母猪利用年限 ……………………………………………… 382
19.6 猪舍对利用年限的影响 ………………………………………… 386
19.7 疾病对母猪利用年限的影响 …………………………………… 389
19.8 季节对母猪利用年限的影响 …………………………………… 389
19.9 人员/管理或饲养管理对母猪利用年限的影响 ……………… 390
19.10　结论 …………………………………………………………… 391
参考文献 ……………………………………………………………… 391

1. 调节后备母猪体况以获得最佳繁殖性能

D. W. Rozeboom

Department of Animal Science, Michigan State University, 474 S. Shaw Lane, 2209 Anthony Hall, East Lansing, MI 48824, USA; rozeboom@msu.edu

摘要：后备母猪的发育对维持生产群至关重要。后备母猪妊娠初期时的发育状况非常重要，因为这会影响后备母猪是否能够正常繁殖和繁殖寿命。研究表明，母猪首次配种的背膘厚度与其利用年限相关性不强。另外，在首配时，后备母猪的理想体重范围为 135~150 kg，并且这是后备母猪管理方案中的一个重要目标。虽然普遍认为后备母猪的体况是最重要的，但我们对母猪体蛋白量或代谢速度与繁殖之间的关系了解还比较少。如何调整后备母猪体况，提高繁殖性能等方面依然具有挑战性。

关键词：母猪，生猪，发育，利用年限

1.1 引言

通过调节后备母猪的体况以获得最佳繁殖性能可以更好地表述为通过调节后备母猪体况以获得最长的繁殖寿命和较高的经济效益。这两个陈述有不同的含义。较高的经济效益主要是在最短的时间内和最少的经济成本下获得最多的后代仔猪，而获得最佳繁殖性能，"最佳"这个词是主观的，并不考虑经济成本。但普遍认为，在某种程度上，后备母猪在发育过程中的"体况调节"影响其繁殖寿命，衡量指标为最大产仔数和繁殖周期中最少的非生产天数。因此，后备母猪的饲养目标是有效的产出，因此衍生出以下问题：是否可以通过调节后备母猪体况从而带来最大数量的断奶仔猪，以及延长母猪繁殖寿命？对于母猪群，每年最佳的更新率为30%，繁殖母猪胎次最好不超过 8 胎次，其提供断奶仔猪 88 头左右为佳。

后备母猪的体况是对体重或身体组成成分的描述或评估。观察评估各种组织（脂肪、肌肉、牙齿、骨骼和结缔组织、神经）和内容物（血液、淋巴、消化液、尿和气体）的数量。实际上关于后备母猪发育，体况主要是指脂肪、蛋白质和骨骼的含量，其增长速度也是衡量体况的重要参数。通常将体况作为衡量青年后备母猪是否为最佳的繁殖寿命做好准备的一个指标。

关于体况，生产者、生产管理者、兽医、遗传学家和营养学家均把其作为后备母猪开始初配行为的身体发育的"目标"指标，以确保后备母猪首次配种和妊娠之前得到充分发育，以防后备母猪配种率低、窝产仔数少、断奶后繁殖性能不一致，以及从猪群中过早被淘汰。有些人认为，后备母猪体况调节也是保证其能够持续、均衡地繁殖大量仔猪而采取的措施。当然，体况只是后备母猪发育的一个方面。后备母猪的健康和外部环境的管理也是影响其繁殖性能的重要因素。

研究表明，后备母猪理想成熟体况（年龄和身体组成共同影响）是由基因型决定的。在养猪研究文献中，常常把年龄、体重、生长速度和背膘厚度作为考察后备母猪发育情况的指标。有时可能把眼肌面积也作为一个参数。许多相关研究资料已经在期刊、杂志出版。值得注意的是，Gill（2007）、Johnston 和 Smits（2007）以及 Bortolozzo 等（2009）在过去 5~7 年中全面总结了体况目标与繁殖成功之间的关系。

身体组成和组织增长率存在遗传变异。现代的后备母猪选择，首先来源于产仔数较大的（每窝 12~13 头活仔猪），以及具有超常的母乳产量（窝乳产量为 2.3~2.7 kg/天）的母系后代。这些母猪通常是杂交的，性格温顺，白色，来自不同的遗传背景，包括来自于杂交群种猪公司以及纯繁种猪生产者的品系和杂交系。在 20 世纪的最后 10 年，市场对精瘦肉需求增加，导致母系猪也逐渐变瘦，并且食欲不佳。目前普遍认为这是一个不利的情况，过去 10 年中，通过采用选育及分子遗传技术，对母系猪进行了选育，使其具有更多脂肪、更大的采食量、更少的肌肉量以及更多的难以测量的优秀特征。

身体组成（脂肪、瘦肉以及骨骼的绝对增长和相对增长）和繁殖性状的基因型差异已经被确认（Johnson 等，2008；Knauer 等，2010；Rozeboom，1999）。大部分种猪生产者认为，每个母猪基因型都会对应独特的身体状况。在此，不再深入讨论这些差异。其讨论是基于体况和基因型相关性分析的结果。

Pinilla 等（2013）表明，最新的母猪正常的生产指标是指第一胎和第二胎总产仔数达 30 头，在 5.2 胎次的繁殖年限中总产 78 头仔猪。希望后备母猪群中：29~31 周龄的母猪超过 60% 被挑选，至 33 周龄时，有超过 90% 的母猪被挑选。操作上而言，后备母猪发育目标相当容易测量和管理。在过去 5 年中，对于后备母猪发育而言，最一致的建议是：

• 在首次配种前与发情活动相关的指标：
—饲养期间的生长速度不低于 550 g/天；
—饲养期间的生长速度不超过 850 g/天；
—在发情期刺激 6 周内有 90% 的后备母猪出现发情；
—记录所有初情不主动配种母猪。

- 关于首次配种的关键指标：
  —体重 135~170 kg；
  —220~270 日龄；
  —饲养期生长速度不高于 850 g/天；
  —背膘厚度为 12~18 mm；
  —体况评分为 3~3.5。

在本书中，对如何通过后备母猪体况调节达到最佳繁殖性能的目标进行了重点阐述。但是过去 10 年，这些目标和重要性有何改变？它们是否已经更有效地最大限度地提高繁殖年限并减少后备猪培育成本？它们的预测能力（可重复实用价值）有改善吗？这些问题需要进一步研究。

1.2 后备母猪在首次配种前的身体状况和发情活动

1991 年，Beltranena 等使用二次曲线描述了在发情期终身生长率与年龄之间的关系。研究表明，日增重由 0.4 kg/天增加到 0.53 kg/天，首次发情日龄随着日增重的增加而降低；日增重从 0.53 kg/天增加到 0.6 kg/天时，首次发情年龄不受影响；但随着日增重超过 0.6 kg/天时，首次发情年龄会随着日增重的增加而增加。因此，较高或较低的增长速度都会使母猪的初情期延迟。Foxcroft 等（2005）称这种发情期生长速率与年龄的曲线反应是极端增长率与初情期边际延迟相关的趋势。Rozeboom 等（1995）报道，通过对后备母猪采取自由采食和大约 120 日龄时公猪诱情使初情期出现，发现初情与身体组织增长的具体比例无关。

但近 10 年来，关于初情期和生长速度的了解进一步深入。Bortolozzo 等（2009）指出，在饲养期间，经历生长速度较快（>700 g/天）的后备母猪较早达到初情期，并且在发情期发情和繁殖周期间不发情情况的发生率较低。他们得出结论，非常快的生长速度（接近 800 g/天）并不会延缓初情期的出现。其他学者也支持这一观点（Amaral Filha 等，2009；Kummer 等，2009；Patterson 等，2010）。

总体而言，最近的研究一致认为，生长速度与首次发情年龄之间几乎没有关系。据推测脂肪和蛋白质积累速率可能影响初情期的发生，这促进了研究生长速度与初情期年龄之间关系的兴趣。最近一些研究和早期的研究结果相一致（Beltranena 等，1993；Patterson 等，2002；Rozeboom 等，1995），这些研究已经肯定了后备母猪的初情期出现不是由特定的身体组成或者特定的脂肪或蛋白质沉积速度所控制的。仍然存在提高将后备母猪成功地转入繁殖母猪群的效率的机会。Gibson 和 Jackson（2012）报道显示，约 7.9% 的后备母猪从未发情。Spörke（2007）报道，每 1.5 头断奶后备母猪中，有 1 头能够成功地繁殖第一窝。因此，研究表明，后备母猪初情期的体况对未来繁殖更加重要。

1.3 后备母猪在首次配种后体况和繁殖表现

母猪体况和终身繁殖性能间的关系已经被广泛研究（Kirkwood 等，1988；Newton 和 Mahan，1993；Rozeboom 等，1996；Young 等，1990）。结论有所不同，但首次配种的身体组成被描述为与随后的繁殖不相关。研究者推荐的第一个育种目标为体重，这是一个安全的推荐，因为体重代表所有的身体组织储存。在最近的研究中，首次配种时的体蛋白质储存的充足性比脂肪沉积更受关注。然而，通常使用基于体重和背膘厚度的预测方程来估计体蛋白质储存。在一些情况下，眼肌面积参数，也用于估算体蛋白质总量。生长速度已被研究作为评价母猪代谢状态的重要参数。

1.3.1 体重

Foxcroft 和 Aherne(2001)研究表明，后备猪群管理必须实施满足繁殖年龄、体重和脂肪目标的营养计划，提出体重125 kg和背膘厚15 mm的繁殖目标。2002年，Foxcroft改变了这一建议，并指出，135 kg作为繁殖体重优于125 kg。这种变化是基于Clowes等(2003a，b)的研究成果，其中包括初产母猪体重在哺乳期下降。研究者基于断奶时身体组成和断奶后卵巢功能之间的关系，根据分娩时的体重和妊娠期体重状况，从而建议首次配种的最小体重为135 kg。Williams 等(2005)的研究结果与之一致，显示出体重为135～150 kg的后备母猪在三胎中可以生产最大数量的仔猪。

Foxcroft(2002)和Williams等(2005)的研究结果，已经成为现代养猪生产广为采纳的目标，得到其他研究者的支持(Amaral Filha等，2009)。Hoving 等(2010)也认为后备猪体重在首次繁殖十分重要，表明当第一次受精时后备母猪体重较轻，其在第二胎的繁殖表现较差(两组平均体重分别为 124 kg 和 145 kg)。Bussières(2013)报道了第一次配种的体重与终生总产仔数之间的关系，与Williams等(2005)研究结果非常相似。Bussières研究组发现，体重145～160 kg的猪终生总产仔数(2013；图1.1)，明显高于最轻(115～130 kg)和最重(175～190 kg)体重组，但与130～145 kg或160～175 kg组无差异。对于后备母猪群体，第一次配种其体重的建议目标范围是130～175 kg。

大量研究发现，第一次繁殖时体重较大的后备母猪更可能发生跛行(Amaral Filha 等，2009；Kummer 等，2006；Patterson 等，2010；Williams 等，2005)。最常提到的体重的关键点约为170 kg。Lyvers-Peffer 等(2003)指出，对于首次配种时体重较大的后备猪，大部分淘汰是因为跛行，而淘汰主要发生在第二胎前。Quinn(2013)也建议后备母猪选育者应将其从65 kg的生长-育肥群中挑出，然后从那时起，饲喂后备猪培育饲料。跛行和蹄甲损伤的减少归因于在后备猪培育饲

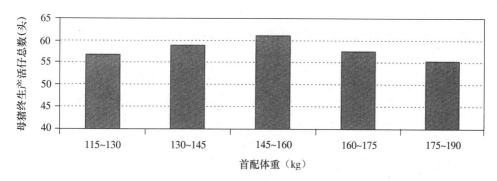

图1.1 母猪首配体重和终生总产仔数之间的关系（Bussières，2013）。

料中使用锌、铜和锰添加剂。De Koning 等（2013）报道了在10~26 周龄进行自由采食的后备猪，与同一生长期限制饲料摄入量（自由采食的80%）的后备猪相比，在26 周龄时屠宰时其关节变形增加，据说这也是骨质发育的重要参数。

相对于骨骼发育来说，生物学家几乎没有可用的经验性措施来评估这一点。通常提到增强骨矿化是在运往屠宰场的商业育肥场而不是针对选留的后备母猪。Rozeboom（2006）报道，为了最大骨矿化，对钙和磷的要求要高出 0.1 个百分点。目前，对生产者继续提出的建议是基于 Nimmo 等（1981）的研究。然而，还有另外三个研究组在同一时间研究了生长猪的骨矿化，他们的工作并不支持类似的结论。迄今为止，还需要对后备母猪骨骼发育的矿物质需求进行更多的研究。

1.3.2 生长速度

关于影响后备猪发育、体况、繁殖力和利用年限的因素研究中，大部分研究者主要聚焦在体重指标上，但同时，一部分研究者也侧重于生长速度或体组织增加的速度。Tummaruk 等（2001）报道，饲养期间生长速度太慢，不利于1~5 胎次的窝产仔数和断奶至发情间隔。Johnston 等（2007）研究发现，后备母猪生长速度过快以及首配体重过重都会使其一生中生产的仔猪数和繁殖寿命减少。Bortolozzo 等（2009）报道，重量超过 150 kg 的后备母猪没有优势。他们断言，在首次繁殖（>150~170 kg）之前，体重增加过快的后备母猪，其运动障碍发生率增加。Kummer 等（2009）报道，生长速度对第一胎繁殖表现（妊娠率、排卵率、总胚胎数、活胚胎和胚胎存活率）无影响。Amaral Filha 等（2010）指出，应该刺激快速增长的后备猪更早的初情期（150~170 日龄），并在185~210 日龄配种，以保证在配种时不会变得太重，以及后期繁殖力下降。最后，Knauer 等（2011）报道了生长速度达到 114 kg 和第一胎分娩的可能性之间存在遗传正相关性（$r=0.52$）；然而，生长速度非常快的后备母猪其成功分娩第一胎可能性降低。

因为首次配种的体重经常与饲养期间的生长速度混淆，分开考虑体重和生长速度对后备母猪的繁殖性能的影响是困难的。Knauer 等（2010）研究表明，生长速度和母猪繁殖年限的关系是不清楚的。Bortolozzo等(2009)写道"很难辨别第一次配种时年龄、体重和背膘厚以及发情次数对繁殖性能和利用年限的影响"。Flowers(2005)同样认为:在后备猪性成熟期间繁殖器官的发育与它们的生长速度呈正相关。换句话说,在相同的年龄段,体重重的后备猪要比体重轻的后备猪的繁殖器官发育得更好。在大多数情况下这可能是正确的,但有一些情形繁殖器官的发育变得与生长速度不相关。在这些情况下,后备母猪在首次配种时其生长速度、年龄和体重是类似的,但是成年后它们的繁殖性能可能差别很大。

1.3.3 体脂

2005年Aherne研究表明,"没有确信的证据表明,背膘厚度直接或间接对繁殖性能有影响"。并且有足够的研究和农场证据表明,尽管母猪在其第一胎时背膘很薄,但依然表现为极佳的生产力(Young等,2004)。因此,背膘(当考虑第一窝时)本身将不是随后母猪繁殖性能的一个可靠预报器。Aherne(2005)指出,可能一个瘦的哺乳母猪,在其增加体重和背膘后,将比一个更肥的母猪而体重或者背膘减少时的繁殖性能更好。

过去5~10年,另一个观点表明,在首次配种时,后备母猪的体脂最低阈值并不影响其终生繁殖性能。Sørensen(2006)展示了丹麦一个商业研究的数据表明,在一个相同的时期,对于100~140 kg的猪,饲喂减少15%氨基酸和增加20%能量的日粮与饲喂典型哺乳料相比,导致背膘厚度更厚,然而,肥胖加重并不影响后备母猪的繁殖性能或者繁殖年限。数据表明,低能量日粮能为后备母猪提供足够的脂肪。因此,本研究在猪日粮中不需特别添加脂肪。Gill(2007)研究也发现,后备猪在首次配种时其背膘厚度和繁殖年限之间没有关系(图1.2)。在他的研究中,采用低蛋白日粮来饲喂后备母猪以在发育期间限制瘦肉增加并增加脂肪沉积。该

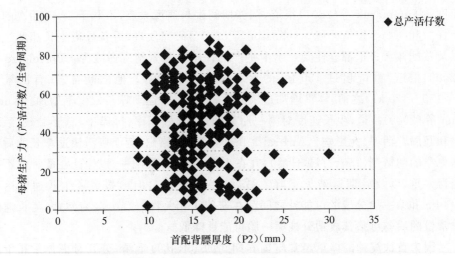

图1.2 首配背膘厚和母猪生产力之间的关系(来源于Gill,2007)。

研究成功地增加首次配种的脂肪厚度；但是与母猪繁殖寿命中所生的仔猪数目不相关。另外，背膘厚度与第一胎断奶到妊娠间隔时间和第二胎完成前因为繁殖失败而造成淘汰母猪数减少有关系。这显示，背膘厚可能依然对使用年限有重要影响。背膘厚度差异一直存在，直到第二胎断奶。

并不是所有的研究支持后备母猪脂肪不重要的观点。一些研究人员主张后备母猪开始繁殖时太少或者太多的体脂可能会使繁殖降低和繁殖稳定性更差的风险增大。确实，Bussières(2013)报道了首配的背膘厚度和终生总产仔数之间存在一个清晰的线性关系(图1.3)。该作者推荐15～16 mm为首配的背膘厚度的目标值，另外一个推荐显示20～26 mm对于高产是更好的，但还不太清楚这个推荐是如何从数据中得出的。

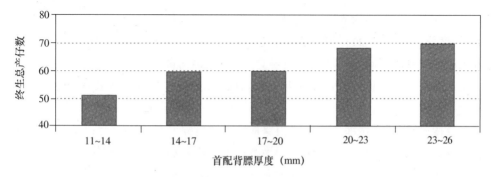

图1.3 首配背膘厚度与母猪终生总产仔数之间的关系(Bussières, 2013)。

1.3.4 体蛋白

最近几年研究发现，体蛋白储存和代谢对于母猪生产力比体脂更重要。Jagger等(2008)报道，后备母猪为了获得最优的繁殖性能达到某一蛋白质量是必需的。关于母猪体蛋白的储备，Clowes等(2003b)报道了如果后备母猪在哺乳期间动用了9%～12%的蛋白储存，在哺乳第20天后泌乳量会减少。另外，为了卵巢正常的功能提供相关蛋白生物合成所需的氨基酸前体，体蛋白量或蛋白代谢率必须达到一定的阈值，如果达不到这些阈值，第二胎的妊娠率和窝产仔数可能会减少。

大量研究者通过研究体重与终生产仔数和母猪使用年限关系，增强了对体蛋白在后备母猪选育中的重要性的认识(The British Pig Executive, 2010a; Williams等, 2005)。体重被解释为蛋白质沉积的结果。Schenkel等(2010)研究了第一胎身体组织代谢对第二胎繁殖性能的影响。身体蛋白沉积评估是依据体重和背膘厚度两个参数计算的。根据Whittemore和Yang(1989)的方程估算体脂肪和体蛋白的百分比。第二胎次窝产仔数的减少与第一胎次时身体损失的阈值相关；体重

＞10％、身体蛋白质＞10％或者体脂＞20％。然而,作者还指出,在一胎次分娩时,体重、体蛋白和体脂的绝对量也是重要的,因为体重大的母猪在第二胎分娩时比体重较轻的母猪生产更多的仔猪,而与体重或体蛋白损失的百分比不相关。

Schenkel 等(2010)的研究结果令人兴奋,认为大多数第一胎母猪(97％)在进行配种时体重达 135 kg 以上,在一胎分娩时体重超过 180 kg,与 Williams 等(2005)的建议一致。这些母猪身体储备足以满足首胎哺乳期,而不会在后续繁殖生产中缺乏营养。然而,实际养猪生产中,大约 25％的母猪在断奶时体重不到 178 kg,并在第二胎出现更少的窝产仔数。在另外一份数据分析评估中,39％的后备母猪出现较少的第二窝产仔数,其在首胎体重损失了 10.1％～25.0％。由此可知,Foxcroft(2002)和 Williams 等(2005)提出的首配目标的实现是不够的。Schenkel 等(2010)研究表明,即使首配时的目标均无法阻止随后母猪繁殖性能的下降,季节、哺乳期采食量减少(未被报告)、身体营养的代谢将增加到一个阈值。首次配种的身体状况是不准确的科学评价指标,因为随后的繁殖性能发生很大的变化。在第一次配种后的繁殖周期阶段,存在关于饲料、分娩、有效环境和饲养管理的所有极端情况。

总体而言,缺乏精准的蛋白沉积测量的长期研究。在评估澳大利亚 4000 多头后备母猪记录之后,Johnston 等(2007)使用回归统计分析,在后备母猪的两个母本遗传系研究中,在 22 周龄时增加眼肌面积可使繁殖寿命略有提高。Knauer 等(2011)估计了初情期眼肌面积与第一胎分娩的可能性、第一胎总产仔数以及第一胎哺乳期后的断奶发情间隔的遗传相关性。三种繁殖性状与初情期(212 日龄,137 kg 体重)时的眼肌面积之间的遗传相关性分别为 0.31、0.01 和 0.10。结论是,初情期更多的肌肉可增加第一胎分娩的可能性。蛋白沉积对第一胎窝产仔数的重要性不太明显。人们常常认为,身体蛋白质储备对于断奶后快速恢复发情活动是重要的,低的正遗传相关性支持这一假设。

相对于体蛋白,常常将饲养期间的瘦肉积累速率作为体况变化的参数,其对母猪的利用年限可能比目前已知的更重要。Johnston 和 Smits(2007)认为,如果瘦肉增长太快,那么利用年限将会缩短。

1.4 结论

最近的研究、综述、评论表明,体况的表型测量对后续繁殖的预测性不会更加精确。换句话说,我们不能通过减少体重、背膘厚度或年龄和体重组合等指标,显著提高后备母猪生产力或母猪使用年限。目前尚无令人信服的研究提供蛋白沉积目标

(精准的眼肌深度或眼肌面积)。

我们需要提高准确性和预测性吗？研究表明，提高目标的准确性和预测能力将增加后备母猪成功和持续繁殖的比例。这将降低后备猪培育的成本，并通过最多产的胎次来保持母猪群。如果能够使体况目标精准化，那么可能阻止代谢状态造成繁殖的、运动的、感染的母猪在生命中过早地被淘汰。具有特殊体况和组织储存发育的后备猪可以满足发情期繁殖周期的营养代谢需要，在母猪第一胎和第二胎期间，身体组织的积累和分解代谢具有良好的平衡。

我们有错误的目标吗？那些在猪肉生产中追求完美效率的人继续认为，如果所有因素都得到精确的控制，那么母猪会及时生产大量的仔猪，在老龄前就不会有母猪损失。最佳的效率包括通过使用体况的繁殖目标将后备母猪高效地纳入繁殖群体。

英国生猪执行委员会(2010b)目前推荐关于后备猪的这些"典型配种目标"：
- 220～270 日龄；
- 135～170 kg 活体重；
- 体况评分 3～3.5。

注意，缺少背膘厚度、眼肌面积和深度等目标。其中一个原因可能是体况评分包括脂肪和蛋白质储存。实际上，后备猪的"合格"被认为是一个更好的指标，比单独考虑的脂肪或蛋白质储存更为全面(Gill, 2007)。合格的定义目前包括饲养和管理后备母猪和青年母猪的身体状况，以及双腿和蹄部的健全性。在将来，它可能包括其他一些目标，它是所有身体组织的增加量或速率的组合或模型。

能够产生明显结果的目标可能会因为各种原因而不同程度避开我们。第一，目标(年龄、体重、生长速度、身体脂肪、身体蛋白质质量和骨骼发育)之间存在"混淆"或"关系"。第二，目标具有固有误差的经验预测因子(Rozeboom 等，1994)。背膘厚度和体重可能无法准确预测全身蛋白质。预测方程不应该用于预测不同母猪群的身体组成。随着用于预测方程式群体的年龄、生理状况、基因型和营养状况有所差异，来自独立变量或测量值的身体成分估计值的准确性降低。第三，我们必须避免方法的错误。不同研究中背膘厚度的测量可能不同。身体上的位置(P2，最后一条肋骨，距离中线 2 cm)可能会有所不同。同样，研究人员可能测量两层或三层脂肪。背最长肌肌肉直径或面积的测量需要技术经验才能确保准确性和一致性。第四，报告之间的时间段有所不同，并造成混乱。Johnston 等(2007)的研究包括断奶中的所有后备母猪的终身性能表现的数学评估。其他只包括从育肥猪舍中选出的后备母猪；还有一些只使用添加到繁殖群中的后备母猪。第五，饲养管理(保育仔猪大小、新生仔猪疾病、空间)，而不是营养或遗传学，有助于培育后备母猪的经验和反应,并可能有助于其后续的繁殖表现(Flowers,2005)。最后,评估猪的性能表现的工作可能会有所不同,因为淘汰标准和用于某些淘汰原因的主观判断存在差异。Pinilla 等(2013)指出,"后备母猪的性能表现和终身生产力表现在健康实践、饲喂/营养和生产原则之间的良性协同作用"。我们对身体状况在后备母猪发育中的重要性的理解有所增加,但一些方面尚未开发并且亟需改进。

参考文献

Aherne, F. 2005. Feeding the gestating sow. Pork Information Gateway Factsheet PIG 07-01-06. Available at: www.extension.org/pages/27438/feeding-the-gestating-sow#.

Amaral Filha, W.S., Bernardi, M.L., Wentz, I. and Bortolozzo, F.P., 2009. Growth rate and age at boar exposure as factors influencing gilt puberty. Livestock Science 120: 51-57.

Amaral Filha, W.S., Bernardi, M.L., Wentz, I. and Bortolozzo, F.P., 2010. Reproductive performance of gilts according to growth rate and backfat thickness at mating. Animal Reproduction Science 121: 139-144.

Beltranena, E., Aherne, F.X. and Foxcroft, G.R., 1993. Innate variability in sexual development irrespective of body fatness in gilts. Journal of Animal Science 71: 471-480.

Beltranena, E., Aherne, F.X., Foxcroft, G.R. and Kirkwood, R.N., 1991. Effects of pre-and postpubertal feeding on production traits at first and second estrus in gilts. Journal of Animal Science 69: 886-893.

Bortolozzo, F.P., Bernardi, M.L., Kummer, R. and Wentz, I., 2009. Growth, body state and breeding performance in gilts and primiparous sows. Society for Reproduction and Fertility Supplement 66: 281-291.

Bussières, D. 2013. Impact of gilt breeding condition on lifetime productivity and performance. Benchmark 8th edition, PigCHAMP and Benchmark.Farms.Com, Ames, IA, USA. Available at: http://benchmark.farms.com/2013_Impact_of_Gilt_Breeding.html.

Clowes, E.J., Aherne, F.X., Foxcroft, G.R. and Baracos, V.E., 2003a. Selective protein loss in lactating sows is associated with reduced litter growth and ovarian function. Journal of Animal Science 81: 753-764.

Clowes, E.J., Aherne, F.X., Schaefer, A.L., Foxcroft, G.R. and Baracos, V.E., 2003b. Parturition body size and body protein loss during lactation influence performance during lactation and ovarian function at weaning in first parity sows. Journal of Animal Science 81: 1517-1528.

De Koning, D.B., Van Grevenhof, E.M., Laurenssen, B.F.A., Van Weeren, P.R., Hazeleger, W. and Kemp, B., 2013. The influence of dietary restriction before and after 10 weeks of age on osteochondrosis in growing gilts. Journal of Animal Science 91: 5167-5176.

Flowers, W.L., 2005. New ideas about gilt development and management. North Carolina State University, Raleigh, NC, USA, 11 pp. Available at: www.prairieswine.com/pdf/2225.pdf.

Foxcroft, G. and Aherne, F., 2001. Rethinking management of the replacement gilt. Advances in Pork Production 12: 197-210.

Foxcroft, G., Beltranena, E., Patterson, J. and Williams, N., 2005. The biological basis for implementing effective replacement gilt management. In: Proceeding of the Allen D. Leman pre-conference reproduction workshop. September 17, 2005. Saint Paul, MN, USA, pp. 5-25.

Foxcroft, G.R., 2002. Fine tuning the breeding program. In: Proceedings of the Saskatchewan Pork Industry Symposium. November 12-14, 2002. Saskatoon, Canada, pp. 49-61. Available at: www.saskpork.com/images/File/symposium_2002_proceedings_2.pdf.

Gibson, S. and Jackson, J., 2012. To improve on lifetime performance we must focus on all aspects to help improve the longevity of our herds. Benchmark 7th edition, PigCHAMP and Benchmark. Farms.Com. Available at: http://benchmark.farms.com/2012_improved_perfomance.html.

Gill, P., 2007. Nutritional management of the gilt for lifetime productivity – feeding for fitness or fatness? In: Proceedings of the London Swine Conference. April 3-4, 2007. London, Ontario, Canada, pp. 83-99. Available at: www.londonswineconference.ca/proceedings/2007/LSC2007_PGill.pdf.

Hoving, L.L., Soede, N.M., Graat, E.A.M., Feitsma, H. and Kemp, B., 2010. Effect of live weight development and reproduction in first parity on reproductive performance of second parity sows. Animal Reproduction Science 122: 82-89.

Jagger, S., Toplis, P. and Wellock, I., 2008. Nutrition of the young breeding pig. AB Agri Ltd, Peterborough, UK. Available at: www.bpex.org.uk/downloads/297224/287981.

Johnson, R.K., Miller, P.S., Anderson, M.W., Perkins, J.M., Rynalds, K.A., Glidden, T.J., McClure, D.R., McGargill, T.E., Barnhill, D.J. and Moreno, R., 2008. Effects of nutrition during gilt development and genetic line on farrowing rates through parity e, causes of culling, sow weights and backfats through parity 4, and factors affecting farrowing rates. Nebraska Swine Report, pp 21-26.

Johnston, L.J., Bennett, C., Smits, R.J. and Shaw, K., 2007. Identifying the relationship of gilt rearing characteristics to lifetime sow productivity. In: Paterson, J.E. and Barker, J.A. (eds.) Manipulating pig production XI. Australasian Pig Science Association, 39 pp.

Johnston, L.J. and Smits, R.J., 2007. Nutrition of the developing gilt for optimal lifetime productivity. In: Proceedings of the 13th Discover Conference on Food Animal Agriculture: sSow productive lifetime. September 9-12, 2007. Nashville, IN, USA. Available at: http://tinyurl.com/oychkvh.

Kirkwood, R.N., Mitaru, B.N., Gooneratne, A.D., Blair, R. and Thacker, P.A., 1988. The influence of dietary energy intake during successive lactations on sow prolificacy. Canadian Journal of Animal Science 68: 283-290.

Knauer, M.T., Cassady, J.P, Newcom, D.W. and See, M.T., 2011. Phenotypic and genetic correlations between gilt estrus, puberty, growth, composition, and structural conformation traits with first-litter reproductive measures. Journal of Animal Science 89: 935-942.

Knauer, M.T., Stalder, K.J., Serenius, T., Baas, T.J., Berger, P.J., Karriker, L., Goodwin, R.N., Johnson, R.K., Mabry, J.W., Miller, R.K., Robison, O.W. and Tokach, M.D., 2010. Factors associated with sow stayability in 6 genotypes. Journal of Animal Science 88: 3486-3492.

Kummer, R., Bernardi, M.L., Schenkel, A.C., Amaral Filha, W.S., Wentz, I. and Bortolozzo, F.P., 2009. Reproductive performance of gilts with similar age but with different growth rate at the onset of puberty stimulation. Reproduction of Domestic Animals 44: 255-259.

Kummer, R., Bernardi, M.L., Wentz, I. and Bortolozzo, F.P., 2006. Reproductive performance of high growth rate gilts inseminated at an early age. Animal Reproduction Science 96: 47-53.

Lyvers-Peffer, P.A., Peng, J.J., Snedegar, J.A. and Rozeboom, D.W., 2003. Effects of a growth-altering pre-pubertal feeding regimen on gilt growth and reproductive longevity. Journal of Animal Science 81(2): 57.

Newton, E.A. and Mahan, D.C., 1993. Effect of initial breeding weight and management system using a high-producing sow genotype on resulting reproductive performance over three parities. Journal of Animal Science 71: 1177-1186.

Nimmo, R.D., Peo, Jr., E.R., Moser, B.D. and Lewis, A.J., 1981. Effect of level of dietary calcium-phosphorus during growth and gestation on performance, blood and bone parameters of swine. Journal of Animal Science 52: 1330-1342.

Patterson, J.L., Ball, R.O., Willis, H.J., Aherne, F.X. and Foxcroft, G.R., 2002. The effect of lean growth on puberty attainment in gilt. Journal of Animal Science 80: 1299-1310.

Patterson, J.L., Beltranena, E. and Foxcroft, G.R., 2010. The effect of gilt age at first estrus and breeding on third estrus on sow body weight changes and long-term reproductive performance. Journal of Animal Science 88: 2500-2513.

Pinilla, J.C., Teuber, R., Thompson, B., Coates, J., Piva, J., Canavate, S. and Molinari, R., 2013. Gilt management for 35 PSY. In: Proceedings of the PIC Boot Camp, Fort Wayne, IN, USA, 31 October 2013. Available at: http://tinyurl.com/p6rur5q.

Quinn, A., 2013. Feeding a gilt developer diet will improve sow longevity and productivity. In: Quinn, A., Calderon Diaz, J.A. and Boyle, L. (eds.) Lameness in pigs. Moorepark Research Dissemination Day. Teagasc, Cork, Ireland. Available at: www.teagasc.ie/publications/2013/2815/LamenessInPigs.pdf.

Rozeboom, D.W. 1999. Feeding programs for gilt development and longevity. Proceedings of the 60th Minnesota Nutrition Conference & Zinpro Technical Symposium. September 22, 1999. Bloomington, MN, USA. 159 pp.

Rozeboom, D.W. 2006. Nutrition of the developing gilt. In: Proceeding of the P.O.R.K. Academy. June 7, 2006. Des Moines, Iowa USA.

Rozeboom, D.W., Pettigrew, J.E., Moser, R.L., Cornelius, S.G. and El Kandelgy, S.M., 1994. *In vivo* estimation of body composition of mature gilts using live weight, backfat thickness and deuterium oxide. Journal of Animal Science 72: 355-366.

Rozeboom, D.W., Pettigrew, J.E., Moser, R.L., Cornelius, S.G. and El Kandelgy, S.M., 1996. Influence of gilt age and body composition at first breeding on sow reproductive performance and longevity. Journal of Animal Science 74: 138-150.

Rozeboom, D.W., Pettigrew, J.E., Moser, R.L., Cornelius, S.G. and El Kandelgy, S.M., 1995. Body composition of gilts at puberty. Journal of Animal Science 73: 2524-2531.

Schenkel, A.C., Bernadi, M.L., Bortolozzo, F.P. and Wentz, I., 2010. Body reserve mobilization during lactation in first parity sows and its effect on second litter size. Livestock Science 132: 165-172.

Sørensen, G., 2006. Feeding gilts in the growth period. The national committee for pig production, danish bacon & meat council. March 20, 2006. Report no. 741. Available at: http://tinyurl.com/oyfdfac.

Spörke, J. 2007. Gilt development techniques that alleviate SPL concerns. In: Proceedings of the 13th Discover Conference on Food Animal Agriculture: sow productive lifetime. September 9-12, 2007. Nashville, IN, USA. Available at: http://tinyurl.com/nr59vyq.

Stalder, K.J., Long, T.E., Goodwin, R.N., Wyatt, R.L. and Halstead, J.H., 2000. Effect of gilt development diet on the reproductive performance of primiparous sows. Journal of Animal Science 78: 1125-1131.

The British Pig Executive (BPEX), 2010a. Gilts: rearing for maximum protein gain or back fat? Research into action 4. Available at: http://tinyurl.com/omft8tr.

The British Pig Executive (BPEX), 2010b. Gilt management: service to farrowing. Action for Productivity 33. Available at: www.bpex.org/2TS/breeding/GiltManagement.aspx.

Tummaruk, P., Lundeheim, N., Einarsson, S. and Dalim, A.M., 2001. Effect of birth litter size, birth parity number, growth rate, back fat thickness and age at first mating of gilts on their reproductive performance as sows. Animal Reproduction Science 66: 225-237.

Whittemore, C.T. and Yang, H., 1989. Physical and chemical composition of the body of breeding sows with differing body subcutaneous fat depth at parturition, differing nutrition during lactation and differing litter size. Animal Production 48: 203-212.

Williams, N. L., Patterson, J. and Foxcroft, G., 2005. Non-negotiables of gilt development. In: Foxcroft, G.R. and Ball, R.O. (eds.) Advances in pork production no. 15. University of Alberta, Agricultural Food and Nutritional Science, Edmonton, Canada, pp. 281-289.

Young, L.G., King, G.J., Walton, J.S., McMillan, I. and Klevorick, M., 1990. Age, weight, backfat and time of mating effects on performance of gilts. Canadian Journal of Animal Science 70: 469-481.

Young, M.G., Tokach, M.D., Aherne, F.X., Main, R.G., Dritz, S.S., Goodband, R.D. and Nelssen, J.L., 2004. Comparison of three methods of feeding sows in gestation and subsequent effects on lactation performance. Journal of Animal Science 82: 3058-3070.

2. 妊娠早期的饲养管理以获得最佳繁殖性能

P. Langendijk

South Australian Research and Development Institute, J. S. Daivies Building, Roseworthy Campus, Roseworthy, SA 5371, Australia; pieter.langendijk@nutreco.com

摘要：在配种后的前 3~5 天，黄体组织分泌孕激素的能力仍然有限，在此期间采食量的增加将使全身孕激素减少。因此，在此期间限制饲喂可能会增加对子宫孕激素的供应，从而改善子宫环境，为着床做准备，从而增加胚胎的存活率。但在着床前，供应到胚胎的能量和蛋白质不太可能是限制性的，胚胎存活主要取决于胚胎发育和胚胎之间差异，后者会造成相对迟滞的胚胎死亡。本章确定了可能促进着床前囊胚发育的具体营养物质，但仍需要在体内进行研究。在妊娠约一周后，黄体功能达到最大，而在此阶段，卵巢-子宫间转移的孕酮在全身孕酮分泌中并不因较高的采食量而减少。在这一阶段，高采食量可产生更多的黄体组织，并且在第 12 天时，可通过胰岛素和胰岛素样生长因子(IGF)途径，增加黄体生成激素而直接增加孕酮分泌。在黄体功能形成期间，较高的采食量可能促进胚胎存活并支持妊娠维持。在妊娠失败风险升高，尤其是在大群饲喂和季节性不孕症等情况下尤其如此。但是胚胎在着床以后，子宫空间变得局限，能量和蛋白质的供应可能成为限制因素。特定营养素如精氨酸可促进胎盘血管化，从而其在早期阶段可促进胚胎存活。

关键词：营养，胚胎存活，妊娠，孕激素

2.1 引言

妊娠的第一个月对妊娠的成功至关重要，因为在这个时期，可能成功妊娠，也可能因为孕体和子宫之间的相互作用不足而不能维持。同样，在这个时期潜在的窝仔数量基本确定，但此时期的管理策略会影响胚胎成活数量进而影响其产仔数。本章主要讨论管理对妊娠建立和胚胎存活的影响，重点是营养方面。本章的目的是描述妊娠早期发生的事件的复杂性，并解释营养策略如何必须考虑到黄体功能和胚胎发育的动态变化。

2.2 早期妊娠问题:妊娠确立和胚胎死亡率

成功妊娠和胚胎存活最大化最重要的是要形成足够的黄体组织。一旦黄体生成素(LH)激增引发在发情期和随后的两周期间排卵前的卵泡黄体化,则形成足够的黄体组织。与其他哺乳动物一样,排卵后猪黄体组织快速发育,在第 10~12 天达到其成熟尺寸,此时后备母猪总黄体量为 6~8 g,经产母猪为 10~15 g(Langendijk 和 Peltoniemi,2013)。影响黄体量的一个主要因素是排卵率,排卵率与黄体量之间的相关性在 0.45~0.62(Almeida 等,2001;Athorn 等,2012a;Willis 等,2003)。这就是经产母猪比后备母猪和第一胎母猪有更多的黄体组织的原因。孕激素的全身浓度与总黄体量有关($r = 0.26 \sim 0.45$;Athorn 等,2012a),但是这种关系可能被低估了,因为它通常用在不同时间点获得的血液样品检测孕激素而不是对黄体量的评估。血液中孕酮大约可以反映出黄体组织的发育情况。

2.2.1 胚胎死亡率

在商业遗传品系中,猪产仔数的出生前损失在 30%~50%(Geisert 和 Schmitt,2002)。大多数这些损失发生在胚胎阶段(第 35 天之前),20%~30% 损失在胚胎发育第三周,剩余 10%~15% 损失是在胚胎期结束时(Ford 等,2002)。大量研究者研究发现,在妊娠期 30~35 天时,通过超排卵(Dziuk,1968)、诱导(Rampacek 等,1975)、子宫角结扎(Chen 和 Dziuk,1993)、卵巢切除(Père 等,1997)和单侧输卵管结扎(Town 等,2004)等技术限制子宫空间,结果发现胚胎损失显著增加,因此子宫容量会影响胚胎成活。大多数胚胎在第 12 天前存活(93%~96%;Anderson 等,1993)。第 12 天后,胚胎发生延长(第 11~13 天),空间增大并着床(第 15~17 天);然而,在这些早期阶段,很少有关胚胎损失的研究报告。对后备母猪的研究表明,在第 25 天,损失范围在 18%~35%,此时胚胎大部分具有独立空间(Dziuk,1968;Pope 等,1972;Webel 和 Dziuk,1974),也就是说,妊娠 25 天后,子宫中空间是限制胚胎成活的一个重要因素。

在营养方面,根据预期,营养物质的竞争仅在妊娠 3 周后开始,此时空间限制开始影响着床区域的大小和胚胎的大小(Père 等,1997;Town 等,2004)。因此,营养物质在 3 周之前不会影响胚胎生长,但特定的营养物质可能会影响胚胎的发育或胚胎发育改变。虽然其具体原因尚不清楚,但普遍认为,妊娠前 3 周胚胎发育的变化被认为是胚胎损失的主要原因(Geisert 和 Schmitt,2002),更确切地说,是在 12~21 天。发育不良的胚胎可能在着床时仅仅获得少量的空间来存活,同时也被雌激素所影响而造成损失,此时发育较好的胚胎也开始分泌雌激素(Geisert 等,2006)。这些雌激素对发育不良的同窝仔猪胚胎直接影响,或通过对子宫环境的间接作用影响产仔数。目前,尚无研究结果表明营养影响胚胎的内在质量或营养管理策略对胚胎变化有影响。例如,减少一胎中发育迟缓的胚胎,或减少或暂时抑制其雌激素分泌的营养管理策略将对胚胎存活的动力学具有相当大的影响,这是一个具有潜力的研究领域。

尽管妊娠前3周的胚胎发育可能不会受到营养方面的直接影响,但是在这些早期阶段营养显然间接影响黄体功能和子宫内环境。黄体功能和子宫内环境是十分重要的,因为随着胚胎发育,胎盘空间越来越受到限制,它们有助于创建更好的着床条件,以减少胚胎损失。毫无疑问,孕酮对于重建子宫内膜是重要的,以为成功的着床、孕体供给营养,因此妊娠早期(排卵后72 h)的全身孕激素与第35天时的胚胎存活相关(Foxcroft,1997,$r=0.48$;Zak等,1998,$r=0.72$)。除了这一阶段,关于胚胎存活与孕酮相关的报道很少(Athorn等,2012a;Gerritsen等,2008)。本章进一步讨论营养对胚胎早期和胚胎中期孕酮浓度的影响,以及营养对功能性胎盘区的影响。

2.2.2 妊娠失败

妊娠失败一般难以度量,因为在胚胎阶段的妊娠失败,母体会对孕体重吸收或在后期引起流产。在商业猪场生产条件下,妊娠失败发生率为5%~40%,但是,其中有一些不是真正意义上的妊娠失败,可能妊娠本身就没有建立。在妊娠大约15天时,如果没有足够的胚胎数,可能会引起黄体溶解,进而引起妊娠失败,这主要可能是受精卵数量或卵子与精子结合失败引起的,可能与营养无关,但是在配种前的营养状况(例如头胎母猪或严重营养不良的情况下)限制了卵母细胞数量或排卵发生(在第17章讨论;Soede和Kemp,2015)。并且在着床后,当功能性胎盘区域变得局限时,所提供的营养可能直接影响胎盘可用的营养物质,在黄体形成过程中,前期营养可能对子宫环境和胚胎存活具有重要作用,从而导致胚胎数量不足,不能维持雌激素信号和黄体正常功能。

妊娠15~25天的胚胎损失可能导致溶黄体信号延迟,导致妊娠失败以及返情延迟,这主要是来自孕体的信号不足和黄体功能失常的结果(Tast等,2002)。在一些气候带,胚胎损失以及妊娠失败的结果可能会在一年的特定时期加剧。这被称为季节性不育症。季节性不孕的机制知之甚少,但有迹象表明营养是重要的一个因素,这将在后面讨论。同样,具体的管理条件,如大群饲喂条件下的饲料竞争可能会导致胚胎损失,可能导致妊娠失败。

排卵后,黄体的发育和孕酮的分泌至少在排卵后10~12天不受垂体LH分泌影响(Peltoniemi等,1995)。因此,在本阶段潜在影响LH分泌的饲喂水平不会影响黄体组织形成。在排卵10~12天之后,LH对黄体功能具有重要的作用,尽管在一些研究中,促性腺激素水平的快速或缓慢降低均导致黄体衰退和妊娠失败。

只有完全抑制 LH 分泌 3~5 天才会导致黄体功能衰竭,导致在妊娠阶段妊娠失败或流产(Langendijk 和 Peltoniemi,2013)。

2.3 配种前营养对妊娠前期黄体功能的影响

配种前营养对黄体功能的影响尚未得到研究。然而,有大量关于卵泡发育和排卵率的研究,其中一些报告了营养对配种后黄体功能的延滞效应。在后备母猪黄体期的第二周进行饲料限制(20%~40%维持需求),导致排卵后孕酮的升高较慢(Almeida 等,2001;Chen 等,2012a)。胰岛素抵抗这种体内效应(Almeida 等,2001),导致体外孕激素分泌增加(Mao 等,2001)。Ashworth 等(1999)同样报道了在配种前饲喂高水平饲料(3.5 kg/天与 1.15 kg/天相比)的后备母猪孕酮分泌量升高,黄体重量增加。在初产母猪模型中,哺乳后期的限制饲喂对排卵后黄体功能具有相似的影响,孕酮浓度在限制性饲喂母猪体内分泌较低(Mao 等,1999)。Chen 等(2012b)研究发现,在哺乳后期饲喂添加淀粉/糖丰富的饲料,与饲喂富含脂肪饲料的母猪相比,在排卵后的第 4 天具有更多的孕酮,并提高了产仔数。在经产母猪断奶时,饲喂后胰岛素特征(峰值和平均浓度)与排卵后孕酮浓度呈正相关(Wientjes 等,2011)。这说明配种前采取营养干预措施,如饲料水平和胰岛素刺激制剂,能够总体上改善卵泡功能和发育,对排卵后孕酮的分泌具有积极的影响。这可能是在黄体生成后影响其分泌能力,进而产生对卵泡细胞数目或者质量的后期影响,并且该作用可能由与代谢相关的因子如 IGF-1 和胰岛素所介导。

2.4 配种后营养、胚胎损失和妊娠维持

2.4.1 饲料水平

为后备母猪提供较多的饲料,由于增大肝脏清除负担,进而造成其全身孕酮浓度减少(Prime 和 Symonds,1993)。因此,大量能量饲喂通常被认为可以降低胚胎存活(Jindal 等,1996)。事实上,Jindal 等(1996)的研究显示,排卵后 3 天饲料水平较高对胚胎存活有负面影响。

在早期妊娠期间(至第 15 天)获得的胚胎存活数据必须谨慎理解(表 2.1)。因为在桑葚胚和囊胚的情况下(Jindal 等,1997;Soede 等,1999),冲洗方法的效率可以影响确定胚胎存活的准确性。一旦胚胎开始延伸,胚胎的脆弱和形态学方面就会使胚胎数量的评估复杂化(Ashworth 等,1999;Athorn 等,2012a;Jindal 等,1997)。

还必须指出的是,通过长期的营养实验发现,在配种前和配种后,营养供应影响

了排卵率,因此,使区别配种后效应还是更早建立的后期作用更加困难(Foxcroft,1997)。

2.4.2 黄体形成、胚胎死亡和妊娠期间的饲料水平

Jindal 等(1996)研究了排卵后不久,不同饲料水平对胚胎存活的影响。但在随后的研究中发现,在整个胚胎期,这种饲喂水平的效应被认为可能是错误的。大量最新研究发现,饲喂水平在妊娠前 3~4 周保持不变,与排卵后的前 3 天相比,饲喂水平的影响不同(表 2.1),饲料效应有很大的变化。较低的饲喂水平通常在每天 1.2~2.5 kg,其中大多数仍然高于维持水平。然而,大多数研究比较了在整个妊娠前 3 周高饲喂水平与对照水平相比,并没有对胚胎存活产生影响,或表现为高饲喂水平使胚胎存活增加。两个研究报道,在整个胚胎期,高水平营养对胚胎存活具有负面影响,同样还报道,相同高水平营养对妊娠率具有正面影响(Dyck 和 Strain,1983;Virolainen 等,2004)。因此,可以假设,在高饲喂水平上,一些胚胎数少的妊娠母猪将在低饲喂水平下不会持久维持妊娠,但实际上高饲喂水平却可以促进妊娠成功。

然而,大部分其他的研究并未发现,在妊娠第 3 天后,高饲喂水平对胚胎存活的负面影响,尽管孕酮一致减少,甚至许多研究还报道了高饲喂水平对胚胎存活具有正面影响(Quesnel 等,2010;Athorn 等,2011a)。Dyck 和 Strain(1983)以及 Virolainen 等(2004)研究发现,在建立的黄体功能期间,更高饲喂水平不仅能促使胚胎存活,而且能够提高妊娠率。同样,Athorn 等(2011b)和 Langendijk 等(2011)研究显示,在妊娠前 25 天,根据其生长速度对第一胎母猪和后备母猪进行分类时,采食量作为测量指标,两项研究结果分别为:生长率最高的 25% 母猪,其妊娠率分别为 100% 和 92%,而生长率最低的 25% 母猪,其妊娠率分别为 92% 和 85%。高饲喂水平对维持妊娠的这些影响,可能是营养对黄体功能产生直接影响,从而降低黄体功能不全母猪的百分比,但也可能是促进较大胚胎存活,导致窝胚胎数变少,窝胚胎数太小从而无法维持足够的雌激素信号来拯救黄体和维持妊娠。

综上所述,配种后的前几天,高饲喂水平可能会降低全身孕激素水平,并对保证后期胚胎存活的子宫环境产生负面影响。之后,高饲喂水平可能对黄体功能、子宫环境并且可能直接对胚胎发育产生不同的影响,并且对胚胎的生存和维持妊娠产生积极影响。妊娠前几周少量黄体分泌有限的孕酮进入全身循环,在妊娠后 10 天变化为完全功能的黄体组织。还有种可能是孕酮从卵巢直接转移到子宫角,绕过全身循环,这取决于子宫内卵巢循环中孕酮的梯度。饲喂水平对孕酮分泌的动力学和孕酮局部转移的影响以前被忽视,仅在最近才得到认可(Langendijk 和 Peltoniemi,2013;Virolainen 等,2005a)。

表 2.1 比较低和高饲喂水平对妊娠率、胚胎存活率和窝产仔数的影响的研究

资源	妊娠状态	饲料添加(kg) 低	饲料添加(kg) 高	妊娠率(%) 低	妊娠率(%) 高		胚胎存活率(%) 低	胚胎存活率(%) 高	胎次[c]
	着床前								
Jindal 等,1997	D3~5	1.7	2.3				87	74	**
Soede 等,1999	D5~10	2.5	4	76	81		74	78	
Athorn 等,2012b	D10	1.4	2.9				77	92	**
Jindal 等,1997	D11~12	1.7	2.3				81	72	
Ashworth 等,1999	D12	1.15	3.5	100	92		86	83	
	着床后								
Condous 等,2013	D25	1.4	2.1				78	88	**
Quesnel 等,2010	D27	2	4	87	100		87	84	
Pharazyn 等,1991	D28	1.2	1.9				86.5	82.5	
Jindal 等,1996	D28	1.9	2.6				84.7	64.5	**
Dyck and Strain 等,1983	D30	1.5	2.5	64	87	**	87	76	**
Toplis 等,1983	D30	2	4	100	100		75	72	MP
Virolainen 等,2004	D35	2.1	4.3	25	100	**	91	70	
Athorn 等,2011a	D35	1.4	2.8	95			65	73	**
Athorn 等,2012a	D35	1.4	2.8	94	91		80	77	
Hoving 等,2012	D35	2.5	3.25	87	95		72	73	P1
	项目						窝产仔数		
Langendijk 等,2011		1.6	3.2	83	91[a]		11.5	11.2	
Athorn 等,2011b		2.0	3.0	97	95[a]		12.1	12.2	P1
Hughes,未发表的		1.9	2.8	70	83[b]	**			P1
Athorn 等,2013		低增加[a]	高增加	92	98[b]	**	11.1	11.6	P1, P2
Sawyer 等,2013									
采食 11~20 天		1.6~2	>2.5	79	90[b]	**			P1
采食 21~30 天		1.6~2	>2.5	50	92[b]	**			
Sawyer 等,2013									
采食 2~10 天		1.6~2	>2.5	76	90[b]	**			P2
采食 11~20 天		1.6~2	>2.5	52	90[b]	**			
采食 20~30 天		1.6~2	>2.5	43	93[b]	**			
Sorensen,2013		2.3	4.6	85	88[b]		17.5	17.5	MP

[a] 根据生长率进行分类时,生长率较低的母猪的妊娠概率较低。[b] 分娩率。[c] P1,第一胎母猪;P2,第二胎母猪;MP,多胎母猪或为后备母猪。** = 差异显著,$P<0.05$。

2.4.3 局部与全身孕酮

营养对子宫孕酮供应的影响是对其全身孕酮影响的结果,其包括分泌和清除,以及孕酮直接从卵巢到子宫角的局部转移。后者的发生是通过卵巢静脉和子宫动脉之间的逆流转运(逆流转运是指从卵巢静脉转运至子宫静脉,即两条静脉血管逆流,类固醇和肽激素能够穿过静脉血、肠液和淋巴至淋巴)和吻合手术进行的(Langendijk 和 Peltoniemi,2013)。孕酮的直接供应对供应子宫孕酮产生了重要的贡献,因为单侧切除一个卵巢可以减少同侧子宫角的胚胎存活,即使两子宫角的初始胚分布保持不变(Athorn 等,2011a)。孕酮的直接供应也解释了为什么子宫动脉中的孕酮显著多于体循环中,并且同样靠近卵巢的子宫动脉中孕酮含量多于卵巢远端的子宫动脉(Stefanczyk-Krzymowska 等,1998)。由于局部的孕酮供应是直接的,它不受肝代谢的影响,如全身孕酮的情况。因此,饲喂促进孕酮可能由于增加卵母细胞分泌孕酮并提供给子宫,而与此同时体循环的清除也增加。这些过程的综合作用结果决定了子宫角可用孕酮量。

有少量研究提供了一些数据,接近分泌源时,孕酮浓度水平及差异如图 2.1 所示。Athorn 等 (2012b) 和 Virolainen 等 (2005a) 研究显示,卵巢分泌大量的孕酮,子宫-卵巢静脉中的浓度(平均为 88 ng/mL)比全身循环中(19 ng/mL;Athorn 等,2012b)高得多。基于有限研究数据,脉冲数(表示孕酮的浓度)似乎随着黄体功能

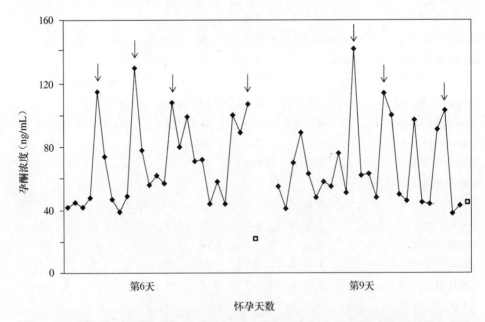

图 2.1 在妊娠第 6 天和第 9 天的 6 h 窗口中,从腔静脉以 15 min 间隔抽取的血浆中的孕酮浓度接近于子宫内膜静脉进入腔静脉位置的浓度。图表右下角的正方形表示在该时间点孕酮的全身浓度。箭头表示作者定义的脉冲(Athorn 等,2012b)。

的建立而增加(Athorn 等,2012b),妊娠第 6 天每 12 h 7 次脉冲,至第 10 天每 12 h 10 次脉冲(Athorn 等,2012b),并且在 11~17 天每 12 h 再次降低 6~8 次脉冲(Brüssow 等,2011),到第 22 天时,降为每 12 h 0~5 个脉冲(Virolainen 等,2005a)。

2.4.4 高饲喂水平对黄体功能的影响

在后备母猪(Athorn 等,2012a)和经产母猪(Gerritsen 等,2008)中,更高的饲喂水平导致在排卵后的第 10 天形成更多的黄体组织(Athorn 等,2012b)。在高饲喂水平下,用脂肪部分替代谷物没有改变黄体组织质量(Athorn 等,2012a)。这些研究结果表明,限制饲喂可能危及孕酮分泌能力,尽管机制了解甚少。对处于发情期的后备母猪使用 GnRH 拮抗剂(Antarelix,安雷利克斯),在黄体期阻断 LH 分泌能够减少黄体组织质量和全身孕酮水平(Quesnel 等,2000)。饲喂量的改变是否具有相似的效果是值得怀疑的。长期限制饲喂可能会降低 LH 分泌(Booth,1990;Booth 等,1996),但短期饲料缺乏可能没有这种作用(Barb 等,2001)。这些对 LH 的影响建立在具有不同 LH 分泌模式的发情期母猪中,而非妊娠猪,在妊娠猪体内孕酮对 LH 分泌产生负反馈。关于饲料水平对 LH 的影响,在后备母猪妊娠期或黄体期间的研究是不明确的,一些研究报告认为对 LH 没有影响(Peltoniemi 等,1997a;Quesnel 等,2000),但是一些研究报道了限制饲喂使妊娠后备母猪中的 LH 分泌减少(Ferguson 等,2003;Peltoniemi 等,1997b)。因此,如果饲喂水平对 LH 有任何影响,对黄体组织质量的影响也是轻微的。然而,可能有其他途径介导饲喂水平对黄体组织形成的影响,例如通过 IGF 和胰岛素。尽管已知猪黄体组织对胰岛素和 IGF-1 有所反应,但几乎没有研究涉及这些途径对黄体组织形成的影响(Miller 等,2003;Ptak 等,2004)。

在黄体期阶段,高饲喂水平也可能直接影响孕酮的分泌。Athorn 等(2012b)研究发现,妊娠后备母猪的饲喂水平从 1.5 kg/天增加到 2.8 kg/天,排卵后第 9 天将尾静脉内的孕酮脉冲数从(3.8±0.7)/6 h 增加至(4.9±1.1)/6 h。这相当于在 24 h 内增加 4.4 个脉冲。尾静脉中的孕酮脉冲平均达 179 ng/mL,比该期间的全身水平高 6 倍,这些脉搏的测量可能更能反映卵巢分泌的孕酮水平。营养对孕酮的这些直接作用可能由 LH 脉冲的改变所介导,然而并不是所有 LH 脉冲都与孕酮脉冲重合(Brüssow 等,2011;Virolainen 等,2005a),如上所述,饲喂水平对 LH 的影响是不清楚的。可以推测,在黄体分泌孕酮方面,低 LH 分泌对于黄体的功能是足够的。事实上,Easton 等(1993)提出,对于黄体功能而言,重要的是 LH 的个别脉冲而不是 LH 的某些基础水平。如果是这样,一些报告(Peltoniemi 等,1997a,b;Prunier 等,1993)中观察到的现象是合理的,即在生理范围内 LH 分泌的变化对黄体维持没有太大的影响。然而,一旦在药理学上使用 GnRH 激动剂刺激 LH 分泌(Peltoniemi 等,1995),孕酮浓度便暂时升高,并且在妊娠第 10 天用 eCG 治疗对孕酮分泌具有直接作用(O'Leary 等,2011)。

胰岛素和 IGF-1 的增加也可以调节高饲喂水平对孕酮的直接作用。Miller 等（2003）报道，在将 IGF-1 注入卵巢后，卵巢的孕酮分泌增加，Langendijk 等（2008）发现在排卵后，全身性 IGF-1 和孕酮之间呈正相关。高饲喂水平通常增加全身的 IGF-1 水平（Ferguson 等，2003）。还有一点值得注意的是，在高饲喂水平的后备母猪中，用脂肪源替代淀粉在日粮中增加了孕酮分泌（在妊娠的第 15 天），而黄体组织质量（第 35 天）没有改变（Athorn 等，2012a）。在这些饲喂之间没有报告 IGF-1 浓度的差异。显然，未来更多研究模型还需要进行研究，如胰岛素、IGF-1、日粮中脂肪以及其他潜在的黄体功能的调节因子如 PGE2（图 2.2）。

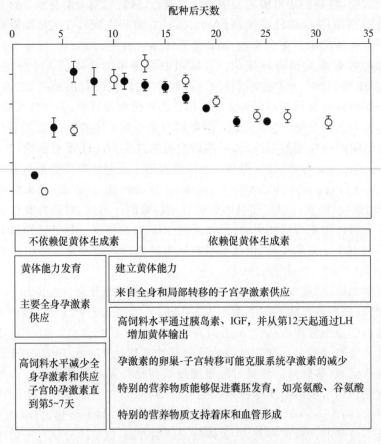

图 2.2 概念图说明了胚胎阶段的黄体质量（黄体的横截面积，实心符号）和全身孕激素（空心符号）的动态变化，解释了营养的不同方面（饲料水平和特定营养）如何影响胚胎期的不同阶段。

2.4.5 营养在妊娠早期风险管理中的作用

在黄体期，上面讨论的低饲喂水平对胚胎存活和维持妊娠的作用在生产上已经被验证,即低饲喂水平使妊娠失败风险增加,例如在大群饲喂情况下,由于饲料竞争或其他原因造成的饲料采食量低增加了妊娠失败的风险。类似地,低饲喂水平可能与其他风险因素(比如季节性不孕症)协同,也会使妊娠失败的风险增加。Kongsted 等(2004)发现,在允许饲料竞争的大群饲养条件下,妊娠前 25 天,背膘增厚最低(甚至损失)的母猪的妊娠失败风险较高。即使在具有母猪饲喂的电子系统中,饲料竞争最小化的情况下,在妊娠 10~30 天,饲料摄入量更低的母猪的受胎率较低(Athorn 等,2013;Sawyer 等,2013)。因此,可以断定,对大群饲养母猪而言,至少一部分妊娠失败是由应激引起营养不良所造成的。Langendijk(未发表的数据)发现,排卵后第 10 天和第 11 天禁食,后备母猪的产仔数从 10.9 降低到 8.8 头,然而,Razdan 等(2004)发现,在第 13 和第 14 天,禁食的后备母猪中没有这样的结果。只有少数几个研究指出了在大群饲养中营养因素是妊娠损失的一个风险因素,因此,仍然不清楚这些妊娠损失在多大程度上是应激相关的,还是营养造成的,或两者兼而有之。为了减少营养不足引起的应激,越来越多农场在大群饲养系统中引入纤维类饲料。与对照日粮相比,在等卡热的日粮中包含 10% 的粗纤维,第一胎母猪(Athorn 等,2011b)和后备母猪(Langendijk 等,2011)中产生了相同的妊娠率和排卵量,不影响后备母猪的胚胎存活率或受胎率(Athorn 等,2012a)。然而,为了达到与自由采食的母猪相当的饱腹感水平,可能必须增加纤维含量以使猪饱腹感接近自由采食(Jensen 等,2012;Knudsen 等,2000)。

与大群饲养竞争的情况一样,对于 LH 分泌和妊娠失败的季节性影响可能会因营养不良而加剧,这是 Love 等(1995)所倡导的一个想法。尽管采食量的变化只能轻度影响 LH 和孕酮,但一些迹象表明,在季节性不孕的时期,这些影响可能与妊娠失败有关。一些报道的数据表明,季节对子宫中孕酮供给有影响(Virolainen 等,2005a)。此外,已经提出,在胚胎着床期间,长时间限制性饲喂,LH 分泌可能减弱,这将导致卵巢孕酮分泌减少,从而影响妊娠第 17 天左右胚胎的雌二醇信号,并导致妊娠中断,在 25~35 天恢复发情,这就是季节性不孕症的典型特征(Tast 等,2002)。相比之下,短时间内高营养水平刺激松果体分泌褪黑激素和垂体分泌 LH 激素,并维持妊娠(Peltoniemi 和 Virolainen,2006)。这些研究结果与充足的饲料可提高后备母猪受胎率结果相一致(Virolainen 等,2004)。

2.4.6 影响胚胎发育和生存的特定营养

在 20 世纪 80 年代和 90 年代,维生素作为与母猪繁殖性能相关的第一个功能性营养物质而被广泛研究,叶酸在那个时代受到了相当大的关注。Lindemann(1993)的研究得出结论,妊娠母猪日粮添加叶酸可以提高产仔数,但其机制尚不清楚。目前,大多数商业饲料中以 0.2~15 mg/kg 的剂量添加叶酸(Gaudre 和 Quiniou,2009),叶酸是否在胚胎期发挥作用尚未得到充分证实。但结果显示叶酸可提高窝产仔猪数(Lindemann,1993;Matte 等,1984)。Tremblay 等(1989)报道,在 eCG 治疗后,排卵率高的母猪中,第 30 天其胚胎存活率增加,而 Harper 等(2003)和 Guay 等(2002)没有发现对胚胎存活的影响,尽管在这个时期通过补充叶酸明显改变了子宫环境(Matte 等,1996)。

在过去 10 年中,特定氨基酸在囊胚阶段、延伸和着床期间对胚胎发育的作用越来越受到关注。已经显示精氨酸通过精氨酸-NO 途径刺激血管生成来影响胎盘的血管形成(Hazeleger 等,2007;Novak 等,2011),并且被认为有利于胎儿存活并提高产仔数。Ramaekers 等(2006)和 Wu 等(2010)报道,当从妊娠 14~28 天或从妊娠第 30 天至分娩时,补充 20~25 g/天精氨酸时,窝产仔数增加。Hazeleger 等(2007)报道,妊娠 16~28 天饲喂 40 g/天精氨酸的母猪妊娠第 30 天胚胎存活率增加,后备母猪主要采用超排处理提高排卵率。Novak 等(2011)报道精氨酸对胚胎存活没有影响,但在第 49 天发现胎儿体重增加。

精氨酸仅在着床后添加(第 15 天起)。据我们所知,只有 Li 等(2010)从配种开始添加 25 天精氨酸,含量为 0.8%,发现第 25 天胎儿数量减少。然而,他们的研究很难解释,因为精氨酸处理后的生长速度远远低于预期,血浆孕酮浓度低于对照组。可能的原因是功能性氨基酸影响着床前的胚胎和滋养层发育,如改善胚胎存活和生长。基于接近着床的子宫和滋养层液体中增加的浓度,Bazer 等(2013)提出,亮氨酸、精氨酸、谷氨酰胺、葡萄糖和果糖可能是影响胚泡发育和滋养层伸长的功能性营养物质。在体外,这些功能性营养素确实刺激了猪滋养外胚层细胞的增殖和特异性细胞信号通路(Bazer 等,2013)。这些功能性营养物质作为组织细胞复合物的一部分转运蛋白和分泌物的表达受孕酮和雌二醇的调节。在牛体内,孕酮明显刺激组织营养素的分泌,据报道在着床前刺激囊胚伸长(Carter 等,2008;Clemente 等,2009)。在猪体内,这些功能性氨基酸在着床前的效应在体内仍有待确认。有趣的是妊娠早期孕酮补充对妊娠胚胎的生存和维持有负面影响(Soede 等,2012)。然而,在着床前,关于上述功能性氨基酸进行体内补充的实验尚未见报道。如果这些营养物质刺激胚泡伸长,胚胎损失将取决于同一窝的胚胎之间的差异是否保持不变、增加或减少,因为胚胎之间的差异已经被认为是胚胎损失的主要原因(Geisert 和 Schmitt,2002)。

2.5 激素干预妊娠维持的作用

因为孕酮是子宫内膜功能的重要驱动因素,被认为是控制胚胎存活的关键因素之一,补充孕酮可以克服黄体功能不全,从而提高窝产仔数或者挽救妊娠,否则妊娠将会失败。在饲喂两倍于维生素需求的后备母猪中,这可能导致全身孕酮的下降,在发情后第 3 天补充孕酮可以在第 28 天将胚胎存活率从 70% 提高到 84%(Jindal 等,1997)。然而,补充孕酮或孕酮类似物有相关风险。事实上,在排卵后过早地补充四烯雌酮似乎减弱了通过输卵管的卵子的运输和受精率，并且在配种后的第 4~6 天补充四烯雌酮使第一胎母猪的产仔数减少（Soede 等, 2012）。

另一个有趣的例子是通过用卵泡刺激剂如 eCG 处理妊娠母猪，随后用 hCG 或 GnRH 类似物诱导排卵或至少使这些卵泡黄体化。开发了这种处理方法以延长后备母猪的黄体期，以便通过在期望的时间点诱导黄体溶解实现同期发情，这种方法使黄体增加了分泌的孕酮总量，因此，维持妊娠和提高了窝产仔数（O'Leary 等, 2011）。单独使用促性腺激素（gonadotrophin）处理还可以增加现有黄体的孕酮分泌，而不需要诱导黄体（O'Leary 等, 2011；Peltoniemi 等, 1995）。然而，尚未对这些模型作为妊娠支持的补救措施进行研究。在卵泡生长的第 12 天，对其诱导可能对维持妊娠具有额外的作用，因为在该妊娠阶段，卵母细胞分泌孕激素可以模拟来自胚胎的内源性孕激素信号，这通常可以防止黄体溶解从而维持妊娠。再者，诱导孕激素分泌的模型已经被用于诱导假性妊娠，从而作为后备母猪同期发情的手段（Noguchi 等, 2010），但是，增加的来自卵泡的孕激素分泌可能用于支持或者维持妊娠，但是这种情况只能在配种后第 12 天使用促性腺激素。然而，这可以营救产仔数少的窝（由于正常死亡或者终止妊娠所造成的）。

2.6　结论

在配种后的前 5 天，限制采食量可能会增加胚胎的存活率。在这段时间之后，似乎没有与更高的饲料供应相关的风险，并且由于高水平营养刺激对孕酮分泌的影响，有可能有益于维持妊娠和胚胎存活，尽管体内孕酮较少，但孕酮通过局部机制直接转移到子宫。因此，在第 5 天之后，后备母猪和初产母猪不需要限制饲喂，并且可以增加饲喂量以匹配所需生长曲线或从先前哺乳期的恢复。在存在妊娠失败风险的母猪中，如大群饲喂情况或季节性不孕时期，维持高采食量，避免厌食事件或低采食量对于减少妊娠失败以及使胚胎存活最大化是关键的。关于特定营养素如亮氨酸和谷氨酰胺在着床之前对囊胚发育的影响是未来的研究方向。在着床时和着床后，精氨酸促进胎盘血管化，从而增加胚胎的存活或发育。

参考文献

Almeida, F.R.C.L., Mao, J., Novak, S., Cosgrove, J.R. and Foxcroft, G.R., 2001. Effects of different patterns of feed restriction and insulin treatment during the luteal phase on reproductive, metabolic, and endocrine parameters in cyclic gilts. Journal of Animal Science 79: 200-212.

Anderson, R.L.H., Christenson, L.K., Christenson, R.K. and Ford, S.P., 1993. Investigations into the control of litter size in swine: II. Comparisons of morphological and functional embryonic diversity between Chinese and American breeds. Journal of Animal Science 71: 1566-1571.

Ashworth, C.J., Beattie, L. and Antipatis, C., 1999. Effects of pre- and post-mating nutritional status on hepatic function, progesterone concentration, uterine protein secretion and embryo survival in meishan pigs. Reproduction, Fertility and Development 11: 67-73.

Athorn, R.Z., Sawyer, K.S., Collins, C.L. and Luxford, B.G., 2013. High growth rates during early pregnancy positively affect farrowing rate in parity one and two sows. In: Pluske, J.R. and Pluske, J.M. (eds.) Manipulating pig production XIV. Australasian Pig Science Association, Melbourne, Australia, 132 pp.

Athorn, R.Z., Stott, P. and Langendijk, P., 2011b. Feeding level and dietary energy source during early pregnancy in first parity sows: effects on pregnancy and litter size. In: Van Barneveld, R.J. (ed.) Manipulating pig production. Proceedings of the Australasian Pig Science Association, Adelaide, Australia, 81 pp.

Athorn, R.Z., Stott, P., Bouwman, E.G., Ashman, R., O'Leary, S., Nottle, M. and Langendijk, P., 2011a. Direct ovarian-uterine transfer of progesterone increases embryo survival in gilts. Reproduction, Fertility and Development 23: 921-928.

Athorn, R.Z., Stott, P., Bouwman, E.G., Chen, T.Y., Kennaway, D.J. and Langendijk, P., 2012b. Effect of feeding level on luteal function and progesterone concentration in the vena cava during early pregnancy in gilts. Reproduction, Fertility and Development 25(3): 531-538.

Athorn, R.Z., Stott, P., Bouwman, E.G., Edwards, A.C., Blackberry, M.A., Martin, G.B. and Langendijk, P., 2012a. Feeding level and dietary energy source have no effect on embryo survival in gilts, despite changes in systemic progesterone levels. Animal Production Science 53: 30-37.

Barb, C.R., Barretta, J.B., Kraeling, R.R. and Rampacek, G.B., 2001. Serum leptin concentrations, luteinizing hormone and growth hormone secretion during feed and metabolic fuel restriction in the prepuberal gilt. Domestic Animal Endocrinology 20: 47-63.

Bazer, F.W., Kim, J., Song, G., Ka, H., Wu, G., Johnson, G.A. and Vallet, J.L., 2013. Roles of selected nutrients in development of the porcine conceptus during pregnancy. In: Rodriguez-Martinez, H., Soede, N.M. and Flowers, W.L. (eds.) Control of pig reproduction IX. Context Products, Packington, UK, pp. 159-174.

Booth, P.J., 1990. Metabolic influences on hypothalamic-pituitary-ovarian function in the pig. In: Cole, D.J.A., Foxcroft, G.R. and Weir, B.J. (eds.) Control of pig reproduction III. Journal of Reproduction and Fertility Supplements 40: 89-100.

Booth, P.J., Cosgrove, J.R. and Foxcroft, G.R., 1996. Endocrine and metabolic responses to realimentation in feed-restricted prepubertal gilts: associations among gonadotropins, metabolic hormones, glucose, and utero ovarian development. Journal of Animal Science 74: 840-848.

Brüssow, K.P., Schneider, F., Wollenhaupt, K. and Tuchscherer, A., 2011. Endocrine effects of GnRH agonist application to early pregnant gilts. Journal of Reproduction and Development 57: 242-248.

Caldwell, B.V., Moor, R.M., Wilmut, I., Polge, C., and Rowson, L.E.A. 1969. The relationship between day of formation and functional lifespan of induced corpora lutea in the pig. Journal of Reproduction and Fertility 18: 107-113.

Carter, F., Forde, N., Duffy, P., Wade, M., Fair, T., Crowe, M.A., Evans, A.C.O., Kenny, D.A., Roche, J.F. and Lonergan, P., 2008. Effect of increasing progesterone concentration from day 3 of pregnancy on subsequent embryo survival and development in beef heifers. Reproduction, Fertility and Development 20: 368-375.

Chen, T.Y., Stott, P., Athorn, R.Z., Bouwman, E.G. and Langendijk, P., 2012a. Undernutrition during early follicle development has irreversible effects on ovulation rate and embryos. Reproduction, Fertility and Development 24: 886-892.

Chen, T.Y., Stott, P., Bouwman, E.G. and Langendijk, P., 2012b. Effects of pre-weaning energy substitutions on post-weaning follicle development, steroid hormones and subsequent litter size in primiparous sows. Reproduction in Domestic Animals 48(3): 512-519.

Chen, Z.Y. and Dziuk, P.J., 1993. Influence of initial length of uterus per embryo and gestation stage on prenatal survival, development, and sex ratio in the pig. Journal of Animal Science 71: 1895-1901.

Clemente, M., De La Fuente, J., Fair, T., Al Naib, A., Gutierrez-Adan, A., Roche, J.F., Rizos, D. and Lonergan, P, 2009. Progesterone and conceptus elongation in cattle: a direct effect on the embryo or an indirect effect via the endometrium? Reproduction 138: 507-517.

Condous, P.C., Kirkwood, R.N. and Van Wettere, W.H.E.J., 2013. Post mating but not pre mating dietary restriction decreases embryo survival of group housed gilts. In: Pluske, J.R. and Pluske, J.M. (eds.) Manipulating pig production XIV. Australasian Pig Science Association, Melbourne, Australia, 132 pp.

Dyck, G.W. and Strain, J.H., 1983. Postmating feeding level effects on conception rate and embryonic survival in gilts. Canadian Journal of Animal Science 63: 579-585.

Dziuk, P.J., 1968. Effect of number of embryos and terine space on embryo survival in the pig. Journal of Animal Science 27: 673-676.

Easton, B.G., Love, R.J., Evans, G. and Klupiec, C., 1993. A role for LH pulses in the establishment of pregnancy. In: Batterham, E.S. (ed.) Manipulating pig production. Australasian Pig Science Association, Canberra, Australia, 246 pp.

Ferguson, E.M., Ashworth, C.J., Edwards, S.A., Hawkins, N., Hepburn, N. and Hunter, M.G., 2003. Effect of different nutritional regimens before ovulation on plasma concentrations of metabolic and reproductive hormones and oocyte maturation in gilts. Reproduction 126: 61-71.

Ford, S.P., Vonnahme, K.A. and Wilson, M.E., 2002. Uterine capacity in the pig reflects a combination of uterine environment and conceptus genotype effects. Journal of Animal Science 80(1): E66-E73.

Foxcroft, G.R., 1997. Mechanisms mediating nutritional effects on embryonic survival in pigs. Journal of Reproduction and Fertility Supplement 52: 47-61.

Gaudré, D. and Quiniou, N., 2009. What mineral and vitamin levels to recommend in swine diets? Revista Brasileira de Zootecnia 38: 190-200.

Geisert, R.D., Ross, J.W., Ashworth, M.D., White, F.J., Johnson, G.A. and Da Silva, U., 2006. Maternal recognition of pregnancy or endocrine disruptor: the two faces of oestrogen during establishment of pregnancy in the pig. Society for Reproduction and Fertility 62: 131-146.

Geisert, R.D. and Schmitt, R.A.M., 2002. Early embryonic survival in the pig: can it be improved? Journal of Animal Science 80: E54-E65.

Gerritsen, R., Soede, N.M., Langendijk, P., Dieleman, S.J., Hazeleger, W., Laurenssen, B.F.A. and Kemp, B., 2008. Feeding level does not affect progesterone levels in intermittently suckled sows with lactational ovulation. Animal Reproduction Science 103: 379-384.

Guay, F., Matte, J.J., Girard, C.L., Palin, M.F., Giguère, A. and Laforest, J.P., 2002. Effect of folic acid and glycine supplementation on embryo development and folate metabolism during early pregnancy in pigs. Journal of Animal Science 80: 2134-2143.

Guthrie, H.D. and Polge, C. 1976. Control of oestrus and fertility in gilts with accessory lutea by prostaglandin analogues, ICI 79,939 and ICI 80,996. Journal of Reproduction and Fertility 48: 427-430.

Harper, A.F., Knight, J.W., Kokue, E. and Usry, J.L., 2003. Plasma reduced folates, reproductive performance, and conceptus development in sows in response to supplementation with oxidized and reduced sources of folic acid. Journal of Animal Science 81: 735-744.

Hazeleger, W., Smits, C. and Kemp, B., 2007. Influence of nutritional factors on placental growth and piglet imprinting. In: Wiseman, J., Kemp, B. and Varley, M. (eds.) Paradigms in pig science. University of Nottingham Press, Nottingham, UK, pp. 309-327.

Hoving, L.L., Soede, N.M., Feitsma, H. and Kemp, B., 2102. Embryo survival, progesterone profiles and metabolic responses to an increased feeding level during second gestation in sows. Theriogenology 77: 1557-1569.

Jensen, M.B., Pedersen, L.J., Theil, P.K., Yde, C.C. and Bach Knudsen, K.E., 2012. Feeding motivation and plasma metabolites in pregnant sows fed diets rich in dietary fiber either once or twice daily. Journal of Animal Science 90: 1910-1919.

Jindal, R., Cosgrove, J.R. and Foxcroft, G.R., 1997. Progesterone mediates nutritionally induced effects on embryonic survival in gilts. Journal of Animal Science 75: 1063-1070.

Jindal, R., Cosgrove, J.R., Aherne, F.X. and Foxcroft, G.R., 1996. Effect of nutrition on embryonal mortality in gilts: association with progesterone. Journal of Animal Science 74: 620-624.

Knudsen, K.E.B., Henry, J. and Canibe, N., 2000. Quantification of the absorption of nutrients derived from carbohydrate assimilation: model experiment with catheterised pigs fed on wheat- or oat-based rolls. British Journal of Nutrition 84: 449-458.

Kongsted, A.G., 2004. Reproduction performances and conditions of group-housed non-lactating sows. The Royal Veterinary and Agricultural University, Frederiksberg, Denmark.

Langendijk, P., Athorn, R.Z. and Stott, P., 2011. Feeding level and dietary fibre content during early pregnancy in gilts and pregnancy rate and litter size. In: Van Barneveld, R.J. (ed.) Manipulating pig production. Proceedings of the Australasian Pig Science Association, Adelaide, Australia, 162 pp.

Langendijk, P., van den Brand, H., Gerritsen, R., Quesnel, H., Soede, N.M. and Kemp, B., 2008. Porcine luteal function in relation to IGF-1 levels following ovulation during lactation or after weaning. Reproduction in Domestic Animals 43: 131-136.

Langendijk, P. and Peltoniemi, O., 2013. How does nutrition influence luteal function and early embryo survival. In: Rodriguez-Martinez, H., Soede, N.M. and Flowers, W.L. (eds.) Control of pig reproduction IX. Context Products, Packington, UK, pp. 145-158.

Li, X., Bazer, F.W., Johnson, G.A., Burghardt, R.C., Erikson, D.W., Frank, J.W., Spencer, T.E., Shinzato, I. and Wu, G., 2010. Dietary supplementation with 0.8% L-Arginine between days 0 and 25 of gestation reduces litter size in gilts. Journal of Nutrition 140: 1111-1116.

Lindemann, M.D., 1993. Supplemental folic acid: a requirement for optimizing swine reproduction. Journal of Animal Science 71: 239-246.

Love, R.J., Klupiec, C., Thornton, E.J. and Evans, G., 1995. An interaction between feeding rate and season affects fertility of sows. Animal Reproduction Science 39: 275-284.

Mao, J., Treacy, B.K., Almeida, F.R.C.L., Novak, S., Dixon, W.T. and Foxcroft, G.R., 2001. Feed restriction and insulin treatment affect subsequent luteal function in the immediate postovulatory period in pigs: progesterone production *in vitro* and messenger ribonucleic acid expression for key steroidogenic enzymes. Biology of Reproduction 64: 359-367.

Mao, J., Zak, L.J., Cosgrove, J.R., Shostak, S. and Foxcroft, G.R., 1999. Reproductive, metabolic, and endocrine responses to feed restriction and gnrh treatment in primiparous, lactating sows. Journal of Animal Science 77: 724-735.

Matte, J.J., Farmer, C., Girard, C.L. and Laforest, J.P., 1996. Dietary folic acid, uterine function and early embryonic development in sows. Canadian Journal of Animal Science 76: 427-433.

Matte, J.J., Girard, C.L. and Brisson, G.J., 1984. Folic acid and reproductive performances of sows.

Journal of Animal Science 59: 1020-1025.

Miller, E.A., Ge, Z., Hedgpeth, V. and Gadsby, J.E., 2003. Steroidogenic responses of pig corpora lutea to insulin-like growth factor I (IGF-I) throughout the oestrous cycle. Reproduction 125: 241-249.

Noguchi, M., Yoshioka, K., Suzuki, C., Arai, S., Itoh, S., and Wada, Y. 2010. Estrus synchronization with pseudopregnant gilts induced by a single treatment of estradiol dipropionate. Journal of Reproduction and Development 56: 421-427.

Novak, S., Paradis, F., Patterson, J.L., Pasternak, J.A., Oxtoby, K., Moore, H.S., Hahn, M., Dyck, M.K., Dixon, W.T. and Foxcroft, G.R., 2011. Temporal candidate gene expression in the sow placenta and embryo during early gestation and effect of maternal Progenos supplementation on embryonic and placental development. Reproduction, Fertility and Development 24: 550-558.

O'Leary, S., Bouwman, E.G., Nottle, M.N. and Langendijk, P., 2011. Increasing endogenous progesterone in early pregnancy increases litter size in pigs. In: Van Barneveld, R.J. (ed.) Manipulating pig production. Proceedings of the Australasian Pig Science Association, Adelaide, Australia, 79 pp.

Peltoniemi, O.A.T., Love, R.J., Klupiec, C. and Evans, G., 1997b. Effect of feed restriction and season on LH and prolactin secretion, adrenal response, insulin and FFA in group housed pregnant gilts. Animal Reproduction Science 49: 179-190.

Peltoniemi, O.A.T., Easton, B.G., Love, R.J., Klupiec, C. and Evans, G., 1995. Effect of chronic treatment with a GnRH agonist (Goserelin) on LH secretion and early pregnancy in gilts. Animal Reproduction Science 40: 121-133.

Peltoniemi, O.A.T., Love, R.J., Klupiec, C., Revell, D.K. and Evans, G., 1997a. Altered secretion of LH does not explain seasonal effects on early pregnancy in gilts. Animal Reproduction Science 49: 215-224.

Peltoniemi, O.A.T. and Virolainen, J.V., 2006. Seasonality of reproduction in gilts and sows. Society for Reproduction and Fertility Supplements 62: 205-218.

Père, M.C., Dourmad, J.Y. and Etienne, M., 1997. Effect of number of pig embryos in the uterus on their survival and development and on maternal metabolism. Journal of Animal Science 75: 1337-1342.

Pharazyn, A., Foxcroft, G.R. and Aherne, F.X., 1991. Temporal relationship between plasma progesterone concentrations in the utero-ovarian and jugular veins during early pregnancy in the pig. Animal Reproduction Science 26: 323-332.

Pope, C.E., Christenson, R.K., Zimmerman-Pope, V.A. and Day, B.N., 1972. Effect of number of embryos on embryonic survival in recipient gilts. Journal of Animal Science 35: 805-808.

Prime, G.R. and Symonds, H.W., 1993. Influence of plane of nutrition on portal blood flow and the metabolic clearance rate of progesterone in ovariectomised gilts. The Journal of Agricultural Science 121: 389-397.

Prunier, A., Martin, C., Mounier, A.M. and Bonneau, M., 1993. Metabolic and endocrine changes associated with undernutrition in the peripubertal gilt. Journal of Animal Science 71: 1887-1894.

Ptak, A., Kajta, M. and Gregoraszczuk, E.L., 2004. Effect of growth hormone and insulin-like growth factor-I on spontaneous apoptosis in cultured luteal cells collected from early, mature, and regressing porcine corpora lutea. Animal Reproduction Science 80: 267-279.

Quesnel, H., Boulot, S., Serriere, S., Venturi, E. and Martinat-Botte, F., 2010. Post insemination level of feeding does not influence embryonic survival and growth in highly prolific pigs.

Animal Reproduction Science 120: 120-124.

Quesnel, H., Pasquier, A., Mounier, A.M. and Prunier, A., 2000. Feed restriction in cyclic gilts: Gonadotrophin-independent effects on follicular growth. Reproduction Nutrition Development 40: 405-414.

Ramaekers, P., Kemp B. and Van der Lende, T., 2006. Progenos in sows increases number of piglets born. Journal of Animal Science 84(1): 394.

Rampacek, G.R., Robison, O.W. and Ulberg, L.C., 1975. Uterine capacity and progestin levels in superinducted gilts. Journal of Animal Science 41: 564-567.

Razdan, P., Tummaruk, P., Kindal, H., Rodriguez-Martinez, H., Hulten, F. and Einarsson, S., 2004. The impact of induced stress during days 13 and 14 of pregnancy on the composition of allantoic fluid and conceptus development in sows. Theriogenology 61: 757-767.

Sawyer, K.S., Athorn, R.Z., Collins, C.L. and Luxford, B.G., 2013. Increasing feed intake in early gestation improves farrowing rate in first and second parity sows. In: Pluske, J.R. and Pluske, J.M. (eds.) Manipulating pig production XIV. Australasian Pig Science Association, Melbourne, Australia, 132 pp.

Soede, N.M., Bouwman, E.G., Van der Laan, I., Hazeleger, W., Jourquin, J., Langendijk, P. and Kemp, B., 2012. Progestagen supplementation during early pregnancy does not improve embryo survival in pigs. Reproduction in Domestic Animals 47: 835-841.

Soede, N.M. and Kemp, B., 2015. Best practices in the lactating and weaned sow to optimize reproductive physiology and performance. Chapter 17. In: Farmer, C. (ed.) The gestating and lactating sow. Wageningen Academic Publishers, Wageningen, the Netherlands, pp. 377-407.

Soede, N.M., Van der Lende, T. and Hazeleger, W., 1999. Uterine luminal proteins and estrogens in gilts on a normal nutritional plane during the estrous cycle and on a normal or high nutritional plane during early pregnancy. Theriogenology 52: 743-756.

Sorensen, G., 2013. Feeding level during early pregnancy in sows: Effects on litter size and farrowing rate. In: Pluske, J.R. and Pluske, J.M. (eds.) Manipulating pig production XIV. Australasian Pig Science Association, Melbourne, Australia, 132 pp.

Stefanczyk-Krzymowska, S., Grzegorzewski, W., Wasowska, B., Skipor, J. and Krzymowski, T., 1998. Local increase of ovarian steroid hormone concentration in blood supplying the oviduct and uterus during early pregnancy of sows. Theriogenology 50: 1071-1080.

Tast, A., Peltoniemi, O., Virolainen, J. and Love, R., 2002. Early disruption of pregnancy as a manifestation of seasonal infertility in pigs. Animal Reproduction Science 74: 75-86.

Toplis, P., Ginesia, M.F.J. and Wrathall, A.E., 1983. The influence of high food levels in early pregnancy on embryo survival in multiparous sows. Animal Production 37: 45-48.

Town, S.C., Putman, C.T., Turchinsky, N.J., Dixon, W.T. and Foxcroft, G.R., 2004. Number of conceptuses in utero affects porcine fetal muscle development. Reproduction 128: 443-454.

Tremblay, G.F., Matte, J.J., Dufour, J.J. and Brisson, G.J., 1989. Days of gestation after folic acid addition to a swine diet survival rate and development of fetuses during the first 30 days. Journal of Animal Science 67: 724-732.

Virolainen, J.V., Love, R.J., Tast, A. and Peltoniemi, O.A., 2005a. Plasma progesterone concentration depends on sampling site in pigs. Animal Reproduction Science 86: 305-316.

Virolainen, J.V., Tast, A., Sorsa, A., Love, R.J. and Peltoniemi, O.A.T., 2004. Changes in feeding level during early pregnancy affect fertility in gilts. Animal Reproduction Science 80: 341-352.

Webel, S.K. and Dziuk, P.J., 1974. Effect of stage of gestation and uterine space on prenatal survival in the pig. Journal of Animal Science 38: 960-963.

Wientjes, J.G.M., Soede, N.M., Van den Brand, H. and Kemp, B., 2011. Nutritionally induced relationships between insulin levels during the weaning-to-ovulation interval and reproductive characteristics in multiparous sows: II. Luteal development, progesterone and conceptus development and uniformity. Reproduction in Domestic Animals 47: 62-68.

Willis, H.J., Zak, L.J. and Foxcroft, G.R., 2003. Duration of lactation, endocrine and metabolic state, and fertility of primiparous sows. Journal of Animal Science 81: 2088-2102.

Wu, G., Bazer, F.W., Burghardt, R.C., Johnson, G.A., Kim, S.W., Li, X.L., Satterfield, M.C. and Spencer, T.E., 2010. Impacts of amino acid nutrition on pregnancy outcome in pigs: mechanisms and implications for swine production. Journal of Animal Science 88: E195-E204.

Zak, L.J., Williams, I.H., Foxcroft, G.R., Pluske, J.R., Cegielski, A.J., Clowes, E.J. and Aherne, F.X., 1998. Feeding lactating primiparous sows to establish three divergent metabolic states: I. Associated endocrine changes and postweaning reproductive performance. Journal of Animal Science 76: 1145-1153.

3. 妊娠母猪的群养

H. A. M. Spoolder* and H. M. Vermeer

Department of Animal Welfare, Wageningen UR Livestock Research, P. O. Box 338, 6700 AH Wageningen, the Netherlands; hans.spoolder@wur.nl

摘要：在全球养猪生产中，妊娠母猪的群养系统逐渐代替单栏限位饲养。现代群养系统的妊娠母猪表现出单栏饲养同等的生产水平。群养系统成功的关键在于能否处理好动物之间的相互争斗行为：群养模式的母猪将为确定自己在群体中的地位或获取资源而争斗，这些行为是母猪的正常行为。但科学的管理、料槽设计和围栏布局能帮助避免这些潜在的负面影响，如受伤、福利减少和生产损失。

通过实践研究，确定了群养系统中成功管理的几个因素。在妊娠的第2～4周期间，应该避免陌生母猪的混养。母猪应该有足够的空间来避免攻击，特别是在引入了不熟悉的母猪之后。母猪群体的规模似乎不那么重要，但是对于第一和第二胎的母猪来说，将其与三胎及以上的母猪分开是很明智的。在母猪群体中混养一头公猪（切除输精管），可能有助于减少母猪之间的攻击性。

以适当的方式调教后备母猪适应群体活动，或者调教后备母猪熟悉即将进入的群体母猪群，这两种方法都能减少混群活动对母猪的影响。料槽的设计和类型对采食争斗行为也有很大影响。使用自动饲喂系统可有效降低采食争斗。

在饲养过程中，母猪的自我保护能力，以及强势的母猪从劣势母猪处偷取食物的程度也是影响母猪争斗程度的重要因素。

饲喂水平也对攻击性行为有一定影响：限量饲喂增加了母猪在料槽周围的活动以及社会交往，因此增大了母猪间争斗的可能性。

圈舍应该设计独立的采食区、休息区和排泄区。

此外，猪舍内还应采取措施控制环境以防暑降温，营造躺卧区的舒适条件能够保证母猪的正常休息，促进其健康和福利。

生产中，有几种类型的群养系统，这些类型之间的主要区别是母猪饲喂方式。不同群养系统的选择取决于养殖场的发展方向：经济方面、动物健康和福利方面。管理水平和劳动力需求也决定了不同群养模式的选择。通过科学的管理，每一种群养系统都能维持群体妊娠母猪较好的状态。

关键词：母猪，群体饲养，动物福利，争斗，母猪生产力，母猪饲喂系统

3.1 引言

在全球的养猪生产中,母猪群养系统逐渐代替了母猪限位系统。欧盟已经禁止了妊娠28天以后的母猪(欧盟,2008)单栏限位饲养。在美国,许多州被强制执行群体饲养(Harris,2014),在澳大利亚和新西兰的立法也包括对妊娠母猪群体饲养的要求(Harris,2014),然而,在实际养猪生产中单栏限位饲喂系统到群体饲养的转变并没有很快被执行:2013年1月1日(欧洲禁令正式发布生效的时间),国际人道协会(HIS)报告说,欧盟14个养猪生产成员国仍然未能通过该法案。由于种种原因,许多农场主推迟到了最后一刻,主要原因是此系统的投资成本和缺乏信心(Tuyttens等,2011)。

关于妊娠母猪是群养还是单栏限位饲养,很多科学家争论十分激烈。母猪群养系统的首次主要倡导者之一是欧洲科学兽医委员会,该委员会于1997年发表了关于集约化养猪福利的科学报告(SVC,1997)。然而,在几年后他们的观点受到了澳大利亚科学家的反对(Barnett等,2001)。关于这个科学争论,美国兽医协会强调必须达成共识。在2005年的报告中,他们在养殖生产中不再强调必须推荐群养畜舍或限位系统,认为"行业发展最急需的是现代猪舍改进和科学管理实践,为养猪企业提供实际可行的方法,以提高母猪的福利"(Rhodes等,2005)。

随后,有关母猪群养系统优于单栏限位饲养系统的研究并没有取得较大进展。Jansen等(2007)证实,群养的母猪比限位母猪更具有攻击性,而且对背膘厚度或繁殖性能并没有显著影响。Harris等(2006)一致认为,妊娠期间一头后备母猪单栏饲养和在4头母猪圈舍群养相比,其繁殖力几乎没有差别,例如,繁殖性能并没有什么不同(尽管攻击性显著不同)。Karlen等(2007)发现,大群饲养的母猪在妊娠早期面临着较大的福利挑战,出现了抓伤率增加、返情率升高以及在妊娠早期皮质醇浓度有升高趋势等现象。这些现象可能都是母猪间相互争斗的结果。与此相反,限位的母猪在妊娠期也会面临较大的福利挑战。因为限位饲养,使母猪出现了较高的蹄趾病发病率,嗜中性粒细胞/淋巴细胞比率增加,这可能是母猪应激增大的原因。总之,这些资料表明,在畜舍系统中,福利的优势和劣势随着时间的推移而改变。最后,Chapinal等(2010)证实,在群养畜舍系统中,呆板行为不那么普遍,而外阴损伤和跛足则更为普遍。他们得出的结论是,在群养畜舍系统中,可能需要更高水平的员工来检测问题,并保持与限位畜舍相同的生产率和组织损伤水平。

对于空怀母猪群养,如果纯粹基于科学证据,似乎不可能得出到底是赞成还是反对的结论。弗雷泽(2003)提出,这可能是对于动物福利如何评估存在不同观点。目前的评估指标主要有:(1)生物功能;(2)没有痛苦;(3)它们的饲养情况的自然性。他说,"当试图评估动物福利时,不同的科学家会选择不同的标准,这反映了他们的价值取向不同。"Weaver 和 Morris（2004）证实了这一观点，并提出，科学不会决定两种方法中哪一种是最好的（群饲或单栏限位饲养）。相反，他们鼓励基于科学研究的道德辩论，另一些人则认为，在一些国家，单栏限位饲养仍然没有大的改变（Caulfield 和 Cambridge，2008）。

从科学的角度，可能不会再讨论妊娠的母猪应该是群养还是单独饲养，也许从来就没有真正的争论过。而公众的争论，尤其是欧洲的争论，已经说服社会采用立法支持母猪群养。此外，还影响大公司改变猪的饲养方式（Mench，2008），以及迫使猪肉零售商从实施这些群养系统的养殖公司购买生猪。目前世界上有数百万种母猪被饲养在一起，并为它们的主人提供了一份收入。还有一个重要的科学问题，那就是如何以一种同时优化母猪福利和农业生产力的方式来完成这一任务。以下的内容概括了当前关于这个问题的最新知识。

3.2 群养系统的母猪行为表现

现代母猪的成功管理可以提高生产效率、提供良好的母猪福利以及猪场工作人员良好的工作环境。单栏限位和母猪群养之间的关键区别是，群养的母猪之间可以相互作用，从而导致积极的和消极的群居互动的增加。母猪之间的攻击性是最显著的负面群居互动形式，它影响着母猪的健康和福利、猪场的生产力，这些又进一步降低了工人的工作满意度。

母猪争斗对动物的福利不利。在争斗期间和争斗之后，争斗失败的动物通常会遭受最严重的伤害，可能是严重的皮肤损伤、脓肿、运动问题（Marchant 等，1990，1995）；与争斗相伴随的急性心理压力导致心率（Svendsen 等，2001）和皮质醇水平增加（Burfoot 等，1992）。此外，有迹象表明，母猪妊娠早期的群居压力可能会导致繁殖率下降（Bokma，1990；Burfoot 等，1997），而在断奶期间发生的侵略行为减少了随后的发情行为表现（Burfoot 等，1993）。在一些情况下，可用资源有限，社会等级较低的母猪可能无法完全适应它们的群养环境，导致严重的体重下降，或者在极端情况下出现死亡。

大多数由侵略行为引起的损伤都是抓伤或皮肤上的伤口。在母猪和后备母猪中，阴户只是偶然的目标。然而，外阴的咬伤并不被认为是一种侵略行为，而是由于缺乏进食机会的一种失望而引起的。攻击性也可能导致腿部问题，特别是当围栏的设计使动物不能轻易地避免事故时（通过空间限制），或者当地板在走动过程中对肢蹄支撑不够时（太滑或质量不佳的漏粪板）。与皮肤损伤相比，腿部疾病的发生率要低得多。最后，攻击性不可避免地与生理和心理上的压力联系在一起（见上文），尽管它未必会影响繁殖能力。Couret 等（2009）在妊娠最后 1/3 的时间里反复地将群养压力应用于妊娠的后备母猪，发现尽管这与重复激活的下丘脑-垂体-肾上腺轴有关,但并没有影响它们的免疫功能和妊娠的结果。Soede 等(2007)之前也获得了类似的结果。

3.2.1 减少负面的社会行为

争夺有限资源和陌生母猪间等级的竞争是母猪争斗的两个主要原因。这两个引起争斗的原因在许多方面有很大的不同。关于食物、水或躺卧的地方的竞争通常是短暂的,但非常频繁(经常是每天)。另一方面,当陌生的母猪混群时,就会通过争斗确定它们的相对等级而建立社会关系,这种类型的攻击并不频繁,但会更加强烈。

建立优势等级

作为一个社会等级高的母猪,不仅具有第一个获得食物、水和更喜欢的卧位空间资源的权利,还有其他更多的优势。Hoy 等(2009b)报告显示,在他们的两个农场中,具有较高排名的母猪的窝产仔数(这两个农场每窝小猪的数量为 12.66 头和 16.14 头)显著高于低等级母猪的产仔数(分别为每窝 12.13 头和 14.83 头小猪)。Tönepöhl 等(2013)研究发现,虽然在混群后作为侵略者的母猪产生更多的总仔猪数和活仔,然而有较高皮肤损伤的母猪的繁殖能力较低(皮肤损伤评分和初生总仔猪数的相关系数:$r=-0.28,P<0.01$)。这些对繁殖性能的影响可能与应激的"可预测性"或"控制"程度有关。

虽然在母猪之间的争斗可能会产生相当严重的后果,但在新猪只引进后的两天内,社会等级制度通常就会建立起来,从而导致攻击性的急剧减少(Bokma 和 Kersjes,1988;Luescher 等,1990),但新引进母猪与生产母猪群完全融合相处需要更长的时间。Moore 等(1993)和 Spoolder 等(1998)研究了小群的后备猪和经产母猪,它们被引入大型的动态群中,完全融合相处它们至少需要 3~4 周。在这个相对稳定的时期,个体的相对社会地位得到确认,要想促进整个群体社会等级形成,可以通过培养饲喂程序(Bressers 等,1993;Hunter 等,1988)或躺卧模式(Moore 等,1993)。与此相反,定位栏饲养的母猪被观察到在长时间内对其邻居保持一定程度的侵略性,可能就是因为它们在身体上被限制而无法建立等级制度的原因(Barnett 等,1987;Vestergaard 和 Hansen,1984)。

当猪只互相不认识对方或在等级制度中互相争抢其位置时,等级秩序的争斗就开始了。Stookey 和 Gonyou(1998)观察了争斗的模式,并提出,在初生仔猪中,相互识别是建立在熟悉的基础上的,它们是在一起长大的,或者是在一起居住的,并且不涉及基因上的亲缘关系。为了减少攻击性,无论是在社会地位还是对资源的获取上,动物倾向于形成所谓的"规避秩序"(Jensen,1982;Jensen 和 Wood-Gush,1984),这一秩序有助于争斗的最小化,并与高等级制度形成紧密联系。母猪经过几周的分离,也可以很快地相互识别。Spoolder 等(1996)观察到在两周间隔期内把两组后备母猪转入一个大的群体中,发现来自一个群体的6头母猪组,很快相互识别,与另外一组后备母猪相比,组内的争斗也相对较少。研究发现,这种分离间隔时间最长不能超过4周。随着年龄的增长,动物等级制度的稳定性似乎有所提高。Parent等(2012)测定了母猪在生长阶段和妊娠阶段等级的稳定性,研究发现妊娠母猪的稳定性持续时间高于生长母猪。并且在妊娠期间的母猪与年轻后备母猪相比,其等级制度会随着时间的延长变得更加稳定(Parent等,2012)。

资源竞争

母猪对诸如空间等资源的竞争是通过躺卧模式来调节的。随着妊娠时间的增加,躺卧模式会随着时间的延长而改变,并且通常母猪会转移到更喜欢的地方(相对安静和舒适的地方)躺卧(Spoolder,1998)。这在群养系统中尤为明显。与此同时,对躺卧模式的评估表明,在妊娠期间邻近的母猪也会发生变化:陌生猪只间的距离减少(Spoolder等,1996,1998)。最近,Krauss和Hoy(2011)证实了这一点,他们发现母猪喜欢与熟悉的群成员躺卧最近,而不是那些进入群养系统之前没有遇到过的母猪。他们计算了混群当天以及3周后最近邻居相互熟悉的场合百分比,并将新母猪融入动态群体(新母猪:第一天94.3%,第21天46.0%;常住母猪:第1天96.8%,第21天74.6%)。正如Spoolder(1998)的研究一样,他们无法得出结论,这种偏好是母猪对私有空间的固定行为还是对同一栏猪只的明确偏好。然而,由于这一效应在新母猪引入后的3周内显著下降,而没有对猪栏中的环境作任何改变,因此,可能是母猪形成了某种形式的"亚群身份"(Moore等,1993)。

对饲料的竞争,特别是在诸如母猪电子饲喂站等连续饲喂系统中,存在所谓的"饲喂顺序"。这些都与社会等级关系密切相关,因为等级最高的母猪通常是第一个在料槽开启后进食的动物(Csermely,1989;Hunter等,1988)。最近,Kirchner等(2012)报告了一项研究,他们支持通过训练母猪响应一个声音信号(喊叫饲喂)来发展一种饲喂顺序。与喊叫饲喂(61.5%)相比,在正常饲喂中,更多的母猪(83.1%)参与了争斗。正常饲喂与喊叫饲喂相比,头部和侧面而不是肩部和后侧严重损伤的数量更大。Kirchner等(2012)提出,将饲喂时间单独地发出信号,可以提高进入饲喂站的可预测性,从而减少母猪之间的竞争。

母猪的侵略程度受到多方面影响:(1)动物的管理;(2)饲养系统;(3)猪舍和建筑物的设计;(4)农场工作人员的技能水平和态度。对于第4点,可以参考第12章(Prunier和Tallet,2015)。其他三个问题将在下面的段落中讨论。

3.2.2 大群管理

混群时间

专家研究发现,应激会对胚胎的存活产生负面影响,特别是在妊娠早期,胚胎贴附在子宫壁的时期(11~16天),以及此后不久的时间。在这一阶段,所谓的"妊娠的母体识别"伴随着许多相关的激素变化。这是否与向一个大群中引入母猪的时机有关,已经引起了很多争论。

荷兰在20世纪80年代后期的研究表明,在妊娠的第一周混群的母猪有20%的返情率,而在妊娠的第四周进行混群,使返情率显著下降,达10%(Bokma,1990)。在本项研究中,采用的是一个母猪电子饲喂站,每个饲喂站有40头母猪,保持动态调整。每周有5~6头新母猪被引入这个群体中,因此可能导致了不好的结果。Kirkwood和Zanella(2005)提出,混群分组的关键时期可能更为有限:他们发现在15头母猪地板式饲喂,重新分组后(第2天)配种表现出最高分娩率,而重新分组第14天显示最低的分娩率,但窝产仔数(那些妊娠到配种的猪)没有明显受分组混群时间的影响。Van der Mheen等(2003)通过观察375头母猪的800胎次记录,在配种后立即被分组混群的母猪具有最短的发情间隔时间以及最大的窝产仔数。然而Knox等(2014)似乎发现了相反的结果,在他们的研究中,他们比较了在第3天到第7天(D3),第13~17天(D14),第35天(D35)混群母猪,以及在整个过程中单栏限位饲喂。结果发现,母猪配种后3~7天混群的繁殖成绩最差。D14和D35处理在繁殖或福利方面几乎没有什么区别,但D14与单栏限位处理相比D14处理在断奶后10天内有更少的返情率(88.8%对96.6%)。

其他研究并没有发现分组混群时间的影响。Van Wettere等(2008)比较了没有混群(但仍在它们的配种前组中)的后备母猪,和妊娠后3~4天或8~9天时混群的母猪。发现妊娠率和胚胎存活率没有差异。同样,Cassar等(2008)将617头母猪分别在配种后2、7、14、21或28天混入15个母猪群中,并没有发现对分娩率或窝产仔数有影响。

因此,目前关于混群时间的研究还没有一致的结果。另外,关于早期胚胎发育与母猪应激之间的直接关系也没有确定的结论。Soede等(2007)对81头发情后备猪给予不同的应激处理,即在卵泡期的应激处理($n=20$)、早期妊娠期的应激处理($n=20$)、两个阶段均进行应激处理($n=21$)和无应激处理($n=20$),并测定了每种处理下的应激参数,如心率、皮质醇浓度、刻板行为、胚胎发育和其他繁殖参数。结果表明,应激对后备母猪在卵泡期和早期妊娠期没有任何影响,也没有观察到对繁殖过程的影响。

因此,与群养相关的因素或所使用的母猪的品种不同,分组混群时间不同对后续繁殖性能的影响也不同。但是,最好避免在妊娠初期发生应激,尤其受精后的2~3周内,以避免母猪生产力的下降。

空间分配

所有的动物不仅需要休息、采食、排便的空间,而且还需要进行社交活动、满足它们探索环境的特性的空间。在没有足够空间的情况下,分组混群会增加攻击性行为,而造成不良后果,导致皮质醇水平长期升高。多项研究报告称,增加每头母猪所占空间将减少争斗和损伤(Barnett 等,2001;Docking 等 2000;Remience 等,2008),这可能主要是由于减少了争斗的机会,例如通过表现顺从的行为(Jensen,1982)。在群养母猪中,空间不足也会导致蹄和腿部受伤的发生率升高,这可能是由于接触了猪舍内设备或地板,或者是由于后备母猪之间的非争斗性相互作用,如母猪之间的相互踩踏(Harris 等,2006)。

  为母猪提供足够的福利所需的最低空间尚未科学地确定。作为进一步量化的起点,EFSA(2005)描述了三种类型的空间:静态空间、行为空间和交互空间。其中静态空间最容易评估,因为它可以应用方程:$A = k \times W^{0.666}$,表示猪只躺卧或站立的状态。在这个方程中,A 是面积,单位为 m^2;W 是体重,单位是 kg;k 是一个常数,根据动物的姿势而定(Baxter,1986)。站立或胸骨躺卧(竖卧)的猪 $k=0.019$,对于休息状态的躺卧(侧卧)$k=0.047$。可以计算出,一个 200 kg 的母猪,侧卧和胸骨躺卧(竖卧)空间分别相当于 $1.61\ m^2$ 和 $0.65\ m^2$。虽然缺乏母猪所需要的行为空间的数据,如采食和排便等行为所需要的空间,但可以根据姿势和动物的数量来估计。有些空间也是和母猪生物学需求相关的。

  母猪相互作用发生的空间需求是最难估计的,例如交配、打架和逃跑。Baxter (1985)估计两头母猪争斗时所需要的空间公式为 $0.11 \times W^{0.667}$,这个公式展示,两头 200 kg 母猪争斗时需要大约 $3.8\ m^2$ 的空间。对母猪空间需求的估计十分复杂,主要受到打斗发生的频率和群体中母猪的认知程度的影响。逃离侵略可能从 2.5 m 以下的距离,直到母猪被追上时已经超过了 20 m(Edwards 和 Riley,1986;Kay 等,1999)。另外,值得注意的是,母猪群体互动的空间需求具有暂时性。争夺主导地位的交配活动和争斗行为都是暂时的活动,这些活动需要专用的混合围栏或交配区提供所需空间。

  有几项研究考察了不同空间对表现和争斗的影响。Barnett 等(2001)建议为获取最佳的母猪生产表现需提供 $1.4 \sim 1.8\ m^2$ 的空间,研究表明,为母猪提供 $3\ m^2$ 的空间与 $2\ m^2$ 相比,其繁殖效果更好。Weng 等(1998)观察了每头猪只占用 2.0、2.4、3.6 和 $4.8\ m^2$ 面积对母猪攻击性和皮肤病变方面的影响。结果显示,对于畜栏中分别饲喂 6 头母猪的母猪群,建议最低的空间面积是每头动物 $2.4\sim3.6\ m^2$。但他们强调,这一结果不应被推广到其他群养规模或饲喂系统中,因为在更大的群体中,所需空间的数量可能更少。Salak-Johnson 等(2012)发现不同的地板空间面积(1.4、2.3 和 $3.3\ m^2$),对 5 头母猪的小群体而言,尽管相互争斗的频率没有明显的差异,但 $3.3\ m^2$ 的母猪与 $1.4\ m^2$ 的母猪相比,进行了更长时间的相互争斗。遗憾的是,在这项研究中没有对病变方面进行比较。在同一课题组的早期工作中,比较相同的饲养密度(Salak-Johnson 等,2007),增加空间面积会减少母猪损伤,这与其他研究结果相一致。因此,2012年出现的明显矛盾的结果可能与增加空间面积条件下互动的强度降低有关。最后,Hemsworth等(2013)也证实,根据攻击性和皮质醇的结果,在将空间面积逐步增加,从1.4、1.8、2.0、2.2、2.4到3.0 m²进行比较时,母猪的幸福感有所改善。研究发现,虽然无法确定对母猪而言,足够的空间面积应该是多少,但他们证实,每头母猪1.4 m²的空间面积太少。

猪群规模

众多课题组已经研究了不同群体规模对争斗行为和性能方面的影响。Barnett等(1986)比较了三组母猪群养时的争斗行为,每组母猪头数分别为2、4、8头,结果发现攻击性(或繁殖变量)没有明显的差异。但在随后的研究结果中发现,在较大的群体中,母猪拥有更大、更丰富的空间,一般来说,相对于在较小的群体中,它们之间的争斗显著减少(Broom等,1995;Mendl,1994)。相比之下,Hemsworth等(2013)比较了每组有10、30和80头母猪群中母猪的皮肤损伤,发现每组含有10头母猪的母猪群中猪只的皮肤损伤在几个观察天内都是最低的。从直觉上看,这更合乎逻辑,因为较小的群体意味着更少的群体互动和潜在的更少的争斗。

在实践中,通过攻击级别评估发现了300头的群养规模对母猪群社会等级没有明显的不利影响。当母猪群体规模增加时,猪可能会改变与群体中的同伴互动策略,而且它们不能够识别所有的个体:它们不太可能参与相互争斗,并表现出更多的逃避行为(Turner和Edwards,2000)。随后,Turner等(2001)在一项研究中证实了这一点,该研究发现,与来自小群体的猪相比,大型群体中的猪对不熟悉的其他猪只的攻击性较低。

是否将后备母猪隔离饲养?

Hodgkiss等(1998)研究了一个农场具有100～110头母猪的大群,发现与同第3胎或者更高胎次的母猪相比,第1胎和第2胎母猪的伤害明显增加。与此类似,Kirkwood和Zanella(2005)发现,同第一胎和第二胎母猪相比,后备母猪更少发生争斗。这种身体强度与支配资源的可能性之间的明显关系提出了一个问题,即是否应该把母猪按照胎次分群,或者至少在群体中老母猪和年轻母猪不混群。

Hoy等(2009a)研究发现,初次分娩的母猪与年轻未分娩母猪一起饲养,相对于初次分娩母猪和老母猪混群饲养而言,其具有明显的更高的社会等级指数:它们在群体中更经常占主导地位。有可能,这意味着它们在妊娠期间会受到更低的攻击和应激。这一点得到了Li等(2012)的证实,他们发现,如果与后备母猪一起混群饲养,争斗次数增加和争斗时间更长,但相对于第一胎母猪与经产母猪在混群饲养时,它们会获得更多胜利,而且更少受伤。因此,建议通过胎次分选,将由混群引起的攻击造成的严重伤害的第一胎母猪挑出,以至于使它们的福利和表现可以在群养系统中得到改善。

在母猪群中饲养一头公猪(输精管切除)可以减少争斗?

有人认为,在母猪群体里,公猪的存在减少了断奶的母猪和后备母猪之间的争斗(Luescher 等,1990)。可能是在这个群体中,一个"超级主导"的公猪,会压制其他个体来主导这个群体的潜在行为愿望。Kirkwood 和 Zanella(2005)证实了这一假设,他们研究了公猪在混群后的影响,Docking 等做出了同样的研究(2000),在专门的混群猪圈中,在没有公猪存在的情况下,对母猪争斗进行打分。Grandin 和 Bruning(1992)通过在猪圈中混养 50 头 100 kg 的猪,在放置和不放置 3 个成熟的公猪情况下,检测公猪的影响。以上研究结果表明,公猪的存在减少了损伤的程度,也降低了争斗的强度和频率。Seguin 等(2006)对有身体接触、通过栅栏接触或没有接触公猪的 5 个含有 15 头母猪的母猪群比较时,发现仅有有限的影响。在他们的研究中,与没有公猪的情况相比,群体中有公猪并且与公猪的身体接触在一定程度上降低了母猪之间的攻击性水平,但同时也增加了唾液皮质醇的水平。总的来说,公猪的存在似乎可以减少争斗,尽管影响很小。

社交技能训练

引入后备母猪到经产母猪群,可通过让后备母猪与经产母猪先进行熟悉来减少其攻击性。Backus 等(1997)研究了一种测试这种假设的精细方法,他们每隔 3 周组织一次大群户外活动将后备母猪暴露给经产母猪。在这种处理之后,后备母猪的发情周期模式更容易预测,而且在引入主要的母猪群的过程中,后备母猪参与的争斗行为很少。Van Putten 和 Buré 等(1997)认为,可能是由于后备母猪拥有更多的社交技能,尤其是在与年龄较大的、占主导地位的母猪接触时表现出顺从的行为。

Kennedy 和 Broom(1994)使用了一种更实用的方法,允许后备母猪在引入至它们将要一起饲喂的母猪群之前先与这些母猪进行 5 天的一定程度的接触。这些后备母猪被安置在母猪群附近的一个猪栏里,能够看到、闻到、听到并偶尔通过它们的围栏接触到经产母猪,与不允许进行这种提前接触的后备母猪相比,提前接触的后备母猪在引入大群后的前几周内受到的攻击较少。似乎动物们建立了一种主导关系,尽管它们实际上并没有参与争斗来确认它。

3.2.3 饲喂系统

同时和顺序饲喂系统

饲喂系统的设计影响到食物竞争的程度。基本上,在这一背景下可以考虑两种饲喂系统:同时和顺序饲喂系统。同时饲喂系统的特征是每天有一两次的饲喂活动,在此期间,动物们想要获取食物并且去采食。一般的同时系统是与自由的大栏圈舍、地板饲喂和长槽(湿)饲喂系统相结合。顺序饲喂系统是指动物以另一种方式饲喂的系统。母猪电子饲喂(electronic sow feeding,ESF)系统在这一类别中是最常用的,但所谓的 Fitmix 系统最近也得到了关注(2008 年 Chapinal 和 2003 年 Van der Mheen 等)。然而,在受限制的饲喂期间受到围栏的保护,例如采用 ESF,是首选。ESF 系统可以为母猪提供足够的福利和性能表现,只要它们的设计和管理良好。

对于同时饲喂系统,特别是当提供固体饲料时,给予个体动物的保护水平尤其重要。因此,如果它们试图吓跑更小、更慢的母猪,那么快速进食就不会很快得到回报。单体栏位通过减少屏障长度以隔离饲喂区域的食槽提供最好的保护(Andersen 等,1999)。

对于顺序饲喂系统来说,更重要的是,保证动物都能吃到料而不能被老母猪顶替。这种偷窃行为在很大程度上是"学习的":每次回访或挑战正在吃食母猪时,都会得到额外的饲喂,这种行为就会得到加强。打破这一学习习惯对所有的顺序饲喂系统都是至关重要的。在没有年龄较大、更占优势的母猪的情况下,后备母猪应该学会如何使用电子母猪饲喂器。一般情况下,它们可以在 1~2 周内学会使用该系统(Nielsen,2008;Spoolder,1998)。

饲喂水平

妊娠母猪一般需要限饲,目的是优化身体状况和繁殖性能。然而,在某些系统中可能会出现过量采食,例如,在自由采食方式饲喂系统中,占统治地位的母猪可以吃掉大量分配的饲料。母猪食用超过其生理需求的饲料会获得更大的体重和更厚的背膘,尽管这不会影响短期的繁殖性能(在妊娠期的分娩率和窝产仔数及随后的哺乳表现),然而,在妊娠早期过度饲喂会增加母猪的胚胎死亡率(Foxcroft,1997)。

Van der Peet-Schwering 等(2004)研究了大群妊娠母猪自由采食含高铁非淀粉多糖(与受限制的传统饲喂相比较)对母猪的繁殖性能的影响。他们跟踪研究母猪超过 3 个繁殖周期。研究发现,在妊娠期间,自由采食的母猪采食量为 1.3 kg/天,超过了限制饲喂母猪的采食量,也增加了体重和背膘。而在哺乳期和繁殖过程中,采食量不受自由采食的负面影响(Van der Peet-Schwering 等,2004)。Stewart 等(2010)也研究了高纤维饲料对母猪行为的影响,并将含有 15% 的粗纤维的日粮与含有 5% 粗纤维的对照日粮相比较。在高纤维的处理组中,相对于限制饲喂的母猪而言,母猪花更多的时间在猪舍休息,更少的时间表现出呆滞的行为,并且攻击性行为减少。Peltoniemi(2010)研究了自由采食饲料(7.7 MJ/kg)和对照组饲料(9.3 MJ/kg)对繁殖性能(受胎率、断奶至发情间隔、产活仔、死胎仔猪和孕激素浓度)的影响。他们没有发现对以上这些指标有显著影响,但报告说,对照组母猪的断奶仔猪数明显多(9.7∶9.4),而自由采食的猪则在断奶时更重(8.8~8.0 kg)。

3.2.4 圈舍设计

功能区域：躺卧区、采食区、排便区

母猪的居住设施应该根据动物的需要而设计，包括躺卧、采食和排便。猪会设法使这些区域保持独立，使它们的采食和躺卧的地方远离粪便和尿液。它们会在猪舍里找到一个最舒服的地方躺卧。提高躺卧区质量要求它应该远离围栏的主要活动区（如走道、食槽和饮水器）、无贼风、提供保护，以及远离吵闹的邻居（例如，用坚固的围栏分区）。一般来说，排便区应该距离躺卧区不远，在一个安静的地方，具有很好的落脚点，这样动物就能舒服地拱起它的背部进行排泄行为。一般来讲，猪会保持它们的躺卧区干净并保持自己的身体干净，但由于环境温度过高不能降低体温时，它们也只会躺在自己的粪尿中。

围栏设计也可能与猪的社会行为有关。Docking 等（2000）总结出一个具有大空间面积的圆形围栏（每头猪 9.3 m^2，由 5 个不熟悉的母猪组成的群体）是最好的减少攻击的方法。然而，对于一个最佳的综合考虑逃跑可能性和建造一个围栏的成本，建议在低密度基础上采用矩形围栏来保障躲避区域。

环境对性能和行为的影响

温度、湿度和光线的变化将影响母猪的行为和繁殖性能。群养畜舍被确定为季节性不孕症的危险因素（Peltoniemi 和 Virolainen，2005；Peltoniemi 等，2000）。季节性不孕症最明显的迹象是所谓的"秋季返情"，即母猪在受精后的 25～30 天内返情现象季节性增加。出现这种现象的原因似乎是由于褪黑激素浓度的增加而抑制了 GnRH/LH 的释放。褪黑激素是由松果体在每天黑暗时期产生的，因此通常在秋季增加。事实上，褪黑激素的注入对分娩率有负面影响（Love 等，1993）。此外，在秋季的光照期中，LH 的浓度下降（Peacock，1991；Love 等，1993），由此导致妊娠前的大约 14 天开始较低的孕酮浓度（因为从那一天起，孕酮的生产依赖于 LH 的水平）。这些较低的孕酮浓度诱导胚胎延迟发育，从而导致黄体的退化并回到发情期。妊娠早期进食量少的母猪显得特别脆弱。Geudeke 和 Gerritsen（2004）还指出，秋季返情现象在妊娠母猪大群饲养的农场中具有更高的发生率，而且还与母猪的平均身体状况和室内气候的质量有关。

猪的行为也与猪舍中的温度和通风有关。正如上面所讨论的，恶劣的气候环境（高温）会导致一些猪通过躺在自己的粪尿中来降温。贼风和其他方面的气流扰乱了休息的行为，可能会增加动物之间的相互干扰，并导致群体的负面表现。

地板

地板的类型及其条件对群养畜舍的健康、行为和性能有直接影响（Vermeer 和 Vermeij，2014）。一个重要的方面是地板上的垫料（通常是稻草）。秸秆供应可以减少蹄部病变（Heinonen 等，2006）以及异常步态（Andersen 等，1999）现象。然而，人们也注意到，稻草并没有降低争斗的程度；稻草的有益效果是由于在交互过程中更好的"抓地力"。Salaün 等（2002）在比较不同的地板类型时，发现了使用 ESF 系统的大群饲喂母猪的攻击性减少。在断奶后，母猪被分成小群，在配种后 6 周进入大群。与漏缝地板系统相比，在垫草上，有更多的活动和更多的打斗，但更少的受伤和擦伤及更少的时间花在寻找一个休息的地方。然而，在研究的不同系统中，繁殖性状并没有什么不同（Salaün 等，2002）。Barnett 等（2001）指出，铺有稻草母猪不仅能更好地抓地，还能提供更多的热量和身体上的舒适感。此外，他们建议用稻草来减少刻板行为和攻击性行为。另一些人则认为，在光滑地板上和有垫料的系统一样会导致蹄子的过度生长从而导致腿病。预防性的蹄部治疗并不能很好地弥补这一点，还会产生大量额外的工作，而蹄部在治疗后迅速恶化。Ehlorsson 等（2003）总结说，改善住宿条件是改善健康的唯一关键。垫草是一种选择，如果它具有良好的质量，则在猪舍有正常磨损的机会。在一些研究中，母猪经常选择具有垫料（例如稻草）的区域，这似乎像能够获得食物一样重要（Matthews 和 Ladewig，1994）。

提供舒服躺卧的替代解决方案也被测试过。Elmore 等（2010）报道称，在铺有垫草的猪舍中，母猪的总损伤比用混凝土猪舍所产生的损伤要低。然而，跛足分数在两个处理上并没有什么不同。他们确实注意到，当提供橡胶垫时，环境温度需要考虑，因为它们可能会增加猪舍的污垢。Diaz 等（2013）也调查了橡胶垫的影响，发现在两个批次研究中，橡胶垫上的母猪会减少肿胀和伤口的风险。这些作者还报道说，用橡胶板条垫的猪舍比没有加垫子的圈舍更脏，身体损伤分数与地板类型之间没有联系。令人惊讶的是，尽管增加的蹄甲损伤与橡胶垫的使用有关，但跛足不受其影响。

3.3 群养系统类型

在不同的系统中，母猪群可以由 6～500 头母猪组成。母猪的群养系统通常根据饲喂系统来命名，这表明饲喂方式对它们的总体设计是至关重要的。在这一段中，我们将介绍一些常见的群体饲养系统：自由出入圈舍、母猪电子饲喂站、地面饲喂、长槽饲养，少喂多餐和自由采食。

3.3.1 自由出入圈舍模式

在自由出入圈舍（free access stalls，FAS）系统中，母猪通常以每个围栏 8～40 头的方式成群饲养。一个猪栏包括 1～2 排带有公共漏缝地板走道的封闭式饲养棚（图 3.1）。

双排式猪舍更适合在粪肥、饮水、探索和社会交往之间提供更多的空间。母猪

妊娠和哺乳母猪

图 3.1　自由进出栏，两列栏之间安装有宽的漏缝板区域。

一天喂一次或两次，每次饲喂 30 min，直到它们的健康和采食量有保障。在食槽中可以提供少量的水来刺激采食量增加，并且可以自由饮水。群体大小是可变的，通过调整猪圈里的围栏。

通常情况下，在同一周内交配的母猪会根据年龄和身体状况进行分类，因为它们是根据群体水平而不是个体水平来饲喂的。由于母猪躺下时不能进行身体接触，低温的临界温度是 18℃。这意味着在 18℃ 以下，母猪需要额外的饲料来维持身体状况。每一头母猪都有独自的宽 65~70 cm，长 200 cm 的栏位，有坚实的地面，也可以作为躺卧的地方。舍内走动的地方宽度至少为 300 cm，可以让母猪自由活动，并降低因攻击而造成的皮肤损伤和跛足的风险。在栏位前建有一个饲喂检查站，对检查、调节料槽和通风有很重要的作用。

然而，为了适应从限位饲养到群体饲养的转变，需要建立饲喂通道，饲喂通道的建立为栏位和更坚固的地板间创造了更多的空间。在配种后，母猪可以在这一系统中生活几天，直到从检测到真正妊娠到预计分娩前一周，然后再把它们移到产床上。带有一个独立的躺卧区域的自由出入圈舍系统（具有或者没有垫料或者休息区）比当饲喂和躺卧区域放在一起的时候需要更多的空间。

3.3.2　母猪电子饲喂系统

在 ESF 系统中，母猪以 40~500 头为一组，每个饲喂站饲喂 40~60 头母猪。在进入饲喂站的时候，母猪会通过电子耳标被识别。母猪可以一次入站就吃完全天日粮，有 95% 的母猪都是这样的。

一旦一头母猪吃完了它的日粮，它就再也不能进入饲喂站了，这样队列里其他母猪就有机会吃食了。每一头母猪需要15～20 min的时间来吃食,这取决于胎次和饲料量。理论上,一个饲喂站可以负责72头母猪饲养。然而,当优势母猪出现时,即使是最可靠的系统也偶尔会出现故障,所以饲料最好不要过量。出于同样的原因,群体母猪的规模常常成倍增加,所以在一个猪栏中有一个以上的饲喂站经济上是可行的。饲喂站提供的饲料数量有限,水也是有限被提供的,而在活动区域的其他地方水可自由饮用。如果在群体中有一段不活跃的时期,这对动物来说是件好事,也可以让饲养员进行一些检查、维护或训练。与其他系统相比,每头母猪所需要躺卧的空间可能要少一些。因为母猪在处于温度适中状态的条件下有大量的身体接触,所以它们需要宽0.5 m、长2.0 m或1 m²的躺卧面积。在有垫料的系统中,躺卧区大约是1.3 m²。在没有铺放垫料的系统中,通常躺卧区供8～10头母猪使用,带有小块的区域用于施展躺卧行为,并在朝向漏粪地板方向有2%的坡度。在铺垫料系统中,躺卧区不能设置障碍。

普通的商业系统使用1.0～2.5 m²用于猪只活动,包括采食和排便,所以每头母猪的总空间限额是2～2.5 m²(图3.2和图3.3)。

图3.2 150头母猪电子饲喂系统(漏缝板无垫料)。

群体的组成可以为静态的,也可以为动态的。在静态组中,所有母猪均在同一周内交配,并在配种结束后的几天或配种后3～4周内进行分组。然而,在妊娠期间,母猪群体组成不会发生变化,而那些返情的母猪则会回到配种舍。

与动态系统相比,静态母猪群对空间和料槽器的使用效率较低,但由于建立了稳定的社会等级结构,因此具有较低的攻击性。动态母猪群的组成个体不断变化,定期引进刚人工授精的母猪,而马上分娩的母猪将会被带到产房。动态母猪群体

妊娠和哺乳母猪

图 3.3 电子饲喂系统(具有大面积铺垫草的躺卧区以及位于前部的排粪区)。

通常比静态组的更大,允许层次冲突和可能从它们中逃离。大多数动态系统不使用铺放垫料的地板,而是一个混凝土的漏粪地板。坚固的地板并添加垫料可以使这些系统在更冷的气候条件和造价更便宜的猪场中使用,而且通常会使蹄甲更健康。

所有的 ESF 系统都是计算机控制的,并且有可能根据胎次、身体状况和妊娠阶段提供特别的日粮和进食速度。该系统还可用于检测健康问题或发情情况。在一个饲喂器或一组饲喂器的前端出口安装自动门,可以将特定的母猪分类到一个隔离区域进行检查、处理或转移到产房。

3.3.3 地面饲喂

地面平养系统相对便宜且简单。它们不需要食槽。饲料可以散布在每头母猪 45~50 cm 的墙边上,也可以分布在水泥地板中间(倒食)或分布在更大的区域(旋转饲喂)。即使在有垫料地板系统中,母猪也能很好地找到最后的颗粒料。本系统的最大挑战在于,为该群体中的每一种母猪提供足够的营养。其中的一个风险是,母猪之间饲料分配不均产生了竞争。建议在每天两次母猪饲喂以 2 h 的间隔,以防止饲料摄入的巨大差异。使用相对较小的静态组(6~15 头母猪)并根据胎次和身体状况进行分选是必须的。

坚固的地板需要稍微倾斜(1%~2%)以允许尿液排出到漏粪板的区域。普通的圈舍尺寸是每头母猪 50 cm 的宽度,深度是 2.5 m 的坚实地面,漏缝地板大约 2.0 m 深。在这个漏缝地板区域的一个侧区上,安装一个饮水器可以自由饮用。

3.3.4 长食槽饲喂

一个长食槽系统通常是用于6～15头母猪组成母猪群,并提供液体饲料。由于不能单独饲喂,因此,分群时必须按照胎次和体重进行排序。每天一餐保证所有母猪不会过量采食。在料槽中,饲料的均等分配是至关重要的,这样的话所有动物都能获得同样数量的饲料。在漏粪地板区域安装有长槽或者碗形饮水器保障饮水。在这种类型的小圈舍中,为提高母猪的福利而提供一些环境上的改进是很困难的。圈舍的设计看起来非常像前一段描述的地面饲喂系统的圈舍设计。在地面饲喂系统中,至少2%的坡度在这些圈舍中很有必要,可以将液体从坚实的地板上移开。

3.3.5 少量多餐模式

少量多餐模式(Biofix饲养)与长槽饲喂类似,但是饲料在每头母猪的食槽中缓慢而连续地供应。分配速度等于进食最慢的母猪的速度。当饲料被连续地投放时,母猪不会打扰其他进食的母猪。在少量多餐饲喂系统中,根据胎次和身体状况进行分类对于减少进食速度的差异至关重要,这样就限制了母猪更换位置的频率,并尽可能减少饲料摄入和体重发育的差异。这种围栏的设计类似于地面饲喂系统。

3.3.6 Fitmix 系统

Fitmix是一种类似于ESF的饲喂系统,但在饲喂过程中没有任何保护。在电子识别出靠近饲喂器的母猪后,一根搅龙开始转动,水和料混合在一起并通过管道分配。母猪把嘴放到管子附近,乳状食物就直接被螺旋转入口中,然后立即被母猪食入。该系统避免了饲料浪费,因为当母猪把它的头从食槽里移开或者被赶离饲喂器,系统就会立即停止投放饲料。

该系统通常适用于含有20头母猪和一个给料器的静态母猪群。在商业化生产实践中使用了各种各样的围栏设计。饲喂系统的软件类似于ESF系统软件,因此母猪可以单独饲喂,自动的数据收集可以成为日常监督管理的辅助工具。就像一个ESF系统一样,Fitmix相比其他饲喂系统需要进行更多的进食训练,并且是一个有竞争力的饲喂系统(Chapinal 等,2010)。

3.3.7 自由采食系统

在自由采食系统中,每组大约10头母猪装配一个简单的喂料器。典型的群体大小在10～30头母猪。这款围栏的设计很简单,给料器就在实心地板前方的角落里,平面图可以与地面饲喂相类似。体积大的饲料通常有很高比例的纤维,如干甜菜渣的形式,以防止吃得过多。

然而,能量消耗一般比限制饲喂的母猪高40%,特别是对于高胎次母猪。与传统饲喂方式相比,采用自由采食的后备母猪在第一次妊娠期间的能量摄入较低。

从福利的角度来看，这个系统是很好的，因为口部刻板行为（与长期的饲喂动机相关联）与受限制的饲喂系统下相比，明显减少了。这个系统的每个母猪的房舍成本很低，但是饲料成本更高。但繁殖性能与在限制饲喂系统中母猪没有区别。

3.3.8 群体饲养系统之间的对比

没有所谓的"最好"的群体饲养系统。系统的好坏取决于它的某些方面的重要性，如劳动力的需求、对饲料摄入的控制程度或母猪群争斗。前面的段落已经确定了几个特定于不同群体饲养系统的具体情况。在表3.1中，提供了所有这些方面的概述。

表 3.1 常用的七种室内群养系统不同方面[1]的比较
（基于 Van der Peet-Schwering 等的研究，2010）

	ESF，稳定组，无稻草	ESF，动态组，无稻草	ESF，动态组，有稻草	自由进出饲喂，稳定组，无稻草	自由采食稳定组，无稻草	地面平饲，稳定组，无稻草	食槽饲喂，稳定组，无稻草
劳动力需求	3	4	5	1	2	2	2
劳动环境	2	2	3	1	1	1	1
技能	3	3	3	1	2	2	2
母猪福利	2	2	1	3	2	3	3
猪只健康	2	3	3	1	2	2	2
技术	2	2	2	1	1	1	1
投资与开发	2	2	2	3	1	2	2
繁殖结果	1	1	1	1	1	1	1
社会认可	2	3	3	4	2	3	3
管理和控制	1	1	1	2	1	1	1

[1]对每个标准进行比较。分值越高，则对该特定标准的性能越低，因此对该系统而言，这个方面是不受欢迎的。

3.3.9 舍内饲养和户外饲养

户外饲养（图3.4）适合于温暖或温带气候条件下的有阳光、自由排水的土壤和不太降雨的地方（<750 mm）。

饲料是通过地面上或在简单的槽里分配给猪群。母猪根据胎次和身体状况进行分组。在寒冷的天气里，与舍内系统相比，需要额外的饲料供给，而且保证饮水不能结冰。母猪通常饲养在每舍包含5~20头的猪群（15~20头/hm²），临时棚舍提供遮蔽物和深草垫，以适应适当的微气候。为了防止晒伤，需要额外的阴凉和遮蔽处。户外散养围场通常由电栅栏隔开，通常由两根地面上200 mm和500 mm

3. 妊娠母猪的群养

图 3.4 母猪户外放养系统。

的电线组成。每天在水槽里提供水,母猪也可以打滚。母猪也可以在室内饲喂,也可以与一群公猪一起奔跑,也可以被引入特定的公猪圈中。户外系统的成本通常较低,但需要专门的技术人员和管理。

3.4 结论

在妊娠期,母猪群体饲喂是一种可行的替代限位栏的方案。但需要有几个关键的成功因素。首先,群体饲养系统的选择需要满足养殖场主的具体要求,包括管理、资产投入、劳动需求及动物健康和福利水平等方面的具体要求。其次,饲养母猪的建筑物和围栏的设计应有助于适当的躺卧、饲喂和排粪行为,以避免行为问题和不良的健康状况。再次,饲喂系统应允许不受干扰的饲喂,为每一头母猪提供适当量的食物,以促进良好的身体状况和繁殖性能。最后,娴熟的管理应该尽量减少由于母猪群内社会等级优势的建立造成的不利影响。

关于母猪混群时,对不熟悉母猪的攻击性,以及造成的负面影响,是不可避免的,但可以降低到动物和农户都可以接受的程度。

参考文献

Andersen, I.L., Boe, K.E. and Kristiansen, A.L., 1999. The influence of different feeding arrangements and food competition at feeding in pregnant sows. Applied Animal Behaviour Science 65: 91-104.

Backus, G.B.C., Vermeer, H.M., Roelofs, P.F.M.M., Vesseur, P.C., Adams, J.H.A.N., Binnendijk, G.P., Smeets, J.J.J., Van der Peet-Schwering, C.M.C., Van der Wilt, F.J., 1997. Comparison of four housing systems for non-lactating sows. Report P1·171. Research Institute for Pig Husbandry, Rosmalen, the Netherlands.

Barnett, J.L., Hemsworth, P.H., Cronin, G.M., Jongman, E.C. and Hutson, G.D., 2001. A review of the welfare issues for sows and piglets in relation to housing. Australian Journal of Agricultural Research 52: 1-28.

Barnett, J.L., Hemsworth, P.H., Winfield, C.G. and Hansen, C., 1986. Effects of social environment on welfare status and sexual behaviour of female pigs. I. Effects of group size. Applied Animal Behaviour Science 16: 249-257.

Barnett, J.L., Hemsworth, PH. and Winfield, C.G., 1987. The effects of design of individual stalls on the social behaviour and physiological responses related to the welfare of pregnant pigs. Applied Animal Behaviour Science 18: 133-142.

Baxter, M., 1985. Social space requirements of pigs. In: Social space for domestic animals. Martinus Nijhoff Publishers, Dordrecht, the Netherlands, pp. 116-127.

Baxter, M.R., 1986. Pig space requirements. Proceedings of the Pig Veterinary Society 15: 56-65.

Bokma, 1990. Housing and management of dry sows in groups in practice: partly slatted systems. In: Electronic Identification in Pig Production, Monograph Series No. 10, Royal Agricultural Society, UK, pp. 37-46.

Bokma, S. and Kersjes, G.J.K., 1988. The introduction of pregnant sows in an established group. Proceedings of the International Congress on Applied Ethology in Farm Animals, Skara, Sweden, pp. 166-169.

Bressers, H.P.M., Te Brake, J.H.A., Engel, B. and Noordhuizen, J.P.T.M., 1993. Feeding order of sows at an individual electronic feed station in a dynamic group-housing system. Applied Animal Behaviour Science 36: 123-134.

Broom, D.M., Mendl, M.T. and Zanella, A.J., 1995. A comparison of the welfare of sows in different housing conditions. Animal Science 61: 369-385.

Burfoot, A., Kay, R.M. and Corning, S., 1995. A scoring method to assess damage caused by aggression between sows after mixing. Animal Science 60: 564.

Burfoot, A., Kay, R.M. and Corning, S., 1997. Reproductive performance and aggression between sows re-mixed into small stable groups at different stages during the embryo implantation period following initial mixing at weaning. Proceedings of the British Society of Animal Science, Scarborough, March 1997: 107.

Calderon Diaz, J.A., Fahey, A.G., Kilbride, A.L., Green, L.E. and Boyle, L.A., 2013. Longitudinal study of the effect of rubber slat mats on locomotory ability, body, limb and claw lesions, and dirtiness of group housed sows. Journal of Animal Science 91(8): 3940-3954.

Cassar, G., Kirkwood, R.N., Seguin, M.J., Widowski, T.M., Farzan, A., Zanella, A.J. and Friendship, R.J., 2008. Influence of stage of gestation at grouping and presence of boars on farrowing rate and litter size of group-housed sows. Journal of Swine Health and Production 16: 81-85.

Caulfield, M.P. and Cambridge, H., 2008. The questionable value of some science-based 'welfare' assessments in intensive animal farming: sow stalls as an illustrative example. Australian Veterinary Journal 86: 446-448.

Chapinal, N., Ruiz de la Torre, J.L., Cerisuelo, A., Gasa, J., Baucells, M.D., Coma, J., Vidal, A. and Manteca, X., 2010. Evaluation of welfare and productivity in pregnant sows kept in stalls or in 2 different group housing systems. Journal of Veterinary Behavior: Clinical Applications and Research 5: 82-93.

Chapinal, N., Ruiz-de-la-Torre, J.L., Cerisuelo, A., Baucells, M.D., Gasa, J. and Manteca, X., 2008. Feeder use patterns in group-housed pregnant sows fed with an unprotected electronic sow feeder (Fitmix). Journal of Applied Animal Welfare Science 11: 319-336.

Couret, D., Otten, W., Puppe, B., Prunier, A. and Merlot, E., 2009. Behavioural, endocrine and immune responses to repeated social stress in pregnant gilts. Animal 3: 118-127.

Csermely, D., 1989. Feeding behaviour in pregnant sows of different social rank. Applied Animal Behaviour Science 22: 84-85.

Docking, C.M., Kay, R.M., Whittaker, X., Burfoot, A. and Day, J.E.L., 2000. The effects of stocking density and pen shape on the behaviour, incidence of aggression and subsequent skin damage of sows mixed in a specialised mixing pen. Winter Meeting of the British Society of Animal Science. Proceedings of the British Society of Animal Science, p. 32.

Edwards, S.A., 2000. Alternative housing for dry sows: system studies or components analyses? Proceedings of the 51st annual meeting of the European Association for Animal Production. August 21-24, 2000. Wageningen Academic Publishers, Wageningen, the Netherlands, pp. 99-107.

Edwards, S.A. and Riley, J.E., 1986. The application of the electronic identification and computerized feed dispensing system in dry sow housing. Pig News and Information 7: 295-298.

Ehlorsson, C.J., Olsson, O. and Lundeheim, N., 2003. Prophylactic measures for improving claw health in dry sows. Svensk Veterinärtidning 13: 11-20.

Elmore, M.R.P., Garner, J.P., Johnson, A.K., Richert, B.T. and Pajor, E.A., 2010. A flooring comparison: the impact of rubber mats on the health, behavior, and welfare of group-housed sows at breeding. Applied animal behaviour science 123(1-2): 7-15.

European Food Safety Authority (EFSA), 2005. The welfare of weaners and rearing pigs: effects of different space allowances and floor types. The EFSA Journal 268: 1-19.

European Union (EU), 2008. Minimum standards for the protections of pigs. Council Directive 2008/120/EC. Available at: http://tinyurl.com/mf9ew2s.

Foxcroft, G.R., 1997: Mechanisms mediating nutritional effects on embryonic survival in pigs. Journal of Reproduction and Fertility Supplement 52: 47-61.

Fraser, D. 2003. Assessing animal welfare at the farm and group level: the interplay of science and values. Animal Welfare 12: 433-443.

Geudeke, M.J. and Gerritsen, C. 2004. Epidemic of early disruption of pregnancy in Dutch sow herds: risk factors on farm level. Proceedings of the 18th IPVS Congress, Hamburg, Germany, pp. 850.

Gonyou, H.W., Hemsworth, P.H. and Barnett, J.L. (1986) Effects of frequent interaction with humans on growing pigs. Applied Animal Behaviour Science 16: 269-278.

Grandin, T. and Bruning, J., 1992. Boar presence reduces fighting in mixed slaughter-weight pigs. Applied Animal Behaviour Science 33: 273-276.

Harris, C., 2014. Moves to group housing systems gather pace globally. The Pig Site, 29 May 2014. Available at: www.thepigsite.com/articles/4741/moves-to-group-housing-systems-gather-pace-globally.

Harris, M.J., Pajor, E.A., Sorrells, A.D., Eicher, S.D., Richert, B.T. and Marchant-Forde, J.N., 2006. Effects of stall or small group gestation housing on the production, health and behaviour of gilts. Livestock Science 102: 171-179.

Heinonen, M., Oravainen, J., Orro, T., Seppä-Lassila, L., Ala-Kurikka, E., Virolainen, J., Tast, A. and Peltoniemi, O.A.T., 2006. Lameness and fertility of sows and gilts in randomly selected loose-housed herds in Finland. Veterinary Record 159: 383-387.

Hemsworth, P.H., Rice, M., Nash, J., Giri, K., Butler, K.L., Tilbrook A.J. and Morrison, R.S., 2013. Effects of group size and floor space allowance on grouped sows: aggression, stress, skin injuries, and reproductive performance. Journal of Animal Science 91: 4953-4964.

Hodgkiss, N.J., Eddison, J.C., Brooks, P.H. and Bugg, P., 1998. Assessment of the injuries sustained by pregnant sows housed in groups using electronic feeders. Veterinary Record 143: 604-607.

Hoy, S., Bauer, J., Borberg, C., Chonsch, L. and Weirich, C., 2009a. Investigations on dynamics of social rank of sows during several parities. Applied Animal Behaviour Science 121: 103-107.

Hoy, S., Bauer, J., Borberg, C., Chonsch, L. and Weirich, C., 2009b. Impact of rank position on fertility of sows. Livestock Science 126: 69-72.

Hunter, E.J., Broom, D.M., Edwards, S.A. and Sibly, R.M., 1988. Social hierarchy and feeder access in a group of 20 sows using a computer-controlled feeder. Animal Production 47: 139-148.

Jansen, J., Kirkwood, R.N., Zanella, A.J. and Tempelman, R.J., 2007. Influence of gestation housing on sow behavior and fertility. Journal of Swine Health and Production 15: 132-136.

Jensen, P., 1982. An analyses of agonistic interaction patterns in group-housed dry sows – aggression regulation through an 'avoidance order'. Applied Animal Ethology 9: 47-61.

Jensen, P. and Wood-Gush, D.G.M., 1984. Social interactions in a group of free-ranging sows. Applied Animal Behaviour Science 12: 327-337.

Karlen, G.A.M., Hemsworth, P.H., Gonyou, H.W., Fabrega, E., Strom, A.D. and Smits, R.J., 2007. The welfare of gestating sows in conventional stalls and large groups on deep litter. Applied Animal Behaviour Science 105(1-3): 87-101.

Kay, R.M., Burfoot, A., Spoolder, H.A.M. and Docking, C.M., 1999. The effect of flight distance on aggression and skin damage of newly weaned sows at mixing. In: Proceedings of the British Society of Animal Science, Scarborough, pp. 14.

Kennedy, M.J. and Broom, D.M., 1994. A method of mixing gilts and sows which reduces aggression experienced by gilts. Proceedings of the 45th Annual Meeting of the EAAP, September 5-8, 1994, p. 333.

Kirchner, J., Manteuffel, G. and Schrader, L., 2012. Individual calling to the feeding station can reduce agonistic interactions and lesions in group housed sows. Journal of Animal Science 90: 5013-5020.

Kirkwood, R. and Zanella, A., 2005. Influence of gestation housing on sow welfare and productivity. National Pork Board Final Report.

Knox, R. J. Salak-Johnson, M. Hopgood, L. Greiner, and Connor, J., 2014. Effect of day of mixing gestating sows on measures of reproductive performance and animal welfare. Journal of Animal Science 92: 1698-1707.

Krauss, V. and Hoy, S., 2011. Dry sows in dynamic groups: an investigation of social behaviour when introducing new sows. Applied Animal Behaviour Science 130: 20-27.

Li, Y.Z., Wang, L.H. and Johnston, L.J., 2012. Sorting by parity to reduce aggression toward first-parity sows in group-gestation housing systems. Journal of Animal Science 90: 4514-4522.

Love, R.J., Evans, G. and Klupiec, C., 1993. Seasonal effects on fertility in gilts and sows. Journal

of Reproduction and Fertility Supplement 48: 191-206.
Luescher, U.A., Friendship, R.M. and McKeown, D.B., 1990. Evaluation of methods to reduce fighting among regrouped gilts. Canadian Journal of Animal Science 70: 363-370.
Marchant, J.N., Whittaker, X. and Broom, D.M., 2001. Vocalisations of the adult female domestic pig during a standard human approach test and their relationships with behavioural and heart rate measures. Applied Animal Behaviour Science 72: 23-39.
Matthews, L.R. and Ladewig, J., 1994. Environmental requirements of pigs measured by behavioural demand functions. Animal Behaviour 47: 713-719.
Mench, J.A., 2008. Farm animal welfare in the U.S.A.: farming practices, research, education, regulation, and assurance programs. Applied Animal Behavior Science 113: 298-312.
Mendl, M., 1994. The social behaviour of non-lactating sows and its implications for managing sow aggression. The Pig Journal 34: 9-20.
Mendl, M., Zanella, A.J. and Broom, D.M., 1992. Physiological and reproductive correlates of behavioural strategies in female domestic pigs. Animal Behaviour 44: 1107-1121.
Moore, A.S., Gonyou, H.W. and Ghent, A.W., 1993. Integration of newly introduced and resident sows following grouping. Applied Animal Behaviour Science, 38: 257-267.
Nielsen, N.P., 2008. Loose housing of sows – current systems. Acta Veterinaria Scandinavica 50(Suppl 1): S8.
Parent, J.P., Meunier-Salauen, M.C., Vasseur, E. and Bergeron, R., 2012. Stability of social hierarchy in growing female pigs and pregnant sows. Applied Animal Behaviour Science 142: 1-10.
Peacock, A.J., 1991. Environmental and social factors affecting seasonal infertility in pigs. PhD thesis, University of Sydney, Sydney, Australia.
Pedersen, L.J., Rojkittikhun, T., Einarsson, S. and Edqvist, L.E., 1993. Postweaning grouped sows: effects of aggression on hormonal patterns and oestrous behaviour. Applied Animal Behaviour Science 38: 25-39.
Peltoniemi, O.A.T., Tast, A. and Love, R.J., 2000. Factors effecting reproduction in the pig: seasonal effects and restricted feeding of the pregnant gilt and sow. Animal Reproduction Science 60-61: 173-184.
Peltoniemi, O.A.T., Tast, A., Heinonen, M., Oravainen, J., Munsterhjelm, C., Hälli, O., Oliviero, C., Hämeenoja, P. and Virolainen, J.V., 2010. Fertility of sows fed *ad libitum* with a high fibre diet during pregnancy. Reproduction in Domestic Animals 45: 1008-1014.
Peltoniemi O.A.T. and Virolainen J.V., 2005. Seasonality of reproduction in female pigs. In: Ashworth, C.J. and Kraeling, R.R. (eds.) Control of pig reproduction VII. Nottingham University Press, Nottingham, UK, pp. 205-218.
Prunier, A. and Tallet, C., 2015. Endocrine and behavioural responses of sows to human interactions and consequences on reproductive performance. Chapter 12. In: Farmer, C. (ed.) The gestating and lactating sow. Wageningen Academic Publishers, Wageningen, the Netherlands, pp. 279-295.
Remience, V., Wavreille, J., Canart, B., Meunier-Salaün, Prunier, A., Bartiaux-Thil, N., Nicks, B. and Vandenheede, M., 2008. Effects of space allowance on the welfare of dry sows kept in dynamic groups and fed with an electronic sow feeder. Applied Animal Behaviour Science 112: 284-296.
Rhodes, R.T., Appleby, M.C., Chinn, K., Douglas, L., Firkens, L.D., Houpt, K.A., Irwin, C., McGlone, J.J., Sundberg, P., Tokach, L. and Wills, R.W., 2005. Task Force Report – a comprehensive review of housing for pregnant sows. JAVMA-Journal of the American Veterinary Medical Association 227: 1580-1590.

Salak-Johnson, J.L., DeDecker, A.E., Horsman, M.J. and Rodriguez-Zas, S.L., 2012. Space allowance for gestating sows in pens: behavior and immunity. Journal of Animal Science 90: 3232-3242.

Salak-Johnson, J.L., Niekamp, S.R., Rodriguez-Zas, S.L., Ellis, M. and Curtis, S.E., 2007. Space allowance for dry, pregnant sows in pens: body condition, skin lesions and performance. Journal of Animal Science 85: 1758-1769.

Salaün, C., Callarec, J., Toudic, M. and Dréan, E., 2002. Effet du type de sol sur le bien-être des truis gestantes en groupe alimentées au distributeur automatique de concentré (DAC). Journées de la Recherche Porcine 34: 217-223.

Scientific Veterinary Committee (SVC), 1997. The welfare of intensively kept pigs. Report of the SVC. European Commission, Brussels, Belgium.

Seguin, M.J., Friendship, R.M., Kirkwood, R.N., Zanella, A.J. and Widowski, T.M., 2006. Effects of boar presence on agonistic behavior, shoulder scratches, and stress response of bred sows at mixing. Journal of Animal Science 84: 1227-1237.

Soede, N.M., Roelofs, J.B., Verheijen, R.J.E., Schouten, W.P.G., Hazeleger, W. and Kemp, B., 2007. Effect of repeated stress treatments during the follicular phase and early pregnancy on reproductive performance of gilts. Reproduction in Domestic Animals 42: 135-142.

Spoolder, H.A.M. 1998. Effects of food motivation on stereotypies and aggression in group housed sows. PhD thesis, Wageningen University, Wageningen, the Netherlands.

Spoolder, H.A.M., Burbidge, J.A., Edwards, S.A, Simmins, P.H. and Lawrence, A.B., 1995. Provision of straw as a foraging substrate reduces the development of excessive chain and bar manipulation in food restricted sows. Applied Animal Behaviour Science 43: 249-262.

Spoolder, H.A.M., Burbidge, J.A., Lawrence, A.B. Edwards, S.A. and Simmins, P.H., 1996. Social recognition in gilts mixed into a dynamic group of sows. Animal Science 62: 630.

Spoolder, H.A.M., Lawrence, A.B., Edwards, S.A., Simmins, P.H. and Armsby, A.W., 1998. The effects of food level on the spatial organisation of dynamic groups of sows. In: Proceedings of the annual meeting of the International Society for Applied Ethology, Clermont-Ferrand, France.

Stewart, C.L., Boyle, L.A., McCann, M.E.E. and O'Çonnell, N., 2010. The effect of feeding a high fibre diet on the welfare of sows housed in large dynamic groups. Animal Welfare 19: 349-357.

Stookey, J.M. and Gonyou, H.W., 1998. Recognition in swine: recognition trough familiarity or genetic relatedness. Applied Animal Behaviour science 55: 291-305.

Svendsen, J., Anderson, M., Olsson, A. Ch., Rantzer, D. and Lundqvist, P., 1990. Group housing of sows in gestation in insulated and uninsulated buildings. Report 66. Swedish University of Agricultural Sciences, Uppsala, Sweden.

Tönepöhl, B., Appel, A.K., Voss, B., König von Borstel, U. and Gauly, M., 2013. Interaction between sows' aggressiveness post mixing and skin lesions recorded several weeks later. Applied Animal Behaviour Science 144: 108-115.

Turner, S.P. and Edwards, S.A., 2000. Housing in large groups reduces aggressiveness of growing pigs. In: Proceedings of the 51st annual meeting of the European Association for Animal Production. August 21-24, 2000. The Hague, the Netherlands.

Turner, S.P., Horgan, G.W. and Edwards, S.A., 2001. Effect of social group size on aggressive behaviour between unacquainted domestic pigs. Applied Animal Behaviour Science 74: 203-215.

Tuyttens, F.A.M., Van Gansbeke, S. and Ampe, B., 2011. Survey among Belgian pig producers about the introduction of group housing systems for gestating sows. Journal of Animal Science 89: 845-855.

Van der Mheen, H.W., Spoolder, H.A.M. and Kiezebrink, M.C., 2003. Stable or dynamic groups for pregnant sows. Report 23. Animal Sciences Group, Lelystad, the Netherlands.

Van der Peet-Schwering, C.M.C., Hoofs, A.I.J., Vermeer, H.M. and Binnendijk, G.P., 2010. Group housing for pregnant sows: characteristics of the different systems. Report 352. Wageningen UR Livestock Research, Lelystad, The Netherlands.

Van der Peet-Schwering, C.M.C., Kemp, B., Plagge, J.G., Vereijken, P.F.G., Den Hartog, L.A., Spoolder H.A.M. and Verstegen, M.W.A., 2004. Performance and individual feed intake characteristics of group-housed sows fed a non-starch polysaccharides diet *ad libitum* during gestation over three parities. Journal of Animal Science 82: 1246-1257.

Van Putten, G. and Buré, R., 1997. Preparing gilts for group housing by increasing their social skills. Applied Animal Behaviour Science 54: 173-183.

Van Wettere, W,H,E.J., Pain, S.J., Stott, P.G. and Hughes, P.E., 2008. Mixing gilts in early pregnancy does not affect embryo survival. Animal Reproduction Science 104: 382-388.

Vermeer, H.M. and Vermeij, I., 2014. Claw health and floor type in group housed sows. CAB Reviews 15: 1-7.

Vestergaard, K. and Hansen, L.L., 1984. Tethered versus loose sows: ethological observations and measures of productivity. I. Ethological observations during pregnancy and farrowing. Annales Recherche Veterinair 15: 245-256.

Weaver, S.A. and Morris, M.C., 2004. Science, pigs, and politics: a New Zealand perspective on the phase-out of sow stalls. Journal of Agricultural & Environmental Ethics 17: 51-66.

Weng, R.C., Edwards, S.A. and English, P.R., 1998. Behaviour, social interactions and lesion scores of group-housed sows in relation to floor space allowance. Applied Animal Behaviour Science 59: 307-316.

4. 乳腺发育

C. Farmer[1*] and W.L. Hurley[2]

[1] *Agriculture and Agri-Food Canada, Dairy and Swine R & D Centre, 2000 College St., Sherbrooke, QC, J1M 0C8, Canada; chantal.farmer@agr.gc.ca*

[2] *University of Illinois, Department of Animal Sciences, 430 Animal Sciences Laboratory, 1207 W. Gregory, Urbana-Champaign, IL 61801, USA*

摘要：乳腺发育状况是母猪产奶能力的一个关键组成部分，因此，必须了解其控制机制。母猪乳腺的快速发育有三个阶段，即90日龄到发情期、妊娠期的最后1/3和整个哺乳期。在这期间内的营养、内分泌状况和后备母猪或母猪的管理可能会影响乳腺发育。更具体地说，生长中的后备母猪在90日龄以前限饲会阻碍乳腺发育，而提供植物雌激素——木黄酮或增加催乳素的体循环浓度则会刺激乳腺发育。在妊娠晚期，松弛素或催乳素的减少会显著抑制乳腺发育，并且过度增加饲料能量会对乳腺发育有不利影响。最近的结果还表明，妊娠母猪的饲养可能会影响其后代进入发情期时的乳腺发育。各种管理因素（如窝产仔数、哺育强度和前一次哺乳期使用或未使用乳头）都将对哺乳结束时存在的乳腺组织的数量产生影响。在断奶时，乳腺退化开始，乳腺的实质组织迅速而剧烈地消退。当乳头不被定期吸吮时，这种退化过程也可能发生在哺乳期早期，但这种早期退化对以后乳腺的发育和产奶量的影响是未知的。显而易见，如何制定后备母猪、妊娠和哺乳期的母猪的最佳管理策略，来使乳腺发育最大化，产奶量增加，在这一方面仍有待研究。

关键词：激素，乳腺发育，产奶量，营养，母猪

4.1 引言

乳汁是仔猪的主要能源，对其生长和生存至关重要。然而母猪不能产生足够多的乳汁来维持其最佳的生长发育。实际上，在断奶前，通过人工饲育自由获得营养的仔猪比由母猪饲育的仔猪断奶时体重明显更高（Harrell等，1993）。在这项研究中，在21日龄时，人工饲养的仔猪比母猪饲喂的仔猪重了53%。最近还显示，为断奶前仔猪补充乳汁显著地增加了断奶时体重（Miller等，2012）。使用高产母猪加剧了仔猪吃奶量不足的问题。因此，必须制定可以提高母猪产奶量的管理策略。

4. 乳腺发育

决定母猪泌乳潜力的一个关键因素是哺乳期开始时的乳腺细胞数量（Head 和 Williams，1991），这应该在开发最佳管理实践中得到更多的关注，以优化生长期后备母猪、妊娠和哺乳母猪的乳腺发育。乳腺快速发育的三个特殊时期内可以通过管理、营养和激素策略来尝试刺激乳腺发育。本章总结了母猪的乳腺发育过程以及影响其发育的各种因素。更具体地说，本章内容包括营养、激素水平、乳头的吮吸对母猪乳腺发育、乳腺退化过程的影响。

4.2 乳腺个体发育

母猪的乳腺位于胸廓到腹股沟，在腹壁中线两侧平行排列。乳腺（胸腹、腹部或腹股沟）通过脂肪和结缔组织附着到腹侧壁。每个乳腺是分开的并与邻近的乳腺不通（Turner，1952），它通常有一个乳头和两个独立的乳头管。每一个导管都会引起静脉窦小幅扩张，并且最终分叉形成其自身的乳泡小叶组织部分，这样每个乳头都有自己的独立导管和腺体系统（Hughes 和 Varley，1980）。乳腺组织来自胚胎中的外胚层，乳腺最终的分化在早期胚胎阶段第一次变得明显，此时在胚胎早期有两个平行的脊出现，被称为"乳线"。沿着这些乳线的结节将使其自身变为乳状芽，每一个都是乳头的祖先。出生时，管道系统发育相对较少，乳腺主要由皮下基质组织组成（Hughes 和 Varley，1980）。乳腺组织和乳腺 DNA（代表细胞数量）直到 90 日龄才缓慢积累。之后乳腺组织和 DNA 的积累速度增加了 4～6 倍（Sorensen 等，2002），使得在后备母猪交配时，虽然乳腺仍然非常小，却已经含有广泛的管道系统，伴随有许多芽状伸出部分（Turner，1952）。最近的数据显示，发情期对乳腺发育具有刺激作用，例如，与年龄相似但没有开始发情的后备母猪相比，已经达到发情期的后备母猪其实质组织块（其含有腺体的上皮细胞组分）增加了 51%，实质外组织块（主要含有脂肪组织）减少了 16%（Farmer 等，2004）。在乳腺发育的所有阶段都由乳腺干细胞提供祖细胞（Borena 等，2013）。乳腺脂肪组织，包括实质和外膜组织内的脂肪组织，对于腺体上皮组分的发育也是重要的（Hovey 和 Aimo，2010）。乳腺干细胞和脂肪干细胞在猪的乳腺发育中的作用迄今仅受到有限的关注。

在妊娠的后备母猪中，在妊娠前 2/3 时乳腺的发育缓慢，而几乎所有乳腺组织和 DNA 的积累发生在最后 1/3（Hacker 和 Hill，1972；Kensinger 等，1982；Sorensen 等，2002）。乳腺组织中 DNA 的浓度在妊娠期的最后 1/3 内急剧增加（King 等，1996）。Ji 等（2006）也报道了在妊娠期 45～112 天乳腺重量明显增加，其在第 75 天后乳腺增生加速发生。从组织学上来说，45～75 天乳腺组织主要由脂肪和基质组成，具有延长的管道和有限的管道分支以形成小叶结构（图 4.1；Ji

等，2006)，类似于后备母猪中发现的针对哺乳动物激素的末端导管小叶单位(Horigan 等，2009)。在第75～112天，由于脂肪和基质组织被腺泡组织广泛代替，腺体经历了较多的组织学变化(图4.1；Hacker 和 Hill，1972；Ji 等，2006；Kensinger 等，1982)。Ji 等(2006)也报道了在妊娠期最后1/3时，乳腺组成从高脂质含量转换到高蛋白质含量，反映了组织中广泛的脂肪。后备母猪的乳腺组织中组织学变化和DNA浓度的差异表明，妊娠期第75～90天上皮细胞分裂增加，第90天存在最大细胞浓度。然后，在第90～105天，与上皮细胞的功能分化相关的细胞器增多和腺泡中分泌物的大量积累，表明泌乳过程的发生(Kensinger 等，1982，1986a)。在分娩时，乳小叶和腺泡完全充满分泌物(图4.1；Turner，1952)。图4.2说明了妊娠的后备母猪和哺乳期母猪的乳腺组织发育情况。在妊娠期乳腺在乳房中的位置会影响其发育。妊娠102天和112天，中间的乳腺(第3、第4和第5对)的重量高于后端乳腺(第6、第7和第8对)(Ji 等，2006)。

妊娠晚期乳腺组织中的这些表型变化与乳腺基因表达的显著变化相吻合。在一项母猪乳腺转录组的研究中，发现许多信号通路和基因调控网络在妊娠80～110天发生变化(Zhao 等，2013)。例如，妊娠后期乳腺细胞中乳脂质合成的增加

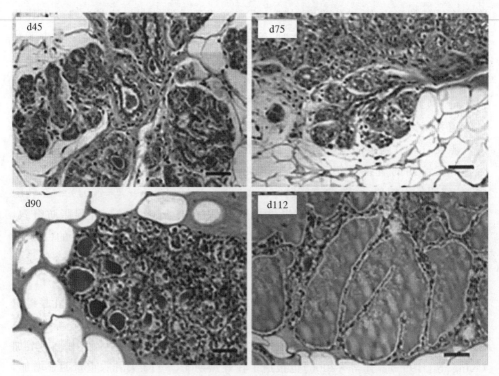

图4.1 怀孕45、75、90和112天后备母猪乳房组织的染色切片。注意在妊娠45天和75天的末端导管小叶单位图片。标尺为50 μm。图片来自Ji 等(2006)关于后备猪的研究。

4. 乳腺发育

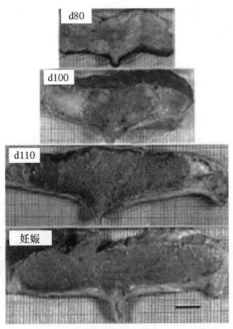

图 4.2　来自妊娠期的最后 1/3 和泌乳期后备母猪的乳腺横切面。图片代表了怀孕 80、100 和 110 天以及妊娠第 3 天。图片来源于 Hurley 等(1991)关于妊娠猪的研究。标尺为 2 cm。

可能是通过一些基因的激活来驱动的，这些基因与脂肪酸生物合成、三羧酸循环及乙醛酸和脱羧酶通量相关。这些分析还表明，妊娠后期可能会出现必需氨基酸的降解降低和其他氨基酸代谢途径的减少，这与乳腺组织蛋白质沉积的急剧增加相一致。细胞间隙连接相关基因的活化、mTOR 信号通路(乳蛋白质合成)和 VEGF 和 MAPK 信号传导(血流调节)都与妊娠后期乳腺组织功能的已知变化一致(Zhao 等,2013)。

乳腺发育并不会在妊娠结束时停止，而是在哺乳期间一直持续。在初产母猪中，乳腺由分布在脂肪和结缔组织之间的细胞芽组成，而在哺乳母猪的乳腺中，结缔组织大部分被腺实质所取代。哺乳母猪的乳腺由分泌单位排列成小叶的复合性管泡状组织组成，由合成乳汁的上皮细胞(乳酸细胞)排列而成。泌乳乳腺的平均重量从哺乳期第 5 天的 381 g 直线增加到第 21 天的 593 g(增加 57%)。在初产母猪中，哺乳期乳腺体积的增加是细胞增生和肥大的结果(Kim 等,1999a)，而在经产母猪中，这似乎主要是由于肥大引起(Manjarin 等,2011)。哺乳期乳腺生长与腺体在乳腺上的位置有关，前五个乳头比在其后的乳头更大(Kim 等，2000)，并且有迹象表明它可能与放乳后按摩的强度有关(Thodberg 和 Sorensen, 2006)。乳腺发育也受到了胎次影响，因为第 1 胎、第 2 胎和第 4 胎的母猪在妊娠期 113 天和哺乳期第 26 天之间乳腺重量分别增加了 63%、21% 和 39%(Beyer 等,1994)。细胞分裂和细胞分化都有助于其他物种如山羊哺乳期早期的产奶量(Knight 和 Peaker,1984)。虽然在哺乳期间母猪的乳腺生长很明显,但在何种程度上猪乳腺细胞分化状态的增加有助于提高产奶量尚未得到充分的探索。

4.3 乳腺发育的控制

如上所述,乳腺发育发生在猪的生长和繁殖的许多阶段。在发情期后期和妊娠期间,各种激素参与猪的乳腺发育控制,最重要的是雌激素(estrogens)、松弛素(relaxin)和催乳素(prolactin)。发情期启动对乳腺发育的剧烈影响证明了雌激素的重要作用(Farmer 等,2004;Sorensen 等,2006)。Farmer 等(2004)报道,与相似年龄的非发情后备母猪相比,发情后备母猪乳房实质组织质量增加了 51%,这又反过来导致了实质性脂肪、蛋白质和 DNA 的增加。

在妊娠期间,母猪血浆中雌激素浓度在 75 天后显著增加(DeHoff 等,1986)。Kensinger 等(1986b,c)研究表明妊娠后期发生的乳腺代谢活动的急剧增加与胎儿来源雌激素的增加有关;确实,在妊娠第 110 天,乳腺 DNA 与母猪雌激素的循环浓度有关。早期的研究还表明,玉米赤霉烯酮(一种具有雌激素样活性的霉菌毒素)影响乳腺发育。在饲喂含有玉米赤霉烯酮的母猪中观察到由于导管增生引起的乳腺腺体元素的增加(Chang 等,1979)。甚至在一些从玉米赤霉烯酮处理的母猪乳房吮吸的 7 日龄仔猪中观察到乳腺发育(Chang 等,1979)。最近,尝试通过提供雌激素的饲料来特异性地刺激后备母猪的乳腺发育。当在 90~183 日龄的后备母猪的饲料中添加 2.3 g/天植物性雌激素木黄酮时,处理期结束后乳腺实质 DNA 增加(代表增生)(Farmer 等,2010a)。在妊娠的最后 1/3 时间,提供类似剂量的木黄酮对后备母猪乳腺发育的影响需要进一步研究。

在繁殖周期和妊娠期的动物中,雌激素与松弛素(relaxin)协同作用以刺激乳腺发育。松弛素是由母猪黄体产生的多肽激素。使用经典替代疗法研究卵巢切除的不孕后备母猪,Winn 等(1994)证实雌激素和松弛素刺激乳腺实质组织的生长。在类似的卵巢切除的妊娠后备母猪研究中,Hurley 等(1991)清楚地表明妊娠期最后 1/3 时松弛素在促进乳腺实质生长中起主要作用。然而,外源性松弛素对完整妊娠后备母猪的乳腺发育的潜在影响尚不清楚。松弛素的这种影响乳腺发育作用可能不会持续到哺乳期。不同泌乳性能的母猪产后 24 h 血浆松弛素浓度没有差异,且激素在 72~120 h 内即检测不到(Porter 等,1992)。

外源性生长激素的研究显示出不同的影响结果。在哺乳后期(12~29 天),猪生长激素的使用显示增加了母猪的产奶量(Harkins 等,1989),而从妊娠第 108 天到哺乳期第 28 天使用猪生长激素则没有使产奶量增加(Cromwell 等,1992)。这些研究都没有评估乳腺生长指标。另外,研究发现生长激素释放因子(growth hormone releasing factor,GRF)对哺乳母猪乳腺发育有潜在刺激作用。在妊娠后期和整个哺乳期,GRF 的使用降低了实质组织重量,但是在哺乳期的第 30 天实质 DNA 浓度增加(Farmer 等,1997)。有趣的是,在乳汁中使用的牛生长激素(bovine somatotropin)的作用主要发生在哺乳期达到顶峰之后(Peel 和 Bauman,1987)。Harkins 等(1989)观察到的积极效应可能主要发生在处理后期。

催乳素(prolactin)是对乳腺发育的影响方面受到最多关注的激素。催乳素影响后备母猪乳腺发育。其第一个迹象来自于一种试验,其中催乳素被提供给后备母猪以试图影响其成长表现(McLaughlin 等,1997)。这些作者报道了持续 28 天对 75 kg 体重的母猪注射 2 mg/天催乳素时乳腺的显著发育。处理后母猪的乳腺的特征在于扩张的腺泡和导管腔以及分泌物质的存在。然而,没有对乳腺组成进行测量。在随后的实验中,将 4 mg/天的重组猪催乳素注射到 29 日龄、体重为 75 kg 的后备母猪中,导致乳腺实质组织块增加 116%,实质 DNA 增加 160.9%(Farmer 和 Palin,2005)。然而,乳腺分泌物也存在,表明有未成熟的乳糖出现。

早在 1945 年,有迹象表明,晚孕母猪的麦角大麦的采食对乳腺发育有不利影响。食用麦角大麦的母猪几乎没有乳腺发育,而所有对照母猪都有正常的乳腺发育(Nordskog 和 Clark,1945)。最近也报道了在分娩前 8 天饲喂麦角碱对乳腺发育的负面影响(Kopinski 等,2007)。与母猪分娩相关的正常催乳素激升发生在产前大约 2 天和产后几天之间(Dusza 和 Krzymowska,1981)。最有趣的是因为发现产后内毒素直接对产后期间催乳素分泌具有抑制作用,从而显示了催乳素的抑制和母猪产奶量不足之间的潜在关系(Smith 和 Wagner,1984)。在 10 年前(译者注:2005 年),使用多巴胺受体激动剂溴麦角环肽抑制催乳素分泌,首次证明了催乳素对已孕后备母猪的乳腺发育具有重要作用(Farmer 等,2000)。妊娠 70~110 天,每天 3 次对后备母猪喂 10 mg 溴麦角环肽,妊娠第 110 天的乳腺实质组织质量为 581 g,对照组为 1011 g,降低了 42.5%(图 4.3)。

图 4.3 妊娠 110 天对照(C)或者处理的(B)后备母猪乳腺的横切面。从妊娠 70~110 天每天 3 次,每次喂食 10 mg 的多巴胺受体激动剂溴麦角环肽。

研究显示,催乳素对乳腺生长发挥其最大刺激作用的具体时间为妊娠90～109天(Farmer和Petitclerc,2003)。在特定时间段内每天给后备母猪喂10 mg溴麦角环肽,在妊娠第110天实质总质量降低46%(918.5对1701.7 g),但是当妊娠期第50～69天或第70～89天饲喂时,治疗无效。最近的数据表明,当在妊娠后期特定的期间出现高催乳素状态时,使用多巴胺拮抗剂多潘立酮(domperidone),对乳腺实质的分泌活性和乳腺上皮细胞分化有显著的有益作用(VanKlompenberg等,2013)。在随后的哺乳期第14天和第21天产奶量也有所改善,仔猪断奶后的体重增加了21%。然而,没有获得乳腺组成成分的测量。

催乳素分泌的刺激程度可能也是重要的考虑因素。King等(1996)从妊娠102天直至哺乳期将高水平的猪催乳素施用于第一胎后备母猪。虽然乳腺组织活检中RNA和DNA的浓度不受催乳素给药的影响,但催乳素给药的母猪的产奶量降低。该研究中的其他证据表明,母猪可能已经开始了过早泌乳发生,并伴随着产后催乳素浓度升高。在另一项研究中,在第三胎母猪哺乳期第2～23天使用重组猪催乳素的,没有观察其到对产奶量或乳腺组成的影响(Farmer等,1999)。这种缺乏效应可能是由于哺乳动物的乳腺受体在对照动物中已经饱和的事实,从而防止外源性催乳素产生其他生物作用(Farmer等,1999)。

寻找使用不是药剂的饲料添加剂来增加循环催乳素浓度是有意义的,它们可以用于猪商业生产中。一种可能是植物提取物水飞蓟素(silymarin,来自乳蓟),在大鼠中具有高催乳素特性(Capasso等,2009)和在妇女(Di Pierro等,2008)和牛(Tedesco等,2004)中具有超半乳糖酶。在最近的一项研究中,证实水飞蓟素可以增加妊娠母猪的催乳素浓度,但是这种增加不足以在乳腺发育方面产生有益效果(Farmer等,2014)。更具体地说,从妊娠90～110天每天提供两次4 g水飞蓟素,导致治疗开始4天后循环催乳素浓度增加51.8%。但是,15天后,这种效果就不再明显。对乳腺发育缺乏有益的影响可能基于催乳素浓度增加不足或持续时间不足的事实。实际上,在VanKlompenberg等的研究中(2013),催乳素浓度增加的积极作用呈现在乳腺组织的分泌活性和上皮细胞分化方面,催乳素浓度在处理24 h内增加了近4倍,并保持较高水平6天。在使用水飞蓟素的研究中(Farmer等,2014),处理24 h后没有获得血液因而不知道催乳素浓度是否先到达峰值。在任何情况下,更大剂量的水飞蓟素可能产生更大的作用。然而,根据处理所需的时间,生产者定期使用这种做法很可能在经济上不可行。

4.4 排乳的作用

乳汁排出对哺乳期的乳腺发育和功能也至关重要。自分泌反馈抑制因子的积累发生在乳腺泡中,作为乳细胞分泌的正常过程的一部分,这种积累会抑制乳汁的进一步分泌(Wilde等,1995)。如果乳汁没有被仔猪从母猪乳腺中吮吸并完全排出,那么腺体会减少进一步的乳汁分泌,并最终开始退化过程(下面讨论)。吮吸也

4.乳腺发育

刺激催乳素和其他激素的分泌(Algers 等,1991;Spinka 等,1999),断奶导致血浆催乳素浓度迅速下降(Bevers 等,1978)。可以预期反馈抑制因子的去除和催乳素分泌的刺激将协同作用刺激乳腺生长,以及维持泌乳功能。

哺乳期中仔猪吮吸和乳汁排出对乳腺发育的影响也可以从研究中看到。妊娠后备母猪卵巢切除和孕激素替代疗法将导致乳腺发育受损(Hurley 等,1991)。对妊娠第 100 天的后备母猪的完整乳腺发育程度与在第 100 天卵巢切除的后备母猪的乳腺发育程度进行对比,然后在第 110 天估计乳腺发育(Hurley 等,1991),发现在第 100~110 天,乳腺似乎在松弛素缺乏的后备母猪中退化,或至少没有进一步发展。Zaleski 等 (1996)研究了妊娠晚期松弛素对随后的泌乳表现的影响。松弛素缺乏的后备母猪没有经历与分娩相关的正常的子宫颈软化(O'Day 等,1989),胎儿在第 114 天被剖腹产(C-section)取出,给予母猪来自于正常分娩健康的窝仔猪,经过 28 天的哺乳期以确定仔猪生长表现。在第 80~114 天缺乏松弛素并因此哺乳期乳腺发育水平较低的后备母猪,预期整窝仔猪生长将明显受损。结果发现,尽管从松弛素缺乏的后备母猪吮吸的仔猪可能会延迟哺乳期早期的生长,但在哺乳期第 21 天,此仔猪重量与对照组没有差异(Zaleski 等,1996)。这表明即使在产前发育受损的情况下,母猪的乳腺也具有广泛的生长潜力。该研究中也提出了证据,即哺乳期的最初几天,由于哺乳和排乳,可能会刺激广泛的乳腺生长。

从窝产仔数的影响可以看出排乳对产奶量的影响,可以提高产奶量和窝产仔数(由 King 综述,2000;同时也在第八章讨论,Quesnel 等,2015)。吮吸强度也影响哺乳期乳腺的生长以及产奶量。例如,窝产仔数明显影响哺乳期乳腺组织总量的增长(下文讨论;Kim 等,1999c)。

从仔猪吸吮的乳腺的大小也同样可以判断出排乳的效果,因为体型较大的仔猪对奶的需求量更大或者说与体型较小的仔猪相比从乳腺中吸取的乳汁更多(The King,2000)。乳腺的大小(以体重或 DNA 的方式)和吮吸乳头的仔猪的生长之间存在显着的相关性,这显示了仔猪与其吮吸的腺体的生长发育之间的关系,即对产奶量的估计(Kim 等,2000;Nielsen 等,2001)。此外,在哺乳期结束时有较高平均仔猪增重的母猪(即 4.46~5.25 kg,哺乳期第 2~21 天)比具有较低仔猪增重的母猪,其每只乳头含有更多的 DNA 和更多的 RNA(Farmer 等,2010b)。

通过对同一个哺乳期的同一日龄的一窝仔猪不同哺乳阶段进行研究,进一步证明仔猪的大小对其吸吮的腺体有影响。King 等 (1997)2 周龄的仔猪吮吸泌乳第 2 天的母猪乳房,与对照组相比在 4~8 天出现更高的产奶量。相反,与对照组相比,2 日龄的仔猪吮吸哺乳期 2 周的母猪,其产奶量有所下降。鉴于产奶量与乳腺细胞数之间的密切关系(Boutinaud 等,2004),这些观察结果表明,哺乳期第 2 天仔猪体型更大的母猪会刺激乳腺发育更大,以满足仔猪的需求,同时较小的仔猪吮吸已建立好的乳腺上,可能导致乳腺组织消退,直到乳腺组织与较小仔猪的需求相平衡。在这两种情况下,对产奶量的影响约持续 2 周,表明此时腺体达到与仔猪需求平衡。对这种效应的另一项研究,曾经使用过表达乳腺特异性转基因的母猪(牛α-乳白蛋白;Marshall 等,2006)。将 7 日龄的仔猪饲养至泌乳第 2 天的母猪处,使得转基因母猪的日产奶量在第 9 天迅速增加至高峰并保持高于对照组直至哺乳期第 15 天。将 7 日龄的仔猪饲养至第 2 天的非转基因母猪处对产奶量的影响更为有限。

仔猪大小如何影响其吸吮的腺体发育的另一个例子是通过比较后备母猪分娩时的乳腺质量与哺乳期第 5 天的腺体质量而得到。比较哺乳期第 5 天的乳腺组织的平均 DNA 含量(Kim 等,1999a)和第 0 天的平均 DNA 含量(Ji 等,2006;Kim 等,1999a)的数据发现,在哺乳期开始的 5 天时间内腺体的发育直接反映了仔猪的乳汁需求水平。在哺乳期开始时,仔猪出生体重保持不变,因此整个乳腺部位的乳汁吮吸强度相似,与分娩相比,第 5 天腺体质量变化减小。在分娩期间最大的腺体(通常为中间的乳腺;Ji 等,2006)可能具有与仔猪吮吸乳汁的能力相对应的过量的组织量,并且可能在哺乳期的最初 5 天内发生消退。相比之下,在哺乳期的最初 5 天内,分娩时最小的腺体(通常为后端乳腺;Ji 等,2006)生长最快,尽管仍然小于前面大的腺体。

4.5 营养对乳腺发育的影响

生长、妊娠或哺乳期猪的营养可能会影响乳腺发育。生长中的后备母猪从 28 天(断奶)到 90 日龄的 34% 限饲对乳腺发育没有显著影响,而从 90 日龄到发情期的 20%(Farmer 等,2004)或 26%(Sorensen 等,2006)限饲分别使乳腺实质量减少了 26.3% 和 34.2%。饲料限制对乳腺发育的影响仅在 90 日龄观察到,这是乳腺快速发育的第一个时期。因此,建议从 90 日龄到发情期保持高饲喂水平,以确保生长中的后备母猪乳腺的发育最佳。另一方面,在同一时期将日粮中粗蛋白质从 18.7% 降低到 14.4% 不影响乳腺发育(Farmer 等,2004),这表明相对于蛋白质吸收,总饲料自身的摄入对后备母猪乳腺发育而言更为重要。Farmer 等(2007)对进食亚麻籽对后备母猪乳腺发育的影响进行了研究,因为它是高分子量的二异辛烯二醇二糖苷(secoisolariciresinol diglycoside),是木质素形成的前体,又表现出雌激素活性(Adlercreutz 等,1987)。然而,第 88~212 天在饲料中补充 10% 亚麻籽,在第 212 天时乳腺发育没有发生显著变化(Farmer 等,2007b)。

生长中的后备母猪的营养可能影响妊娠结束时的乳腺发育。Lyvers-Peffer 和 Rozeboom(2001)研究了在发情期前改变饲养方案对妊娠结束时乳腺发育的影响。他们使用膳食纤维(35% 向日葵壳)来实现中等生长阶段与最大生长阶段之间的过渡。研究发现,9~12 周龄和 15~20 周龄的中度饲养方式的后备母猪在妊娠的第

4.乳腺发育

110天比对照组后备母猪具有较少的乳腺实质。在使用类似方法的后续实验中，特定时期限饲（提供对照饲料的70%蛋白质和DE含量）然后再超额补给（提供对照饲料的115%蛋白质和DE），在发情期之后这种饲养方式对生长中的后备母猪的乳腺发育没有任何有益的影响。事实上，这种饲喂方式会导致发情期乳腺实质组织的减少（Farmer等，2012a）。同样的营养处理也不会在妊娠结束时影响乳腺实质质量，但会出现乳腺实质中蛋白质百分比降低的趋势（Farmer等，2012b）。

妊娠期间的营养无疑会影响妊娠结束时的乳腺发育。一项通过在妊娠期间控制蛋白质和能量摄入来改变母猪的身体组成的早期研究表明，在妊娠结束时，过度肥胖的后备母猪（背膘厚度36 mm）和更瘦的后备母猪（背膘厚度24 mm）具有相似的乳腺重量，但与更瘦的后备母猪相比，过度肥胖的后备母猪的乳腺DNA浓度（即细胞数）急剧下降（约3倍）（Head和Williams，1991）。然而，这些身体状况并不代表商业生产中看到的，并且当根据现行标准比较时，不知道肥胖、胖瘦适宜和瘦的后备母猪是否具有这种乳腺DNA的差异。这是为了确定妊娠结束时最佳乳腺发育所需的理想身体状况而需要考虑的。从妊娠第75天到妊娠结束时日粮能量的增加（5.76对10.5 Mcal ME/天），降低了妊娠105天时的乳腺实质重量和DNA（Weldon等，1991）。另一方面，增加蛋白摄入量（330对216 g CP/天）对乳腺发育没有影响（Weldon等，1991）。这一发现后来被Kusina等（1999）证实，他们表明，妊娠25～105天的赖氨酸摄取量为4，8或16 g/天，并没有改变妊娠结束时乳腺的发育。在妊娠前10周限饲（提供对照饲料的70%蛋白质和DE），然后10周后超额补给（提供对照饲料的115%蛋白质和DE）直到妊娠结束，在妊娠结束时实质组织较少，实质组织组成没有变化（Farmer等，2014b）。该研究的目标是考虑补偿性饲喂对乳腺发育的影响，尽管过度饲喂期的增长率有所增加，但这一增加不足以补偿在早孕限饲期间的体重减轻。需要开发更好的适应性饲喂方式，以便能够真正评估补充饲喂对妊娠后备母猪的乳腺发育的影响。最近有报道显示妊娠和哺乳期母猪的营养可能影响其后代的乳腺发育。的确，从妊娠第63天到哺乳期结束时，饲料中补充10%亚麻籽增加了后代在发情期时乳腺实质量（Farmer和Palin，2008）。这是第一次证明了这种猪子宫内膜效应，并且在促进乳腺发育的饲养策略上开发了新的途径。

在哺乳期乳腺快速增生的最后阶段，其营养也会影响乳腺发育，但是关于这个问题的信息很少。Kim等（1999b）饲喂哺乳期初产母猪不同蛋白质（32或65 g/天赖氨酸）和能量（12或17.5 Mcal ME/天）水平组合四种的饲料。吮吸乳腺的湿重和干重均受能量和蛋白质摄入量的积极影响。结果表明，当母猪每天平均消耗16.5 Mcal ME和950 g粗蛋白质时，乳腺的湿重和干重出现最大化，后者相当于每天消耗52.3 g的赖氨酸。因此，显而易见的是，哺乳期的营养摄入量对于乳腺发育而言是非常重要的。

4.6 乳腺退化

4.6.1 断奶期

乳腺特别有趣，因为它具有生长、哺乳期和退化的重复周期。在仔猪断奶时，吮吸行为的突然停止导致乳腺退化。该过程的特征在于在断奶后的前7天，乳腺实质的快速消退（Ford等，2003）。乳腺组织的变化非常显著（图4.4）。断奶后7天，乳腺实质组织湿重和实质DNA分别降低了68.8%和66.8%。在早期断奶2天后就可以看到明显的变化，其特征在于截面积、腺体湿重和实质DNA急剧减少。从第2天到第4天，乳腺退化降到最小，直到断奶后第7天再次显著下降（Ford等，2003）。在断奶后发现每块组织DNA的比例不被改变，表明在退化过程中每毫克乳腺组织的细胞数不变。然而，在断奶后的乳腺实质中观察到其他组成变化，即蛋白质百分比的降低和脂肪百分比的增加。可能是由于组织内的脂质增加，这很好地反映了乳脂质的暂时性积累（Ford等，2003）。早期的组织学研究还显示，在断奶后的前几天乳腺变得充血，并且内腔中的乳汁似乎被重新吸收（Cross等，1958）。

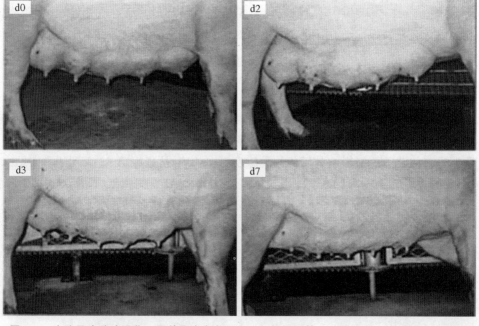

图4.4 头胎母猪乳腺退化。图片取自断奶（0天）和断奶后第2、3和7天。注意第2天乳腺明显充血，并且第3天减少。

Ford 等（2003）提出断奶母猪乳腺退化分为三个阶段,即:①断奶至第 2 天;②从第 2 天起至断奶后第 4 或 5 天;③从断奶后第 4 或 5 天至断奶后至少 7 天。在初始阶段,突然断乳导致乳汁淤滞,并通过在囊泡腔中积累自分泌反馈抑制哺乳作用,从而来抑制乳汁分泌(Wilde 等,1995)。组织液剧烈损失,伴随着组织脂肪含量的增加和由于凋亡导致的相当大的细胞损失。退化过程中的细胞损失至少部分通过一组控制凋亡的基因的表达发生,但也涉及巨噬细胞对凋亡细胞的吞噬作用(Motyl 等,2001)。在断奶后的最初 16 h 内,乳腺血流量下降了 40%（Renaudeau 等,2002)。第二阶段的特征是乳腺组分质量的变化更为有限,与乳汁代谢物的有限变化一致(Atwood 和 Hartmann,1995)。在最后一个退化阶段,只有有限的乳腺分泌物可以被收集而且非常黏稠(Atwood 和 Hartmann,1995)。在这个阶段,囊泡结构残留非常少(Cross 等,1958),乳腺实质组织和 DNA 最终也减少(Ford 等,2003)。

乳腺退化的过程受哺乳期长短的影响。当比较 22 天和 44 天的哺乳期长度时,乳汁中的 Na/K 比率在 44 天更大,表明随着哺乳期的进行乳腺上皮细胞连接不再紧密(Farmer 等,2007a)。乳腺上皮完整性的破坏加上血浆乳糖浓度的增加表明,退化过程开始于断奶前(哺乳期的第 44 天)。乳腺退化的发生可能与吮吸强度的降低有关,因为在 44 天断奶的仔猪在哺乳期的第 22 天时可以接受教槽料。或者,这一时期标志着母猪高哺乳峰期后下降的阶段。反刍动物高哺乳期峰期下降的阶段与乳腺上皮细胞的丧失有关(Stefanon 等,2002)。

4.6.2 泌乳期

泌乳早期未吮吸的乳腺以相似的速率复原,并遵循与断奶后退化的乳腺相似的回归模式(Kim 等,2001)。当乳腺未被吮吸时,乳腺组织的重量在哺乳期的第 7～10 天下降了 2/3,而其后的复原速度则慢得多(Kim 等,2001)。与之相应的,哺乳期未吮吸的乳腺在断奶后则没有显示实质组织有进一步的损失(Ford 等,2003)。当比较窝仔数为 6～12 头仔猪时,在哺乳期第 21 天未吮吸乳腺的组成或大小没有差异,表明窝产仔数大小对乳腺退化没有影响(Kim 等,2001)。另一方面,未吮吸乳腺的复原率受泌乳期日粮营养水平的影响。Kim 等(2001)报道,与饲喂低能量低蛋白日粮的母猪相比,哺乳期的第 5 天饲喂高能量(17.5 对 12 Mcal ME/天)高蛋白(65 对 32 g 赖氨酸/天)日粮的母猪的未吮吸腺体的湿重高 91%。因此,在促进乳腺生长的日粮条件下,未吮吸的腺体的复原率似乎最慢。然而,哺乳期未哺乳乳腺的消退程度和速度的影响对其未来发育和产奶量的影响尚不清楚。

母猪乳腺复原的程度是可逆的这一点尤为重要,因为仔猪的交叉哺乳是一种常见的做法。Kim 等的研究结果(2001)显示未哺乳腺体中有大量的湿重和 DNA 损失,表明在哺乳期的最初几天之后,哺乳期功能的丧失是不可逆转的。在 2005年,Theil 等专门针对这个问题进行了研究,为了防止仔猪哺乳,在分娩后 24 或

72 h 使奶头不可见。结果发现,哺乳期乳头不可见 24 h 时乳腺发育与常规乳腺腺体相似,而乳头不可见 72 h 的乳腺发育较哺乳期第 6 天更小。泌乳早期,在前 24 h 内未吮吸的乳腺复原是可逆的(腺体拯救),但在 3 天后则不可逆转(Theil 等,2005)。在 24 h 内未被吮吸的复原腺体的产奶量在哺乳期间仍然较低。培育母猪分娩后 24 h 交叉培育的仔猪在断奶时的重量比对照仔猪重 900 g 进一步证明了这一点(Thorup,1998)。吮吸强度对于乳腺复原也很重要。Theil 等(2006)比较没有乳汁、暂时性乳汁(直到产后 12~14 h)或乳腺正常哺乳,观察到定期吸吮乳腺保持泌乳,而瞬时哺乳和非吮吸腺体在哺乳期退化复原。

4.7 影响乳腺发育的管理策略

窝产仔数的大小不仅会影响产奶量(King,2000),而且也对乳腺发育有影响(Kim 等,1999c)。窝产仔数较多(12 头猪)的第一胎母猪比窝产仔数少(6 头猪)的母猪在哺乳期第 21 天总乳腺块更大,但产仔数较多的窝中每个个体腺体的重量较低(Kim 等,1999c)。

总哺乳腺体湿重和干重随窝产仔数呈线性增加,窝产仔数从 6 头增加到 12 头导致乳腺总重量增加了 65%,乳腺总 DNA 增加了 67%,总蛋白质增加了 63%。由于在单个哺乳期的乳腺湿重和干重以及无脂肪组织、蛋白质、DNA 和脂肪的量随着仔猪数量的增加而减少,所以这些增加是由于哺乳期腺体数量较多。结果表明,不同窝仔数的母猪在其哺乳腺的细胞密度或细胞大小没有差异,而在具有较小窝仔数的母猪的腺体中,实质组织质量较大。一个重要的发现是,乳腺大小的增加或乳腺蛋白质的数量对仔猪体重增加有积极的影响。此发现也说明在哺乳期间通过饲喂和其他策略使乳腺生长最大化的重要性。

另一个管理方面可以影响乳腺发育的是乳头使用或哺乳历史。也就是说,一个哺乳期腺体生长和哺乳的程度可能会影响在随后泌乳中腺体的生长和功能。Ford 等(2003)指出,在哺乳期间吮吸乳腺大于未吮吸的腺体,这表明吮吸可能对在下一次妊娠期间乳腺的再次发育产生有益的影响。研究结果表明,在第一胎使用或不使用乳头对其后续哺乳期生产力的影响可能有所不同,有吮吸与乳头位置的混杂效应(Fraser 等,1992)。最近可靠的数据表明,第一胎未吮吸的乳头会损害其第二窝的发育(Farmer 等,2012c)。在第一次和第二次哺乳期间相同的乳头或不同的瞎乳头和第一胎时未吮吸的乳头,其都在第二次哺乳期结束时比提前已经吮吸的乳头具有较少的实质组织和较少的实质 DNA 和 RNA。这表明之前使用过的腺体实质细胞既有增生又有代谢活动发生。此外,吮吸以前使用过的乳头的仔猪会比吮吸以前未使用过的乳头的仔猪在 56 日龄时更重(1.12 kg)(Farmer 等,2012c)。

4.8 结论

乳腺发育受到众多因素影响,包括后备母猪或母猪的营养和内分泌状况,但仍然需要学习最佳管理策略以最大限度地去提高后备母猪、妊娠母猪和哺乳期母猪的产奶量。表4.1和表4.2分别总结了会显著影响后备母猪和妊娠动物乳腺发育的各种处理方法。总而言之,限饲90日龄生长的后备母猪阻碍了乳腺发育,并且有迹象表明,供应植物性雌激素木黄酮或增加催乳素浓度可以刺激乳腺发育。在妊娠期间,证明了松弛素和催乳素对于乳腺发育的重要作用,然而,不知道使用这些外源性激素是否会刺激乳腺发育。妊娠期营养研究主要表现为负面影响,因此,过量采食能量或限饲,会减少乳腺实质质量。哺乳期间的管理,如改变吮吸腺体数量和吮吸腺体的持续时间,也会影响哺乳期末期乳腺的发育,第一窝仔猪使用乳头会提高其后续哺乳期的生产力和发育。从我们目前的知识可以看出:优化生长、妊娠和哺乳期的后备母猪或母猪的乳腺发育的理想饲喂方式尚未开发。

表4.1 在发情期或妊娠结束时,各种处理对生长后备母猪的乳腺发育的影响

处理[1]	显著作用于	参考文献
发情期后缺乏雌激素和松弛素	↓实质的横截面面积	Winn等(1994)
催乳素注射29天(代谢体重75 kg BW)	↑实质质量和DNA	Farmer and Palin (2005)
90~183天饲喂木黄酮	↑实质的DNA	Farmer等(2010a)
88~212天饲喂10%亚麻籽	—	Farmer等(2007)
90~202天14.4%与18.7%粗蛋白	—	Farmer等(2004)
28~90天的饲料限制为34%	—	Sorensen等(2006)
90~202天的饲料限制为20%	↓实质质量	Farmer等(2004)
90~170天的饲料限制为26%	↓实质质量	Sorensen等(2006)
9~12周龄和15~20周龄的生长限制:妊娠期的效果	↓实质质量	Lyvers-Peffer and Rozeboom(2001)
在发情期限制饲喂,然后过度饲喂:在发情期的效果	↓实质质量	Farmer等(2012a)
在发情期限制饲喂,然后过度饲喂:在哺乳期的效果	↓实质中的蛋白比例(%)	Farmer等(2012b)

[1]对处理最后一天收集的腺体作用,除非另有说明。

表 4.2　各种处理对妊娠后备母猪的乳腺发育或其后代的影响。

处理[1]	显著作用于	参考文献
80～110 天抑制松弛素	↓实质的横截面面积	Hurley 等（1991）
从妊娠第 100 天到哺乳期的第 29 天注射 GRF	↓实质质量， ↑实质 DNA	Farmer 等（1997）
70～110 天抑制催乳素	↓实质质量，DNA，RNA 和蛋白比例（%）	Farmer 等（2000）
90～110 天抑制催乳素	↓实质质量和蛋白比例（%）	Farmer and Petitclerc（2003）
第 90～110 天每天两次喂 4 g 植物提取物水飞蓟素	—	Farmer 等（2014）
从妊娠第 63 天到哺乳期结束时喂 10% 亚麻籽：对发情期后代的影响	↑实质质量	Farmer and Palin（2008）
第 75～105 天喂 10.5 与 5.76 Mcal ME/天	↓实质质量 DNA	Weldon 等（1991）
第 75～105 天喂 330 与 216 g/天粗蛋白	—	Weldon 等（1991）
25～105 天喂 16 与 4 g/天的赖氨酸	—	Kusina 等（1999）
饲料限制，然后过度饲喂：在妊娠期的效果	↓实质质量	Farmer 等（2014）
喂食限制，然后过度饲喂：在哺乳期效果		Farmer 等（2014b）

[1] 对处理最后一天收集的腺体作用，除非另有说明。

参考文献

Adlercreutz, H., Höckerstedt, K., Bannwart, C., Bloigu, S., Hämäläinen, E., Fotsis, T. and Ollus, A., 1987. Effect of dietary components, including lignans and phytoestrogens, on enterohepatic circulation and liver metabolism of estrogens and on sex hormone binding globulin (SHBG). Journal of Steroid Biochemistry 27: 1135-1144.

Algers, B., Madej, A., Rojanasthien, S. and Uvnas-Moberg, K., 1991. Quantitative relationships between suckling-induced teat stimulation and the release of prolactin, gastrin, somatostatin, insulin, glucagon and vasoactive intestinal polypeptide in sows. Veterinary Research Communications 15: 395-407.

Atwood, C.S. and Hartmann, P.E., 1995. Assessment of mammary gland metabolism in the sow. III. Cellular metabolites in the mammary secretion and plasma following weaning. Journal of Dairy Research 62: 221-236.

Bevers, M.M., Willemse, A.H. and Kruip, T.A.M., 1978. Plasma prolactin levels in the sow during lactation and the postweaning period as measured by radioimmunoassay. Biology of Reproduction 19: 628-634.

Beyer, M., Jentsch, W., Hoffmann, L., Schiemann, R. and Klein, M., 1994. Studies on energy and nitrogen metabolism of pregnant and lactating sows and sucking piglets. 4. Chemical composition and energy content of the conception products, the reproductive organs as well as liveweight gains or losses of pregnant and lactating sows. Archives Animal Nutrition 46: 7-36.

Borena, B.M., Bussche, L., Burvenich, C., Duchateau, L. and Van de Walle, G.R., 2013. Mammary stem cell research in veterinary science: An update. Stems Cells and Development 22: 1743-1751.

Boutinaud, M., Guinard-Flament, J. and Jammes, H., 2004. The number and activity of mammary epithelia cells, determining factors for milk production. Reproduction, Nutrition, Development 44: 499-508.

Capasso, R., Aviello, G., Capasso, F., Savino, F., Izzo, A.A., Lembo, F. and Borrelli, F., 2009. Silymarin BIO-C®, an extract from *Silybum marianum* fruits, induces hyperprolactinemia in intact female rats. Phytomedicine 16: 839-844.

Chang, K., Kurtz, H.J. and Mirocha, C.J., 1979. Effects of the mycotoxin zearalenone on swine reproduction. American Journal of Veterinary Research 40: 1260-1267.

Cromwell, G.L., Stahly, T.S., Edgerton, L.A., Monegue, H.J., Burnell, T.W., Schenck, B.C. and Schricker, B.R., 1992. Recombinant porcine somatoptropin for sows during late gestation and throughout lactation. Journal of Animal Science 70: 1404-1416.

Cross, B.A., Goodwin, R.F.W. and Silver, I.A., 1958. A histological and functional study of the mammary gland in normal and agalactic sows. Journal of Endocrinology 17: 63-74.

DeHoff, M.H., Stoner, C.S., Bazer, F.W., Collier, R.J., Kraeling, R.R. and Buonomo, F.C., 1986. Temporal changes in steriods, prolactin and growth hormone in pregnant and pseudopregnant gilts during mammogenesis and lactogenesis. Domestic Animal Endocrinology 3: 95-105.

Di Pierro, F., Callegari, A., Carotenuto, D. and Tapia, M.M., 2008. Clinical efficacy, safety and tolerability of BIO-C® (micronized Silymarin) as a galactagogue. Acta Biomedica 79: 205-210.

Dusza, L. and Krzymowska, H., 1981. Plasma prolactin levels in sows during pregnancy, parturition and early lactation. Journal of Reproduction and Fertility 61: 131-134.

Farmer, C., Knight, C. and Flint, D., 2007a. Mammary gland involution and endocrine status in sows: effects of weaning age and lactation heat stress. Canadian Journal of Animal Science 87: 35-43.

Farmer, C., Lapointe, J. and Palin, M.F., 2014a. Effects of the plant extract silymarin on prolactin concentrations, mammary gland development, and oxidative stress in gestating gilts. Journal of Animal Science 92: 2922-2930.

Farmer, C. and Palin, M.F., 2005. Exogenous prolactin stimulates mammary development and alters expression of prolactin-related genes in prepubertal gilts. Journal of Animal Science 83: 825-832.

Farmer, C. and Palin, M.F., 2008. Feeding flaxseed to sows during late-gestation affects mammary development but not mammary expression of selected genes in their offspring. Canadian Journal of Animal Science 88: 585-590.

Farmer, C., Palin, M.F., Gilani, G.S., Weiler, H., Vignola, M., Choudhary, R.K. and Capuco, A.V., 2010a. Dietary genistein stimulates mammary hyperplasia in gilts. Animal 4: 454-465.

Farmer, C., Palin, M.F. and Hovey, R., 2010b. Greater milk yield is related to increased DNA and RNA content but not to mRNA abundance of select genes in sow mammary tissue. Canadian Journal of Animal Science 90: 379-388.

Farmer, C., Palin, M.F. and Martel-Kennes, Y., 2012a. Impact of diet deprivation and subsequent over-allowance during prepuberty. Part 1. Effects on growth performance, metabolite status, and mammary gland development in gilts. Journal of Animal Science 90: 863-871.

Farmer, C., Palin, M.F. and Martel-Kennes, Y., 2012b. Impact of diet deprivation and subsequent over-allowance during prepuberty. Part 2. Effects on mammary gland development and lactation performance of sows. Journal of Animal Science 90: 872-880.

Farmer, C., Palin, M.F. and Martel-Kennes, Y., 2014b. Impact of diet deprivation and subsequent over-allowance during gestation on mammary gland development and lactation performance. Journal of Animal Science 92: 141-151.

Farmer, C., Palin, M.F., Theil, P.K., Sorensen, M.T. and Devillers, N., 2012c. Milk production in sows from a teat in second parity is influenced by whether it was suckled in first parity. Journal of Animal Science 90: 3743-3751.

Farmer, C., Pelletier, G., Brazeau, P. and Petitclerc, D., 1997. Mammary gland development of sows injected with growth hormone-releasing factor during gestation and(or) lactation. Canadian Journal of Animal Science 77: 335-338.

Farmer, C., Petit, H.V., Weiler, H. and Capuco, A.V., 2007b. Effects of dietary supplementation with flax during prepuberty on fatty acid profile, mammogenesis, and bone resorption in gilts. Journal of Animal Science 85: 1675-1686.

Farmer, C. and Petitclerc, D., 2003. Specific window of prolactin inhibition in late gestation decreases mammary parenchymal tissue development in gilts. Journal of Animal Science 81: 1823-1829.

Farmer, C., Petitclerc, D., Sorensen, M.T., Vignola, M. and Dourmad, J.Y., 2004. Impacts of dietary protein level and feed restriction during prepuberty on mammogenesis in gilts. Journal of Animal Science 82: 2343-2351.

Farmer, C., Sorensen, M.T. and Petitclerc, D., 2000. Inhibition of prolactin in the last trimester of gestation decreases mammary gland development in gilts. Journal of Animal Science 78: 1303-1309.

Farmer, C., Sorensen, M.T., Robert, S. and Petitclerc, D., 1999. Administering exogenous porcine prolactin to lactating sows: milk yield, mammary gland composition, and endocrine and behavioral responses. Journal of Animal Science 77: 1851-1859.

Ford, Jr., J.A., Kim, S.W., Rodriguez-Zas, S.L. and Hurley, W.L., 2003. Quantification of mammary gland tissue size and composition changes after weaning in sows. Journal of Animal Science 81: 2583-2589.

Fraser, D., Thompson, B.K. and Rushen, J., 1992. Teat productivity in second lactation sows: influence of use or non-use of teats during the first lactation. Animal Production 55: 419-424.

Hacker, R.R. and Hill, D.L., 1972. Nucleic acid content of mammary glands of virgin and pregnant gilts. Journal of Dairy Science 55: 1295-1299.

Harkins, M., Boyd, R.D. and Bauman, D.E., 1989. Effects of recombinant porcine somatotropin on lactational performance and metabolite patterns in sows and growth of nursing pigs. Journal of Animal Science 67: 1997-2008.

Harrell, R.J., Thomas, M.J. and Boyd, R.D., 1993. Limitations of sow milk yield on baby pig growth. In: Cornell University (ed.) Proceedings of the Cornell Nutrition Conference for Feed Manufacturers. October 19-21, 1993. Rochester, NY, USA, pp. 156-164.

Head, R.H. and Williams, I.H., 1991. Mammogenesis is influenced by pregnancy nutrition. In: Batterham, E.S. (ed.) Manipulating pig production III. Australasian Pig Science Association, Atwood, Australia, 33 pp.

Horigan, K.C., Trott, J.F., Barndollar, A.S., Scudder, J.M., Blauwiekel, R.M. and Hovey, R.C., 2009. Hormone interactions confer specific proliferative and histomorphogenic responses in the porcine mammary gland. Domestic Animal Endocrinology 37: 124-138.

Hovey, R.C. and Aimo, L., 2010. Diverse and active roles of adipocytes during mammary gland growth and function. Journal of Mammary Gland Biology and Neoplasia 15: 279-290.

Hughes, P.E. and Varley, M.A., 1980. Lactation. In: Hughes, P.E. and Varley, M.A. (eds.) Reproduction in the pig. Butterworth & Co., London, UK, pp. 136-158.

Hurley, W.L., Doane, R.M., O'Day-Bowman, M.B., Winn, R.J., Mojonnier, L.E. and Sherwood, O.D., 1991. Effect of relaxin on mammary development in ovariectomized pregnant gilts. Endocrinology 128: 1285-1290.

Ji, F., Hurley, W.L. and Kim, S.W., 2006. Characterization of mammary gland development in pregnant gilts. Journal of Animal Science 84: 579-587.

Kensinger, R.S., Collier, R.J. and Bazer, F.W., 1986a. Ultrastructural changes in porcine mammary tissue during lactogenesis. Journal of Anatomy 145: 49-59.

Kensinger, R.S., Collier, R.J. and Bazer, F.W., 1986b. Effect of number of conceptuses on maternal mammary development during pregnancy in the pig. Domestic Animal Endocrinology 3: 237-245.

Kensinger, R.S., Collier, R.J., Bazer, F.W. and Kraeling, R.R., 1986c. Effect of number of conceptuses on maternal hormone concentrations in the pig. Journal of Animal Science 62: 1666-1674.

Kensinger, R.S., Collier, R.J., Bazer, F.W., Ducsay, C.A. and Becker, H.N., 1982. Nucleic acid, metabolic and histological changes in gilt mammary tissue during pregnancy and lactogenesis. Journal of Animal Science 54: 1297-1308.

Kim, S.W., Easter, R.A. and Hurley, W.L., 2001. The regression of unsuckled mammary glands during lactation in sows: the influence of lactation stage, dietary nutrients, and litter size. Journal of Animal Science 79: 2659-2668.

Kim, S.W., Hurley, W.L., Han, I.K. and Easter, R.A., 1999a. Changes in tissue composition associated with mammary gland growth during lactation in sows. Journal of Animal Science 77: 2510-2516.

Kim, S.W., Hurley, W.L., Han, I.K. and Easter, R.A., 2000. Growth of nursing pigs related to the characteristics of nursed mammary glands. Journal of Animal Science 78: 1313-1318.

Kim, S.W., Hurley, W.L., Han, I.K., Stein, H.H. and Easter, R.A., 1999b. Effect of nutrient intake on mammary gland growth in lactating sows. Journal of Animal Science 77: 3304-3315.

Kim, S.W., Osaka, I., Hurley, W.L. and Easter, R.A., 1999c. Mammary gland growth as influenced by litter size in lactating sows: impact on lysine requirement. Journal of Animal Science 77: 3316-3321.

King, R.H., 2000. Factors that influence milk production in well-fed sows. Journal of Animal Science 78(3): 19-25.

King, R.H., Mullan, B.P., Dunshea, F.R. and Dove, H., 1997. The influence of piglet body weight on milk production in sows. Livestock Production Science 47: 169-174.

King, R.H., Pettigrew, J.E., McNamara, J.P., McMurty, J.P., Henderson, T.L., Hathaway, M.R. and Sower, A.F., 1996. The effect of exogenous prolactin on lactation performance of first-litter sows given protein-deficient diets during the first pregnancy. Animal Reproduction Science 41: 37-50.

Knight, C.H. and Peaker, M., 1984. Mammary development and regression during lactation in goats in relation to milk secretion. Quarterly Journal of Experimental Physiology 69: 331-338.

Kopinski, J.S., Blaney, B.J., Downing, J.A., McVeigh, J.F. and Murray, S.A., 2007. Feeding sorghum ergot (*Claviceps africana*) to sows before farrowing inhibits milk production. Australian Veterinary Journal 85: 169-176.

Kusina, J., Pettigrew, J.E., Sower, A.F., Hathaway, M.R., White, M.E. and Crooker, B.A., 1999. Effect of protein intake during gestation on mammary development of primiparous sows. Journal of Animal Science 77: 925-930.

Lyvers-Peffer, P.A. and Rozeboom, D.W., 2001. The effects of a growth-altering pre-pubertal feeding regimen on mammary development and parity-one lactation potential in swine. Livestock Production Science 70: 167-173.

Manjarin, R., Trottier, N.L., Weber, P.S., Liesman, J.S., Taylor, N.P. and Steibel, J.P., 2011. A simple analytical and experimental procedure for selection of reference genes for reverse-transcription quantitative PCR normalization data. Journal of Dairy Science 94: 4950-4961.

Marshall, K.M., Hurley, W.L., Shanks, R.D. and Wheeler, M.B., 2006. Effects of suckling intensity on milk yield and piglet growth from lactation-enhanced gilts. Journal of Animal Science 84: 2346-2351.

McLaughlin, C.L., Byatt, J.C., Curran, D.F., Veenhuizen, J.J., McGrath, M.F., Buonomo, F.C., Hintz, R.L. and Baile, C.A., 1997. Growth performance, endocrine, and metabolite responses of finishing hogs to porcine prolactin. Journal of Animal Science 75: 959-967.

Miller, Y.J., Collins, A.M., Smits, R.J., Thomson, P.C. and Holyoake, P.K., 2012. Providing supplemental milk to piglets preweaning improves the growth but not survival of gilt progeny compared with sow progeny. Journal of Animal Science 90: 5078-5085.

Motyl, T., Gajkowska, B., Wojewodzka, U., Wareski, P., Rekiel, A. and Ploszaj, T., 2001. Expression of apoptosis-related proteins in mammary gland of sow. Comparative Biochemistry and Physiology Part B 128: 635-646.

Nielsen, O.L., Pederson, A.R. and Sorensen, M.T., 2001. Relationships between piglet growth rate and mammary gland size of the sow. Livestock Production Science 67: 273-279.

Nordskog, A.W. and Clark, R.T., 1945. Ergotism in pregnant sows, female rats and guinea pigs. American Journal of Veterinary Research 6: 107-116.

O'Day, M.B., Winn, R.J., Easter, R.A., Dziuk, P.J. and Sherwood, O.D., 1989. Hormonal control of the cervix in pregnant gilts. II. Relaxin promotes changes in the physical properties of the cervix in ovariectomized hormone-treated pregnant gilts. Endocrinology 125: 3004-3010.

Peel, C.J. and Bauman, D. E., 1987. Somatotropin and lactation. Journal of Dairy Science 70: 474-486.

Porter, D.G., Friendship, R.M., Ryan, P.L. and Wasnidge, C., 1992. Relaxin is not associated with poor milk yield in the postpartum sow. Canadian Journal of Veterinary Research 56: 204-207.

Quesnel, H., Farmer, C. and Theil, P.K., 2015. Colostrum and milk production. Chapter 8. In: Farmer, C. (ed.) The gestating and lactating sow. Wageningen Academic Publishers, Wageningen, the Netherlands, pp. 173-192.

Renaudeau, D., Lebreton, Y., Noblet, J. and Dourmad, J.Y., 2002. Measurement of blood flow through the mammary gland in lactating sows: methodological aspects. Journal of Animal Science 80: 196-201.

Smith, B.B. and Wagner, W.C., 1984. Suppression of prolactin in pigs by *Escheriscia coli* endotoxin. Science 224: 605-607.

Sorensen, M.T., Farmer, C., Vestergaard, M., Purup, S. and Sejrsen, K., 2006. Mammary development in prepubertal gilts fed restrictively or *ad libitum* in two sub-periods between weaning and puberty. Livestock Science 99: 249-255.

Sorensen, M.T., Sejrsen, K. and Purup, S., 2002. Mammary gland development in gilts. Livestock Production Science 75: 143-148.

Spinka, M., Illmann, G., Stetkova, Z., Krejc, P., Tomanek, M., Sedlak, L. and Lidicky, J., 1999. Prolactin and insulin levels in lactation sows in relation to nursing frequency. Domestic Animal Endocrinology 17: 53-64.

Stefanon, B., Colitti, M., Gabai, G., Knight, C.H. and Wilde, C.J., 2002. Mammary apoptosis and lactational persistency in dairy animals. Journal of Dairy Research 69: 37-52.

Tedesco, D., Tava, A., Galletti, S., Tameni, M., Varisco, G., Costa, A. and Steidler, S., 2004. Effects of silymarin, a natural hepatoprotector, in periparturient dairy cows. Journal of Dairy Science 87: 2239-2247.

Theil, P.K., Labouriau, R., Sejrsen, K., Thomsen, B. and Sorensen, M.T., 2005. Expression of genes involved in regulation of cell turnover during milk stasis and lactation rescue in sow mammary tissue. Journal of Animal Science 83: 2349-2356.

Theil, P.K., Sejrsen, K., Hurley, W.L., Labouriau, R., Thomsen, B. and Sorensen, M.T., 2006. Role of suckling in regulating cell turnover and onset and maintenance of lactation in individual mammary glands of sows. Journal of Animal Science 84: 1691-1698.

Thodberg, K. and Sorensen, M.T., 2006. Mammary development and milk production in the sow: Effects of udder massage, genotype and feeding in late gestation. Livestock Science 101: 116-125.

Thorup, F., 1998. Kuldudjaevningens betydning for fravaenningsvaegten. Erfaring fra Landsudvalget for svin Copenhagen, Denmark.

Turner, C.W., 1952. The anatomy of the mammary gland of swine. In: Turner, C.W. (ed.) The mammary gland. I. The anatomy of the udder of cattle and domestic animals. Lucas Brothers, Columbia, MO, USA, pp. 279-314.

VanKlompenbeg, M.K., Manjarin, R., Trot, J.F., McMicking, H.F. and Hovey, R.C., 2013. Late-gestational hyperprolactinemia accelerates mammary epithelial cell differentiation that leads to increased milk yield. Journal of Animal Science 91: 1102-1111.

Weldon, W.C., Thulin, A.J., MacDougald, O.A., Johnston, L.J., Miller, E.R. and Tucker, H.A., 1991. Effects of increased dietary energy and protein during late gestation on mammary development in gilts. Journal of Animal Science 69: 194-200.

Wilde, C.J., Addey, C.V.P., Boddy, L.M. and Peaker, M., 1995. Autocrine regulation of milk secretion by a protein in milk. Biochemistry Journal 305: 51-58.

Winn, R.J., Baker, M.D., Merle, C.A. and Sherwood, O.D., 1994. Individual and combined effects of relaxin, estrogen, and progesterone in ovariectomized gilts. II. Effects on mammary development. Endocrinology 135: 1250-1255.

Zaleski, H.M., Winn, R.J., Jennings, R.L. and Sherwood, O.D., 1996. Effects of relaxin on lactational performance in ovariectomized gilts. Biology of Reproduction 55: 671-675.

Zhao, W., Shahzad, K., Jiang, M., Graugnard, D.E., Rodriguez-Zas, S.L., Luo, J., Loor, J.J. and Hurley, W.L., 2013. Bioinformatics and gene network analyses of the swine mammary gland transcriptome during late gestation. Bioinformatics and Biology Insights 7: 193-216.

5. 妊娠期高纤维饲喂

M.C. Meunier-Salaün[1,2]* and J.E. Bolhuis[3]

[1] INRA, UMR1348 PEGASE, 35590 Saint Gilles, France; marie-christine.salaun@rennes.inra.fr

[2] AgrocampusOuest, UMR1348 PEGASE, 35000 Rennes, France

[3] Wageningen University, Department of Animal Sciences, Adaptation Physiology Group, P.O. Box 338, 6700 AH Wageningen, the Netherlands

摘要：妊娠母猪饲喂水平通常是较低的,不能提供足够的饱腹感,也不允许母猪充分发挥它们的采食动机来表现出觅食和采食行为。因此,限饲可能导致非进食性口腔活动发生率高。包括在群养母猪中出现刻板、躁动和攻击行为,这被解释为持续采食动机和挫折的信号。日粮纤维的添加降低了饲料中的能量密度,从而使得在不增加能量供给的情况下获得更大的进食量。此外,日粮纤维能够增加摄食和吸收后的饱腹感和满足感。本章回顾了日粮纤维对妊娠母猪的行为和福利的影响,并且描述它们对生产性能的潜在作用。日粮纤维一般能够降低刻板行为的发生,减少烦躁行为和活动,还有一些研究报告说能够减少侵略性。这些影响很可能与日粮纤维对饱腹感和采食的影响的潜在行为和生理机制有关。然而,对日粮纤维的反应程度取决于纤维性日粮的特性(包含率、纤维来源、理化特性)、猪舍和饲喂条件以及母猪的特性,特别是胎次,且这种反应对年轻母猪有更大的影响。在妊娠期间提供的日粮纤维通常会导致母猪在哺乳期间饲料摄入增加,这可能是由于它们对胃肠道的大小和能力造成的影响。纤维对繁殖性能的影响的相关研究很少,且研究结果比较多变,这可能是由于对妊娠期间饲料能量含量的高估或低估引起的。总之,日粮纤维对限制饲喂的母猪的行为和福利有积极的影响。妊娠期高纤维日粮多个连续的周期对母猪繁殖性能的影响值得进一步研究。

关键词：母猪,纤维饲料,行为,饱腹感

5.1 引言

为了保持良好的健康和最佳的生产表现,能繁母猪在整个繁殖周期中通常被限饲以维持相对恒定的机体状况(Dourmad 等 1994,1996)。妊娠期间,母猪过量

5. 妊娠期高纤维饲喂

的体重增加和脂肪沉积,可能会延长分娩过程,并增加运动障碍,因此,妊娠母猪通常应限制饲喂。提供的饲料水平相当于母猪自由采食量的40%～60%(Lawrence 等,1988;Van der Peet-Schwering 等,2004),这导致了低水平的饱足感和食欲的降低以及进食行为减少等表现。这种饲喂水平可能无法满足妊娠母猪的行为需要,尤其是它们表达觅食和采食行为的动机。因此,饲料限制引起了广泛的行为问题,比如采食时的刻板行为(Lawrence 和 Rushen,1993),以及在群体饲养的母猪因饲养竞争而出现的侵略行为(Terlouw 等,1991)。这两种行为问题都被认为反映了一种未实现的采食动机和福利的受损(De Leeuw 等,2008,2001,2001;Philippe 等,2008)。此外,过度的饲料限制也可能会对仔猪的出生体重和生存能力造成不利影响,以及对断奶至发情间隔或受胎率的影响,这两种情况都与分娩时较低体脂有关(Campos 等,2012,Van der Peet-Schwering 等,2003b)。

目前关于集约化动物生产系统的社会争论较多的是一些对农场动物福利状况的怀疑(Eurobarometer,2005)。这些质疑尤其关注的是,母猪无法表达自由采食行为,未能满足立法要求的五项自由中的其中一项:自由表达正常行为的自由(Council Directive 1991 630/EEC;EC,1991)。一种满足妊娠母猪饲喂动机的方法,是为母猪提供纤维材料作为基质种类或提供包含纤维成分日粮,同时维持有限的能量供应以防止过量的体重增加和脂肪沉积。

在欧洲关于猪福利的立法(Council Directive 2001/88/EC;EC,2001)中,这一替代方法被考虑。日粮中纤维成分的供给或作为支持的物质,可能会促进营养性饱腹感,例如,通过增加饲料量而不增加能量和营养的供给,以及通过它们对采食行为的食欲和完成顺序的影响而产生的行为饱腹感。

本章的目的是描述为妊娠母猪所使用的日粮纤维的主要影响。我们将讨论纤维的特性,对饱腹感的影响,以及纤维饲料对动物福利的表现和行为活动的影响。

5.2 日粮纤维:定义、来源和特征

Hipsley(1953)引入了植物细胞壁的非消化成分的定义。此后,人们广泛讨论了专家的新提议,并在2009年的食品法典委员会会议(Codex Alimentarius Commission)上通过了最终的定义(Berridge,2009,Phillips and Cui,2011)。

日粮纤维被定义为含有10个或更多单体的碳水化合物聚合物,且不能由人类小肠内的内源性酶来水解。在动物生产中,日粮纤维通常被描述为植物的可食用部分或其提取物的组成部分,或合成的类似物(Bach Knudsen,2001)。

日粮纤维包括来自植物细胞壁的4种主要成分:纤维素(cellulose)、半纤维素(hemi-cellulose)、果胶和木质素(pectin and lignin),以及一些多糖细胞质,如抗性淀粉(直链淀粉和支链淀粉,amylose and amylopectin)、瓜尔胶(guar gum)和黏质

物(mucilage,如海藻酸)。日粮纤维分析根据 Van Soest 等(1991),再加上最新的分析方法描述了不同的纤维组分,相应的有抗性淀粉,非消化性的低聚糖、木质素和非淀粉多糖(non-starch polysaccharides,NSP),其进一步细分为中性洗涤纤维(neutral detergent fibre,NDF)、酸性洗涤纤维(acid detergent fibre,ADF;Champ 等,2003;参见图 5.1 综述)。

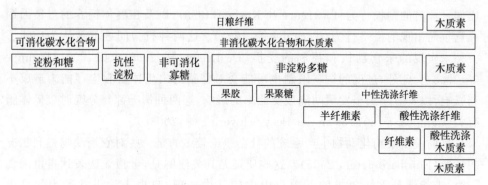

图 5.1 日粮纤维图示(De Leeuw 等,2008,改编于 Van Soest 的分析,1991)。

日粮纤维的来源是植物、水果、蔬菜、坚果和谷物(表 5.1)。日粮纤维的主要特点是不能被小肠消化和吸收,其通常是在大肠内完全或部分发酵。日粮纤维可以其物理化学性质为特征,包括其在水中的溶解性(由此分类为可溶性或不溶性)、黏度、凝胶的形成、结合水的能力、膨胀和发酵能力(Blackwood 等,2000)。

表 5.1 妊娠母猪的主要日粮纤维的营养和物理化学特性[1]

纤维资源	CF	NDF	ADF	ADL	可发酵性[2]	系水力[3]	容重[4]
小麦秸秆	38.2	72.1	45.8	7.5			
大豆壳	34.2	56.4	40.4	2.2	+++	++	
燕麦壳	29.0	73.5	35.3	6.9	+	+	—
脱水苜蓿	28.4	46.1	32.6	8.3	+++	++	
葵花籽粕	28.7	46.3	33.0	11.4	++		
脱水甜菜渣	19.4	45.5	23.1	2.1	++++	+++	——
葵花籽	16.7	31.0	20.1	6.1	+		
大麦芽	15.4	52.8	20.4	5.4	++		
燕麦	14.0	35.4	16.2	2.7	+		
菜籽饼	13.9	31.7	20.6	9.9	+	+	
大豆壳					+++	++	—
小麦麸	10.6	45.5	13.7	3.9	++	+	———

续表 5.1

纤维资源	CF	NDF	ADF	ADL	可发酵性[2]	系水力[3]	容重[4]
玉米蛋白粉	8.5	38.4	10.0	1.3	+	+	
玉米粒	8.3	39.2	13.9	2.3	++	+	
小麦粒	7.9	23.7	10.8	4.6	+	+	
豌豆	6.0	14.3	7.1	0.4	++	++	
小麦	2.5	12.6	3.2	0.9	+	−	
抗性淀粉					+++	+	−

[1] 在纤维素(CF)、中性洗涤纤维(NDF)、酸性洗涤纤维(ADF)和酸性洗涤木质素(ADL)中干物质含量(%)(Le Gall 等，2011；Sauvant 等，2004)。
[2] 数据来自于 Evapig (www.evapig.com/x-home-fr)
[3] +，++，+++，++++ 相对于其他纤维资源，增加的更高程度的可发酵性和 WHC，−，−−，−−− 相对于其他纤维资源，增加的更高程度的体积密度。根据一些文献资源，来自 De Leeuw 等(2008) 和 Le Gall 等(2011)的对比进行估计。
[4] 低容重表示高膨松度。

尽管纤维的膨胀性不同，膨胀性的高低可能取决于它们的结构(体积)和水的结合能力，但所有的纤维都有膨胀的特性，因为它们包含在等热日粮中会降低饲料能量密度(Wanders 等，2011)。

5.3 日粮纤维和饱腹感：新陈代谢和生理影响

根据日粮纤维的来源及其物理化学性质，包含它们已经被证明会影响饲料的可用能量含量(Noblet 和 Le Goff，2001)和影响饱腹感的过程，包括饱腹感关联的三个主要阶段：感觉和认知、摄食后和吸收后阶段(Benelam，2009；Blundell 和 Burley，1987)。许多研究都提到了日粮纤维的能量价值，并估算其对能量供应的贡献范围(Noblet 等，2003)。猪能够在合理的范围内利用日粮纤维，而纤维性饲料对能量供应的贡献在更成熟的动物中也会增加，特别是在能繁母猪中。因此，日粮纤维可以用于妊娠母猪的饲喂方式中，由于低能量能允许更大的采食量而不增加饲料日粮能量(Noblet 和 Shi，1993)。记录整个饱腹感关联的日粮纤维的作用已被广泛研究和描述(Burton-Freeman，2000；De Leeuw 等，2008)。在口腔消化水平上，日粮纤维的作用与它们的感官特性有关，能够调节食物的味道、气味和食物的质地以及对特定食物的适口性，尤其是在开始采食的时候。

关于以前采食的信息（大小和时间间隔）以及之前对特定食物的经验也有助于日粮纤维的饱腹信号(Bach Knudsen，2001；Forbes，1998)。日粮纤维作为一种主要的纤维对饱腹感的感觉阶段的影响可以通过食用甜菜渣中增加的咀嚼活动来说明，而甜菜渣具有很高的水结合能力(Ramonet 等，1999)。在人体内，咀嚼活动增强会在口腔中产生更长的感觉暴露时间,从而提高饱腹感,或终止用餐(Zijlstra 等,2009)。餐后短时间内的进食动机减少,也可以说明与日粮纤维有关的饱腹感(Meunier-Salaün 等,2001;Souza da Silva 等,2012)。已经报道,在胃肠道中与饱腹后期相关的日粮纤维的几种推定效应(Brownlee,2011):增加唾液和胃液的产生从而促进胃扩张,延迟胃排空,丰富传入的迷走神经信号,长时间的流动时间可能导致饱腹感相关的激素分泌增加和延长的化学感受,最后,短链脂肪酸(short-chain fatty acids,SCFA)产量的增加与远端小肠部分(主要在结肠)的生产发酵相关(Sappok,2012)。吸收后的因素与采食的代谢变化有关。研究表明,摄入的日粮纤维影响饱腹感的生物标志物是葡萄糖、胰岛素和 SCFA。的确,饲料中纤维的掺入导致采食后血糖降低和胰岛素反应迟缓,反映出酶消化葡萄糖减少,以及采食后 SCFA 生产增加(Farmer 等,2002;Ramonet 等,2000a),这可能刺激饱腹感相关的激素的释放如 PYY 或 GLP-1(De Leeuw 等,2008)和延长身体的能源供应,从而促进饱腹感的持续时间。

目前关于妊娠母猪日粮纤维,研究发现纤维来源和日粮纤维水平显著影响采食后诱导的代谢和激素分布变化(De Leeuw 等,2004;Farmer 等,2002;Guerin 等,2001;Quesnel 等,2009;Ramonet 等,2000a;Serena 等,2008)。例如,含甜菜渣为主要纤维源饲料的研究表明,通过增加咀嚼时间,在感觉阶段有强烈的影响,同时在吸收后的阶段,与含麦麸饲料相比葡萄糖和胰岛素峰值延迟,葡萄糖和胰岛素水平较低(Ramonet 等,2000a)。Brouns 等(1995)也报告说,与使用其他纤维来源的日粮相比,含有甜菜渣的饲料适口性比较低。

总之,日粮纤维对饱腹过程中涉及的机制产生了影响。然而,影响作用取决于纤维来源及其物理化学性质,包括膨胀、黏度、胶凝或可发酵性。而妊娠母猪饲料中日粮纤维诱导饱腹感是可变的(Le Gall 等,2009;Serena 等,2008,2009;Slavin and Green,2007)。对主要理化性质的影响研究表明,膨胀作用在采食时和采食后具有更强的效果,而可发酵纤维在采食后延长餐后饱腹感多个小时(Benelam,2009;De Leeuw 等,2008;Souza da Silva 等,2012)。除了饱腹效果之外,日粮纤维还可能对结肠健康和微生物群的构成有好的作用(Haenen 等,2013a,b)。这些影响超出了本章的范围,但将在第 15 章中讨论(Le Huërou-Luron 和 Ferret-Bernard,2015)。

5.4 日粮纤维对行为活动的影响

日粮纤维常常作为饱腹感的信号,关于其对猪行为活动影响的研究多集中在饲喂模式和进食动机。其他与进食行为有关的活动,如没有进食时无聊的口腔咀嚼活动,也被认为是由于限制饲喂而引起的沮丧导致的,而好斗的行为可能反映了群体饲养的母猪的食物竞争。然而,如果母猪被关在铺有可用作日粮纤维的营养来源丰富垫料的地板上的猪舍中,不仅仅花费在饲养行为上的时间有很大变化,而且纤维(如抗性淀粉、RS)也可能使各种活动减少(De Leeuw 等,2008),包括觅食和探索性行为(D'Eath 等,2009;De Leeuw 等,2008;Meunier-Salaun 等,2001;Le Gall 等,2009;Philippe 等,2008)。尽管这些研究使用了不同的品系、饲养条件、日粮纤维和记录时期(采食前、采食后或全天),但可以得出关于纤维日粮对母猪行为的影响的一般结论,即日粮纤维主要影响体现在饲喂模式的满足感增加,非进食口腔活动的发生率减少,包括群养母猪的呆板和侵略行为减少以及一般活动的下降。然而,响应的强度取决于与动物相关的因素、饲喂条件和所提供的日粮的特征,如下所述。

5.4.1 饲喂模式和摄取食物的动机

与传统的、高能量的日粮相比,妊娠母猪饲喂纤维日粮通常花费更多的时间来消耗增加的供应量以便提供相同的能量。根据提供的纤维类型,单独饲养的母猪和在室外群体饲养的母猪可以花费最多 5 倍的时间来采食和消化(Brouns 等,1994,1997)(Brouns 等,1995;Martin 和 Edwards,1994;Ramonet 等,1999;表 5.2)。这是因为母猪花更多的时间来咀嚼纤维食物。Guillemet 等(2006)发现当母猪饲喂 31% 的 NDF 饲料时,在采食中会休息一次,相比之下当饲喂 17% 的 NDF 饲料时则很少休息。这些饲喂模式的变化也与较低的进食率相联系(Brouns 等,1997;Mroz 等,2006;Mroz 等,1986;菲利普等,2008,Ramonet 等,2000,Guillemet 等,2003)。然而,纤维日粮对进食率的影响可能取决于猪舍条件。例如,6 头猪在单体限位栏饲养和群养两种饲养模式相比,均提供 6 个饲喂槽,自由采食,结果显示,单体限位栏饲养效果显著好于群养。这可能与群养时的竞争情况有关,这种情况下母猪在料槽边由于竞争而浪费了大量时间。

饲喂模式的改变也可能与纤维来源的适口性有关。例如,研究表明,与麦麸相比,含有甜菜渣或脱水苜蓿的饲料母猪优先选择性较低(Brouns 等,1995,Brouns 等,2007a,2011)。饲料的纤维来源、添加比率和新鲜度可能都会影响适口性,因此,可能通过饱腹感和采食终止而对短期饲料摄入产生影响,其中口腔感官刺激将起重要作用。与胃膨胀、延迟胃排空(Guerin 等,2001)相关的代谢作用,葡萄糖和胰岛素峰值的减少以及 SCFA 产量的增加均可以通过促进饱腹感来形成不同的饲喂模式,从而推迟新的饲喂周期的开始(De Leeuw 等,2008;Meunier-Salaün 等,2001;Philippe 等,2008))。然而,这些代谢变化似乎在一定程度上是暂时的,这是由于供应高水结合和可发酵的日粮纤维而导致胃肠器官(胃,盲肠,结肠)的体积增加,从而促进更高的摄取能力(Jorgensen 等,1996;Whittaker 等,2000)。对饲喂纤维性饲料的母猪采食动机进行了研究,目前主要关注使用混合纤维来源的研究(Meunier-Salaün 等,2008;Philippe 等,2008)。然而,比较 Brouns 等(1997)关于

表 5.2　日粮纤维对采食前后测量的妊娠母猪的饲喂模式和饲喂动机的影响
（即采食前 90 min 至采食后 23 h）

纤维资源	粗纤维占干物质比例[1](%)	胎次	圈舍/饲喂	餐数[2]	饲喂环境		条件模型		参考文献
					采食时长[3]	饲喂率[3]	程序[4]	奖励次数[3]	
小麦麸皮/玉米缨糖	10	P1-P2	I/R	2	↑ P1				Robert 等，1993
小麦麸皮/燕麦	20	P1-P2	I/R	2	↔ P2				Robert 等，1993
甜菜浆 (SBP)	500 g/kg	P1	I/R	1	↔				Brouns 等，1994
SBP	500 g/kg	多胎	I/AL	1	↔				Brouns 等，1995
SBP	500 g/kg	多胎	G/R, FS	1					Brouns 等，1997
混合物[5]	13, 26	多胎	I/R	1	↑				Ramonet 等，1999
混合物[5]	18	多胎	I/R	1	↔				Ramonet 等，2000a
燕麦壳/苜蓿	18.2	多胎	I/R	2	↑				Bergeron 等，2002
燕麦壳	23.5	多胎	I/R	2	↔				Bergeron 等，2002
SBP	20/30	多胎	G/T, R	—					Rijnen 等，2003
混合物	12.4	多胎	I/R	1	↔		PR, GFD		Guillemet 等，2006
燕麦壳	20.4	后备猪	I/R	2			BFM1, BFM2	↓	Robert 等，1997
小麦麸皮	10.1	后备猪	I/R	2			PR, GSD; AFM 4 h; AFM 23 h	↓	Robert 等，1997
SBP	10	多胎	I/R	1			PR, GFD; BFM1, AFM2	↓	Ramonet 等，2000b
燕麦壳和苜蓿	18.2	P1	I/FS	2			PR, LR, AFM1; BFM2	→	Robert 等，2002
燕麦壳	18.2/23.5	多胎	I/R	2	→/↑		PR, BFM1, AFM2		Bergeron 等，2000
果胶渣，马铃薯浆，SBP	35	P1	G/R	2					Jensen 等，2012

[1] 圈舍设计：G = 群养，I = 单体栏位。饲喂：AL = 自由采食饲喂，FS = 饲喂料槽，R = 限饲，T = 常见料槽。
[2] 每天餐数。
[3] ↓，↑，→：当纤维日粮与对照饲料对比时，分别减少、增加以及不受潜食处理影响。
[4] 条件程序。奖励计划：FR = 固定比率，PR = 渐进率。食物奖励：GSD = 妊娠标准料，LD = 哺乳料。条件测试时间表（每日两餐）：BFM1 = 第 1 餐前；BFM2 = 第 2 餐前；AFM1 = 第 1 餐后；AFM2 = 第 2 餐后。
[5] 混合物 = 甜菜浆 + 小麦麸 + 麦花粉 + 大豆壳 + 玉米粉。

用甜菜渣、大麦秸秆、燕麦壳、麦芽或大米中两者之一等能饲料的研究,显示出具有高水平甜菜渣饲料的自由采食量较低,说明母猪的采食自我限制,当添加物水平从 400 g/kg 到 650 g/kg 的饲料时,采食量从 5.0 kg/天降低到 3.0 kg/天。这种饲料摄入量的降低可归因于该纤维源的具体性质,例如适口性差、高水结合和发酵特性。这也可以解释为由于采食动机减少而导致的,进食动机的测定是使用一个电子操作的饲喂器,或者一个装有轮子的饲料分配器,在那里猪必须按下按钮或转动轮子来获得食物奖励。按压的次数反映了采食动机的水平,从而评估了饥饿和饱腹感程度(Bergeron 等,2002;Lawrence 和 Illius,1989;Ramonet 等,2000b;Robert 等,1997;Souza da Silva 等,2012)。在食欲阶段期间,有关采食行为中的食物搜索行为也在一条跑道上进行了测试,母猪必须单独走一条固定的 U 形轨道才能在其末端获得食物奖励(Souza da Silva 等,2012))。无论什么样的模式,饲喂纤维日粮的母猪在进食后的短期测试时往往具有较少的奖励数量,而 Souza da Silva 等(2012)报告说,与黏性纤维(pectin,果胶)相比,较大的木质纤维素(lignocellulose)或可发酵纤维(resistant starch,抗性淀粉)具有较强的作用。此外,同一作者关于可发酵纤维的比较研究还表明,与瓜尔胶和菊糖相比,饲料中可发酵纤维水平升高则动物进食动机降低,特别是在木薯淀粉(抗性淀粉)具有最显著的作用(Souza da Silva 等,2013)。然而,各种测试结果的差异很大,主要是可能由于操作程序差异,包括执行的任务(操作中的固定或渐进比例),所提供的报酬的性质(对照日粮,适口或纤维性日粮),相对于之前采食的时间点。事实上,采食大量饲料的母猪在进一步采食的时候可能表现出动机较少,尽管仍然处于"代谢饥饿"或胃部充盈却愿意采食。这表明,大量日粮的供应可能会在短期内增加饱腹感,但可能无法解决进食中、长期持续的进食动机。在这些研究中,诸如母猪的年龄或胎次的因素各不相同,表明纤维性日粮对于母猪与母猪的进食动机有较强的影响。

5.4.2 非进食口腔活动和刻板行为

妊娠母猪针对圈舍中的垫物(例如嘴掘地、舔或咬的杆、链条、进料器、地板、墙壁或垫料)或不自主的口腔活动(假咀嚼、玩舌头、磨牙、摇头),一般与未满足的进食动机有关,由于饲料限制(饲料不足以产生饱腹感)并且由于缺乏垫料而导致觅食行为的挫败。另外,饲料限制也会导致水的消耗高于正常生理需求(Douglas 等,1998)。这种口腔活动是母猪应激和福利受损的迹象(Mason,1991;Lawrence 和 Rushen,1993)。在许多情况下,这些活动主要伴随着采食活动进行(从采食前至采食后 3 h),并且在饲喂前后发现了不同形式的刻板行为(Blackshaw and McVeigh,1985;Robert 等,2002;Rushen,1985)。例如,已经描述了啃咬物体几乎完全在进食之前发生,并且以较高的频率反应对饲喂的期望,而真空咀嚼和啃咬链条通常是进食后最常见的刻板行为。这些结果与先前的观点一致,即饲喂前的刻板行为可能与觅食挫败相关(例如觅食行为),而饲喂后的刻板行为与采食动机的持续性相关(Robert 等，2002)。

许多研究报告显示，在妊娠母猪日粮中添加纤维，进食后立即产生的口腔活动发生率降低。然而，纤维量降低口腔的刻板活动的有效性取决于纤维的性质、添加比率和饲喂条件(De Leeuw 等，2008；Le Gall 等，2009；Meunier-Salaün 等，2001；Philippe 等，2008)。实际上，与较低水平相比，当饲料中以约 300 g NDF/kg 的比例掺入时刻板行为的减少程度似乎更多(Bergeron 等，2000；Ramonet 等，1999；Robert 等，1993，1997)，同时保持净能量(net energy，NE)在相似的水平。在摄入高 NSP 日粮之后，已经报道了对刻板行为发生率的主要影响因素，特别是添加甜菜渣、燕麦壳或小麦麸的情况下(Cerneau 等，1997；Courboulay and Gaudré，2002；Paboeuf 等，2000；Ramonet 等，2000a；Robert 等，1997)。Ramonet 等 (2000a)强调指出，饲喂纤维日粮的母猪减少刻板活动可能某种程度上是由于进食时间较长，无进食活动的时间较短的原因，这可能解释了与麦麸相比，日粮甜菜渣的影响更强的原因。然而，De Leeuw 等 (2008)强调，即使纠正采食时间，进食后，日粮纤维的添加也可以立即减少非进食口腔活动。日粮纤维的这种采食后效应似乎与蓬松性的来源或者程度无关(De Leeuw 等，2008)。即使具有较低膨松度的抗性淀粉，最近也被报道可以减少母猪的刻板行为(Souza da Silva 等，2013)。然而，重要的是，只有在满足能量和营养需求的情况下，纤维日粮才能有效地减少刻板行为，这是考虑到纤维日粮中较低净能(NE)的一个重要因素(Ramonet 等，2000a)。

日粮纤维提供的方式也可以调节进食后对口腔或刻板行为的影响。例如，在早上进食后 20 min 内，接受粉状而不是颗粒状的高纤维日粮(50% NDF 或 43% NDF 对 20% NDF 日粮的对照)的母猪显著降低了啃咬链条的时间(Bergeron 等，2002)。当后备母猪每天饲喂两次时，它们在采食之前和之后表现出比每天饲喂一次时更多的刻板行为，同时在用纤维性日粮和对照日粮。每次饲喂两次并饲喂高纤维饲料的后备母猪，与对照组后备母猪相比，早上饲喂前 5 min 内紧张度降低(Robert 等，2002)。

个别因素，如胎次，也在日粮纤维减少了口腔活动的有效性上发挥了作用，与经产母猪相比，对青年母猪的影响更大(Meunier-Salaün 等，2001)。这可能是由于刻板的行为在连续胎次的情况下逐渐变得更加死板和频繁(Dantzer，1986)。

综上所述，日粮纤维在减少进食后口腔活动方面，如果从第一胎就开始提供时其作用效果更高(Paboeuf 等，2000)，并且受到所添加的纤维类型的影响。De Leeuw 等 (2008)发现，尽管蓬松类型的纤维具有更强的效应但膨胀性较低但可发酵的纤维进食后也对口腔活动有影响。因此，膨胀特性在多大程度上对母猪产生影响仍然没有定论。

5.4.3 身体行为(休息、探索和社会行为)

许多研究报告显示,在母猪饲料中加入含有各种纤维成分后的 24 h 内,一般躁动不安的情况有所减少(De Leeuw 等,2008;Meunier-Salaün 等,2001;Van der Peet-Schwering 等,2003a)。行为反应以较低的站立时间或较低的姿势变化频率来表示,效果的大小取决于个体和饲料因素。例如,饲喂纤维性饲料的母猪,行为姿势变化的减少在进食前更加明显(Paboeuf 等,2000),并且在妊娠母猪的站立时间和口腔活动之间呈现正相关(Ramonet 等,1999)。日粮纤维对身体活动的影响也可以被胎次调节,站立时间的缩短在第二胎母猪中比第一胎母猪更为明显(Robert 等,1993)。此外,日粮纤维对身体活动的影响可能取决于纤维成分,例如与添加麦麸和玉米的饲料相比时,当母猪饲喂添加燕麦壳和燕麦的饲料具有更强的作用(Robert 等,1993)。De Leeuw 等(2004,2008)强调了纤维饲料发酵对母猪进食后几个小时内身体活动的影响的重要作用,而在 24 h 周期内膨胀性本身对身体活动的影响尚未得出确凿的结论。身体活动似乎是进食后饱腹感延长的可靠指标,但是需要进一步研究,以确定饲喂大量的纤维饲料的母猪躺卧时间的增加是否代表了进食动机水平的降低,或者由于日粮中水分结合能力升高相关的胃粘连性增高引起的腹部不适。

5.4.4 侵略行为

在群养母猪中采用限制饲喂产生了竞争现象,其中侵略行为的程度可由饲养者设计和饲养调教调节(Meunier-Salaün 等,2001)。纤维日粮可能减少侵略行为的表现,如通过电子饲喂站(Vantat Putten and Van de Burgwal,1990)或者地板饲喂高纤维饲料可以使母猪外阴咬伤减少(Whittaker 等,1999)。当与限制水平的常规日粮相比,母猪自由采食高纤维饲料,在电子饲喂站中引入母猪后的第一天,动态群体中的母猪侵略行为严重程度显著降低(Whittaker 等,1999)。通过饲喂调教也可以调节日粮纤维对母猪相互争斗的影响,这可能有助于提高竞争水平。例如,通过比较使用饲喂料槽与电子饲喂器(electronic self-feeder,ESF)的母猪时发现,饲喂纤维日粮的母猪采食率较低,而料槽饲喂的母猪由于竞争被挤走的频率更高,另外,在 ESF 组非进食性活动更多(Courboulay 等,2001)。

日粮纤维对攻击性行为的影响也可能取决于纤维来源。事实上,Danielsen 和 Vestergaard(2001)报道,在含有甜菜渣(50%)或混合纤维来源(20%燕麦,15%麦麸和15%干草粉)的不同能量日粮的母猪中,而仅有采食添加甜菜渣日粮的母猪减少了争斗的频率。

总之,纤维日粮可能潜在地减少与饲料竞争有关的侵略行为,但是这种影响取决于纤维来源和饲喂条件,并且如果不提供觅食垫料,纤维日粮就可能更重要。

5.5 日粮纤维对生产性能的影响

关于日粮纤维对妊娠和哺乳期间的生产表现的影响有许多研究但通常是不确定的,因为不同研究人员所使用的纤维来源的巨大差异、添加比率、饲养条件(例如,单独围栏或群体猪舍,常规或富集环境,饲喂策略)或母猪的个体特征(例如,胎次、体重、品系),如表 5.3 所示。

当能量供应相似时妊娠期间的体重或背膘厚度增加似乎不受日粮补充纤维的影响,无论纤维来源或母猪的胎次(Cerneau 等,1997;Farmer 等,1996;Guillemet 等,2007b;Paboeuf 等,2000)如何。

然而,应该注意的是,在几项比较标准日粮与纤维日粮热能的研究中[基于估计的消化能(digestible energy, DE)、代谢能(metabolizable energy, ME)或净能(NE)],饲喂高纤维日粮的母猪体重较高或较低或背膘厚度增加,这部分是由于过度或低估了饲料中能量及其 NE 供应(Guillemet 等,2006,2007b;Matte 等,1994;Mroz 等,1986;Noblet 等,2003,Van der Peet-Schwering 等,2003b)。这些对母猪不同身体条件的影响反过来也会影响其后续哺乳期间的表现。此外,饲喂纤维性日粮的妊娠母猪体重增加的变化某种程度上可以解释为:胃肠道的重量较大或与日粮纤维的消化作用的肠道充实有关(Danielsen and Vestergaard,2001)。这种假设通过在分娩后对精料的快速消化适应来加强,由于消化器官体积较小和食物的充实,活体重减少 5~6 kg(Van der Peet-Schwering 等,2003b;Vestergaard 和 Danielsen,1998)。

许多研究报道了在妊娠期间提供纤维饲料对哺乳期的采食量(通常为浓缩日粮)的有益作用,在哺乳期间(Courboulay and Gaudré,2002;Farmer 等,1996;Matte 等,1994;Quesnel 等,2009;Van der Peet-Schwering 等,2003b;表 5.3),特别是在哺乳期第一周同样具有有益作用(Farmer 等,1996;Guillemet 等,2006)。这种效应可以通过由妊娠期间提供的纤维日粮引起的消化道的长度和容量增加来解释,从而促进采食,特别是在哺乳期开始时。然而,随着哺乳期的进展由于消化道的适应性过程,这种作用往往会易于消失(Farmer 等,1996)。一些作者提出了另外的解释,并将高纤维妊娠日粮对哺乳期采食量的有益影响归因于:在妊娠时纤维饲料饲喂的母猪在分娩后母猪的背膘厚度较少,身体状况较差,进而引起哺乳期母猪采食量增加(Matte 等,1994;Danielsen 和 Vestergaard,2001)。然而,妊娠期纤维性饲料对母猪背膘厚的影响与哺乳期间的饲料摄入量之间似乎并不存在关系。

妊娠期饲料的膨胀度越高,哺乳期摄入量增加似乎更大。这主要在基于甜菜渣而不是基于混合纤维来源的饲料中观察到这种效果(麦麸、燕麦和干草粉;

5. 妊娠期高纤维饲喂

表 5.3 在不同研究中，日粮纤维对母猪生殖性能和窝生长的影响[1]

妊娠纤维源[2]	粗纤维占干物质比例(%)	研究的胎次[3]	圈舍[4]	妊娠期间应用的时期[5]	妊娠期间的餐数	哺乳期的采食	产活仔	窝产仔数 断奶数	窝 出生体重 仔猪	断奶体重 窝 仔猪	参考文献
燕麦壳	4.0~6.2~14.9	P2	I	整个妊娠期	1	—	NS	—	↑	NS —	Mroz 等,1986
燕麦壳	15	P1	I	整个妊娠期	2	↑	NS	—	NS	NS NS	Farmer 等,1996
小麦麸皮/玉米缠糖	10.07	P1-P2	I	妊娠第1天直到产后第2天	2	NS	NS	NS	NS (P1); ↑ (P2)	—	NS (P1); ↑ (P2)	Matte 等,1994
燕麦壳/燕麦	20.41	P1-P2	I	妊娠第1天直到产后第2天	2	↑	NS	NS	NS (P1); ↑ (P2)	—	NS (P1); NS (P2)	Matte 等,1994
小麦秸秆	7	P1-Pn	G	妊娠第2~109天	1	NS	NS	NS	—	NS	NS NS	Renteria-Flores 等,2008
大豆壳	12	P1-Pn	G	妊娠第2~109天	1	↑	NS	NS	—	NS	NS NS	Renteria-Flores 等,2008
大豆壳	14.8	多胎; 1-6	I	断奶后第1天到妊娠109天	1 or 2	NS	↓	NS	—	NS	NS NS	Holt 等,2006
大豆壳	19.3% NDF as-fed basis	小母猪; 多胎	I	妊娠第1天到分娩第4天	—	NS	NS	NS	—	NS	NS —	Darroch 等,2008
大豆壳	7.9	P1	I	第91~105天	1	NS	NS	NS	—	NS	NS NS	Loisel 等,2013
小麦麸皮	5	P1-P2	I	第1~90天	2	NS	↓ P1	N	↓ (P1); NS (P2)	↑ P1; NS P2	↓ (P1); NS (P2)	— NS	Che 等,2011
小麦秸秆	8.26	3胎; P0-P4	I	整个妊娠期	1	↑ P1,P2, P3,overall	↑ overall	↑ P2, overall	↑ overall	—	↑ P2, overall	—	Veum 等,2009
甜菜渣	600 g/kg	P1 P2	G	妊娠28天至产前1周	自由采食限饲	NS (P1) (P2)	NS	NS	NS	NS	NS	NS	Whittaker 等,2000
甜菜渣	12.4	P1	G	25天至分娩	1	↑	NS	NS	NS	NS	NS	NS	Guillemet 等,2007b
甜菜渣	12.4	P1	G	25天至分娩	1	NS	NS	NS	NS	NS	NS	↑ 0	Quesnel 等,2009
甜菜渣	300 g/kg	3胎; 多胎	25天至分娩	1天至分娩	1个电子饲喂站；2栏位	↑ P1,P2, P3	↓ P1,P2, NS	↓ P1,P2	NS	↓ P1 P2	—	NS	Van der Peet-Schwering 等,2003b

续表 5.3

妊娠粗纤维资源[a]	粗纤维占干物质比例(%)	研究的胎次[3]	圈舍[4]	妊娠期间应用的时期[5]	妊娠期间的餐数[6]	哺乳期的采食	窝产仔数 产活仔	窝产仔数 断奶数	出生体重 窝	出生体重 仔猪	断奶体重 窝	断奶体重 仔猪	参考文献
甜菜渣	15.9	3胎；多胎	G	1天至分娩	1个电子饲喂站；2栏位	↑	NS	—	NS	NS	—	NS	Van der Peet-Schwering 等，2004
甜菜渣	12.7	P1-5	I	0~112天	2	↑	NS	NS	—	↓	—	NS	Danielsen 和 Vestergaard，2001
混合 -	13.5	P1-5	I	0~112天	2	NS	NS	NS	—	—	—	NS	Danielsen 和 Vestergaard，2001
甜菜渣	13.2	3胎后备母猪多胎	I	0~112天	2	↑ P1 P2 overall	NS	NS	—	↓ P1 P2 P3,overall	NS	NS	Vestergaard 和 Danielsen，1998
混合 2	14.1	3胎后备母猪多胎	I	0~112天	2	NS	NS	NS	—	—	—	NS	Vestergaard 和 Danielsen，1998
混合 3	13.2	多胎	I/G	断奶至分娩	2个栏位；1个饲喂站	↑	NS	NS	NS	↓	↓	↓	Courboulay 和 Gaudré，2002

[1] 日粮纤维饮食与标准饮食相比的影响：↑ = 上升；↓ = 下降；NS = 无影响。
[2] 混合 1：干草粕＋小麦麸皮＋燕麦壳；混合 2：绿草粕＋小麦麸皮＋燕麦壳；混合 3：甜菜渣＋麸皮＋葵花粕。
[3] 圈舍设计：I 单体栏位，G 群养圈。

Danielsen and Vestergaard，2001）。据报道，与添加麦麸的饲料相比，添加燕麦的饲料其采食量增加更多，但是添加燕麦的饲料其粗纤维也高得多（20.41 对 10.07 mg/kg 干物质分别对应燕麦和麦麸饲料），说明这种效应与纤维水平相关（Matte 等，1994）。

只有很少的研究是针对妊娠期纤维性饲料对分娩过程的影响，这些研究在日粮纤维的特性、添加比率以及遗传品种方面结果有所不同。一些研究表明，日粮添加纤维素可以使母猪分娩时间减少，例如，在妊娠晚期饲喂纤维性日粮的母猪，分娩时间下降了 9%～29%（Bilkei，1990；Kurcman-Przedpelska，1989；Morgenthum and Bolduan，1988），而其他人则（Guillemet 等，2007b；Loisel 等，2013）报道对妊娠或分娩的持续时间没有显著影响。一项研究报告说，在妊娠晚期饲喂高纤维日粮的母猪，第一头和第三头仔猪的出生间隔较短（Loisel 等，2013）。同时也发现了纤维性妊娠饲料对母猪围产期行为的不同影响。Farmer 等（1995）研究表明，在两个连续繁殖周期中，饲喂纤维日粮的母猪在围产期会花更多的时间躺在一边而很少时间站立，而其他人则发现其对母性行为没有影响（Guillemet 等，2007b）。

大多数研究报告说，无论纤维来源、胎次或繁殖周期如何，日粮纤维对断奶至发情间隔（weaning-to-oestrus interval，WEI）和受胎率无影响（Che 等，2011；Darroch 等，2008；Guillemet 等，2007b；Matte 等，1994；Paboeuf 等，2000，Van der Peet-Schwering 等，2003b；Veum 等，2009；Whittaker 等，2000），除了一项研究（Farmer 等，1996）报道，在妊娠期间饲喂高纤维日粮的母猪中，WEI 低于 10 天的比例趋于升高。

许多研究表明，无论胎次或繁殖周期如何，妊娠期间日粮纤维对母猪繁殖表现没有重大的积极或消极影响（Meunier-Salaün 等，2001；Philippe 等，2008；表 5.3）。

当母猪被饲喂多于一个繁殖周期的纤维饲料时，对窝产仔数具有潜在的有益影响，但是在第二周期后则效果较小（Che 等，2011；Reese 等，2008；Van der Peet-Schwering 等，2003b）。在妊娠期间，每天提供 350～450 g NDF 的饲料（Reese，1997），或者添加燕麦壳（40% NDF；Mroz 等，1986），玉米蛋白粉饲料（34% NDF，Honeyman 和 Zimmerman，1990），甜菜渣（47% NSP；Van der Peet-Schwering 等，2003b）或小麦秸秆（Veum 等，2009）为基础的饲料，可以使窝仔数量增加。与之相反的是，大豆壳日粮对窝仔数具有负面影响（Holt 等，2006），其他报告显示没有影响（表 5.3）。Mroz 等（1986）提出，当饲料中添加日粮纤维时，可能由于纤维素会减少大肠吸收毒素的作用而提高胚胎的着床率和存活率（Bergner，1981）。最近，Renteria-Flores 等（2008）报告说，饲喂富含可溶性纤维（燕麦麸）、不溶性纤维（麦秆）或两者（大豆皮）的饲料的母猪具有较低妊娠子宫和胎盘重量，且饲喂后两种来源纤维时活胚数量较少。然而，应该指出的是，在同一篇论文的第二个试验中,对产(活)猪数没有影响,而其他人也报道了早期妊娠期间高纤维日粮对胚胎存活的影响(Athorn 等,2013)。

在妊娠期间供应日粮纤维对哺乳仔猪的出生体重和生长速度在不同研究之间差异极大,并且仍然是不确定的。然而,一些研究表明纤维饲料的饲喂减少了仔猪出生体重(Courboulay 和 Gaudré,2002;Van der Peet-Schwering 等,2003b),特别是如甜菜渣来源的纤维(Danielsent Vestergaard,2001;表 5.3)。Matte 等(1994)研究发现,母猪胎次与所用纤维来源之间的相互作用,仔猪的出生体重在饲喂添加燕麦壳的日粮的第一胎母猪中较低,而在以玉米为饲料的第二胎母猪中则较高。其他因素,如饲喂水平的变化和母猪的 NE 摄入量的不准确测量,特别是在胎儿生长可能受影响的妊娠最后阶段,都是可能造成研究结果之间的差异的原因。最近的一项研究(Loisel 等,2013)表明,在妊娠晚期饲喂高纤维日粮会影响初乳组成(但不是产量),而低出生体重仔猪的初乳消耗增加了 76%(<900 g)。这些影响可能是由于在围产期间改变行为(Farmer 等,1995),或与小猪活力的变化有关。Che 等(2011)还报道说,提供高纤维妊娠期饲料导致新生仔猪的肝脏、心脏和肾脏的重量相对较高,并且一窝之间的出生体重均匀度提高。

日粮纤维对断奶仔猪数量和体重的影响是不确定的(Meunier-Salaün 等,2001;Philippe 等,2008)。Loisel 等(2013)发现断奶前死亡率降低,Veum 等(2009)报告说,在三个生产周期内断奶的仔猪数量较多。Guillemet 等(2007b)和 Quesnel 等(2008)报道了对断奶重量的积极影响,一些研究表明,断奶前的窝生长率将增加 13% 和 20%(Evert,1991;Matte 等,1994)。与此形成对比的是,Che 等(2011)发现,在第一胎时饲喂高纤维日粮的母猪在第 22 天发现对仔猪体重有负面影响。

总之,日粮纤维对母猪繁殖性能的影响是可变的,这取决于所使用的日粮纤维及其添加比率,也取决于母猪的生殖周期数。对母猪哺乳期摄食量存在正面影响,但对仔猪性能的影响尚不明确。

5.6 结论

本文强调了日粮纤维对限制饲喂的妊娠母猪的行为和福利的有益影响。纤维性饲料通常会减少刻板行为的发生,减少在采食时的不安和侵略行为。这很可能是因为日粮纤维在行为性(长时间进食)和生理性的影响作用,这些纤维能够促进满足感和饱腹感。此外,在大多数研究中,在妊娠期间提供的日粮纤维增加了母猪在哺乳期的采食量,这可能是由于其对胃肠道的大小和容积的影响。关于纤维对繁殖性能的影响的研究相当少,并且显示了可变的结果,部分原因可能与妊娠期间 NE 供应的测量不准确有关。然而,日粮纤维的影响程度取决于与纤维本身(添加比率、纤维来源、物理化学性质)，环境条件(饲喂和圈舍系统)以及个体因素(胎次)有关的许多因素。在大多数关于日粮纤维对生产性能影响的研究中，在一个生殖周期母猪被单独饲养，表明对最近的欧洲动物福利监管政策和通过多次生殖连续循环是有道理的。日粮纤维来源对行为的影响不确定，尽管和物理化学特性密切，但仍有待研究。最近的研究揭示了发酵和膨胀度促进饱腹感的生理机制，以及纤维采食对采食量和繁殖性能的长期影响。然而，仍然需要进一步的研究来充分认识具有不同物理化学性质的纤维之饱腹效应。

日粮纤维的使用也有一些实际的缺点，可能限制其使用。例如，高纤维饲料可能对环境产生影响，这是由于生产的粪肥数量增加(Philippe 等，2008)。这也说明纤维饲料的最佳配方的重要性，不仅要确保改善母猪福利，同时满足环境和经济要求。

致谢

作者们想感谢 M. Legall(Provimi France)，感谢她的批判性阅读，她的贡献，尤其是对表 5.1 的贡献，以及她对提高书稿质量的总体评价。

参考文献

Athorn, R.Z., Stott, P., Bouwman, E.G., Chen, T.Y., Kennaway, D.J. and Langendijk, P., 2013. Effect of feeding level on luteal function and progesterone concentration in the vena cava during early pregnancy in gilts. Reproduction, Fertility and Development 25: 531-538.

Bach Knudsen, K.E., 2001. The nutritional significance of 'dietary fiber' analysis. Animal Feed Science and Technology 90: 3-20.

Beattie, V.E. and O'Connell, N., 2002. Relationships between rooting behaviour and foraging in growing pigs. Animal Welfare 11: 295-303.

Benelam, B., 2009. Satiation, satiety and their effects on eating behavior. Nutrition Bulletin 34: 126-173.

Bergeron, R., Bolduc, J., Ramonet, Y., Meunier-Salaün, M.C. and Robert, S., 2000. Feeding motivation and stereotypies in pregnant sows fed increasing levels of fibre and/or food. Applied Animal Behaviour Science 70: 27-40.

Bergeron, R., Meunier-Salaün, M.C. and Robert, S., 2002. Effects of food texture on meal duration and behaviour of sows fed high-fibre or concentrate diets. Canadian Journal of Animal Science 82: 587-589.

Bergner, H.,1981. Chemically treated straw meals as a new source of fibre in the nutrition of pigs. Pigs News Information 2: 135-140.

Berridge, V., 2009. Dietary fibre: an evolving definition. Nutrition Bulletin 34: 122-125.

Bilkei, G.P., 1990. The effect of increased fiber content fed on the previous week on the parturition of sows. Magyar Allatorvosok Lapja 45: 597-601.

Blackshaw, J.K. and McVeigh, J.F., 1985. Stereotype behaviour in sows and gilts housed in stalls, tethers, and groups. In: Fox, M.W. and Mickley, L.D. (eds.) Advances in animal welfare science 1984. Martinus Nijhoff Publishers, Boston, MA, USA, pp 163-174.

Blackwood, A.D., Salter, J., Dettmar, P.M. and Chaplin, M.F., 2000. Dietary fiber, physicochemical properties and their relationship to health. The journal of the Royal Society for the Promotion of Health Royal Society of Health 120: 242-247.

Blundell, J.E. and Burley, V.J., 1987. Satiation, satiety and the action of fibre on food intake. International Journal of Obesity 11: 9-25.

Brouns, F. and Edwards, S.A., 1994. Social Rank and feeding behaviour of group-housed sows fed competitively or *ad libitum*? Applied Animal Behaviour Science 39: 225-235.

Brouns, F., Edwards, S.A. and English, P.R., 1995. Influence of fibrous feed ingredients on voluntary feed intake. Animal Feed Science and Technology 54: 301-313.

Brouns, F., Edwards, S.A. and English, P.R., 1997. The effect of dietary inclusion of sugar beet pulp on the feeding behaviour of dry sows. Animal Science 65: 129-133.

Brownlee, I.A., 2011. The physiological roles of dietary fibre. Food Hydrocolloids 25: 238-250.

Burton-Freeman, B, 2000. Dietary fiber and energy regulation. Journal of Nutrition 130: 272S-275S.

Campos, P.H., Silva, B.A., Donzele, J.L., Olieveira, R.F. and Knol, E.F., 2012. Effects of sow nutrition during gestation on within-litter birth weight variation: a review. Animal 6: 797-806.

Cerneau, P., Meunier-Salaün, M.C., Lauden, P. and Godfrin, K., 1997. Incidence du mode de logement et du mode d'alimentation sur le comportement de truies gestantes et leurs performances de reproduction. Journée Recherche Porcine en France 29: 175-182.

Champ, M., Langkilde, A.M., Brouns, F., Kettlitz, B. and Le Bail-Collet, Y., 2003. Advances in dietary fibrecharacterisation.2 Consumption, chemistry, physiology and measurements of resistant starch; implication for health and food labelling. Nutrition Research Review 16: 143-161.

Che, L., Feng, D., Wu, D., Fang, Z., Lin, Y. and Yan, T., 2011. Effect of dietary fibre on reproductive performance of sows during the first two parities. Reproductive Domestic Animal 46: 1061-1066.

Courboulay, V., Aude, D. and Meunier-Salaün, M.C., 2001. La distribution d'aliment riche en fibres affecte l'activité alimentaire des truies gestantes logées en groupe. Journée Recherche Porcine en France 33: 307-312.

Courboulay, V. and Gaudré, D., 2002. Faut-il distribuer des aliments riches en fibres aux truies en groupe. Journée de la Recherche Porcine en France 34: 225-232.

D'Eath, R.B., Tolkamp, B.J., Kyriasakis, I. and Lawrence A.B., 2009. Freedom fromhunger and preventing obesity: the animal welfare implication of reducing food quantity or quality. Animal Behaviour 77: 275-288.

Danielsen, V. and Vestergaard, E.M., 2001. Dietary fibre for pregnant sows: effect on performance and behaviour. Animal Feed Science and Technology 90: 71-80.

Dantzer, R., 1986, Behavioural, physiological and functional aspects of stereotyped behavior: a review and re-interpretation. Journal Animal Science 62: 1776-1786.

Darroch, C.S, Dove, C.R., Maxwell, C.V., Johnson, Z.B., and Southern, L.L., 2008. A regional evaluation of the effect of fiber type in gestation diets on sow reproductive performance. Journal of Animal Science 86: 1573-1578.

De Leeuw, J.A., Bolhuis, J.E., Bosch, G. and Gerrits, W.J.J., 2008. Effects of dietary fibre on behaviour and satiety in pigs. The Proceedings of the Nutrition Society 67(4): 334-342.

De Leeuw, J.A., Jongbloed, A.W. and Verstegen, M.W.A., 2004. Dietary fiber stabilizes blood glucose and insulin levels and reduces physical activity in sows (Sus scrofa). Journal of Nutrition 134: 1481-1486.

Douglas, M.W., Cunnick, J.E., Pekas, J.C., Zimmerman, D.R. and Von Borell, E.H., 1998. Impact of feeding regimen on behavioral and physiological indicators for feeding motivation and satiety, immune function, and performance of gestating sows. Journal of Animal Science 76: 2589-2595.

Dourmad, J.Y., Etienne, M. and Noblet, J., 1996. Reconstitution of body reserves in multiparous sows during pregnancy: effect of energy intake during pregnancy and mobilisation during the previous lactation. Journal of Animal Science 74: 2211-2219.

Dourmad, J.Y., Etienne, M., Prunier, A. and Noblet, J., 1994. The effect of energy and protein intake of sows on their longevity: a review. Livestock Production Science 40: 87-97.

Eurobarometer, 2005. Attitudes of consumers towards the welfare of farmed animals. Special Eurobarometer 229 / Wave 63.2 – TNS Opinion and social, 138 pp.

European Commission (EC), 1991. Council Directive 91/630/EEC of 19 November 1991 laying down minimum standards for the protection of pigs. Official Journal of the European Union L340: 33-38.

European Commission (EC), 2001. Council Directive 2001/88/EC of 23 October 2001 amending Directive 91/630/EEC laying down minimum standards for the protection of pigs. Official Journal of the European Union L316: 1-4.

Everts, H., 1991. The effect of feeding different sources of crude fiber during pregnancy on the reproductive performance of sows. Animal Production 52: 175-184.

Farmer, C., Meunier-Salaün, M, C., Bergeron, R. and Robert, S., 2002. Hormonal responses of pregnant gilts fed a high-fiber or a concentrate diet once or twice daily. Canadian Journal of Animal Science 82: 159-164.

Farmer, C., Robert, S. and Matte, J.J., 1996. Lactation performance of sows fed a bulky diet during gestation and receiving growth hormone-releasing factor during lactation. Journal of Animal Science 74: 1298-1306.

Forbes, J.M., 1998. Dietary awareness. Applied Animal Science 57: 287-297.

Guerin, S., Ramonet, Y., Le Cloarec, J., Meunier-Salaün, M.C., Bourguet, P. and Malbert, C.H., 2001. Changes in intragastric meal distribution are better predictors of gastric emptying rate in conscious pigs than are meal viscosity or dietary fibre concentration. British Journal of Nutrition 85: 343-350.

Guillemet, R., Comyn, S., Dourmad, J.Y. and Meunier-Salaün, M.C., 2007a. Gestating sows prefer concentrate diets to high fibre diet in two choice tests. Applied Animal Behaviour Science 108: 251-262.

Guillemet, R., Dourmad, J.Y. and Meunier-Salaün, M.C., 2006. Feeding behaviour in primiparous lactating sows: impact of a high fiber diet during pregnancy. Journal of Animal Science 84: 2474-2481.

Guillemet, R., Hamard, A., Quesnel, H., Père, M.C., Etienne, M., Dourmad, J.Y. and Meunier-Salaün, M.C., 2007b. Dietary fibre for gestating sows: effects on parturition progress, behaviour, litter and sow performance. Animal 1: 872-880.

Haenen, D., Souza da Silva, C., Zhang, J., Koopmans, S.J., Bosch, G., Vervoort, J., Gerrits, W.J.J., Kemp, B., Smidt, H., Müller, M. and Hooiveld, G.J., 2013b. Resistant starch induces catabolic but suppresses immune and cell division pathways and changes the microbiome in the proximal colon of male pigs. Journal of Nutrition 143: 1889-1898.

Haenen, D., Zhang, J., Souza da Silva, C., Bosch, G., Van der Meer, I.M., Van Arkel, J., Van den Borne, J.J., Pérez Gutiérrez, O., Smidt, H., Kemp, B., Müller, M. and Hooiveld, G.J., 2013a. A diet high in resistant starch modulates microbiota composition, SCFA concentrations, and gene expression in pig intestine. Journal of Nutrition 143: 274-283.

Hipsley, E.H., 1953. Dietary Fibre and pregnancy toxaemia. British Medical Journal 2: 420-422.

Holt, J.P; Johnston, L.J., Baidoo, S.K. and Shurson, G.C., 2006. Effects of high-fiber diet and frequent feeding on behavior, reproductive performance, and nutrient digestibility in gestating sows. Journal of Animal Science 84: 946-955.

Honeyman, M.S. and Zimmerman, D.R., 1990. Long-term effects of corn glute, feed on the reproductive performance and weight of gestating sows. Journal of Animal Science 68: 1329-1336.

Jensen, M.B., Pedersen, L.J., Theil, P.K., Yde, C.C. and Bach Knudsen, K.E., 2012. Feeding motivation and plasma metabolites in pregnant sows fed diets rich in dietary fiber either once or twice daily. Journal of Animal Science 90: 1910-1919.

Jorgensen, H., Zhao, X.Q. and Eggum, B.O., 1996. The influence of dietary fibre and environmental temperature on the development of the gastrointestinal tract, digestibility, degree of fermentation in the hindgut and energy metabolism in pigs. British Journal of Nutrition 75: 365-378.

Kurcman-Przedpelska, B., 1989. Duration of pregnancy and farrowing of sows fed on concentrate mixtures of different fibre content. Acta Academic Agricukltural Technlogy 32: 65-74.

Lawrence, A. and Rushen, J., 1993. Stereotypic animal behaviour: fundamentals and applications to welfare. CAB International, Wallington, UK, 210 pp.

Lawrence, A.B., Appleby, M.C. and Macleod, H.A., 1988. Measuring hunger in the pig using operant conditioning: the effect of food restriction. Animal Production 47: 131-137.

Lawrence, A.B. and Illius, A.W., 1989. Methodology for measuring hunger and food needs using operant conditioning in the pig. Applied Animal Behaviour Science 24: 273-285.

Le Gall, M., Montagne, L., Jaguelin-Peyraud, Y., Pasquier, A. and Gaudré, D., 2011. Prédiction de la teneur en fibres totales et insolubles de matières premières courantes dans l'alimentation du porc à partir de leur composition chimique. Journée de la Recherche Porcine 43: 117-123.

Le Gall, M., Montagne, L., Meunier-Salaün, M.C. and Noblet, J., 2009. Valeurs nutritives des fibres, conséquences sur la santé des porcelets et le bien-être de la truie. INRA Productions Animales 22: 17-24.

Le Huërou-Luron, I. and Ferret-Bernard, S., 2015. Development of gut and gut-associated lymphoid tissues in piglets. Role of maternal environment. Chapter 15. In: Farmer, C. (ed.) The gestating and lactating sow. Wageningen Academic Publishers, Wageningen, the Netherlands, pp. 335-355.

Loisel, F., Farmer, C., Ramaekers, P. and Quesnel, H., 2013. Effects of high fiber intake during late pregnancy on sow physiology, colostrum production, and piglet performance. Journal Animal Science 91: 5269-5279.

Martin, J.E. and Edwards, S.A., 1994. Feeding behaviour in outdoor sows: the effect of diet quantity and type. Applied Animal Behaviour Science 41: 63-74.

Mason, G.J., 1991. Stereotypies: a critical review. Animal Behaviour 41: 1015-1037.

Matte, J.J., Robert, S., Girard, C., Farmer, C. and Martineau, G.P., 1994. Effects of bulky diets based on wheat bran or oats hulls on reproductive performance of sows during their first 2 parities. Journal of Animal Science 72: 1754-1760.

Meunier-Salaün, M.C., Edwards, S.A. and Robert, S., 2001. Effect of dietary fibre on the behaviour and health of the restricted fed sow. Animal Feed Science and Technology 90: 53-69.

Morgenthum, R. and Bolduan, G., 1988. Effects of live weight and dietary factors of sows on length of parturition. Monatshefte Fur Veterinarmedizin 43: 194-196.

Mroz, Z., Partridge, I.G., Mitchell, G. and Keal, H.D., 1986. The effect of oat hulls, added to the basal ration for pregnant sows, on reproductive-performance, apparent digestibility, rate of passage and plasma parameters. Journal of the Science of Food and Agriculture 37: 239-247.

Noblet, J., Bontems, V. and Tran, G., 2003. Estimation de la valeur énergétique des aliments pour le porc. INRA Productions Animales 16: 197-210.

Noblet, J. and Le Goff, G., 2001. Effect of dietary fibre on the energy value of feeds for pigs. Animal Feed Science and Technology 90: 35-52.

Noblet, J. and Shi, X.S., 1993. Comparative digestibility of energy and nutrients I growing pigs fed *ad libitum* and adult sows fed at maintenance. Livestock Production Science 34: 137-152.

Paboeuf, F., Ramonet, Y., Corlouër, A., Dourmad, J.Y., Cariolet, R. and Meunier-Salaün, M.C., 2000. Impact de l'incorporation de fibres dans un régime de gestation sur les performances zootechniques et le comportement des truies. Journée de la Recherche Porcine en France 32: 105-113.

Philippe, F.X., Remience, V., Dourmad, J.Y., Cabaraux, J.F., Vandenheede, M. and Nicks, B., 2008. Les fibres dans l'alimentation des truies gestantes: effets sur la nutrition, le comportement, les performances et les rejets dans l'environnement. INRA Productions Animales 21: 277-290.

Phillips, G.O. and Cui, S.W., 2011. An introduction: evolution and finalisation of the regulatory definition of dietary fibre. Food Hydrocolloids, 25: 139-143.

Quesnel, H., Meunier-Salaün, M.C., Hamard, A., Guillemet, R., Etienne, M., Farmer, C., Dourmad, J.C. and M.C. Père, 2009. Dietary fiber for pregnant sows: influence on sow physiology and performance during lactation. Journal of Animal Science 87: 532-543.

Ramonet, Y., Bolduc, J., Bergeron, R., Robert, S. and Meunier-Salaün, M.C., 2000b. Feeding motivation in pregnant sows: effects of fibrous diet in an operant conditioning procedure. Applied Animal Behaviour Science 66: 21-29.

Ramonet, Y., Meunier-Salaün, M.C. and Dourmad, J.Y, 1999. High-fiber diets in pregnant sows: digestive utilization and effects on the behavior of the animals. Journal of Animal Science 77: 591-599.

Ramonet, Y., Robert, S., Aumaître, A., Dourmad, J.Y. and Meunier-Salaün, M.C., 2000a. Influence of the nature of dietary fiber on digestive utilization, some metabolite and hormone profiles and behaviour of pregnant sows. Animal Science 70: 275-286.

Reese, D., Prosch, A., Travnicek, D.A. and Eskridge, K.M., 2008. Dietary fiber in sow gestation diets – an update review. Nebraska Swine Reports, University of Nebraska – Lincoln, Lincoln, NE, USA, pp. 14-18.

Reese, D.E., 1997. Dietary fiber in sow gestation diets: a review. Nebraska Swine Reports, University of Nebraska – Lincoln, Lincoln, NE, USA, 4 pp.

Renteria-Flores, J.A., Johnston, L.J., Shurson, G.C., Moser, R.L. and Webel, S.K., 2008. Effect of soluble and insoluble dietary fiber on embryo survival and sow performance. Journal of Animal Science 86: 2576-2584.

Rijnen, M., Verstegen, M.W.A., Heetkamp, M.J.W. and Schrama, J.W., 2003. Effect of dietary fermentable carbohydrates on behavior and heat production in group-housed sows. Journal of Animal Science 81: 182-190.

Robert, S., Bergeron, R., Farmer, C. and Meunier-Salaün, M.C., 2002. Does the number of daily meals affect feeding motivation and behaviour of gilts fed high-fibre diets? Applied Animal Behaviour Science 76: 105-117.

Robert, S., Matte, J.J., Farmer, C., Girard, C.L. and Martineau, G.P., 1993. High-fibre diets for sows: effects on stereotypies and adjunctive drinking. Applied Animal Behaviour Science 37: 297-309.

Robert, S., Rushen, J. and Farmer, C., 1997. Both energy content and bulk of food affect stereotypic behaviour, heart rate and feeding motivation of female pigs. Applied Animal Behaviour Science 54: 161-171.

Rushen, J., 1985. Stereotypies, aggression and the feeding schedules of tethered sows. Applied Animal Behaviour Science 14: 137-147.

Salovaara, H., Gates, F. and Tenkanen, M., 2007. Dietary fibre components and functions. Wageningen Academic Publishers, Wageningen, the Netherlands, 345 pp.

Sappok, M.A., 2012. Fermentation capacity of hindgut micobiota in pigs related to dietary fibre. Ph.D. thesis, Wageningen University, Wageningen, the Netherlands.

Sauvant, D., Perez J.M. and Tran, G., 2004. Tables de composition et de valeur nutritive des matières premières destinées aux animaux d'élevage porcs, volailles, bovins, ovins, caprins, lapins, chevaux, poissons, 2nd ed. INRA Editions, Versailles, France, 304 pp.

Serena, A., Jorgensen, H. and Bach Knudsen, K.E., 2008. Digestion of carbohydrate and utilization of energy in sow fed diets with contrasting levels and physicochemical properties of dietary fiber. Journal of Animal Science 86: 2208-2216.

Serena, A., Jorgensen, H. and Bach Knudsen, K.E., 2009. Absorption of carbohydrate-derived nutrients in sows as influenced by types and contents of dietary fiber. Journal of Animal Science 87: 136-147.

Slavin, J. and Green, H., 2007. Dietary fibre and satiety. Nutrition Bulletin 32: 32-42.

Sola-Oriol, D., Roura, E. and Torrallardona, D., 2011. Feed preference in pigs: effect of selected protein, fat and fiber sources at different inclusion rates. Journal of Animal Science 89: 3219-3227.

Souza Da Silva, C., Bolhuis, J.E., Gerrits, W.J.J., Kemp, B. and Van den Borne, J.G.C., 2013. Effects of dietary fibres with different fermentation characteristics on feeding motivation in adult females pigs. Physiology and Behaviour 110-111: 148-157.

Souza Da Silva, C., Van den Borne, J.G.C., Gerrits, W.J.J., Kemp, B. and Bolhuis, J.E., 2012. Effects of dietary fibres with different physicochemical properties on feeding motivation in adult females pigs. Physiology and Behaviour 107: 218-230.

Terlouw, E.M.C., Lawrence, A.B. and Illius, A.W., 1991. Influences of feeding level and physical restriction on development of stereotypies in sows. Animal Behaviour 42: 981-991.

Van de Putten, G. and Van de Burgwal, J.A., 1990. Vulva biting in group-housed sows: a preliminary report. Applied Animal Behaviour Science 26: 181-186.

Van der Peet-Schwering, C.M.C., Kemp, B., Binnendijk, G.P., Den Hartog, L.A., Spoolder, H.A.M. and Verstegen, M.W.A., 2003a. Development of stereotypic behaviour in sows fed a starch diet or a non-starch polysaccharide diet during gestation and lactation over three parities. Applied Animal Behaviour Science 83: 81-97.

Van der Peet-Schwering, C.M.C., Kemp, B., Binnendijk, G.P., Den Hartog, L.A., Spoolder, H.A.M. and Verstegen, M.W.A., 2003b. Performance of sows fed high levels of non starch polysaccharides during gestation and lactation over three parities. Journal of Animal Science 81: 2247-2258.

Van der Peet-Schwering, C.M.C., Kemp, B., Plagge, J.G., Vereijken, P.F.G., Den Hartog, L.A., Spoolder, H.A.M. and Verstegen, M.W.A., 2004. Performance and individual feed intake characteristics of group-housed sows fed a nonstarch polysaccharides diet *ad libitum* during gestation over three parities. Journal of Animal Science 82: 1246-1257.

Van Putten, G. and Van de Burgwal, J.A., 1990. Vulva biting in group-housed sows: preliminary report. Applied Animal Behaviour Science. 26, 181-186.

Van Soest, P.J., Robertson, J.B. and Lewis, B.A., 1991. Methods for dietary fiber, neutral detergent fiber and nonstarch polysaccharides in relation to animal nutrition. Journal of Dairy Science 74: 3583-3597.

Van Soest, P.J., Robertson, J.B. and Lewis, B.A., 1991. Methods for dietary fiber, neutral detergent, and nonstarch polysaccharides in relation to animal nutrition. Journal of Dairy Science 74: 3583-3597.

Vestergaard, E.M. and Danielsen, V., 1998. Dietary fibre for sows: effects of large amounts of soluble and unsoluble fibres in the pregnancy period on the performance of sows during three reproductive cycles. Animal Science 67: 355-362.

Veum, T.L., Crenshaw, J.D., Crenshaw, T.D., Cromwell, G.L., Easter, R.A., Ewan, R.C., Nelssen, J.L., Miller, E.R., Pettigrew, J.E. and Ellersieck, M.R., 2009. The addition of ground wheat straw as a fiber source in the gestation diet of sows and the effect on sow and litter performance for three successive parities. Journal of Animal Science 87: 1003-1012.

Wanders, A.J, Van den Borne, J.J.G.C., Mars, M, Schols, H.A. and Feskens E.J.M., 2011. Effects of dietary fibre on subjective appetite, energy intake and body weight: a systematic review of randomzed controlled trails. Obesity Review 12: 724-739.

Whittaker, X., Edwards, S.A., Spoolder, H.A.M., Corning, S. and Lawrence, A.B., 2000. The performance of group-housed sows offered a high fibre diet *ad libitum*. Animal Science 70: 85-93.

Whittaker, X., Edwards, S.A., Spoolder, H.A.M., Lawrence, A.B. and Corning, S., 1999. Effects of straw bedding and high fibre diets on the behaviour of floor group-housed sows. Applied Animal Science 63: 25-39.

Zijlstra, N., De Wijk, R.A., Mars, M., Stafleu, A. and De Graaf, C., 2009. Effects of bite size and oral processing of a semisolid food on satiation. American Journal Clinical Nutrition 90: 269-275.

6. 母猪氨基酸和能量供给

N. L. Trottier[1*], L. J. Johnston[2] and C. F. M. de Lange[3]

[1] Michigan State University, Department of Animal Science, Monogastric Animal Nutrition, Anthony Hall, East Lansing, MI 48824, USA; trottier@msu.edu

[2] University of Minnesota, West Central Research and Outreach Center, 46352 State Highway 329, Morris, MN 56267, USA

[3] University of Guelph, Animal and Poultry Science, 50 Stone Road East, Guelph, ON, N1G 2W1, Canada

摘要：妊娠和哺乳母猪的营养优化策略取决于几个因素，包括能量和氨基酸需要的有效估算，饲料原料品质和有效营养物质方面的知识，以及饲养方式。这些因素对于确保母猪健康、福利和长期生产力至关重要。决定妊娠能量和氨基酸需要的因素包括身体维持、孕体生长、母体能量和蛋白质储存的变化；而对于泌乳，仔猪窝生长率则是目前认为最重要的决定因素。这些过程的能量需要已经被量化，并且估算了有效代谢能需要。与净能相比，有效代谢能解释了日粮能量来源对能量效率的影响。氨基酸需要现在普遍以标准回肠可消化氨基酸表示，其代表了小肠吸收的氨基酸，并且考虑和计入了基础内源性氨基酸损失。维持氨基酸需要的主要决定因素包括肠基础性内源氨基酸损失和来自皮肤与毛发的氨基酸损失，这两者分别与采食水平和母猪代谢体重大小有关。对于妊娠，由来自于母体、胎体、子宫、胎盘和相关体液以及乳房的五个蛋白质库，一起组成母猪的全身体蛋白库，每个库的独特的氨基酸谱（即氨基酸：赖氨酸）和预估的氨基酸沉积效率，被用于计算妊娠每个氨基酸的总体标准回肠可消化氨基酸需要。对于泌乳，氨基酸需要由两个库（母体和乳）的氨基酸组成，由预估的氨基酸利用效率和它们在泌乳过程中对蛋白质合成的相对贡献来定义。确保妊娠和哺乳母猪获得满足预期生产需要（分别为孕体和仔猪生长）所需的能量和氨基酸，取决于饲料原料品质、妊娠和哺乳饲养管理方式。优化饲养管理方式如改进料槽设计和饲料传送方法等均可提高饲料采食，减少行为规癖（也译作：古板行为、刻板行为、呆板行为）和确保动物福利。

关键词：母猪，能量，氨基酸

6.1 引言

优化妊娠和哺乳母猪的营养对于确保它们的健康、福利和长期生产力至关重要。在本章中,将讨论如何为妊娠和哺乳母猪的能量和氨基酸需要提供最佳的饲养管理方案。本章还介绍了关于母猪日粮能量和氨基酸需要的最新推荐标准,以及 NRC(2012)中对能量和氨基酸需要预测的关键点和概念。

6.2 总体原则

母猪饲养中需要了解的一个重要概念是繁殖周期中各个阶段是相互关联的。任何饲养计划的首要目标都是要满足母猪当时的营养需要。但是,同样重要的是,养殖者必须考虑母猪在未来繁殖周期各阶段的繁殖性能,并相应地规划当时的营养程序。当前的日粮和饲养管理系统必须为未来最佳生产奠定基础。

妊娠饲养管理状况可以影响下一次哺乳性能,以及随后的繁殖性能。妊娠期的采食水平似乎与哺乳期母猪的自由采食量相反。Weldon 等(1994)证明,在妊娠期最后 40 天,母猪自由采食消耗的饲料是母猪营养需要量的两倍。然而,在随后的哺乳期中,这些自由采食母猪的采食量是在之前妊娠期限制饲喂母猪采食量的 50%左右。其他研究者(Dourmad,1991;Sinclair 等,2001)也报道了妊娠期自由采食引起的哺乳期自由采食量下降的类似结果。由于哺乳期采食量和断奶后母猪生产性能之间是密切相关的,那么哺乳期采食量的下降就会导致问题发生。

母猪哺乳期采食量不仅会影响营养平衡,也将影响随后的繁殖性能。一些研究结果(Johnston 等,1989;Koketsu 等,1996a;Yoder 等,2013)清晰地表明,由于泌乳和哺乳期低采食导致的哺乳期体重过度损失,将使断奶后发情间隔延长,并增加母猪乏情的发生率。在哺乳期的任何阶段(早期、中期或晚期),母猪都容易受到这种负营养平衡的影响(Koketsu 等,1996b;Zak 等,1997a)。这些营养缺乏会影响卵母细胞的质量(Yang 等,2000;Zak 等,1997b),胚胎存活(Zak 等,1997b),从而影响下一次妊娠的窝产仔数。

营养对繁殖群中的母猪使用年限具有重要影响。母猪使用年限具有重要的经济和动物福利影响。由于与母猪饲养的成本相关,当后备母猪进入繁殖猪群中并不能立刻产生利润,而直到大约第 3 个胎次才能产生(Stalder 等,2003)。因此,在第 3 胎次之前的部分母猪会因为繁殖性能差或功能丧失而被淘汰,从而减少了母猪使用年限,并对生产者造成经济损失。由于种种原因,营养不良可能会加剧母猪的早期淘汰。营养不良对母猪使用年限的影响可能在母猪生活的早期就已开始。一些研究者认为,母猪在发育过程中营养和生理因素会影响其终身繁殖力和使用年限(Hoge 和 Bates,2011;Johnston 等,2007;Nikkila 等,2013)。这些在本书的第 1 章(Rozeboom,2015)和第 19 章(CaldéronDíaz 等,2015)中有更多介绍。如果让妊娠期母猪和断奶仔数较少的母猪自由采食,从而摄入过多营养时,会导致母猪肥胖和超重。与体况控制较好母猪相比,超重母猪通常会增加分娩问题和运动障碍,从而缩短了使用年限(Dourmad 等,1994)。然而,若由于能量和营养摄入不足而变得过瘦的母猪,也会减少其在繁殖群体中的时间(Anil 等,2006;Knauer 等,2010),并增加淘汰率(Hughes 等,2010)。骨骼适当的发育和维持对于延长母猪的使用年限很重要。营养,特别是矿物质营养,对骨骼、关节软骨和关节的正常发育与维持有重要的影响(Van Riet 等,2013)。生物可利用的 Ca、P、维生素 D 以及其他营养物质的供应不足可能会使母猪罹患蹄病和跛行(Van Riet 等,2013)(另见第 18 章;Friendship 和 O'Sullivan,2015),从而被淘汰(Anil 等,2007)。

通常,营养不当的负面影响似乎对母猪生产性能和使用年限具有累积效应。随着每个连续的繁殖周期,即使之前胎次的轻微营养缺乏,也会逐渐加重,最终导致母猪的功能受损,进而被淘汰。这种现象有助于建立这样的概念:在母猪繁殖周期中的一个阶段的营养状况,都将影响当前和未来的母猪生产力。

6.3 能量供给

6.3.1 能量体系和能量测定方法:消化能、代谢能和净能

基于消化能(DE)和代谢能(ME)的日粮配方体系正逐渐被净能(NE)体系所代替。NE 体系对合成和沉积蛋白质与脂肪的饲料原料提供更接近"真实"能量的估算(Payne 和 Zijlstra,2007)。图 6.1 展示了妊娠母猪乳成分(乳糖、蛋白质和脂肪)中蛋白质、脂肪和胎儿糖原从饲料能量到沉积能量的能量流。DE 是基于全肠道表观总消化和吸收的能量,其来源于蛋白质、碳水化合物和脂肪消化所产生的能量。因此,饲料能量和 DE 之间的差值就是排泄的粪便中未消化的营养物质(即粪能)的能量。蛋白质、碳水化合物和脂肪消化的主要产物,在血液循环中作为吸收营养物质出现的分别为氨基酸、单糖及(挥发性)有机酸和脂肪酸。这些营养物质被用作蛋白质、脂肪和乳糖(哺乳母猪)合成的底物,或者提供能量以支持与身体功能维持或繁殖相关的一系列代谢过程。当氨基酸不用于维持或增殖所需的蛋白质合成时,其将会被脱氨基,所得到的氨基氮主要以尿素的形式排出,因此,在尿中损失的能量主要归因于尿素。代谢能是从 DE 中减去尿能和气体能后剩余的能量。由于猪的气体能损失非常小,所以常从计算中将其忽略。因此,在提供给母猪的饲料原料中,ME 通常占 DE 的 92%~98%(NRC,2012)。例如,脂质(脂肪和油)的 ME 含量是 DE 的 98%(NRC,2012)。而高蛋白原料如豆粕的 ME 含量则为 DE 的 91%(NRC,2012)。

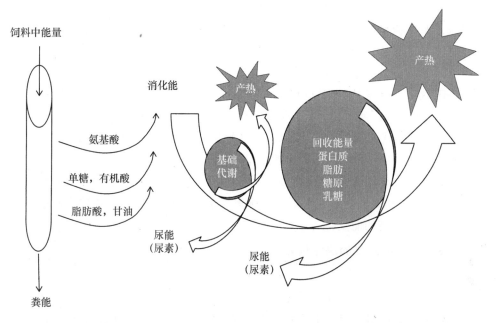

图 6.1 妊娠母猪中从饲料到蛋白质、脂肪和胎儿糖原的能量流,和哺乳母猪中乳成分(乳糖、蛋白质和脂肪)的能量流。消化能等于摄入饲料能量减去粪能。代谢能等于消化能减去尿能。净能等于吸收的能量减去与蛋白质、脂肪、糖原和乳糖沉积相关的能量。

ME 和 NE 之间的差异与饲料消化、吸收和吸收营养物质的代谢产热有关,这些统称为采食或进食热增耗。因此,NE 代表妊娠期和哺乳期中蛋白质、脂肪、糖原和乳糖在增生组织和代谢产物中回收的能量(图 6.1)。饲料原料的 NE 值由于淀粉、纤维、蛋白质和脂肪组成的不同而具有较大的差异。与纤维和蛋白质的消化和吸收相关的热增耗大于淀粉和脂肪的热增耗。因此,含有较高纤维或蛋白质含量的原料具有较低的 NE 含量。例如,玉米(黄色,马齿型)和豆粕(脱壳,溶剂浸提)含有类似水平的 ME,分别为 3395 和 3294 kcal,但是玉米的 NE 明显高于豆粕,两者分别为 2672 和 2087 kcal。ME 体系高估了高纤维和高蛋白质含量的饲料原料"真正有效"的能量浓度。相对于 DE 和 ME 体系,NE 体系对纯化的氨基酸添加剂或高纤维含量的副产品的评价更为准确。在母猪中,高纤维饲料备受重视,因为其在妊娠日粮中应用可增加饲料膨松度,控制食欲(第 5 章;Meunier-Salaün 和 Bolhuis,2015),并减少便秘(第 10 章;Peltoniemi 和 Oliviero)、难产和子宫炎-乳房炎无乳综合征(MMA)的发生率。

当前使用 NE 体系仅考虑了日粮能量来源(即日粮营养)对能量利用效率的影响。但这难以将 NE 体系与估算用于各种身体功能(如维持体蛋白质和体脂增加、泌乳)的能量需要结合起来,而这些能量利用效率确实是不同的,是受使用能量的各身体功能影响的。因此,NRC(2012)引入了有效代谢能(effective ME)的概念,其考虑了日粮能量源对能量效率的影响,并且基于各种身体功能的能量效率,并基于这些来估算有效代谢的能量需要。

在本章的下一节中,将详细讨论妊娠和哺乳期能量需要的决定因素。妊娠母猪的 ME 需要等于维持(即基础代谢)所需的能量,妊娠产物和母体沉积能量,而哺乳母猪的 ME 需要等于维持能量和乳汁中的能量(NRC,2012)。如图 6.1 所示,ME 包括与基础代谢相关的热量,以及与妊娠期间蛋白质、脂肪和糖原沉积相关的热增耗,或是与哺乳期间蛋白质、脂肪和乳糖产生相关的热量。因此,与孕体发育及乳合成相关的热增耗,就成了妊娠与哺乳母猪的 ME 和 NE 需要之间的唯一差异。因此,妊娠和哺乳期的 NE 需要分别相当于在孕体和乳中的基础代谢能量和沉积(或恢复)所需的能量。

6.3.2 能量需要的决定因素

妊娠

妊娠母猪能量需要的主要决定因素是身体功能维持、孕体的生长和母体能量存储(包括子宫和乳腺组织)的变化。在 NRC(2012)中,基于对现有文献的广泛回顾,对这些生理过程的能量需要进行了量化,发现其与胎次、母猪体重、母猪生产力水平(窝产仔数和平均仔猪初生重)和环境状况(有效环境温度、地板类型和母猪运动)相关。NRC(2012)妊娠模型可用于估算个体母猪的能量需要和利用率。根据 NRC(2012),机体将优先满足身体功能维持、孕体的生长和母体体蛋白沉积的能量需要。当能量摄入超过这三个过程的需求后才用于体脂沉积。如果能量摄入不足以支持维持需要、孕体生长和母体体蛋白沉积,那么母体体脂将被动员提供能量。能量分配的优先次序一般情况下是不变的,仅在能量摄入严重受限时,孕体的生长受阻(Dourmad 等,1999)。2 胎母猪的每日有效代谢能摄入量的典型分配方式如图 6.2 所示。

对处于相对无应激压力和适宜温度环境中的母猪,维持能量需要每千克代谢体重($W^{0.75}$)估算为 100 kcal/天有效代谢能。然而,应仔细考虑环境条件对妊娠母猪维持能量需要的影响。由于妊娠母猪的采食量通常受到限制,因此与自由采食的哺乳母猪或生长猪相比,其体热产量相对较低。当处于较低的临界温度时,例如有效环境温度低于临界温度时,需要额外的饲料提供能量来维持体温。当母猪在混凝土地板上单独饲养时,这种情况尤其明显,因为这将导致母猪身体的热量通过高速传导而损失。根据 NRC(2012),单栏饲养和群养母猪的低临界温度分别为 20℃ 和 16℃;对于在稻草上群养的母猪,比低临界温度要低 4℃(即 12℃)。当低于最低临界温度时,每下降 1℃,单栏饲养和群养母猪维持 ME 需要每千克代谢体重分别增加 4.30 和 2.39 kcal。猪站立和行走的能量消耗比其他动物高,反映了猪具有相对较大的躯干长度。因此,提高运动水平将增加母猪的维持能量需要。如 NRC (2012)所述,每天站立或行走超过 4 h 的母猪,站立或行走每增加 1 min,其维持 ME 需要量每千克代谢体重增加 0.0717 kcal/天。

如图 6.2 所示,孕体中的能量沉积随时间呈指数增长。为此,在妊娠晚期可以增加饲料供给量,以避免母猪动员体脂。当每日采食量在整个妊娠期间保持恒定时,母猪在妊娠晚期很可能处于能量负平衡,这可能加重哺乳期间负能量平衡对母猪长期繁殖性能的负面影响。据 NRC(2012)估算,在妊娠 90 天后,母猪通常需要额外补充 400 g/天的玉米豆粕型日粮,以满足支持孕体的生长增加所需的能量需要。这些补充量随着预期的窝产仔数和平均仔猪初生重而变化。如本章前面所述,增加采食量超过孕体生长所需时,将会增加母体脂质沉积,这将使分娩时体况过肥,因此增加分娩问题,减少哺乳期食欲,并减少母猪繁殖力和使用年限。

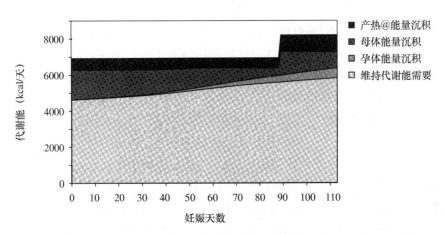

图 6.2 基于代谢能(ME)单位和表 6.2 中生产性能水平的第 2 胎妊娠母猪的摄入能量(E)分配图。

如 Dourmad 等(1999,2008)所述,妊娠期母体体蛋白沉积有两个主要组成部分,即时间依赖性和能量摄入依赖性体蛋白沉积(NRC,2012)。时间依赖性体蛋白沉积发生在妊娠早期,在能量沉积转向胎儿生长之前。能量摄入依赖性体蛋白沉积反映的是随着能量摄入的增加,体蛋白沉积将线性增加。这种能量摄入依赖性体蛋白沉积在 1 胎母猪中最高。事实上,随着母猪成熟,能量摄入依赖性体蛋白沉积和能量摄入之间的斜率下降,并且估算在 4 胎母猪(NRC 2012)中斜率逐渐达到 0。目前已建立了时间依赖性和能量摄入依赖性体蛋白沉积的一般模式,但是实际的沉积率可能在不同基因型母猪之间变化并且受到哺乳期体重减轻的影响。因此,用于估算妊娠母猪营养需要的 NRC(2012)模型允许用户在基于妊娠期间观测到的体重和背膘厚变化,以及在限定的每日能量摄入水平来调整母体体蛋白沉积。

哺乳

泌乳是哺乳母猪能量需要的最大决定因素。由于维持能量需要占总能量需要的比例相对较小，因此准确估算哺乳母猪的维持能量需要，或过多考虑与维持能量需要有关的环境因素并不太重要。此外，哺乳母猪通常处于明显高于其低临界温度的环境温度下。母猪每千克代谢体重（$W^{0.75}$）维持需要的合理估算值为100 kcal/天有效代谢能（NRC，2012）。

泌乳的有效代谢能需要可以从窝平均日增重（ADGl）和窝产仔数（LS；NRC，2012）进行合理准确的估算，如下式所示：

泌乳的有效代谢能需要（kcal/天）＝ 7.03×ADGl(g)－129×LS

在实际应用中，当使用具有 3300 kcal/kg 的有效代谢能含量的玉米豆粕型日粮时，体重为 200 kg 的哺乳母猪将需要 1.6 kg/天用于维持和 1.7 kg/天用于仔猪生长的饲料。

在现代高产母猪中，能量摄入通常不足以支持身体功能维持和泌乳的能量需要。因此，这些母猪将从储存的体蛋白质和体脂肪中调动能量，如前所述，这可能影响母猪长期的繁殖性能。因此，哺乳母猪管理的要旨在于如何最大限度地提高能量摄入，这将在本章后面讨论。

6.3.3 能量来源

日粮能量来自蛋白质、脂质和碳水化合物消化和吸收后氧化产生的能量。来自谷物和豆类的淀粉是碳水化合物的主要成分，其所带来的葡萄糖氧化供能，是母猪日粮能量的主要来源。母猪日粮中的常见谷物包括黄玉米、大麦、燕麦、高粱和小麦，其添加水平会根据谷物的供应情况和价格，以及生产阶段而变化。谷物粗蛋白质含量较低（玉米中 8.3%，硬红小麦中 14%），并且通常缺少赖氨酸或苏氨酸；然而，它们的淀粉含量很高，通常介于燕麦的 39% 和高粱的 70% 之间。这些谷物的能量和淀粉组成列于表 6.1。在所列的谷物中，玉米和高粱能量最高，部分原因在于它们的脂肪和淀粉含量较高。对于哺乳母猪，通常以植物油或动物脂肪的形式添加额外的能量，以改善适口性并增加日粮的能量浓度。源自日粮中脂质的脂肪酸氧化产生的能量是淀粉的两倍以上（表 6.1）。根据日粮的原料组成和环境温度，通常可添加 2%～7% 的油脂。Schoenherr 等（1989）报道了当处于 32℃ 的热应激环境时，饲喂含有 10.6% 精选白色动物油脂的饲料，可使母猪窝重增加。然而，较高的脂肪含量可能会降低饲料的流动性。由乙醇工业产生的高质量的酒精糟及其可溶物（DDGS），含有超过 27% 的粗蛋白质、33% 的中性洗涤纤维和超过 6% 的脂肪，可以用作妊娠和哺乳期母猪的能量、纤维和蛋白质来源。尽管相关的文献有限，但一些研究者建议高质量 DDGS 的最高添加水平在妊娠日粮中可达 40%，在哺乳日粮中可达 30%（Hill 等 2008；Stein，2007；Stein 和 Shurson，2009）。

表 6.1 妊娠和哺乳母猪日粮中常用谷物和脂肪原料的能量构成(NRC,2012)

谷物和油脂原料来源	能值(kcal/kg)			淀粉(%)
	消化能	代谢能	净能	
谷物				
大麦	3150	3073	2327	50.2
玉米	3451	3395	2672	63.0
燕麦	2627	2551	1893	39.0
高粱	3596	3532	2780	70.0
小麦	3313	3215	2472	59.5
动物脂肪				
牛油	7995	7835	6895	0
精选白色动物油脂	8290	8124	7149	0
猪油	8288	8123	7148	0
植物油				
玉米油	8754	8579	7549	
大豆油	8749	8574	7545	0
动植物混合油脂[1]	8393	8225	7238	0
DDGS,脂肪>10%[2]	3620	3434	2384	6.7
DDGS,6%<脂肪<9%[2]	3582	3396	2343	9.6
DDGS,脂肪<4%[2]	3291	3102	2009	10.0

[1] 25%猪油,25%家禽脂肪,25%牛油和25%玉米油。
[2] DDGS:酒精糟及其可溶物。

6.3.4 日粮的能量需要

妊娠和哺乳母猪推荐的日有效代谢能摄入量,以及基于含有 3300 kcal/kg 有效代谢能的玉米豆粕型日粮摄入量分别列于表 6.2 和表 6.3 中。对于妊娠母猪,这些推荐量是针对两个后续阶段的,以便适应妊娠晚期孕体生长增加的能量需要。表 6.2 中关于 4 胎母猪的值表明母猪体重对能量需要的影响,这反映了母猪体重和维持能量需要之间的密切关系,以及维持能量需要占据了妊娠母猪能量需要的很大比例。胎次对母体体重增加和组成的影响也有关。随着胎次的增加,身体脂肪增加对母体体重增加的贡献提高,每单位母体体重增加需要更多的能量。对于实际饲养管理,应考虑如热环境等决定能量需要的因素,根据分娩时预期实现的体况目标来调整每头母猪的饲喂水平。

对于哺乳母猪,仔猪窝生长率是衡量产奶量的重要指标,也是目前能量和饲料需要量最重要的决定因素。在需要大量泌乳的现代母猪中,应当使饲料和能量摄入最大化,以尽量减少哺乳期所有阶段中的母体减重。

表6.2　妊娠母猪日粮的能量、关键氨基酸、钙和磷的推荐水平(90%饲料干物质含量)[1]

胎次	1		2		3		4+					
配种时体重(kg)	140		165		185		205		205		230	
预期窝产仔数	12.5		13.5		13.5		13.5		15.5		13.5	
预期妊娠增重(kg)	65		60		52.5		40		45		40	
妊娠天数	<90	>90	<90	>90	<90	>90	<90	>90	<90	>90	<90	>90
预期采食量,含5%饲料浪费(kg/天)	2.13	2.53	2.21	2.61	2.21	2.61	2.05	2.45	2.08	2.48	2.21	2.61
有效代谢能(kcal/kg)[2]	3300	3300	3300	3300	3300	3300	3300	3300	3300	3300	3300	3300
净能(kcal/kg)[2]	2518	2518	2518	2518	2518	2518	2518	2518	2518	2518	2518	2518
SID 赖氨酸(%)[3]	0.57	0.76	0.48	0.67	0.41	0.58	0.35	0.53	0.36	0.55	0.34	0.50
SID 蛋氨酸(%)	0.17	0.23	0.13	0.19	0.11	0.17	0.10	0.14	0.10	0.15	0.09	0.14
SID 蛋氨酸+半胱氨酸(%)	0.37	0.50	0.32	0.44	0.29	0.40	0.25	0.36	0.26	0.39	0.24	0.35
SID 苏氨酸(%)	0.41	0.53	0.36	0.47	0.32	0.43	0.30	0.40	0.31	0.42	0.29	0.39
SID 色氨酸(%)	0.10	0.14	0.09	0.13	0.08	0.12	0.08	0.11	0.08	0.12	0.07	0.11
SID 异亮氨酸(%)	0.33	0.40	0.28	0.35	0.24	0.31	0.21	0.26	0.22	0.29	0.20	0.26
SID 缬氨酸(%)	0.41	0.54	0.35	0.47	0.31	0.43	0.28	0.40	0.29	0.41	0.26	0.37
总钙(%)	0.67	0.91	0.59	0.86	0.54	0.79	0.51	0.78	0.51	0.82	0.50	0.75
STTD 磷(%)[4]	0.30	0.40	0.26	0.37	0.23	0.34	0.22	0.34	0.22	0.36	0.21	0.33

[1] 数值来源于NRC(2012),但是考虑到饲料配制的不准确性和估算母猪生产性能造成的误差,氨基酸、钙和STTD磷的安全系数增加了10%。

[2] 日粮能量含量基于玉米和豆粕型日粮。有效代谢能(effective ME)含量是根据净能(NE)含量,使用母猪的修订转换值计算的。对于基于玉米和豆粕的日粮,有效代谢能含量与实际代谢含量相近。最佳日粮能量含量随当地饲料原料的供应情况和成本而变化。当使用替代原料时,建议基于净能含量配制日粮,并调整营养物质水平以维持恒定的养分与能量比。

[3] SID,标准回肠消化率。

[4] STTD,标准全消化道消化率。

表6.3　哺乳母猪在21天泌乳期日粮中的能量、关键氨基酸、钙和磷的推荐水平(90%饲料干物质含量)[1]

胎次	1			2+		
产后体重(kg)	175			210		
窝产仔数	11			11.5		
哺乳仔猪平均日增重(g)	190	230	270	190	230	270
预期妊娠母猪体重变化(kg)	1.5	−7.7	−17.4	3.7	−5.8	−15.9
预期采食量,含5%饲料浪费(kg/天)	5.95	5.95	5.93	6.61	6.61	6.61
有效代谢能(kcal/kg)[2]	3300	3300	3300	3300	3300	3300

续表 6.3

胎次	1			2+		
产后体重（kg）	175			210		
窝产仔数	11			11.5		
净能（kcal/kg）²	2518	2518	2518	2518	2518	2518
SID 赖氨酸（%）³	0.83	0.89	0.96	0.79	0.87	0.92
SID 蛋氨酸（%）	0.22	0.23	0.25	0.21	0.23	0.24
SID 蛋氨酸＋半胱氨酸（%）	0.43	0.47	0.52	0.42	0.45	0.50
SID 苏氨酸（%）	0.52	0.56	0.61	0.51	0.54	0.58
SID 色氨酸（%）	0.15	0.17	0.19	0.14	0.17	0.18
SID 异亮氨酸（%）	0.45	0.50	0.54	0.44	0.47	0.52
SID 缬氨酸（%）	0.70	0.76	0.81	0.67	0.73	0.78
总钙（%）	0.69	0.78	0.88	0.66	0.75	0.84
STTD 磷（%）⁴	0.34	0.40	0.44	0.33	0.37	0.42

[1] 数值来源于 NRC(2012)，但是由于饲料制备的不准确性和估算母猪性能造成的误差，氨基酸，钙和 STTD 磷的安全系数增加了 10%。

[2] 日粮能量含量基于玉米和豆粕型日粮。有效代谢能（effective ME）含量是根据净能（NE）含量，使用母猪的修订转换值计算的。对于基于玉米和豆粕的日粮，有效代谢能含量与实际代谢能含量相近。最佳日粮能量含量随当地饲料原料的供应情况和成本而变化。当使用替代原料时，建议基于净能含量配制日粮，并调整营养物质水平以维持恒定的养分与能量比。

[3] SID 为标准回肠消化率。

[4] STTD 为标准全消化道消化率。

6.4 氨基酸营养

6.4.1 氨基酸需要的表达方式：总、表观回肠可消化，标准回肠可消化和生物可利用氨基酸需要

总氨基酸需要

在玉米和豆粕构成常用能量和蛋白质来源的地区，与使用其他谷物和蛋白质来源的地区相比，以可消化基础来表达需要量，比用总需要量表示的优势要少得多。这主要归因于以下事实：许多评估母猪赖氨酸需要量的研究是基于玉米豆粕型日粮进行的。因为不同饲料原料的氨基酸的生物利用率，特别是在蛋白质质量较低原料中差异很大，所以用饲料原料的总氨基酸来确定氨基酸需要量时，将根据用于特定日粮的饲料原料而不同。因此，现在通常的做法是在"回肠可消化"氨基酸的基础上表示母猪的氨基酸需要。营养学家认为可以假定回肠消化率可提供氨基酸生物利用率的合理估算（Stein 等，2007）。

回肠可消化氨基酸需要

为什么是回肠？在本书中，"回肠"是指小肠的最远端，为可在结肠和小肠结合处之前收集小肠食糜的部位。通过测定在小肠消化道中和饲料中氨基酸浓度，从而计算出氨基酸的消化量，称为"消化率"（Sales and Janssens，2003）。这种方法的生物学原理是，在这个位置取样，而不是侵入性较低的传统粪便收集方法，可避免大肠（盲肠和结肠）中微生物代谢对日粮氨基酸消化造成的混淆。大肠中大量的微生物活动会改变摄入蛋白质的氨基酸组成（Metges，2000）。大肠有比小肠中多得多的微生物数量和更长的食糜通过时间（Hovgaard 和 Brondsted，1996），这有利于发酵蛋白质和氨基酸降解，以及微生物通过来自非特异性和日粮氮源产生氨基酸（Sauer 等，1975）。因此，粪便氨基酸消化率可能高估（对于大多数氨基酸）或低估回肠消化率，这取决于日粮组成和特定的氨基酸。由于没有计入大肠吸收的日粮氨基酸（Darragh 等，1994；Moughan 和 Stevens，2012；Torrallardona 等，2003），回肠取样的方法可以更好地测量动物消化和吸收的氨基酸，即氨基酸生物利用率，因此比粪便消化率更为准确。

虽然回肠氨基酸消化率不一定等于生物利用率，但它提供了可以常规测量的估算值。然而，以回肠可消化的表达氨基酸需要的方法也有其局限性。内源性氨基酸是饲料在通过消化道期间被分泌或在消化道中损失的氨基酸，而回肠可消化法的关键问题是无法把在消化道中未被消化的日粮氨基酸和内源性氨基酸之间进行精确区分。另一个问题是，在热处理过的原料中，回肠消化率可能高估氨基酸（特别是赖氨酸）的生物利用率（Stein 等，2007）。

表观、标准和真回肠氨基酸消化率

内源性氨基酸损失与消化酶、消化道黏液和由消化道分泌的其他蛋白质有关，而在常规回肠消化率测定中对回肠食糜进行取样时，未消化的日粮氨基酸和内源性氨基酸损失之间没有区分开来。因此，常规消化率值应被称为表观消化率。猪内源性蛋白质损失有两个主要部分：①基础损失，其不受日粮影响，可以通过给猪饲喂无蛋白质日粮进行评估；②日粮或原料特异性内源性损失，因饲料和日粮组成的差异而不同。如果没有校正内源损失，日粮氨基酸的真实消化率就会被低估，此时被称为"表观"消化率。基础内源损失校正后，可用来估算标准消化率，而对基础内源损失和原料特异性损失校正后可用来估算"真"消化率。由于用于估算原料特异性内源性氨基酸损失的方法有限，成本高昂和通常带有侵入性，目前没有足够的数据来估算常用的猪饲料原料中的"真"回肠氨基酸消化率。因此，对表观消化率进行基础回肠内源性损失校正后的的标准回肠消化率，被普遍用于猪日粮的制定。使用标准回肠消化率的主要好处是，标准回肠消化率的测量在很大程度上不受日粮蛋白质含量的影响，并且在猪饲料原料混合配制饲料时具有可加性。这些概念在 Stein 等（2007）的文章中有更详细的描述。

6.4.2 氨基酸需要的决定因素

维持需要

维持氨基酸需要的主要决定因素包括基础肠内源性氨基酸损失,其与采食量水平有关,皮肤和毛发氨基酸损失,其与代谢体重($W^{0.75}$)大小的成函数关系(Moughan,1999;NRC,2012)。皮肤和毛发的氨基酸组成已由 Van Milgen 等(2008)报道。母猪肠内源性损失的氨基酸谱被认为与生长育肥猪的相似(NRC,2012)。这些氨基酸谱与回肠赖氨酸损失有关,回肠赖氨酸损失对于妊娠(0.522 g/kg 干物质摄入)和泌乳(0.292 g/kg 干物质摄入)是不同的,并其可被用于计算每种必需氨基酸损失。另外,大肠的氨基酸内源损失大约占回肠内源损失的10%。除了这些机体氨基酸损失,氨基酸最小代谢损失也对维持氨基酸需要有影响。然而,关于与机体维持功能相关的氨基酸分解代谢的文献非常欠缺。因此,当需要计算氨基酸维持需要的估算值时,NRC(2012)采用了一个计算损失的因子,既考虑了机体氨基酸损失的需要,也反映了氨基酸的分解代谢。由于母猪的皮肤和毛发有关的氨基酸损失非常少,因此在妊娠期和哺乳期氨基酸的维持需要差别主要基于采食量,哺乳母猪为妊娠母猪的2~3倍(表6.4)。值得注意的是,当对妊娠母猪饲喂含有较高纤维水平的日粮时,内源性氨基酸损失可能略高,从而增加维持需要。

表 6.4 妊娠期和哺乳期的维持氨基酸需要,以 $mg/W^{0.75}$ 为单位(改编自 NRC,2012)

氨基酸	妊娠期	哺乳期
精氨酸	17.09	38.45
组氨酸	12.09	19.74
异亮氨酸	29.78	38.76
亮氨酸	36.03	53.41
赖氨酸	34.79	43.26
蛋氨酸	9.17	10.16
总硫	36.80	36.33
苯丙氨酸	25.03	35.66
芳香族氨基酸总量	45.49	62.14
苏氨酸	44.53	56.78
色氨酸	10.94	13.97
缬氨酸	69	65.81
N × 6.25	979.33	1320.51

妊娠蛋白质库

妊娠氨基酸需要的决定因素包括:母体的氨基酸组成和与繁殖直接相关的四个主要动态蛋白质库,以及它们在妊娠过程中对机体总蛋白质周转的相对贡献。

5个蛋白质质库，即母体、胎儿、子宫、胎盘和羊水，以及乳房，它们一起组成了母猪的全身体蛋白库。这些蛋白质库中的每一个都具有独特的氨基酸谱（即氨基酸：赖氨酸），如表6.5所示。

表6.5 赖氨酸含量和以赖氨酸含量的百分比表示的母体、胎儿、胎盘、子宫、绒毛尿囊液、乳房和乳汁的氨基酸组成（改编自 NRC,2012）。

氨基酸	母体	胎儿	子宫	胎盘＋羊水	乳房
赖氨酸(g/100 g CP)[1]	6.74	4.99	6.92	6.39	6.55
氨基酸：赖氨酸 × 100					
精氨酸	105	113	103	101	84
组氨酸	47	36	35	42	35
异亮氨酸	54	50	52	52	24
亮氨酸	101	118	116	122	123
赖氨酸	100	100	100	100	100
蛋氨酸	29	32	25	25	23
蛋氨酸＋半胱氨酸	45	54	50	50	51
苯丙氨酸	55	60	63	68	63
苯丙氨酸＋酪氨酸	97	102			
苏氨酸	55	56	61	66	80
色氨酸	13	19	15	19	24
缬氨酸	69	73	75	83	88

[1] CP,粗蛋白质。

与生命周期的许多其他阶段相反，妊娠的特征在于动态代谢过程，其中不同的组织以非常不同的速率吸引蛋白质。图6.3展示了每个蛋白质库中不同妊娠天数对应的蛋白质沉积速率。胎儿和乳房生长在妊娠期间有类似的蛋白质沉积模式，但它们在妊娠第60～80天的沉积率却发生了明显变化。而胎盘和绒毛尿囊液中的蛋白质沉积从第40天一直增加到约第60天，并在余下的妊娠期保持相对恒定。子宫对身体蛋白质沉积的量影响最小但却有显著的贡献，其随妊娠过程线性增加。尽管在本章中并未提及，但在前面章节已指出，对于初产母猪，其母体蛋白质沉积与妊娠天数（具有时间依赖性）成函数关系，并且从数量上能量摄入比胎儿蛋白质沉积更为重要。

了解了每个相应蛋白质库中的氨基酸谱与蛋白质的每日沉积率，就可以计算每天母猪体内和妊娠产物中的总氨基酸沉积量，这对计算氨基酸需要意义重大。氨基酸不能以100％效率沉积到蛋白质中；按NRC(2012)估算，妊娠期赖氨酸的沉积效率值为49％。该数值与来自妊娠晚期母猪根据不同日粮赖氨酸水平测定氮平衡的试验结果一致(NRC,2012)。由于其他必需氨基酸的此类研究数量有限，NRC(2012)必须估算每个氨基酸的单独的效率值，以计算妊娠母猪的可消化氨基酸需要。尽管如此,这些估算的效率值仍需要与妊娠母猪的经验值进行互相验证。这些效率估算值见表 6.6。

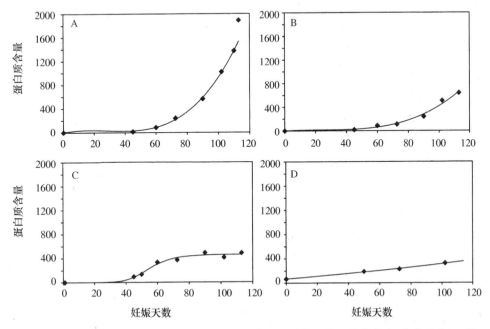

图 6.3　妊娠天数与胎儿($n = 12$ 仔猪)(A),乳房(B),胎盘和绒毛尿囊液(C)和空子宫(D)的蛋白质含量之间的关系。符号(♦)表示为实测值,实线表示为预测值(改编自 NRC,2012)。

表 6.6　日粮回肠可消化氨基酸对妊娠蛋白质增加和泌乳期乳蛋白生产的利用效率(改编自 NRC,2012)

氨基酸	妊娠期	哺乳期
精氨酸	0.960	0.816
组氨酸	0.636	0.722
异亮氨酸	0.491	0.698
亮氨酸	0.588	0.723
赖氨酸	0.490	0.670
蛋氨酸	0.495	0.675
总硫	0.402	0.662
苯丙氨酸	0.542	0.733
芳香族氨基酸总量	0.537	0.705
苏氨酸	0.527	0.764
色氨酸	0.467	0.674
缬氨酸	0.549	0.583
N×6.25	0.555	0.759

泌乳蛋白质库

哺乳期的氨基酸需要由两个库(母体和乳)的氨基酸组成及其对泌乳过程中蛋白质积累的相对贡献来定义。同样,这两个库具有各自独特的氨基酸谱(即氨基酸:赖氨酸),如表 6.7 所示。乳蛋白质生产是占哺乳母猪氨基酸需要的最大部分,这意味着哺乳期间的氨基酸需要模式是相对恒定的,并且主要由乳蛋白质的氨基酸谱特征决定。

表 6.7 赖氨酸含量,以赖氨酸含量的百分比表示的母体和乳蛋白质的氨基酸组成(改编自 NRC,2012)

氨基酸	母体蛋白质	乳蛋白质
赖氨酸(g/100 g CP) [1]	6.74	7.01
氨基酸:赖氨酸×100		
精氨酸	105	69
组氨酸	47	43
异亮氨酸	54	56
亮氨酸	101	120
赖氨酸	100	100
蛋氨酸	29	27
总硫	45	50
苯丙氨酸	55	58
芳香族氨基酸总量	97	115
苏氨酸	55	61
色氨酸	13	18
缬氨酸	69	71

[1] CP,粗蛋白质。

乳蛋白质中可消化氨基酸沉积效率

根据 NRC(2012)中描述的许多严格筛选和发表的研究结果,可以直接计算出标准回肠可消化赖氨酸沉积到乳蛋白质中的效率为 67%。目前其他必需氨基酸用于母猪产奶需要的数据不多,还没有充足到能像赖氨酸那样可以直接计算出氨基酸沉积到乳蛋白质中的效率。除赖氨酸外,表 6.6 中所示的其他氨基酸的标准回肠氨基酸用于泌乳的效率值都是由模型推导的(NRC,2012),并且仍然需要通过稳健实证研究(robust empirical studies)进行验证。然而,这些用于泌乳的效率值,对于计算泌乳的标准回肠氨基酸需要是必不可少的。

6.4.3 氨基酸营养与能量营养之间的关系

在母猪中,能量和氨基酸营养之间存在重要的相互作用。例如,在妊娠母猪中,尤其在 1--3 胎次期间,此时母猪尚未达到其体成熟蛋白质总量,在典型能量摄入水平的基础上,每日能量摄入量的少量增加都将会增加母体蛋白质沉积,这反过来也增加了日粮的氨基酸需要(Dourmad 等,2008;NRC,2012)。哺乳期母猪一般处于负能量平衡,能量摄入的增加可以减少机体蛋白质和体脂的动员(Dourmad 等,2008;NRC,2012)。其结果是,动员机体蛋白质的氨基酸向乳蛋白质的流动减少,而需要更多的日粮氨基酸摄入以满足乳蛋白质生产的需要(表6.3)。Tokach 等(1992)证明了哺乳母猪的能量摄入和每日氨基酸需要之间的这种相互作用,其研究结果表明每日赖氨酸需要随哺乳母猪的每日能量摄入量增加而提高。这可能也适用于肥育猪,或者说当能量摄入与肌肉生长(即机体蛋白质沉积)之间没有关系时的情况。基于这些概念可以推测,将每日氨基酸摄入量作为目标,并简单地从估算的每日氨基酸需要除以饲料采食量而推导出日粮的目标氨基酸浓度来配制母猪日粮是不合适的。对于母猪,通过估算氨基酸与能量的比率,进而估算日粮的目标氨基酸浓度是更合理的。然而,建立日粮氨基酸浓度的优先方法是析因法(建模法),此方法综合考虑了每日能量摄入量测定值,母猪生产性能和环境条件的情况(例如NRC,2012)。

6.4.4 蛋白质来源和质量

对人或动物来说,蛋白质源的营养质量可以被描述为该蛋白质满足氨基酸和氮需要的能力(Schaafsma,2005)。蛋白质质量在饲料原料中差异很大,并且由其氨基酸组成和生物利用率来评价。谷物的蛋白质常缺乏关键的必需氨基酸,因此,常用高蛋白质饲料原料与谷物组合,来补充谷物中缺乏的必需氨基酸。尽管如此,在母猪日粮中最常用的谷物,如玉米、大麦、小麦和高粱,却提供了60%的总氨基酸需要(NRC,2012)。用于母猪日粮的常见植物蛋白原料包括豆粕、豌豆(field peas)、双低菜粕(canola meal)、向日葵粕(sunflower meal)和谷物副产品,如DDGS、玉米麸质饲料(corn gluten feed)和玉米蛋白粉(corn gluten meal)。

豆粕是从大豆中提取豆油后的副产物。在脱壳和随后的提油过程中,大豆蛋白被加热,从而降低对热敏感的抗营养因子,尤其是存在于生大豆中的胰蛋白酶抑制因子(trypsin-inhibitors)。有两类豆粕:经去皮和溶剂浸提,含有49%~50%粗蛋白质(CP)的高蛋白豆粕,以及含有44%~46%粗蛋白质和更高纤维含量的低蛋白豆粕。正确加工处理的豆粕是一种较为稳定一致的产品,通常是母猪最经济的氨基酸来源。此外,豆粕是唯一可提供与动物蛋白具有相似氨基酸质量的常用植物蛋白原料。由于其具有很高的氨基酸消化率和高赖氨酸含量,豆粕可以作为妊娠和哺乳母猪日粮中唯一的蛋白原料,并且最小化尿和粪便中的氮排泄(Pettigrew 等,2008)。然而,如果可以适当地设计和制备日粮,也可以在母猪日粮中添加多种其他蛋白质来源,而不损害母猪的生产性能。因此,各种蛋白质来源的使用应主要由其营养价值和成本驱动,并使用最低成本饲料配方系统进行评估(Patience 等,1995)。

当生长条件有利于豌豆(Pisum sativum L.)时,这种原料是母猪日粮中豆粕或玉米的良好部分替代物。与大豆不同,豌豆的抗营养因子含量非常低,因此可以直接饲喂(Thaler 和 Stein,2003)。热处理还可增加其氨基酸的回肠消化率,以及淀粉和能量的回肠消化率(Stein 和 Bohlke,2007)。对某些品种而言,豌豆是良好的蛋白质(23.4%)和赖氨酸(1.7%)来源,但是蛋氨酸和色氨酸稍微缺乏,与豆粕相比,其色氨酸和含硫氨基酸的标准回肠消化率较低(Stein 等,2004)。妊娠和哺乳母猪日粮推荐添加比例分别为16%和24%。

溶剂浸提的双低菜粕,也称为溶剂浸提的低硫代葡萄糖苷(glucosinolate)和芥酸菜籽粕(erucic acid rapeseed meal),即使它包含比豆粕更多的纤维和更少的蛋白质,对母猪而言,也是一种有价值的蛋白质来源。菜籽粕含有硫代葡萄糖苷,其在酶促水解后会产生具有抗甲状腺(致甲状腺肿大)性质的副产物(Bell 和 Baker,1957)。在妊娠期间,这些化合物会导致母猪和胎儿的甲状腺肥大,尤其胎儿会特别敏感。胎儿的血浆甲状腺素浓度和体重减小与母猪日粮中的硫代葡萄糖苷水平相关。如果菜籽粕是通过选育获得的新油菜品种所制,其与传统菜籽粕相比具有非常低硫代葡萄糖苷含量,则可以在繁殖母猪中使用,并且可以添加非常高的水平(King 等,2001)。当日粮中硫代葡萄糖苷浓度低于 2 μmol/g 时,即可避免不利于健康的后果。Etienne 等(1975)报道了在大麦基础日粮中添加11%豆粕或21%双低菜粕,并补充合成蛋氨酸时,两组母猪的采食量,产奶量和乳成分均相似。对于日粮中分别添加 0,101 和 202 g/kg 双低菜粕的母猪,哺乳期的平均饲料采食量分别为 5.08、5.50 和 5.67 kg/天(King 等,2001)。

根据脱壳和提油的程度不同(未脱壳的 CP 为29%～33%,脱壳的 CP 为35%～39%),向日葵粕在质量等级上差异很大。幸运的是,向日葵粕不含内源性抗营养因子,因此不需要任何热处理。饲料级向日葵粕能量相对较低,赖氨酸不足,但富含蛋氨酸,因此与用于哺乳日粮相比,其更适合作为妊娠日粮蛋白质的来源(CETION,2003)。当给哺乳母猪饲喂豆粕时,添加向日葵粕可以很好地补充豆粕所不足的蛋氨酸(译者注:原著为赖氨酸,推测应为蛋氨酸)需要。

玉米麸质饲料和玉米蛋白粉是玉米湿磨工业的产品,如同向日葵粕,不含抗营养因子。玉米麸质饲料能量较低,纤维含量高,蛋白质含量在20%～25%,可消化赖氨酸和色氨酸含量低,并且与豆粕相比,母猪的适口性较差。由于玉米麸质饲料较为蓬松,不推荐用于哺乳日粮,但是非常适合作为妊娠母猪日粮中主要或唯一的蛋白质来源,其添加比例可在50%～70%(Honeyman 和 Zimmerman,1990;Meunier-Salaun,2001)。玉米蛋白粉含有高能量和高可消化氨基酸,因此非常适合作为妊娠和哺乳日粮中的主要蛋白质来源,添加水平可为20%。

如前所述,高品质 DDGS 是一种优良的蛋白质来源,由于其高纤维含量,对妊娠母猪来说尤其适合。在妊娠母猪中,DDGS 可以替代所有的豆粕,而在哺乳母猪日粮中，DDGS可以替代大部分玉米和豆粕，添加量可达30%（Stein和Shurson，2009）。

动物副产品也可有限度地用于母猪日粮，例如喷雾干燥鸡蛋粉（spray-dried egg）、血液副产品（blood co-products）、禽肉粉（poultry meal）、肉骨粉（meat and bone meal）和鱼粉（fish meal），鱼粉可能比较常见。这些产品在母猪日粮中的添加量通常限制在3%或更低（Patience等，1995）。然而，这些动物副产品中的蛋白质质量非常好，具有高消化率和高蛋氨酸浓度。同时，这些产品往往又相当昂贵，因此从经济角度考虑往往阻碍了它们进入最低成本的母猪日粮配方中（Patience等，1995）。

晶体氨基酸，包括 L-赖氨酸，L-苏氨酸，DL-蛋氨酸和 L-色氨酸目前都可以方便以纯晶体形式购得。每种结晶体氨基酸都可以单独用于哺乳母猪日粮中。其中，L-赖氨酸是目前使用最多的，其次可能是 L-苏氨酸，这取决于所用的谷物情况。结晶形式的氨基酸是100%可消化的，并且可以以非常精确的量来补充日粮氨基酸缺乏。

6.4.5 应用日粮氨基酸需要

当饲喂玉米豆粕基础日粮时，不同生产水平的妊娠和哺乳母猪的关键氨基酸日推荐需要量列于表6.2和表6.3。这些值来自NRC（2012），但通常增加了10%的安全余量，以防止日粮配方和制备过程中，以及估算营养需要时不够精确。因此，表6.2和表6.3中提出的氨基酸需要量被称为"推荐"需要量或推荐量。

对于妊娠母猪，妊娠晚期（第85天后）的每日氨基酸需要大大高于妊娠早期和中期（第1~90天）。这些更高的需要反映了妊娠晚期妊娠产物的蛋白质积累和氨基酸需要呈指数增加（图6.3）。当将妊娠晚期与早期和中期进行比较时，会发现前者氨基酸需要的增加大于能量需要的增加。因此，在妊娠晚期，最佳日粮氨基酸与能量比将升高。基于这些概念，应该考虑将妊娠母猪分阶段饲养，在妊娠90天左右，其饲养水平和日粮成分都应有所变化。妊娠母猪的氨基酸需要随胎次的增加而明显降低，反映了随着母猪接近成熟体蛋白质量，母体体蛋白质沉积在逐渐减少。在大规模母猪实际生产中，可以考虑使用分胎次的妊娠日粮，以便更贴合满足不同胎次群体母猪的氨基酸需要，并且在保持母猪生产力的同时降低饲料成本和进入环境的营养物质损失。随着计算机控制投喂设备和母猪个体鉴定技术的快速发展，未来将可以实现两种或多种基础日粮进行优化组合，从而满足每头母猪独特的氨基酸需要。在这种情况下，基础日粮可以配制成满足具有最高（例如1胎母猪的妊娠晚期）和最低（例如4胎以上母猪妊娠早期）的氨基酸需要。

与能量需要的情况一样，仔猪窝生长率是哺乳母猪氨基酸需要的最重要决定因素。对哺乳母猪特别是在1胎母猪而言，若不能满足能量和氨基酸需要将导致产奶量减少，增加母体体蛋白质的动员，并将对仔猪生长性能和母猪随后的繁殖性能产生不利影响。为了满足具有超过平均窝生长率的1胎和更高胎次哺乳母猪相对较高的氨基酸需要，应当使用包含高蛋白质和良好适口性原料配制而成的优质饲料。在窝生长率一定的情况下，母猪自由采食量的增加将导致每日氨基酸需要的增加（表 6.3）。这种情况可在较高的饲料采食水平下减少体蛋白动员，并且可减少氨基酸用于乳蛋白质合成的体蛋白动员。

6.5 饲养管理

6.5.1 饲料加工的考虑

饲料加工方法对母猪繁殖性能的影响研究很少受到关注。当将哺乳日粮中玉米粒度从 1200 μm 减小至 400 μm 时，除窝重增加 11% 外，哺乳母猪的自由采食量、干物质、氮和总能的表观消化率均得到提高，母猪繁殖性能也有小幅提高（Wondra 等，1995）。在对单个哺乳期的研究中发现，粒度的减小增加了胃溃疡和胃角质化的发生率，但无明显有损母猪健康的迹象。小粒度日粮是否会对母猪的胃健康有长期影响的多胎次研究尚未开展。Pettigrew 等（1985）报道了当玉米豆粕型基础日粮的粒度由 619 μm 减小到 444 μm 时，对哺乳母猪的自由采食量、母猪繁殖性能或仔猪生长性能均无影响。粒度减小可提高日粮消化率，由此可减少营养物质排泄，有益于环境。但是，减小饲料的粒度可能会导致在商业饲料传送系统中流动性下降的问题（Ganesan 等，2005）。如果饲料流动性下降导致不易通过饲料传送系统和进入母猪料槽，这将使母猪的自由采食量受到限制。可以通过制粒来减轻精细研磨日粮的流动性下降的问题。制粒后可改善日粮中干物质、能量和氮的表观消化率，而对自由采食量或母猪繁殖性能没有影响（Johnston 等，1998）。在选择母猪日粮的适宜粒度时，一方面可能有改善母猪采食量和营养物质消化率的益处；另一方面可能会对母猪肠道健康、管理效率和饲料传送带来挑战，需要在利弊之间取得平衡。

6.5.2 妊娠期饲养方式

通常会限制妊娠母猪采食量，以防止妊娠期体重过度增加。妊娠母猪有限的采食量可引起称为"行为规癖"的行为恶习（Lawrence 和 Terlouw，1993），这被认为可能是母猪福利受损的迹象。饲养管理需要每天提供所需的营养物质，但也必须尽量减少有损母猪福利的可能性。饲喂的频率会影响行为规癖的发生。Holt 等（2006）报道，与母猪每天饲喂一次相比，每天饲喂两次发生行为规癖的总时间增加，但是分娩和泌乳性能不受影响。每天两次而非一次的母猪显示出较多饲喂诱导的行为规癖。每天饲喂 6 次，比每天两次的母猪减少了围栏母猪的互相攻击，但对分娩或哺乳性能没有影响（Schneider 等，2007）。后备母猪的行为和分娩性能不受每日饲喂 6 次和 2 次的影响。因此，与单次饲喂相比，母猪每天多次饲喂似乎没什么优势。此问题的一个主要解决方案是给妊娠母猪饲喂高纤维日粮，这在第 5

章中有详细描述(Meunier-Salaün 和 Bolhuis,2015)。

任何妊娠期饲养管理的一个关键措施都是要确保每头母猪可以接受到它全部的饲料配给,这是特别重要的,因为妊娠母猪通常只能吃到相对少量的营养丰富的日粮。当母猪没有得到其完全分配的饲料(即使很短的时间),繁殖性能可能受损(Kongsted,2006),并可能导致过早淘汰。这在母猪群养的情况下尤其重要,在这种情况下,侵略性强的母猪可能会对社会阶层较低端的侵略性较弱的母猪产生不利影响。畜栏设计、饲养管理方式和饲养人员都在确保母猪妊娠期间获得应得饲料中发挥重要作用。这在第3章关于妊娠群养母猪中有详细讨论(Spoolder 和 Vermeer,2015)。

继续提高猪肉生产效率的动力促使人们对妊娠母猪更有针对性的饲养方法产生兴趣。妊娠期间的能量和营养需要受体重和母猪状况的影响很大。传统上,母猪饲喂基础水平的日粮,然后根据对母猪体况的视觉评估而增加或减少营养供给。然而,Young 等(2004)根据测量母猪的体重和背膘厚来调整饲喂量,从而可以使母猪在分娩时实现目标背膘厚。他们更有针对性的目标管理饲养法成功地在分娩时轻微减少了母猪肥胖的发生率,但是没有显著降低瘦母猪分娩时肥胖的发生率。对分娩和哺乳性能来说,传统和目标管理饲养法之间没有差异。然而,目标管理饲养方式可使有相同哺乳性能的妊娠母猪消耗更少的饲料,这将能更好地利用饲料资源。需要特别指出的是,根据个体体重和背膘厚制定目标饲养的妊娠母猪,会比传统方法饲养的母猪获得的体脂和体蛋白质要少。由于这项研究只针对一个胎次,因此未来长期的研究结果值得关注。

初产母猪和经产母猪的能量和氨基酸需要在整个妊娠期不是恒定的(NRC,2012)。青年母猪(后备母猪和1胎母猪)和成年母猪的需要也不同(NRC,2012)。由于妊娠产物的快速生长,能量和营养需要在妊娠后期会明显增加(Noblet 等,1985;Ullrey 等,1965)。这种需要的重要变化表明,日粮中能量和营养物质的供应应该去适应妊娠母猪的年龄和各阶段的生理需要。然而,大多数妊娠期饲养方式中并未在整个妊娠期提供设定多个营养水平的饲料,也不论母猪年龄,而仅根据体况的视觉评估进行调整。通常,在妊娠最后3周会考虑到胎儿和乳腺发育的增加,而提高标准日粮的采食量,但是日粮的组成却未作改变。近来研究人员建议采用胎次分离和阶段饲养的方法来提高妊娠饲养规划的效率和有效性(Ball 和 Moehn,2013;Kim 等,2009;见本章前面部分)。Ball 和 Moehn(2013)总结了大量证据,提出妊娠应分为早期(0~84天)和晚期(85~114天)两个阶段。他们提出每个阶段应制定不同的日粮,由于在妊娠晚期氨基酸的需要比能量增加更多,因此使用单一日粮是不合适的。除了对妊娠期不同阶段需要调整,他们还建议应对不同胎次,如1胎,2胎,3胎和更高胎次的母猪施以不同的饲养水平。从生理学角度来看,妊娠母猪的胎次分离,分阶段饲养是有道理的;但是,由于需要重新配置饲料处理系统和增加管理难度,在商业条件下的实施会受到限制。猪肉生产者需要先实现经济回报以支付这些附加费用,然后这种方法才能广泛采用。当妊娠母猪群养模式变得更普遍,母猪电子饲喂设备被纳入群养系统,分胎次和分阶段饲养的商业应用就将变得更加可行。

6.5.3 哺乳期饲养方式

哺乳期是母猪生殖周期中最艰巨的阶段。支持哺乳期的能量和营养需要远远超过生殖周期的任何其他阶段(NRC,2012)。鉴于如此高的营养需要,人们普遍认为,必须以最大限度地提高采食量和营养摄入量来饲养和管理哺乳母猪(Eissen等,2003)。然而哺乳母猪的自由采食量会受许多因素影响。许多研究者(Eissen等,2000;Johnston,1992;Eissen等,2000;O'Grady等,1985)提出过哺乳母猪采食量最大化的方法,并进行了详尽的讨论。

料槽设计可对哺乳母猪的自由采食量产生重要影响。在20世纪90年代初,猪设备制造商根据伊利诺伊大学的研究成果重新设计了哺乳母猪的料槽(Taylor,1990)。高效的料槽是宽敞的,可以允许母猪吃料时进行小幅的身体运动,并舒适地站在料槽前。容易获得饲料可以促进哺乳母猪取得最大的采食量。然而,即使使用设计良好的料槽,如果管理当中不能为母猪提供足够的饲料,则采食量仍可能受到限制。当饲养员每天手工饲喂母猪两次时,这个问题常常出现。一个有效的解决方案是给料槽加装一个容器(料斗),该容器可以储存比母猪在下一次加料之前所能采食的更多的饲料。这种方法比人工饲喂增加了哺乳母猪的自由采食量(Peng等,2007;Peterson等,2004)。最近,已经为哺乳母猪开发了采用自动化螺旋推运器和小料斗,可以每天24 h提供不间断供料的饲喂系统。

多年来,哺乳期的采食问题一直是讨论的主题。Koketsu等(1996c)报道,在哺乳早期达到采食量高峰的母猪,其整个哺乳期的平均日采食量更大。他们还报道,哺乳早期的高采食量是哺乳后期采食量短暂下降的一个重要的风险因素。因此,母猪管理者必须在母猪产后自由采食时平衡这两种相反的效应。在许多养殖场普遍采用了妥协的方案,即在分娩后的头几天控制限制采食量,但是在分娩后4或5天使母猪完全自由采食(Trottier和Johnston,2001)。这种饲养实践既可满足母猪分娩后由产奶量上升推动的快速增加的营养需要,也可能会降低饲料成本。

6.6 结论

随着世界人口和肉类需要的增长,以及可用土地减少和全球变暖的加剧,母猪的精准饲养越来越重要。本章介绍了妊娠期和哺乳期母猪氨基酸和能量需要的最新发现,以及最大限度地提高营养摄入量和效率的方法。然而,未来仍需要更多的工作来持续验证日粮氨基酸和能量预测模型。

参考文献

Almeida, F.N., Htoo, J.K., Thomson, J. and Stein, H.H., 2014. Digestibility by growing pigs of amino acids in heat-damaged sunflower meal and cottonseed meal. Journal of Animal Science 92: 585-593.

Anil, S.S., Anil, L., Deen, J., Baidoo, S.K. and Walker, R.D., 2006. Association of inadequate feed intake during lactation with removal of sows from the breeding herd. Journal of Swine Health Production 14: 296-301.

Anil, S.S., Anil, L., Deen, J., Baidoo, S.K. and Walker, R.D., 2007. Factors associated with claw lesions in gestating sows. Journal of Swine Health Production 15: 78-83.

Ball, R.O. and Moehn, S., 2013. Feeding pregnant sows for optimum productivity: past, present and future perspectives. In: Pluske, J.R. and Pluske, J.M. (eds.) Manipulating Pig Production XIV. Proceedings of the Fourteenth Biennial Conference of the Australasian Pig Science Association, pp. 151-169.

Bell, J.M. and Baker, E., 1957. Growth depressing factors in rapeseed oil meal. Canadian Journal of Animal Science 37: 21-30.

Calderón Díaz, J.A., Nikkilä, M.T. and Stalder, K., 2015. Sow longevity. Chapter 19. In: Farmer, C. (ed.) The gestating and lactating sow. Wageningen Academic Publishers, Wageningen, the Netherlands, pp. 423-452.

Centre Technique Interprofessionnel des Oléagineux et du Chanvre (CETIOM), 2003. Sunflower meal: quality protein and fibre. Fiches Techniques Octobre 2003. Edition CETIOM.

Darragh, A.J., Cranwelland, P.D., and Moughan, P.J., 1994. Absorption of lysine and methionine from the proximal colon of the piglet. British Journal of Nutrition 71: 739-752.

Dourmad, J.Y., 1991. Effect of feeding level in the gilt during pregnancy on voluntary feed intake during lactation and changes in body composition during gestation and lactation. Livestock Production Science 27: 309-319.

Dourmad, J.Y., Etienne, M., Prunier, A., and Noblet, J., 1994. The effect of energy and protein intake of sows on their longevity: a review. Livestock Production Science 40: 87-97.

Dourmad, J.Y., Etienne, M., Valancogne, A., Dubois, S., Van Milgen, J. and Noblet, J., 2008. InraPorc: a model and decision support tool for the nutrition of sows. Animal Feed Science and Technology 143: 372-386.

Dourmad, J.Y., Noblet, J., Pere, M.C. and Etienne, M., 1999. Mating, pregnancy and prenatal growth. In: Kyriazakis, I. (ed.) Quantitative Biology of the Pig. CABI, Wallingford, UK, pp. 129-152.

Eissen, J.J., Apeldoorn, E.J., Kanis, E., Verstegen, M.W.A. and De Greef, K.H., 2003. The importance of a high feed intake during lactation of primiparous sows nursing large litters. Journal of Animal Science 81: 594-603.

Eissen, J.J., Kanis, E. and Kemp, B., 2000. Sow factors affecting voluntary feed intake during lactation. Livestock Production Science 64: 147-165.

Etienne, M., Duée, P.H., and Pastuszewska, B., 1975. Nitrogen balance in lactating sows fed on diets containing soybean oil meal or horsebean (*Vicia faba*) as a protein concentrate. Livestock Production Science 2: 147-156.

Friendship, R.M. and O'Sullivan, T.L., 2015. Sow health. Chapter 18. In: Farmer, C. (ed.) The gestating and lactating sow. Wageningen Academic Publishers, Wageningen, the Netherlands, pp. 409-421.

Ganesan, V., Rosentrater, K.A. and Muthukumarappan, K., 2005. Flowability and handling characteristics of bulk solids and powders – a review. American Society of Agricultural and Biological Engineers Paper No. 056023.

Hill, G.M., Link J.E., Rincker, M.J., Kirkpatrick, D.L., Gibson, M.L. and Karges, K., 2008. Utilization of distillers dried grains with solubles and phytase in sow lactation diets to meet the phosphorus requirement of the sow and reduce fecal phosphorus concentration. Journal of Animal Science 86: 112-118.

Hoge, M.D. and Bates, R.O., 2011. Developmental factors that influence sow longevity. Journal of Animal Science 89: 1238-1245.

Holt, J.P., Johnston, L.J., Baidoo, S.K. and Shurson, G.C., 2006. Effects of a high-fiber diet and frequent feeding on behavior, reproductive performance, and nutrient digestibility in gestating sows. Journal of Animal Science 84: 946-955.

Honeyman M.S. and D.R. Zimmerman., 1990. Long-term effect of corn gluten feed on the reproductive performance and weight of gestating sows. Journal of Animal Science 68: 1329-1336.

Hovgaard, L. and Brondsted, H., 1996. Current applications of polysaccharides in colon targeting. Critical Review in Therapeutic Drug Carrier Systems 13: 185-223.

Hughes, P.E., Smits, R.J., Xie, Y. and Kirkwood, R.N., 2010. Relationships among gilt and sow live weight, P2 backfat depth, and culling rates. Journal of Swine Health and Production 18: 301-305.

Johnston, L.J., 1992. Maximizing feed intake of lactating sows. Compendium on Continuing Education for the Practicing Veterinarian 15: 133.

Johnston, L.J., Bennett, C., Smits, R.J. and Shaw, K., 2007. Identifying the relationship of gilt rearing characteristics to lifetime sow productivity. In: Paterson, J.E. and Barker, J.A. (eds.) Manipulating Pig Production XI. Australasian Pig Science Association, pp. 39.

Johnston, L.J., Fogwell, R.L., Weldon, W.C., Ames, N.K., Ullrey, D.E. and Miller, E.R., 1989. Relationship between body fat and postweaning interval to estrus in primiparous sows. Journal of Animal Science 67: 943-950.

Johnston, S.L., Hancock, J.D., Hines, R.H., Behnke, K.C., Kennedy, G.A., Maloney, C.A., Traylor, S.L. and Sorrell, S.P., 1998. Effects of expander conditioning of corn- and sorghum-based diets on pellet quality and performance in finishing pigs and lactating sows. Kansas State University Swine Day Report, pp. 213-225.

Kim, S.W., Hurley, W.L., Wu, G. and Ji, F., 2009. Ideal amino acid balance for sows during gestation and lactation. Journal of Animal Science 87: E123-E132.

King, R.H., Eason, P.E., Kerton, D.K. and Dunshea, F.R., 2001. Evaluation of solvent-extracted canola meal for growing pigs and lactating sows. Australian Journal of Agricultural Research 52: 1033-1041.

Knauer, M., Stalder, K. J., Serenius, T., Baas, T.J., Berger, P.J., Karriker, L., Goodwin, R.N., Johnson, R.K, Mabry, J.W., Miller, R.K., Robison, O.W. and Tokach, M.D., 2010. Factors associated with sow stayability in 6 genotypes. Journal of Animal Science 88: 3486-3492.

Koketsu, Y., Dial, G.D., Pettigrew, J.E. and King, V.L., 1996a. Feed intake pattern during lactation and subsequent reproductive performance of sows. Journal of Animal Science 74: 2875-2884.

Koketsu, Y., Dial, G.D., Pettigrew, J.E., Marsh, W.E. and King, V.L., 1996c. Characterization of feed intake patterns during lactation in commercial swine herds. Journal of Animal Science 74: 1202-1210.

Koketsu, Y., Dial, G.D., Pettigrew, J.E., Marsh, W.E. and King, V.L., 1996b. Influence of imposed feed intake patterns during lactation on reproductive performance and on circulating levels of glucose, insulin, and luteinizing hormone in primiparous sows. Journal of Animal Science 74: 1036-1046.

Kongsted, A.G., 2006. Relation between reproduction performance and indicators of feed intake, fear and social stress in commercial herds with group-housed non-lactating sows. Livestock Science 101: 46-56.

Lawrence, A.B. and Terlouw, E.M.C., 1993. A review of behavioral factors involved in the development and continued performance of stereotypic behaviors in pigs. Journal of Animal Science 71: 2815-2825.

Metges, C.C., 2000. Contribution of microbial amino acids to amino acid homeostasis of the host. American Journal of Nutrition 130: 1857S-1864S.

Meunier-Salaün, M.C. and Bolhuis, J.E., 2015. High-Fibre feeding in gestation. Chapter 5. In: Farmer, C. (ed.) The gestating and lactating sow. Wageningen Academic Publishers, Wageningen, the Netherlands, pp. 95-116.

Moughan, P.J., 1999. Protein metabolism in the growing pig. In: Kyriazakis, I. (ed.) Quantitative biology of the pig. CABI, Wallingford, UK, pp. 299-331.

Moughan, P.J. and Stevens, B.R., 2012. Digestion and Absorption of Protein. In: Stipanuk, M.H. and Caudill, M.A. (eds.) Biochemical, physiological and molecular aspects of human nutrition. Elsevier, St Louis, MO, USA, pp. 162-178.

National Research Council (NRC), 2012. Nutrient requirements of swine, 11th revised edition. The National Academy Press, Washington, DC, USA.

Nikkila, M.T., Stalder, K.J., Mote, B.E., Rothschild, M.K., Gunsett, F.C., Johnson, A.K., Karriker, L.A., Boggess, M.V. and Serenius, T.V., 2013. Genetic associations for gilt growth, compositional, and structural soundness traits with sow longevity and lifetime reproductive performance. Journal of Animal Science 91: 1570-1579.

Noblet, J., Close, W.H., Heavens, R.P. and Brown, D., 1985. Studies on the energy metabolism of the pregnant sow. 1. Uterus and mammary tissue development. British Journal of Nutrition 53: 251-265.

O'Grady, J.F., Lynch, P.B. and Kearney, P.A., 1985. Voluntary feed intake by lactating sows. Livestock Production Science 12: 355-365.

Patience, J.F., Thacker, P.A. and De Lange, C.F.M., 1995. Swine Nutrition Guide, 2nd edition. Prairie Swine Centre Inc., Saskatoon, Canada.

Payne, R.L. and Zijlstra, R.T., 2007. A guide to application of net energy in swine feed formulation. Advances in Pork Production 18: 159-165.

Peltoniemi, O.A.T. and Oliviero, C., 2015. Housing, management and environment during farrowing and early lactation. Chapter 10. In: Farmer, C. (ed.) The gestating and lactating sow. Wageningen Academic Publishers, Wageningen, the Netherlands, pp. 231-252.

Peng, J.J., Somes, S.A. and Rozeboom, D.W., 2007. Effect of system of feeding and watering on performance of lactating sows. Journal of Animal Science 85: 853-860.

Peterson, B.A., Ellis, M., Wolter, B.F. and Willams, N., 2004. Effects of lactation feeding strategy on gilt and litter performance. Journal of Animal Science 82(1): 35.

Pettigrew, J.E., Miller, K.P., Moser, R.L. and Cornelius, S.G., 1985. Feed intake of lactating sows as affected by fineness of corn grind. University of Minnesota Swine Research Report AG-BU-2300, pp. 54-55.

Pettigrew, J.E., Soltwedel, K.T., Miguel, J.C. and Palacios, M.F., 2008. Fact sheet – soybean use – swine. Soybean Meal Information Center, Ankeny, IA, USA. Available at: www.soymeal.org/.

Rozeboom, D.W., 2015. Conditioning of the gilt for optimal reproductive performance. Chapter 1. In: Farmer, C. (ed.) The gestating and lactating sow. Wageningen Academic Publishers, Wageningen, the Netherlands, pp. 13-26.

Sales, J. and Janssens, G.P.J., 2003. The use of markers to determine energy metabolizability and nutrient digestibility in avian species. World Poultry Science Journal 59: 314-327.

Sauer, F.D., Erfle, J.D. and Mahadevan, S., 1975. Amino acid biosynthesis in mixed rumen cultures. Biochemistry Journal 50: 357-372.

Schaafsma, G., 2005. The protein digestibility-corrected amino acid score (PDCAAS) – a concept for describing protein quality in foods and food ingredients: a critical review. Journal of Association of Official Analytical Chemist International 88: 988-994.

Schneider, J.D., Tokach, M.D., Dritz, S.S., Nelssen, J.L., DeRouchey, J.M. and Goodband, R.D., 2007. Effects of feeding schedule on body condition, aggressiveness, and reproductive failure in group-housed sows. Journal of Animal Science 85: 3462-3469.

Schoenherr, W.D., Stahly, T.S. and Cromwell, G.L., 1989. The effects of dietary fat or fiber addition on yield and composition of milk from sows housed in a warm or hot environment. Journal of Animal Science 67: 482-495.

Sinclair, A.G., Bland, V.C. and Edwards, S.A., 2001. The influence of gestation feeding strategy on body composition of gilts at farrowing and response to dietary protein in a modified lactation. Journal of Animal Science 79:2397-2405.

Spoolder, H.A.M. and Vermeer, H.M., 2015. Gestation group housing of sows. Chapter 3. In: Farmer, C. (ed.) The gestating and lactating sow. Wageningen Academic Publishers, Wageningen, the Netherlands, pp. 47-71.

Stalder, K.J., Lacy, R.C., Cross, T.L. and Conatser, G.E., 2003. Financial impact of average parity of culled females in a breed-to-wean swine operation using replacement gilt net present value analysis. Journal of Swine Health and Production 11: 69-74.

Stein, H., Fuller, M., Moughan, P.J., Seve, B. and De Lange, C.F.M., 2007. Amino acid availability and digestibility in pig feed ingredients: terminology and application. Journal of Animal Science 85: 172-180.

Stein, H.H., 2007. Distillers dried grains with solubles (DDGS) in diets fed to swine. University of Illinois Swine Focus-001: 1-8.

Stein, H.H., Benzoni, G., Bohlke, R.A. and Peters, D.N., 2004. Assessment of the feeding value of South Dakota-grown field peas (*Pisum sativum* L.) for growing pigs. Journal of Animal Science 82: 2568-2578.

Stein, H.H. and Bohlke, R.A., 2007. The effects of thermal treatment of field peas (*Pisum sativum* L.) on nutrient and energy digestibility by growing pigs. Journal of Animal Science 85: 1424-1431.

Stein, H.H. and Shurson, G.C., 2009. The use and application of distillers dried grains with solubles in swine diets. Journal of Animal Science 87: 1292-1303.

Tanksley, T.D., 1990. Cottonseed meal. In: Thacker, P.A. and Kirkwood, R.N. (eds.) Nontraditional feed sources for use in swine production. Butterworth Publishers, Stoneham, MA, USA. pp. 139-152.

Taylor, I., 1990. Design of the sow feeder: a systems approach. University of Illinois, Urbana-Champaign, Il, USA.

Thaler, B. and Stein, H., 2003. Using South Dakota grown field peas in swine diets. Cooperative Extension Service, South Dakota State University, Animal and Range Science, Extension Extra 2041.

Tokach, M.D., Pettigrew, J.E., Crooker, B.A., Dial, G.D. and Sower, A.F., 1992. Quantitative influence of lysine and energy intake on yield of milk components in the primiparous sow. Journal of Animal Science 70: 1864-1872.

Torrallardona, D., Harris, C.I. and Fuller, M.F., 2003. Lysine synthesized by the gastrointestinal microflora of pigs is absorbed, mostly in the small intestine. American Journal of Physiology 284: E1177-E1180.

Trottier, N.L. and Johnston, L.J., 2001. Feeding gilts during development and sows during gestation and lactation. In: Lewis, A.J. and Southern, L.I. (eds.) Swine Nutrition, 2nd edition. CRC Press, Washington, DC, USA. pp. 725-770.

Ullrey, D.E., Sprague, J.I., Becker, D.E. and Miller, E.R., 1965. Growth of the swine fetus. Journal of Animal Science 24: 711-717.

Van Milgen, J., Noblet, J., Valancogne, A., Dubois, S. and Dourmad, J.Y., 2008. InraPorc: a model and decision support tool for the nutrition of growing pigs. Animal Feed Science and Technology 143: 387-405.

Van Riet, M.M.J., Millet, S., Aluwe, M. and Janssens, G.P.J., 2013. Impact of nutrition on lameness and claw health in sows. Livestock Science 156: 24-35.

Weldon, W.C., Lewis, A.J., Louis, G.F., Kovar, J.L., Giesemann, M.A. and Miller, P.S., 1994. Postpartum hypophagia in primiparous sows: I. effects of gestation feeding level on feed intake, feeding behavior, and plasma metabolite concentrations during lactation. Journal of Animal Science 72: 387-394.

Wondra, K.J., Hancock, J.D., Kennedy, G.A., Hines, R.H. and Behnke, K.C., 1995. Reducing particle size of corn in lactation diets from 1,200 to 400 micrometers improves sow and litter performance. Journal of Animal Science 73: 421-426.

Yang, H., Foxcroft, G.R., Pettigrew, J.E., Johnston, L.J., Shurson, G.C., Costa, A.N. and Zak, L.J., 2000. Impact of dietary lysine intake during lactation on follicular development and oocyte maturation after weaning in primiparous sows. Journal of Animal Science 78: 993-1000.

Yoder, C.L., Schwab, C.R., Fix, J.S., Stalder, K.J., Dixon, P.M., Duttlinger, V.M. and Baas, T.J., 2013. Estimation of deviations from predicted lactation feed intake and the effect on reproductive performance. Livestock Science 154: 184-192.

Young, M.G., Tokach, M.D., Aherne, F.X., Main, R.G., Dritz, S.S., Goodband, R.D. and Nelssen, J.L., 2004. Comparison of three methods of feeding sows in gestation and the subsequent effects on lactation performance. Journal of Animal Science 82: 3058-3070.

Zak, L.J., Cosgrove, J.R., Aherne, F.X. and Foxcroft, G.R., 1997a. Pattern of feed intake and associated metabolic and endocrine changes differentially affect postweaning fertility in primiparous lactating sows. Journal of Animal Science 75: 208-216.

Zak, L.J., Xu, X.D., Hardin, R.T. and Foxcroft, G.R., 1997b. Impact of different patterns of feed intake during lactation in the primiparous sow on follicular development and oocyte maturation. Journal of Reproduction and Fertility 110: 99-106.

7. 围产期母猪饲养管理

P.K. Theil

Department of Animal Science, Aarhus University, Research Centre Foulum, Blichers Álle 20, P.O. Box 50, 8830 Tjele, Denmark; peter.theil@agrsci.dk

摘要：从妊娠晚期到哺乳早期的围产期时间相当短,但是对于高产母猪的生产力具有重要意义。围产期,定义为妊娠的最后10天和哺乳期的前10天,这期间母猪经历了重大变化。具体来说,妊娠晚期胎儿生长、乳腺发育、初乳产生和母猪维持都需要大量的营养物质。分娩后,营养物质主要用于乳汁合成和母猪维持,但子宫恢复的过程也在向血液回供大量的氨基酸。母猪的生理要确保营养物质向后代的转移,并且营养物质优先在分娩前分配到子宫组织,并在分娩后分配到乳房。在围产期,以当前的饲养模式下,母猪会由于泌乳的高度优先性而转变为分解代谢。事实上,对于大多数母猪来说,饲料从妊娠日粮改变为哺乳日粮,饲料供应也通常从限制供应到自由采食。此外,围产母猪常常面临猪舍环境的变化,在欧洲,这种变化与现在正从松散的群养向单栏饲养的转变相关。在分娩前后,初乳已开始分泌,并且乳腺中已开始乳的合成。在泌乳开始后,特别是在最初几天,乳成分发生变化,并且在整个围产期中产奶量持续增加,成为营养需要最重要的决定因素。因此,围产母猪的营养需要受许多内在因素的影响,这些需要迅速变化,但母猪饲养实践并未很好回应这些变化,因此开发专门适应围产母猪的新饲养策略,可能对满足母猪快速变化的营养需要很重要。

关键词:初乳合成,乳腺分泌,营养需要,围产母猪,蛋白质

7.1 引言

过去,在妊娠期和哺乳期通常只给母猪饲喂一种日粮,这在一些蛋白质饲料相当便宜且饲养密度低的国家仍然是普遍的饲养方式。然而,现在母猪在大部分(或整个)妊娠期被饲喂能量和蛋白质低的日粮,但在围产期改为饲喂能量和蛋白质较高的哺乳期日粮。围产期,定义为妊娠期的最后10天和哺乳期的前10天,这期间母猪经历了重大变化。

对待进入围产期的母猪,有很多显著不同的饲养制度。事实上,许多养殖者、

养殖顾问和饲料专家仅仅基于试验和误差,而不是科学知识开发了多种饲养制度。理想状态下,在围产期的母猪饲养应该适应每个个体,同时考虑生理阶段(即妊娠天数或哺乳天数)、仔猪活重和生产水平(例如产奶量),以满足其对于营养物质的需要。然而,在实践中,饲养母猪似乎更注重人力最小化和规避可能的错误,而不是母猪生产力最大化,并且在大多数国家,母猪饲养时并不注意诸如母猪的胎次、体重、繁殖阶段和生产力水平。

能量、蛋白质和必需氨基酸的营养需要在围产期快速变化。然而,在大多数养殖场的饲养方式仅为每个母猪栏中投放一种日粮,因此从妊娠日粮转为哺乳日粮,通常正好发生在分娩前不久将母猪转移到产房的时刻。因此,妊娠和哺乳日粮的营养物质含量,需要在胎儿生长、乳腺发育、初乳生产、产奶、母猪维持、母猪体增重和身体动员之间互相平衡而达到最佳。本章将重点关注实际母猪饲养的简单性和母猪生理学的复杂性之间的矛盾,以及确保最佳母猪表现所需的营养物质的快速变化之间的差异。本章将揭示在饲养围产母猪时哪些是要重点考虑的特征,也将概述当今应如何饲养高产母猪,以及应根据析因法计算能量、赖氨酸和氮的需要,从而获得最优饲养方案。

7.2 围产期的重要性

母猪围产期尽管持续时间短,但其非常重要,主要是因为断奶仔猪的数量是母猪生产力的主要决定因素,而仔猪的死亡大多发生在分娩后的前3天(Rootwelt等,2013)。在围产期发生了许多重要的、剧烈的与繁殖产出相关的生理变化。这些特征是(或可能潜在地)受母猪营养的影响。以下将介绍已知或预期将影响母猪营养的最重要的特征。

7.2.1 新生仔猪死亡和新生仔猪能量供应

新生仔猪死亡包括分娩之前和期间分娩仔猪的损失(死胎)及产后活仔猪的死亡。这两种死亡都是复杂的问题,有多种因素,并都涉及许多深层原因。死胎至少部分与分娩过程相关(Oliviero等,2010)。大多数活产仔猪的死亡发生在产后前3天(Rootwelt等,2013),尽管仔猪健康和缺乏免疫也可能发挥作用(Pedersen等,2010),但新生仔猪死亡在很大程度上是由能量供应不足导致(Quesnel等,2012;Theil等,2014a)。刚出生后,新生仔猪依赖于来自肝脏和肌肉储存的糖原氧化,以及来自摄入的初乳中营养物质的氧化来获取能量。在初乳期结束后不久的第2天,母猪就开始大量产奶(Hartmann等,1984a)。然而,从第1头仔猪出生到开始哺乳,时间为23~39 h(Vadmand等,未发表的数据)。产奶发生时间的这种差异对新生仔猪具有巨大的影响,因为在分娩15 h后,几乎不可获得母猪初乳,从而导致仔猪能量摄入下降(Theil等,2014a),此时的仔猪将极易因饥饿而致死(Baxter

等,2013;Quesnel 等,2012)。Jean 和 Chiang(1999)报道,出生时体重小于 1100 g 的仔猪,当母猪补充了 10%的大豆油后,其出生第 3 天的存活率为 48%,而如果补充的是 10%的椰子油和中链脂肪酸,则存活率分别提高到 80%和 98%。新生仔猪死亡是一个复杂的问题,详见第 11 章(Edwards 和 Baxter,2015)。本章将仅关注与新生仔猪死亡有关的潜在生物学特征,并介绍母猪营养对其的影响。这些潜在的生物学特征包括胎儿生长、窝产仔数、糖原储备、胎盘和子宫组织的生长、乳腺发育、分娩过程、初乳合成、泌乳开始时间、乳合成、生产相关疾病、母猪产热、生理适应、体况和身体动员。其中许多特征在本书其他章节有更详细的描述,在这里只涉及在围产期母猪的营养。

7.2.2 胎儿生长、仔猪初生重和窝产仔数

胎儿生长在整个妊娠期呈指数增长,并且主要发生在分娩前的最后 10 天。事实上,胎儿 1/3 的体重增加几乎是在这么短的时间内完成(Noblet 等,1985)。胎儿生长(和其后的出生重)对于哺乳期和断奶后的仔猪存活和生长性能都很重要(Akdag 等,2009;Cabrera 等,2012)。在母猪营养方面,高胎儿生长率增加了妊娠晚期的蛋白质和氨基酸需要。通常营养物质优先于后代(Theil 等,2012),如果母体饲料供应不足,或者如果饲料组成对于妊娠晚期母猪不是最佳的,则将动员身体脂肪和蛋白质库以确保胎儿和其他生殖组织的生长。与此相应,增加采食量或增加能量供应,对妊娠晚期的母猪通常没有,或只是对仔猪出生重有轻微效果(Campos 等,2012)。然而,L-肉碱似乎能够在妊娠期刺激胎儿生长,其作用模式被认为是通过增加胎盘生长,导致更多的营养转移给仔猪(Eder,2009)。最近经遗传改良提高了现代高产母猪品系的窝产仔数(Baxter 等,2013;Pedersen 等,2010),而增加窝产仔数可能降低了仔猪平均出生重,但还是增加了母体对胎儿生长和其他繁殖组织生长的营养物质供应(Andersen 等,2011)。从营养学角度来看,最好应根据实际的窝产仔数,来为围产期母猪提供更优的营养供应。然而,从实践的角度来看,这是不可能的,因为在分娩前窝产仔数是未知的。但在分娩后,最终窝产仔数通常等于哺乳第 1 天或第 2 天的活仔数,可用此指标来代表实际窝产仔数。

7.2.3 新生仔猪的糖原储备

新生仔猪非常容易因能量供应不足而死亡(Theil 等,2014a)。在初乳期,新生仔猪通过动员来自肝和肌肉的大量糖原(Pastorelli 等,2009;Theil 等,2011),以应对子宫外的低环境温度,并且由于仔猪的高体表面积与体积比率,而导致非常大的热损失。糖原在妊娠晚期储存在胎儿中(不仅在最后 10 天),并且这些储备的主要目的确实是为新生仔猪提供能量。在胎儿生命期间,增加仔猪存活率的遗传选择似乎增加了肝脏和肌肉组织中的糖原沉积(Leenhouwers 等,2002)。保留在一窝

17头新生仔猪中的糖原总计约为 1 kg(Theil 等,2011),该量大致相当于 2 kg 标准母猪妊娠饲料中的淀粉含量。由于糖原储备库在分娩前的最后 2~4 周期间积累(Père,2003),并且妊娠期日粮中淀粉非常丰富,新生仔猪的糖原储备从营养数量角度看可能并不缺乏。然而,从福利和经济的角度来看,确实有必要研究是否可以通过改变母猪营养来改善糖原储备。Seerley 等(1974)报道了当母猪在妊娠第 109 天开始以玉米淀粉额外补充能量时,新生仔猪肝脏中糖原浓度较高。根据 Jean 和 Chiang(1999)报道,与大豆油相比,从妊娠第 84 天直到分娩饲喂 10%中链脂肪酸或 10%椰子油,可提高出生后 4 h 仔猪的肝糖原含量。然而,Boyd 等(1978)报道,当母猪从妊娠的第 100 天直到分娩额外添加玉米淀粉或牛油时,新生仔猪的肝糖原浓度并未提高。同样,Newcomb 等(1991),在母猪妊娠期的最后 14 天,饲喂淀粉、大豆油和中链甘油三酯,均未对新生仔猪的肝糖原含量产生影响。

7.2.4 胎盘、子宫、羊水和胎膜

胎盘和子宫也在妊娠晚期以指数增长,因此,在妊娠的最后 10 天营养物质沉积量很大。相反的,羊水体积在妊娠的第 80 天左右达到峰值(Noblet 等,1985),并且胎膜增长相当缓慢,因此羊水和胎膜在妊娠晚期的营养物质需要量可以忽略。胎盘和子宫(肌肉增长)增长所需的能量相当低,可以忽略,但是它们的肌肉增长所需的氨基酸的量是相当大的(Noblet 等,1985)。在分娩过程中,胎盘、羊水和胎膜被母猪排出,因此,保留在这些繁殖组织中的营养物质也随之从母猪丢失。相反,子宫恢复会在分娩后的第一周开始,这一过程中释放了可用于泌乳的营养物质。从子宫恢复回供给母猪循环的能量很少并且可以忽略,但是大量的蛋白质和必需氨基酸被释放回血液中,因此围产母猪营养最好考虑到这一点。目前尚不清楚母猪营养是否影响胎盘和子宫组织的增长和复旧,但因为这些生理过程与后代存活相关,而且是高度优先的,因此推测营养的影响可能相当小。

7.2.5 乳腺发育

大部分乳腺发育发生在妊娠期的最后 1/3,但分娩前 10 天乳房仍然相当小。在妊娠晚期乳腺发育速度的模式尚不清楚,但毫无疑问,通过目测妊娠晚期母猪的乳房变化,就可知在妊娠期的最后 10 天里乳腺发育的速度在加快。乳腺持续发育到哺乳期第 10 天左右(Kim 等,1999a),但在分娩后的发育速度比分娩前慢(Noblet 等,1985),并且对营养需要产生影响。此外,如果乳腺由于某些原因在分娩后没有被吮吸,这些乳腺将在分娩后的第一周内退化(Kim 等,2001;Theil 等,2006),此时,大量的氨基酸和少量能量将被再循环到血液中。然而,在现代高产母猪中,由于有大量的哺乳仔猪,非哺乳乳腺的数量较少,因此常常忽略了来自复旧乳腺的氨基酸动员。从生产力的角度来看,若能洞悉乳腺发育是否可以通过营养手段来增强,将很有意义。这方面在第 4 章(Farmer 和 Hurley,2015)中有所阐述,但很显然,该章也缺乏关于围产期母猪营养对乳腺发育影响方面的信息。

7.2.6 分娩过程:持续时间、胎间隔、死胎率和仔猪活力

最佳母猪饲养管理应以缩短分娩持续时间、降低死胎率和提高仔猪出生活力为目标,然而这些却没有受到很多的科学关注。显然,分娩前最后几天的饲养管理对于分娩过程是重要的,因此,改变营养状况可能是增加母猪生产力(例如降低死胎率)的一条途径。2007—2013 年在丹麦 Foulum 研究中心进行的 5 项研究数据显示,126 头母猪平均分娩持续时间为(343±15)min(Vadmand 等,未发表的数据)。针对每头母猪,分娩时间长度为 87~935 min,这意味着最快的母猪只花了 1.5 h 分娩,而最慢的要花 15.5 h。所有母猪均产下超过 10 头仔猪,平均为 16.1 头,而每窝有 1.1 头为死胎。分娩持续时间与仔猪出生总数并不相关,但分娩持续时间越长,死胎率越高(图 7.1)。然而,尚不清楚是死胎的存在增加了分娩持续时间,还是延长的分娩时间增加了死胎的发生率。在上述研究中的产仔间隔平均为(20±1)min,但从个体来看,生产最快的母猪每 7 min 产生一头仔猪,而最慢的母猪每 52 min 产生一头仔猪。

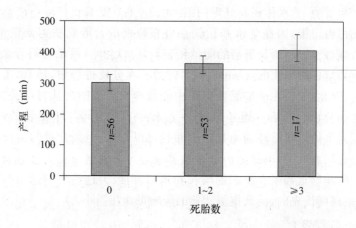

图 7.1 母猪产 0、1~2、3 头或更多死胎对应的产程。

饲养可以通过几种方式对分娩过程起作用。在分娩前使用包含纤维的日粮可能会减轻便秘,否则便秘可能会阻碍胎儿通过产道排出。饲养策略对围产期便秘的影响将在第 10 章(Peltoniemi 和 Oliviero,2015)中讨论。Oliviero 等(2010)发现低便秘发生率与较短的分娩时间有关。日粮纤维的另一个益处是可以从胃肠道吸收更多和更持续的能量(Serena 等,2009)。事实上,即使在最后一次采食的 24 h 后,母猪也能从胃肠道大量纤维中吸收能量,这对于分娩期间的母猪来说可能是特别重要的,因为它们在分娩前和分娩期间的采食量可能会被调低。分娩时过度肥胖也会延长分娩时间(Oliviero 等,2010),这表明母猪在围产期的营养状况可能对分娩过程起作用。日粮矿物质的供应也可能是重要的,并且已证实钙对肌肉收缩是必需的。母猪分娩时可能需要适当的钙,以使子宫周围的肌肉进行适当收缩,从

而将胎儿向前推动经产道排出。Vallet 等（2014）最近的研究显示，在妊娠晚期通过添加适宜量的锌，可以降低死胎率，尽管锌的作用方式目前仍是未知的。分娩过程中的并发症可能导致死胎率增加，通过营养手段降低死胎率的研究方向非常值得进一步关注。Coffey 等（1987）报道，在妊娠晚期母猪日粮中添加 10% 的脂肪降低了死胎率，但是应当强调的是，他们的结论是建立在每个日粮处理仅 8 头母猪的基础上，这对研究死胎来说是非常低的重复数。另一项围绕母猪分娩前后营养有积极影响的研究结果是，营养可能有增强仔猪活力的潜力，尽管这将更有利于改善活产仔猪的生存，而不是降低死胎率（Baxter 等，2008）。

7.2.7 初乳合成

母猪在分娩前开始合成初乳，尽管不确切知道在什么时候发生。在妊娠期第 80 天左右，可在母猪血浆中检测到第一种已知的初乳成分（β-乳球蛋白）（Dodd 等，1994），而乳猪血浆中 α-乳白蛋白（催化乳糖合成的酶复合物的一部分）增加，是在妊娠期的最后一周。与此一致，Hartmann 等（1984b）报道，母猪血浆中乳糖含量在分娩前 4 天开始显著增加，这些研究表明，初乳乳糖主要在妊娠的最后几天产生。与之相反，目前尚不知道初乳中脂肪何时产生，也不知免疫球蛋白和生长因子何时合成。从营养学角度来看，生产乳糖、通过初乳分泌产生的能量并不是母体的主要投入，但是就蛋白质和氨基酸而言，初乳生产显然是一项重大投入。最近发现，实际初乳产量比以往报道的平均要高 30%（Theil 等，2014b），因此，就蛋白质和氨基酸而言，母体的投入甚至比以前认为的更大。据报道，母猪的初乳产量平均为（5.9±0.1）kg（n = 126 头母猪），但以个体计，其范围为 2.7～8.5 kg（Theil 等，2014b；Vadmand 等，未发表的数据）。目前认为大多数的泌乳启动发生在妊娠的最后 7～10 天（Theil 等，2014a）。然而，当初乳由仔猪吸出后，还可能在分娩开始后产生初乳。妊娠晚期的营养状况会影响母猪的初乳产量，关于饲料组成对初乳产量影响的更多细节可见第 8 章（Quesnel 等，2015）和最近的一篇综述（Thei 等，2014a）。要重点强调的是，初乳生产似乎不会对母猪的能量需要有很大的影响，因为它是在长时间里生产，而初乳产量会对母猪蛋白质和氨基酸平衡（在本文后面讨论）产生较大影响。除了初乳的数量之外，初乳质量对于仔猪可能也是很重要的，因为母猪的生产力，是要由初乳的体积和成分，来向仔猪传送营养物质和生物活性成分。初乳含有乳糖和脂肪，具有向新生仔猪供应能量的重要作用。此外，初乳含有非常丰富的蛋白质，主要是由于高浓度的免疫球蛋白构成，但也有生长因子存在（Hurley 和 Theil，2011）。有关初乳组成的更详细介绍，请参阅第 9 章（Hurley，2015）。母猪营养可以增加脂肪的浓度，或者可能改变免疫球蛋白和生长因子的浓度，而可能不会改变乳糖浓度，因为乳糖会通过渗透作用将水吸入乳腺腔。母猪营养对初乳数量的影响将在第 8 章中详细描述（Quesnel 等，2015）。从营养学角度来看，初乳的产量比其组成更重要，因为在各个母猪之间，初乳产量可能有 3 倍的差异,而初乳蛋白质浓度变化的程度要小得多(Theil 等,2014b;Vadmand 等,未发表数据)。

7.2.8 泌乳开始时间

在分娩后的第一个 48 h,乳房分泌物(初乳和常乳)数量的变化相当大。在第 1 头仔猪出生后的最初 12～15 h 内,初乳可大量获得,但随后分泌初乳的量将下降 (Krogh 等,2012)。此时乳腺分泌物的量较少,直到第 1 头仔猪出生后平均 33～34 h 的时候,大量乳开始分泌(Krogh 等,2012)。尚不清楚为什么初乳在第 1 头仔猪出生的 15 h 后变得几乎不可得,可能是与当分娩停止时,母猪血浆中催产素浓度的下降相关。有趣的是,我们发现一些母猪的首次泌乳开始时间,可早在产后 23 h,而有的母猪,要晚到产后 39 h。目前,我们只能推测是什么原因导致了母猪之间的差异。关于这种差异,我们已经考查了首次泌乳时间是否会与围产期母猪的饲料组成或饲料采食量,母猪分娩时的体重或背膘厚,分娩持续时间或初乳期时的窝产仔数这些指标有关,但都不能给出解释(Vadmand 等,未发表的数据)。因此,母猪之间哺乳启动时间的差异原因尚待阐明。这也许与同窝仔猪建立奶头偏好性有关,仔猪们会协调一致行动,成功地按摩乳房和诱导乳汁排出。另外有趣的是,哺乳早期的泌乳启动与哺乳期第 1、2、3 和 4 周更高的产奶量有关(Vadmand 等,未发表的数据)。泌乳启动对母猪的生理有很大的影响,因为一旦泌乳开始,能量和蛋白质的平衡会突然打破(Hansen 等,2012b,2014)。

7.2.9 乳合成

泌乳是哺乳母猪的主要投入,在哺乳高峰期,一半来自饲料的氮(52%)和能量(50%)通过乳汁转移到仔猪(Theil 等,2004)。因此,乳汁的合成能力是对每日母猪营养需要具有最大影响的繁殖性状。乳汁合成的营养需要取决于乳汁的产量和组成,并且这两个特征都受哺乳阶段的影响(Theil 等,2002,2004;Hansen 等,2012b),然而,无疑产奶量对营养需要的影响最大。

最近有研究者使用数学模型来描述母猪的泌乳曲线(Hansen 等,2012b)。所开发的数学模型基于仔猪窝增重和窝产仔数来预测母猪产奶量,并且通过这些简单的输入,可以估算哺乳期每天的产奶量。该数学模型参考了自 1980—2012 年所有公布的,而且在哺乳期至少测量了两次的乳汁产量数据。该哺乳曲线的特点在于,从泌乳开始(在第 2 天时的泌乳量约为 5.7 kg)到围产期结束(泌乳的第 10 天)的产奶量不断增加,平均值为 11～14 kg/天。围产期后,泌乳量在第 17～19 天达到峰值,这取决于窝产仔数和窝增重(Hansen 等,2012b),最好的母猪产量为 15～17 kg/天。作为乳腺分泌活动的结果,乳腺血浆流量从分娩前第 10 天的约 3500 L/天,增加到产后第 3 天的 6000 L/天,并且在哺乳第 17 天时增加到 9500 L/天(Krogh 等,未发表)。影响母猪产奶量的各种因素见第 8 章(Quesnel 等,2015)。

以常量化学成分计,猪乳由乳糖、脂肪、蛋白质、矿物质、维生素和水组成(Theil 等,2012)。有关乳汁成分的更详细描述,请参见第 9 章(Hurley,2015)。在泌乳开始后,虽然不如初乳期那么快,但乳成分变化也相当快(Jackson 等,1995)。Klobasa 等(1987)和 Csapo 等(1996)的研究证明了在围产期乳成分发生的变化,发现乳成分在泌乳第 10 天后趋于恒定。这与 Hansen 等(2012b)开发的数学模型一致。在围产期中乳脂肪含量的增加,要么是对哺乳阶段的反应,要么是对日粮脂肪补充的反应(Lauridsen 和 Lauridsen,2004 年),在干物质的比例上也有类似的增加。相比之下,乳糖浓度在 5%~6%时相当恒定,因为乳糖会将水吸收到乳汁中。在哺乳期的第 2~4 天时,蛋白质浓度略高(6.0%~6.5%),而此后相当恒定地保持在 5.5%(Klobasa 等,1987)。伴随着在第 5 天的乳蛋白质的下降,氨基酸的相对浓度(每 100 g 蛋白质的氨基酸克数)发生改变(Csapo 等,1996)。一些氨基酸(谷氨酸、脯氨酸、异亮氨酸和赖氨酸)在第 5 天之后变得更丰富,一些氨基酸变得不太丰富(苏氨酸、丝氨酸、甘氨酸、丙氨酸、半胱氨酸、缬氨酸、蛋氨酸、亮氨酸和苯丙氨酸),还有一些氨基酸(天冬氨酸、酪氨酸、组氨酸、色氨酸和精氨酸)当以氨基酸与蛋白质比例表示时,相对不变。

7.2.10 围产期生产相关疾病

分娩前后母猪的健康对于出生后母猪总体生产力和仔猪生长性能很重要,但在分娩前后,疾病的发病率似乎会上升(Blaney 等,2000;Oliviero 等,2010)。其原因是未知的,但是可以推测,母猪的一般免疫力受到诸如免疫球蛋白损失到初乳中的影响,这使得母猪更易受到与分娩期间疾病相关风险因素的影响。分娩过程涵盖了范围广泛的疾病,包括紊乱、乳腺炎和产后泌乳障碍综合征[PPDS;定义为在第一头仔猪出生后 72 h 内,母猪中初乳和乳产量不足(Klopfenstein,2006)]受到最多的关注,因为它们对泌乳有重大影响。第 18 章(Friendship 和 O'Sullivan,2015)将对母猪健康和疾病进行概述。尽管与生产相关疾病的发病原因是多样的,并且常常没有研究透彻,但是涉及许多因素,其中母猪营养是重要的方面。例如,使用麦角菌感染的小麦(和黑麦)所生产的饲料会显著降低初乳(Blaney 等,2000)和常乳产量(Kopinski 等,2008)。在围产期每天保证纤维摄入似乎对减少分娩时便秘风险也很重要。实际上,如果纤维摄入减少,则后肠发酵下降,并且后肠和粪便中的食糜量也会相应地减少。纤维摄入量由母猪日粮的饲料摄入量和纤维含量决定。通常,在妊娠晚期纤维摄入减少,原因可能是在妊娠晚期饲料供应减少,或者由于哺乳日粮(其通常喂给妊娠晚期母猪)中的纤维含量降低,或同时由于这两个原因。相反,过高的饲养水平是发生产后泌乳障碍综合征的危险因素(Papadopoulos 等,2010),甚至可能增加母猪的死亡率(Abiven 等,1998)。此外,长时间的过量饲料供应可能导致母猪肥胖,而这将会延长分娩持续时间(Oliviero 等,2010)。据报道,与来自鱼粉和肉骨粉的蛋白质相比,在妊娠晚期日粮中添加植物蛋白质可减少母猪无乳症的发生(Göransson,1990)。同样,降低日粮能量浓度(Göransson,1989b)或在分娩前最后几天减少能量供应(Göransson,1989a),也将减少无乳症的发生率。一般来说,经历过分娩相关疾病的母猪,常常会减少初乳或常乳的合成,或两者同时减少。目前还不清楚为什么,但部分解释可能是营养优先级的显著变化(例如氨基酸可优先用于产生新的免疫球蛋白以抗击疾病,而不是用于蛋白质合成,并分泌到初乳或常乳中)。另一个可能的解释是,在分娩前后的水摄入量不足,这也可能会降低乳产量。

7.2.11 围产母猪产热

围产期母猪产热量增加主要是由于产奶量的升高。分娩时除外,维持能量需要(ME)的数量在以每单位代谢体重($W^{0.75}$)的形式表示时是恒定的。事实上,哺乳母猪(460 kJ/$W^{0.75}$)的维持能量需要高于妊娠晚期母猪(405 kJ/$W^{0.75}$;NRC 2012)。在妊娠期的最后10天,即使胎儿生长迅速,母猪的代谢体重也几乎不变,因此其维持能量需要是相当恒定的。对于体重200 kg的青年母猪,维持能量需要为21.5 MJ/天,而对于体重为300 kg的经产母猪,维持能量需要约为29.2 MJ/天。这些数据说明,母猪的体重对能量需要起着关键作用。如果使用标准母猪饲料,那么体重为200 kg的青年母猪需要1.7 kg饲料以满足维持能量需要,而体重为300 kg的成熟母猪需要2.2 kg饲料。在分娩时,母猪的体重通常会因为仔猪出生,胎盘和子宫羊水的排出而下降约20 kg,但是相关的母猪代谢体重的下降,几乎被从分娩前的405 kJ/$W^{0.75}$增加至分娩后的460 kJ/$W^{0.75}$维持能量需要所抵消。因此,维持能量需要在整个围产期也可以被认为是恒定的。然而,母猪额外产生的热量,与繁殖消耗和日粮诱导的产热有关(Noblet等,1985;Van Milgen等,1997)。日粮诱导产热对母猪额外热损失的贡献是未知的,但是在妊娠晚期母猪(妊娠的第104天,采食3.5 kg/天)中,通过减去计算的维持能量需要(Theil等,2002),可得出总额外热损失估算为7.5 MJ/天。在刚刚分娩后,额外的热损失可能非常低,因为母猪还没有产生大量的初乳或常乳。相比之下,从第2天以后,大量乳汁将开始分泌,额外的热损失每天显著增加。与乳汁合成相关的额外热损失可以如下公式估算:

$$额外的热损失 = \frac{乳汁中分泌的能量}{k} - 乳汁中分泌的能量 \quad (1)$$

其中在乳汁中分泌的能量以 MJ/天为单位,k是将代谢能转换成乳汁中能量的效率。在通过重水(也称氘化水,deuterated water,DO)稀释技术评估产奶量的研究中,k被报道为0.78(Theil等,2004)。分娩后,分泌乳汁的能量从第2天的大约26 MJ增加到第10天的45 MJ,并且与之相随,额外的产热从7.3 MJ/天增加到12.7 MJ/天。从第2天到第10天,由乳合成增加的产热量相当于大约0.5 kg饲料中的能量。这种额外热量的增加仅与乳合成产生的额外热量相关,并不包括与日粮诱导相关的额外产热,可以预期将增加所需的饲料量。

7.2.12 围产期肝脏的生理适应和作用

在围产期,母猪的能量和蛋白质平衡以及中间代谢将发生显著变化(Hansen等,2012a;Mosnier等,2010;Theil等,2002,2004,2013)。在分娩前后,母猪从合成代谢转变为分解代谢,并伴随着蛋白质平衡显著降低(Theil等,2002,2004)。此外,肝代谢也在改变,例如,在哺乳期间,门静脉血流量和动脉血浆流量分别为妊娠晚期的2倍和3倍多(Flummer等,未发表的数据)。肝脏重量仅占母猪体重的2%~4%,但在哺乳期,肝脏氧气消耗占心脏输出量的40%(Kristensen和Wu,2012)。有趣的是,随着围产期肝脏代谢活动显著增加,动脉和门静脉血液向肝脏的供应显著增加,然而,母猪的平均总产热量仅从分娩前第10天的31 MJ/天(Theil等,2002),温和地在分娩后第10天增加到37 MJ/天(Theil等,2004)。伴随着乳腺代谢活动的增加,乳汁合成需要大量的能量和营养物质。肝动脉血液的供应增加3倍,表明肝脏执行了与泌乳相关的代谢负荷的一部分。肝脏的另一个作用是维持葡萄糖稳态,并且在哺乳高峰期,肝脏动用糖原储备,来补充采食后4 h直至下一餐之间的血浆葡萄糖(Flummer等,未公开)。此外,与从胃肠道吸收的量相比时,肝脏可以非常高效地(>95%)提取丙酸盐。肝脏可将丙酸盐转化为乳酸盐,后者可以用作肌肉或乳腺组织中的能量来源(Flummer等,未发表的数据)。

低 n-6∶n-3 比例的日粮对于围产期母猪的炎症反应和分娩不久后的采食量都是有益的(Papadopolous等,2009)。这些近期研究的作者进一步指出,日粮的 n-6∶n-3 比例,与分娩前后母猪的代谢生理适应相关性更强。代谢适应可能涉及肝脏代谢的改变,因为已知日粮 n-6∶n-3 比例会影响断奶猪肝脏基因的表达(Theil和Lauridsen,2007)。在围产期通过饲料提供的营养物质对母猪的中间代谢很重要。在最近的一项研究中,母猪在直到预产期7天之前,一直饲喂低脂肪和高纤维或低纤维日粮,然后在整个围产期和哺乳期饲喂哺乳日粮。从血浆中脂肪酸的含量可以看出,在预产期7天之前饲喂高纤维日粮显著降低了发酵,并且削弱了初乳期仔猪的生产性能(Hansen等,2012a)。另一项最近的研究指出母猪的初乳产量与血浆尿素呈正相关(Loisel等,2014),由于尿素在肝脏中产生,这一观察表明肝脏代谢和氧化模式(即蛋白质,碳水化合物和脂肪的氧化)会影响围产期母猪的生产力。

7.2.13 体况和身体动员

身体状况是围产母猪的另一个重要方面。当母猪进入产房时,它们的体况不应该太高,也不能太低。如果母猪过肥,它们在分娩过程中可能会遇到问题(Oliviero等,2010),此外,肥胖母猪所产的新生仔猪死亡率可能更高(Hansen等,2012a)。然而,高体况有利于持久泌乳,也就是有利于哺乳高峰期的乳汁产量(Hansen等,2012a)。太瘦的母猪也是不理想的,因为会增加肩部损伤和延长断奶后发情间隔的可能性(King 和 Williams,1984)。

在妊娠晚期动员身体脂肪和蛋白质储备,被证明与母猪初乳产量正相关(Decaluwé 等,2014)。Loisel 等(2013)发现的初乳产量与分娩时血浆尿素呈正相关(表明了蛋白质的氧化)支持了这一观点。此外,Hansen 等(2012a)报道,分娩前最后几天的负能量平衡有益于提高泌乳期第 7~10 天的产奶量。这反证了之前的发现,即在分娩前母猪自由采食对随后的产奶量是有害的(Danielsen,2003),然而,潜在的机制尚不清楚。

7.3 现代高产母猪围产期饲养方法

7.3.1 围产期何时更换日粮

母猪通常饲喂妊娠日粮,然后是哺乳日粮,并且日粮变化通常与母猪从妊娠舍转移到产房的时间相一致。然而,在不同养殖场之间更换日粮的时间差异很大(图 7.2)。一些养殖场在分娩前大约一周更换日粮,这在生物学上是有意义的,因为胎儿与乳腺发育及初乳合成都需要大量的蛋白质和赖氨酸。其他养殖场有相同的意愿,但由于产房的空间不足,转舍被推迟到分娩前几天。这些养殖场可能经历了产后泌乳障碍综合征的流行增加,然而并不知道具体原因,是否与饲养改变、太接近分娩时的身体运动或其他原因导致的应激有关(Papadopoulos 等,2010)。然而,有些养殖者选择仅在分娩期,分娩几天后或甚至 7 天后更换日粮。使用这些策略的养殖场声称这样的操作是最佳的,因为此阶段仔猪的吮乳能力相当低,这种做法可降低哺乳早期母猪乳房的压力。这种更换日粮的时间选择,更多地是基于固执的观念而不是科学证据。这种固执的观念还表现在围产期饲喂曲线的选择上(稍后讨论)。可以推测,有一些妊娠晚期母猪饲喂了哺乳日粮,而一些哺乳早期母猪却饲喂了妊娠日粮。目前,给母猪饲喂特定的围产期日粮,或是妊娠和哺乳日粮以不同比例混合作为围产日粮的情况还很少,然而,这是一个应该考虑的,能满足围产母猪快速变化营养需要的方法。

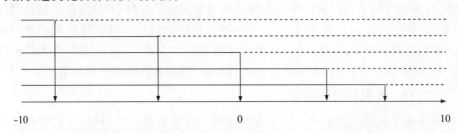

图 7.2 养殖场在围产期期间,母猪饲料从妊娠日粮(实线)到哺乳日粮(虚线)改变时使用不同的策略(箭头显示不同的策略;0 代表分娩当天)。

7.3.2 围产期母猪日粮中能量、赖氨酸和氮的含量

日粮中能量源于饲料中的碳水化合物、蛋白质和脂肪,通常这三种营养物质占饲料的87%~89%,而水与少量矿物质和维生素占剩余的11%~13%。在提供能量的饲料原料中,碳水化合物总是最为丰富,并且淀粉是母猪日粮中最主要的能量来源。通常,淀粉占母猪饲料干物质部分的50%以上。以能量(总能)计算,碳水化合物(淀粉和纤维)分别构成妊娠和哺乳日粮能量的76%和64%。相比之下,蛋白质分别占妊娠和哺乳日粮能量的16%和22%,而脂肪通常分别占8%和14%。以代谢能基础计,来自不同国家的母猪日粮中能量含量非常相似(表7.1),其范围为11.9~13.6 MJ ME/kg,而哺乳日粮的能量浓度通常比妊娠日粮高6%~10%(为13.0~14.0 MJ ME/kg)。添加更多富含纤维的原料会减少日粮代谢能含量,这通常在配制妊娠日粮时采用。相反,日粮代谢能含量可通过添加更多的脂肪而增加,并且在大多数哺乳日粮中,实际添加有3%~5%的脂肪。然而,一般来说,妊娠和哺乳日粮的能量含量变化小于20%(表7.1)。2014年春季进行的调查显示,目前在加拿大、丹麦、荷兰、法国和美国,均尚未使用特定的围产期日粮。

妊娠日粮中的氮含量约为2%,而哺乳日粮中的氮含量接近3%(表7.1)。这些值分别相当于在妊娠和哺乳日粮中的粗蛋白质约为13%和18%。日粮中氮主要来源于蛋白质饲料原料,例如豆粕、小麦和大麦(欧洲)及玉米(美国和加拿大)。哺乳日粮通常用22%~25%的豆粕配制,而妊娠日粮常含有15%~18%的豆粕。在欧洲,可能是由于粪便中的氮污染问题公众广为关注,其哺乳日粮的氮含量低于美国和加拿大。

日粮中标准回肠可消化(SID)赖氨酸含量的差异远远大于日粮中能量和蛋白质含量的差异。一般来说,哺乳日粮中的赖氨酸水平是妊娠日粮中的2倍。在不同国家之间,日粮SID赖氨酸含量变化很大,在妊娠日粮中,SID赖氨酸的含量在0.31%~0.58%,而在哺乳日粮中,其范围为0.63%~1.0%(表7.1)。日粮赖氨

表7.1 妊娠和哺乳日粮中能量、氮和赖氨酸的典型含量

	妊娠日粮			哺乳日粮		
	代谢能 (MJ/kg)[1]	氮,SID (日粮中%)[2]	赖氨酸,SID (日粮中%)[3]	代谢能 (MJ/kg)	氮,SID (日粮中%)	赖氨酸,SID (日粮中%)
加拿大	11.9	1.9	0.52	13.4	3.0	1.0
丹麦	13.0	2.2	0.31	14.0	2.9	0.63
荷兰	12.1	2.2	0.50	13.0	2.9	0.75
法国	12.1	2.2	0.50	12.9	2.6	0.89
美国	13.6	2.2	0.58	13.8	3.1	1.0

[1]代谢能,等于总能减去粪便、尿液和气体中损失的总能量。[2]标准回肠可消化氮(含量等于日粮氮的表观回肠消化率经基础内源性损失校正后的数值)。[3]标准回肠可消化赖氨酸(含量等于日粮赖氨酸的表观回肠消化率经基础内源性损失校正后的数值)。

酸通常来源于豆粕、小麦/大麦和玉米(与日粮中氮来源相似),另外,通常会将晶体赖氨酸加入哺乳日粮中,以特异性地提高赖氨酸水平。

7.3.3 围产期饲喂曲线和母猪食欲

大多数母猪在整个妊娠期间(也包括围产期),被限制采食,以避免母体肌肉和脂肪组织过度沉积,然而,这也可能会导致分娩问题和其他健康相关问题,例如分娩前后的乳腺炎、子宫炎和无乳症。在妊娠晚期,与妊娠早期和中期相比,饲料供给量通常会增加,并且通常在分娩前2~4周实施。这种做法可看作是一个简单的,为满足胎儿生长氨基酸需要增加所做的尝试,但它没有考虑到胎儿生长速度在这一阶段成指数变化(Noblet等,1985),并且胎儿在稳定地增加氮和赖氨酸沉积。直到目前,有些养殖场选择在整个妊娠期每天给母猪提供相同的饲料量,而有些养殖场使用高纤维饲料,为方便管理,允许母猪自由采食饲料。增加日粮纤维水平确实降低了一个母猪群内的平均能量摄入量(Danielsen 和 Vestergaard,2001),然而,不推荐妊娠母猪自由采食,因为随着时间的推移,其会导致母猪个体之间产生巨大的体况差异。在第5章(Meunier-Salaün 和 Bolhuis,2015)中详细描述了在妊娠期饲喂高纤维日粮对母猪的各种影响。

在妊娠晚期,大多数母猪按能量需要或超出能量需要来饲喂(图7.3)。在分娩时,母猪通常饲喂与分娩前相同的饲料量,然后从产后第3天开始,每天逐步增加饲料供给量,以满足增加泌乳的需要(Hansen等,2012b)。当在哺乳早期阶段为母猪增加饲料供应时,母猪的食欲似乎是营养摄入的限制因素,因此常常使母猪饲料摄入大大低于其能量需要。理想情况下,饲料采食量应该增加得相当快,以避免体脂和蛋白质的过度动员,但如果在哺乳早期饲料喂量增加过快,后期母猪的采食量通常会下降(Hansen,2012)。加拿大、丹麦、法国、荷兰和美国推荐的饲养水平似乎相差不大。尽管如此,在不同的养殖场可以看到饲料供应的巨大差异,并且围

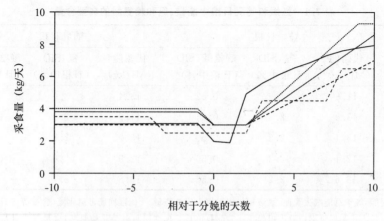

图7.3 围产期母猪需要的饲料量(实线)和通常采用的饲养曲线(点线和虚线)。

产母猪的饲养水平似乎仅仅基于当前的观念和实践的可能性,而不是营养需要。

许多养殖者试图在妊娠早期和中期恢复母猪的体况。相比之下,围产期的饲喂曲线(甚至日粮推荐量)对于所有母猪是相同的,而不用考虑胎次、体重、猪舍情况和环境条件(如温度和湿度)。这里有一个例外,是在法国的推荐标准中,经产母猪比初产母猪从妊娠第 101 天每天多饲喂 0.4 kg/天饲料,直到分娩。在荷兰,通常在冬季时多提供 0.15~0.25 kg/天饲料给妊娠母猪,但不限于围产期。

7.4 最新进展

7.4.1 使用析因法研究围产期营养需要

在妊娠晚期每天所需的营养物质可以通过析因法计算,包括对母猪维持,胎儿生长,乳腺发育,子宫组织(子宫、胎盘、羊水和胎膜),初乳生产和与这些繁殖性状相关的热损失(Feyera 和 Theil,2014;图 7.4)。从营养学角度来看,即使在妊娠后期,分娩前繁殖所需的能量需要也相当小(占摄入量的 12%),而所需蛋白质的量要高得多(占摄入量的 41%;Noblet 等,1985)。关于母猪体重和其后代体重与营养需要和动态变化的有关详细描述,请参阅本书第 6 章(Trottier 等,2015)。举例来说,妊娠母猪每千克代谢体重的代谢能需要被认为是是恒定的($405\ kJ/W^{0.75}$;NRC,2012),但众所周知,胎儿生长遵循指数增长曲线。子宫组织的动态变化也被研究得很透彻(Noblet 等,1985)。然而,对初乳生产营养需要的研究仍然缺乏,目

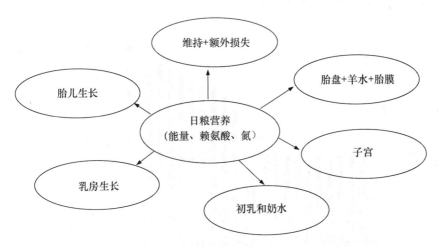

图 7.4 描述围产母猪营养需要的数学模型。对于能量,额外的损失是指与繁殖产出相关的超出维持能量的部分,对于氮,它们是指由如氨基酸不平衡等导致的额外的蛋白质氧化。需要注意的是,分娩后,胎儿生长、胎盘、羊水和胎膜将不再需要提供营养,并且子宫复旧过程中会向循环系统提供营养物质,并可用于产乳。粪便中的损失未显示。

前尚不知道初乳真正在何时产生,并且以何种速率生产(Theil 等,2014b)。为简单起见,假定初乳在妊娠的最后 10 天期间被均匀地合成。并且,尽管已知乳腺发育在分娩前比在后更快,但妊娠晚期乳腺发育的速率和营养需要都还是未知的(Kim 等,1999b;Noblet 等,1985)。对于妊娠晚期母猪,从能量的角度来看,这些未知的营养需要变化不是主要的,因为大多数能量需要来自母猪身体,而不是繁殖组织。相比之下,初乳的生产对于评估蛋白质和必需氨基酸的需要是重要的。

分娩后,需要营养物质用于与泌乳、维持、乳腺发育和与这些繁殖性状相关的热损失。子宫恢复对母猪氨基酸和蛋白质需要量具有相当大的影响。由于在文献中没有关于子宫恢复速度的信息,因此在数学模型中假设子宫恢复发生在哺乳的第 1 周。然而,初乳在妊娠最后 10 天内均匀合成,子宫恢复在泌乳的前 7 天,这些假设显然太简单,需要进一步研究以更好地评估这些过程的时间。

7.4.2 能量需要

母猪维持是妊娠后期能量需要的主要部分(图 7.5)。从妊娠的 105～115 天,母猪需要大约 39 MJ/天的代谢能(Feyera 和 Theil,2014),其中最高的比例(79%)作为热能而损失掉(30.5 MJ/天;Noblet 等,1985,Theil 等,2002)。剩余的 21% 保留在生殖组织或产物中,例如初乳(3.6 MJ/天)、胎儿生长(2.6 MJ/天)、乳腺发育(1.6 MJ/天)以及子宫、胎盘、羊水和胎膜(0.3 MJ/天)。热损失是维持、初乳生产、胎儿生长、乳腺发育和子宫组织生长所需要的。能量需要在妊娠晚期相当恒定,并且从妊娠第 105 天到分娩,由于母猪的代谢体重($W^{0.75}$)仅稍微增加,因此能量需要几乎没有增加。分娩后,能量需要有几天较低,主要是因为不再需要将营养

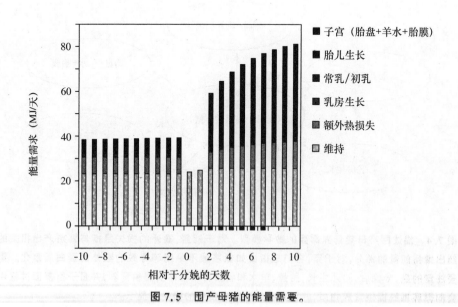

图 7.5 围产母猪的能量需要。

物质保留在胎儿、胎盘或子宫组织,并且大量的初乳和常乳很可能在分娩后的前1.5天还没有开始合成。从哺乳期的第2天开始,由于更高的产奶量(Hansen等,2012b)和与泌乳相关的热量增加,每天的能量需要大幅增加。在哺乳期的第10天,母猪需要约81 MJ/天的代谢能,其中53%分泌在乳汁中(42.8 MJ/天),另外,12 MJ/天作为泌乳引起的热量损失,剩余25.7 MJ/天为维持热量损失。子宫恢复在哺乳期的第1周仅提供少量的能量,这种贡献略微降低了每日能量需要(-1.6 MJ/天)。最后,乳腺发育所需的能量(0.6 MJ/天)是可以忽略的(Kim等,1999b)。值得注意的是,额外热损失(超过维持所需的热损失)的比例在分娩前为19%~20%,而在分娩后仅为13%~15%。在分娩前热损失的比例较大的原因,可能是由于妊娠后期比哺乳早期的身体活动更多(两者站立时间分别为6 h和4 h;Theil等,2002,2004),但是也可以推测,代谢能转化为初乳的效率要低于代谢能用于分泌常乳的效率。

7.4.3 赖氨酸需要

在围产期内,赖氨酸主要用于繁殖需要,而母猪维持需要少量赖氨酸(Feyera和Theil,2014;图7.6)。在妊娠期的最后10天,由于胎儿生长成指数增加,赖氨酸需要增加大约24%(Noblet等,1985)。如果乳腺发育也像妊娠晚期胎儿生长那样遵循指数生长曲线,赖氨酸需要的增加甚至可能大于24%。然而,目前尚不知道乳腺发育和初乳生产的需要在分娩前后如何转变。如果我们假设所有初乳在妊娠的最后10天期间均匀产生,则可以计算出乳腺发育和初乳合成的每日赖氨酸需要多达59%(第106天)和48%(第115天)。在分娩时,赖氨酸需要突然下降,因为

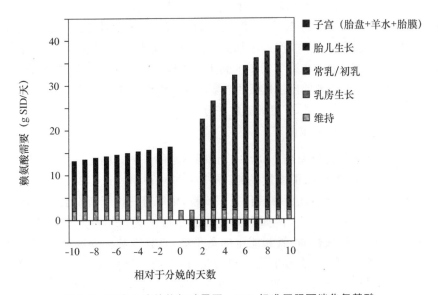

图7.6 围产母猪的赖氨酸需要。SID,标准回肠可消化氨基酸。

大多数赖氨酸被用于繁殖，只需要很少的量用于母猪维持。赖氨酸需要从分娩第2天和整个围产期显著增加，因为母猪对于泌乳具有非常高的赖氨酸需要。Wu和Knabe(1994)报道，乳汁中含有29 mmol/L赖氨酸，相当于每千克乳汁含4.12 g赖氨酸(假设乳汁密度为1.029 kg/L)。然而，分娩后子宫恢复提供约2.6 g/天的赖氨酸，该数量高于母猪维持所需的量。总的来说，子宫供应占哺乳期第2天赖氨酸总需要量的大约13%，第7天约为8%。这些数字仅在分娩后子宫恢复均匀发生(这是不可能的情况)时有效，但是这些数据说明，在优化母猪营养供应时应考虑子宫的贡献，因为它提供了大量的赖氨酸和其他氨基酸。

7.4.4 氮需要

氮(作为必需和非必需氨基酸)也需要用于母猪维持和繁殖性状，但是与能量和赖氨酸相比，母猪维持的氮需要没有很好地定义。这样做的原因是许多必需氨基酸的需要是未知的。此外，任何过量(相对于赖氨酸的不平衡)的日粮氨基酸将在肝脏中脱氨基，转化为尿素，分泌于尿液中，而碳骨架可用作能量源或生糖前体。因此，更合适的是量化尿中损失的氮量，而不是评估维持所需的氮量(Hansen等，2014)，但氮的尿损失取决于氮摄入量。这里给出的数字(Feyera和Theil，2014)基于这样一个事实，尿中的氮损失量取决于日粮摄入量(36%在妊娠后期损失，28%在哺乳期损失；Theil等，2002，2004)，因为广泛接受的观点是蛋白质周转随日粮氮水平提高而增加。即使维持需要不能精确地量化，"每日氮需要量"这个术语还是被用来描述对氮的需要。

尿液中氮损失的百分比范围，为妊娠后期计算氮需要量的36%~48%(Feyera和Theil，2014)。胎儿生长、初乳生产和乳腺发育所需的氮量大致相似，并且总计占妊娠晚期的每日氮需要量的50%~62%。分娩后，随着采食量和泌乳量的增加，在尿中损失的氮量也将在整个围产期增加。同时，在子宫恢复第2天贡献计算每日氮需要的19%，而子宫供应在第7天减少到11%(图7.7)。值得注意的是，计算的每日氮需要从妊娠期至哺乳期比赖氨酸的需要增加要更少。这表明蛋白质可能是妊娠晚期母猪生产力的限制因素，并且赖氨酸在分娩后可能是具有限制性的。

7.4.5 围产期营养平衡

当给围产母猪直到分娩前1周，一直饲喂标准的妊娠日粮时，会出现少量的赖氨酸和氮的负平衡，但略微正能量平衡(图7.8；Feyera和Theil，2014)。如果在围产期一开始时就用标准哺乳日粮替代妊娠日粮，则所有三种平衡都会变为正平衡，直到最终饲料供应降低为止(例如在分娩前的最后3天)，如丹麦推荐标准所示。减少饲料供应会导致赖氨酸、氮和能量的负平衡，直到分娩。分娩后，氮平衡恒定为正，赖氨酸平衡接近零，能量平衡最初为负，但在约一周后变为零。在妊娠后期日粮变化后，能量、赖氨酸和氮的轻微正平衡表明，母猪的营养需要在分娩前通过饲料组成和饲喂水平得到很好的满足。相反，在哺乳早期，正氮平衡与负能量平衡

7.围产期母猪饲养管理

表明,哺乳日粮不能很好地平衡和提供泌乳所需的营养物质。

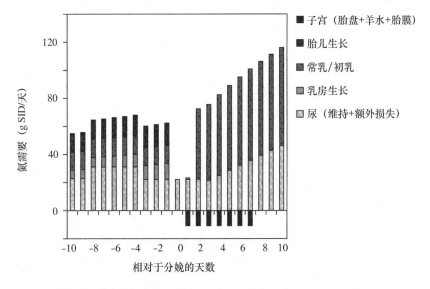

图 7.7 围产母猪的氮需要。SID,标准回肠可消化氨基酸。

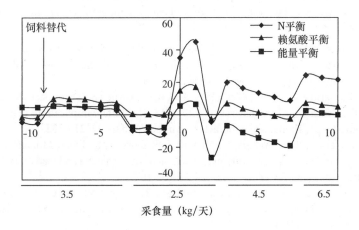

图 7.8 围产母猪的能量、赖氨酸和氮平衡,母猪饲喂一般妊娠料直到产前一周,然后饲喂泌乳料。饲料分步供应情况如图所示。

7.5 未来展望

理想情况下,母猪的营养需要应该以其所处妊娠期或哺乳期的时间表示,而不是基于饲料。在围产期经历营养需要急剧而又快速地变化时,尤其如此,因此不可

141

能由一种单一日粮就可以达到最优化的饲喂母猪。由于对能量和蛋白质的需要是彼此在独立地变化，因此以饲料为基础表达营养需要，无论是以千克表示，还是以能量为基础表示，都是没有意义的。未来应当考虑，开发新的饲养策略以满足围产母猪快速变化的营养需要，以提高母猪的饲料利用率和母猪生产力。这可能会增加母猪的使用年限，降低仔猪甚至母猪的死亡率。毫无疑问，分娩前后的能量、粗蛋白质和赖氨酸供应都必须要能满足母猪生产力和营养需要，但也必须强调，其他日粮营养物质（其他氨基酸，维生素和矿物质）可能也是真正影响围产期的母猪生产力的限制因素。

参考文献

Abiven, N., Seegers, H., Beaudeau, F., Laval, A. and Fourichon, C., 1998. Risk factors for high sow mortality in French swine herds. Preventive Veterinary Medicine 33: 109-119.

Akdag, F., Arslan, S. and Demir, H., 2009. The effect of parity and litter size on birth weight and the effect of birth weight variations on weaning weight and pre-weaning survival in piglet. Journal of Animal and Veterinary Advances 8: 2133-2138.

Andersen, I.L., Nævdal, E. and Bøe, K.E., 2011. Maternal investment, sibling competition, and offspring survival with increasing litter size and parity in pigs (Sus scrofa). Behavioral Ecology and Sociobiology 65: 1159-1167.

Baxter, E.M., Jarvis, S., D´Eath, R.B., Ross, D.W., Robson, S.K., Farish, M., Nevison, I.M., Lawrence, A.B. and Edwards, S.A., 2008. Investigating the behavioural and physiological indicators of neonatal survival in pigs. Theriogenology 69: 773-783.

Baxter, E.M., Rutherford, K.M.D., D'Eath, R.B., Arnott, G., Turner, S.P., Sandoe, P., Moustsen, V.A., Thorup, F., Edwards, S.A. and Lawrence, A.B., 2013. The welfare implications of large litter size in the domestic pig II: management factors. Animal Welfare 22: 219-238.

Blaney B.J., McKenzie, R.A., Walters, J.R., Taylor, L.F., Bewg, W.S., Ryley, M.J. and Maryam, R., 2000. Sorghum ergot (Claviceps africana) associated with agalactia and feed refusal in pigs and dairy cattle. Australian Veterinary Journal 78: 102-107.

Boyd, R.D., Moser, B.D., Peo, E.R. and Cunningham, P.J., 1978. Effect of energy source prior to parturition and during lactation on piglet survival and growth and on milk lipids. Journal of Animal Science 47: 883-892.

Cabrera, R.A., Lin, X., Campbell, J.M., Moeser, A.J. and Odle, J., 2012. Influence of birth order, birth weight, colostrum and serum immunoglobulin G on neonatal piglet survival. Journal of Animal Science and Biotechnology 3: 42.

Campos, P.H., Silva, B.A., Donzele, J.L., Oliveira, R.F. and Knol, E.F., 2012. Effects of sow nutrition during gestation on within-litter birth weight variation: a review. Animal 6: 797-806.

Coffey, M.T., Yates, J.A. and Combs, G.E., 1987. Effects of feeding sows fat or fructose during late gestation and lactation. Journal of Animal Science 65: 1249-1256.

Csapo, J., Martin, T.G., Csapo-Kiss, Z.S. and Hazas, Z., 1996. Protein, fats, vitamin and mineral concentrations in porcine colostrum and milk from parturition to 60 days. International Dairy Journal 6: 881-902.

Danielsen, V., 2003. Fodringsstrategier for diegivende søer (Feeding strategies for lactating sows). In Danish. Grøn Viden, Husdyrbrug 33: 1-8.

Danielsen, V. and Vestergaard, E.M., 2001. Dietary fibre for pregnant sows: effect on performance and behaviour. Animal Feed Science and Technology 90: 71-80.

Decaluwé, R., Maes, D., Cools, A., Wuyts, B., De Smet, S., Marescau, B., De Deyn, P.P. and Janssens, G.P., 2014. Effect of peripartal feeding strategy on colostrum yield and composition in sows. Journal of Animal Science 92: 3557-3567.

Dodd, S.C., Forsyth, I.A., Buttle, H.L., Gurr, M.I. and Dils, R.R., 1994. Milk whey proteins in plasma of sows – variation with physiological state. Journal of Dairy Research 61: 21-34.

Eder, K., 2009. Influence of l-carnitine on metabolism and performance of sows. British Journal of Nutrition 102: 645-654.

Edwards, S.A. and Baxter, E.M., 2015. Piglet mortality: causes and prevention. Chapter 11. In: Farmer, C. (ed.) The gestating and lactating sow. Wageningen Academic Publishers, Wageningen, the Netherlands, pp. 253-278.

Farmer, C. and Hurley, W.L., 2015. Mammary development. Chapter 4. In: Farmer, C. (ed.) The gestating and lactating sow. Wageningen Academic Publishers, Wageningen, the Netherlands, pp. 73-94.

Feyera, T. and Theil, P.K., 2014. Nutrient balances of energy, lysine and nitrogen in late gestating and early lactating sows. In: Proceedings of 65st Annual EAAP meeting. August 25-28, 2014. Copenhagen, Denmark, p. 342

Friendship, R.M. and O'Sullivan, T.L., 2015. Sow health. Chapter 18. In: Farmer, C. (ed.) The gestating and lactating sow. Wageningen Academic Publishers, Wageningen, the Netherlands, pp. 409-421.

Göransson, L., 1989a. The effect of feed allowance in late pregnancy on the occurrence of agalactia post partum in the sow. Zentralblatt für Veterinarmedizin 36: 505-513.

Göransson, L., 1989b. The effect of dietary crude fiber content on the frequency of post partum agalactia in the sow. Zentralblatt für Veterinarmedizin 36: 474-479.

Göransson, L., 1990. The effect of protein source in late pregnancy feed on the occurrence of agalactia post partum in the sow. Acta Veterinaria Scandinavica 31: 117-120.

Hansen, A.V., 2012. Feed intake in reproducing sows. In: Bach, K.E., Knudsen, N.J., Kjeldsen, H.D. and Jensen, B.B. (eds.) Nutritional physiology of pigs. Danish Pig Research Center, Copenhagen, Denmark. Available at: http://tinyurl.com/lw2h5sh.

Hansen, A.V., Lauridsen, C., Sorensen, M.T., Bach Knudsen, K.E. and Theil, P.K., 2012a. Effects of nutrient supply, plasma metabolites, and nutritional status of sows during transition on performance in the next lactation. Journal of Animal Science 90: 466-480.

Hansen, A.V., Strathe, A.B., Kebreab, E. and Theil, P.K., 2012b. Predicting milk yield and composition in lactating sows – a Bayesian approach. Journal of Animal Science 90: 2285-2298.

Hansen, A.V., Strathe, A.B., Theil, P.K. and Kebreab, E., 2014. Energy and nutrient deposition and excretion in the reproducing sow: model development and evaluation. Journal of Animal Science 92: 2458-2472.

Hartmann, P.E., McCauley, I., Gooneratne, A. and Whitely, J., 1984a. Inadequacies of sow lactation: survival of the fittest. In: Peaker, M., Vernon, R.G. and Knight, C.H. (eds.) Physiological stategies in lactation. Academic Press, London, UK, pp. 301-326.

Hartmann, P.E., Whitely, J.L. and Willcox, D.L., 1984b. Lactose in plasma during lactogenesis, established lactation and weaning in sows. The Journal of Physiology 347: 453-463.

Hurley, W.L., 2015. Composition of sow colostrum and milk. Chapter 9. In: Farmer, C. (ed.) The gestating and lactating sow. Wageningen Academic Publishers, Wageningen, the Netherlands, pp. 193-229.

Hurley, W.L. and Theil, P.K., 2011. Perspectives on immunoglobulins in colostrum and milk. Nutrients 3: 442-474.

Jackson, J.R., Hurley, W.L., Easter, R.A., Jensen, A.H. and Odle, J., 1995. Effects of induced or delayed parturition and supplemental dietary fat on colostrum and milk composition in sows. Journal of Animal Science 73: 1906-1913.

Jean, K.B. and Chiang, S.H., 1999. Increased survival of neonatal pigs by supplementing medium-chain triglycerides in late-gestating sow diets. Animal Feed Science and Technology 76: 241-250.

Kim, S.W., Easter, R.A. and Hurley, W.L., 2001. The regression of unsuckled mammary glands during lactation in sows: the influence of lactation stage, dietary nutrients, and litter size. Journal of Animal Science 79: 2659-2668.

Kim, S.W., Hurley, W.L., Han, I.K. and Easter, R.A., 1999a. Effect on nutrient intake on mammary gland growth in lactating sows. Journal of Animal Science 77: 3304-3315.

Kim, S.W., Hurley, W.L., Han, I.K. and Easter, R.A., 1999b. Mammary gland growth as influenced by litter size in lactating sows: impact on lysine requirement. Journal of Animal Science 77: 3316-3321.

King, R.H. and Williams, I.H., 1984. The effect of nutrition on the reproductive performance of first-litter sows. 2. Protein and energy intakes during lactation. Animal Production 38: 249-256.

Klobasa, F., Werhahn, E. and Butler, J.E., 1987. Composition of sow milk during lactation. Journal of Animal Science 64: 1458-1466.

Klopfenstein C., Farmer, C. and Martineau, G., 2006. Diseases of the mammary glands. In: Straw, B.E., Zimmerman, J.J., Allaire, S. and Taylor, D.J. (eds.) Diseases of Swine, 9th edition. Blackwell Publishing, Oxford, UK, pp. 57-86.

Kopinski J.S., Blaney, B.J., Murray, S.A. and Downing, J.A., 2008. Effect of feeding sorghum ergot (*Claviceps africana*) to sows during mid-lactation on plasma prolactin and litter performance. Journal of Animal Physiology and Animal Nutrition 92: 554-561.

Kristensen, N.B., and Wu, G., 2012. Metabolic functions of the porcine liver. In: Bach, K.E., Knudsen, N.J., Kjeldsen, H.D. and Jensen, B.B. (eds.) Nutritional physiology of pigs. Danish Pig Research Center, Copenhagen, Denmark. Available at: http://tinyurl.com/q73v6c5.

Krogh, U., Flummer, C., Jensen, S.K. and Theil, P.K., 2012. Colostrum and milk production of sows is affected by dietary conjugated linoleic acid. Journal of Animal Science 90(4): 366-368.

Lauridsen, C. and Danielsen, V., 2004. Lactational dietary fat levels and sources influence milk composition and performance of sows and their progeny. Livestock Production Science 91: 95-105.

Leenhouwers, J.I., Knol, E.F., De Groot, P.N., Vos, H. and Van der Lende, T., 2002. Fetal development in the pig in relation to genetic merit for piglet survival. Journal of Animal Science 80: 1759-1770.

Loisel, F., Farmer, C., Ramaekers, P. and Quesnel, H., 2013. Effects of high fiber intake during late pregnancy on sow physiology, colostrum production, and piglet performance. Journal of Animal Science 91: 5269-5279.

Loisel, F., Farmer, C., Ramaekers, P. and Quesnel, H., 2014. Colostrum yield and piglet growth during lactation are related to gilt metabolic and hepatic status prepartum. Journal of Animal

Science 92: 2931-2941.
Meunier-Salaün, M.C. and Bolhuis, J.E., 2015. High-Fibre feeding in gestation. Chapter 5. In: Farmer, C. (ed.) The gestating and lactating sow. Wageningen Academic Publishers, Wageningen, the Netherlands, pp. 95-116.
Mosnier, E., Etienne, M., Ramaekers, P. and Pére, M.C., 2010. The metabolic status during the peri partum period affects the voluntary feed intake and the metabolism of the lactating multiparous sow. Livestock Science 127: 127-136.
National Research Council (NRC), 2012. Nutrient requirements of swine,. 11th revised edition. The National Academies Press, Washington, DC, USA, 400 pp.
Newcomb, M.D., Harmon, D.L., Nelssen, J.L., Thulin, A.J. and Allee, G.L., 1991. Effect of energy source fed to sows during late gestation on neonatal blood metabolite homeostasis, energy stores and composition. Journal of Animal Science 69: 230-236.
Noblet, J., Close, W.H., Heavens, R.P. and Brown, D., 1985. Studies on the energy metabolism of the pregnant sow. 1. Uterus and mammary tissue development. British Journal of Nutrition 53: 251-265.
Oliviero, C., Heinonen, M., Valros, A. and Peltoniemi, O., 2010. Environmental and sow-related factors affecting the duration of farrowing. Animal Reprodtion Science 119: 85-91.
Papadopoulos, G.A., Maes, D.G., Van Weyenberg, S., Van Kempen, T.A., Buyse, J. and Janssens, G.P., 2009. Peripartal feeding strategy with different n-6: n-3 ratios in sows: effects on sows' performance, inflammatory and periparturient metabolic parameters. British Journal of Nutrition 101: 348-357.
Papadopoulos, G.A., Vanderhaeghe, C., Janssens, G.P., Dewulf, J. and Maes, D.G., 2010. Risk factors associated with postpartum dysgalactia syndrome in sows. Veterinary Journal 184: 167-171.
Pastorelli, G., Neil, M. and Wigren, I., 2009. Body composition and muscle glycogen contents of piglets of sows fed diets differing in fatty acids profile and contents. Livestock Science 123: 329-334.
Pedersen, L.J., Berg, P., Jørgensen, E., Bonde, M.K., Herskin, M.S., Knage-Rasmussen, K.M., Kongsted, A.G., Lauridsen, C., Oksbjerg, N., Poulsen, H.D., Sorensen, D.A., Su, G., Sørensen, M.T., Theil, P.K., Thodberg, K. and Jensen, K.H., 2010. *Pattegrisedødelighed i DK. Muligheder for reduktion af pattegrisedødeligheden i Danmark* (Piglet mortality in Denmark. Possibilities for reducing piglet mortality in Denmark). In Danish. Aarhus Universitet, Det Jordbrugsvidenskabelige Fakultet, Aarhus, Denmark, report Husdyrbrug 86: 1-77.
Peltoniemi, O.A.T. and Oliviero, C., 2015. Housing, management and environment during farrowing and early lactation. Chapter 10. In: Farmer, C. (ed.) The gestating and lactating sow. Wageningen Academic Publishers, Wageningen, the Netherlands, pp. 231-252.
Père, M.C., 2003. Materno-foetal exchanges and utilisation of nutrients by the foetus: comparison between species. Reproduction, Nutrition, Development 43: 1-15.
Quesnel, H., Farmer, C. and Devillers, N., 2012. Colostrum intake: Influence on piglet performance and factors of variation. Livestock Science 146: 105-114.
Quesnel, H., Farmer, C. and Theil, P.K., 2015. Colostrum and milk production. Chapter 8. In: Farmer, C. (ed.) The gestating and lactating sow. Wageningen Academic Publishers, Wageningen, the Netherlands, pp. 173-192.
Rootwelt, V., Reksen, O., Farstad, W. and Framstad, T., 2013. Postpartum deaths: piglet, placental, and umbilical characteristics. Journal of Animal Science 91: 2647-2656.

Seerley, R.W., Pace, T.A., Foley, C.W. and Scarth, R.D., 1974. Effect of energy-intake prior to parturition on milk lipids and survival rate, thermostability and carcass composition of piglets. Journal of Animal Science 38: 64-70.

Serena, A., Jorgensen, H. and Bach Knudsen, K.E., 2009. Absorption of carbohydrate-derived nutrients in sows as influenced by types and contents of dietary fiber. Journal of Animal Science 87: 136-147.

Theil, P.K., Cordero, G., Henckel, P., Puggaard, L., Oksbjerg, N. and Sørensen, M.T., 2011. Effects of gestation and transition diets, piglet birth weight, and fasting time on depletion of glycogen pools in liver and 3 muscles of newborn piglets. Journal of Animal Science 89: 1805-1816.

Theil, P.K., Flummer, C., Hurley, W.L., Kristensen, N.B., Labouriau, R.L. and Sørensen, M.T., 2014b. Mechanistic model to predict colostrum intake based on deuterium oxide dilution technique data and impact of gestation and lactation diets on piglet intake and sow yield of colostrum. Journal of Animal Science, in press. DOI: http://dx.doi.org/10.2527/jas2014-7841.

Theil, P.K., Jørgensen, H. and Jakobsen, K., 2002. Energy and protein metabolism in pregnant sows fed two levels of dietary protein. Journal of Animal Physiology and Animal Nutrition 86: 399-413.

Theil, P.K., Jørgensen, H. and Jakobsen, K., 2004. Energy and protein metabolism in lactating sows fed two levels of dietary fat. Livestock Production Science 89: 265-276.

Theil, P.K. and Lauridsen, C., 2007. Interactions between dietary fatty acids and hepatic gene expression in livers of pigs during the weaning period. Livestock Science 108: 26-29.

Theil, P.K., Lauridsen, C. and Quesnel, H., 2014a. Neonatal piglet survival: Impact of sow nutrition around parturition on fetal glycogen deposition and production and composition of colostrum and transient milk. Animal 8: 1021-1030.

Theil P.K., Nielsen, M.O., Sørensen, M.T. and Lauridsen, C., 2012. Lactation, milk and suckling. In: Bach, K.E., Knudsen, N.J., Kjeldsen, H.D. and Jensen, B.B. (eds.) Nutritional physiology of pigs. Danish Pig Research Center, Copenhagen, Denmark. Available at: http://tinyurl.com/luywh5g.

Theil, P.K., Olesen, A.K., Flummer, C., Sørensen, G. and Kristensen, N.B., 2013. Impact of feeding and post prandial time on plasma ketone bodies in sows during transition and lactation. Journal of Animal Science 91: 772-782.

Theil, P.K., Sejrsen, K., Hurley, W.L., Labouriau, R., Thomsen, B. and Sørensen, M.T., 2006. Role of suckling in regulating cell turnover and onset and maintenance of lactation in individual mammary glands of sows. Journal of Animal Science 84: 1691-1698.

Trottier, N.L., Johnston, L.J. and De Lange, C.F.M., 2015. Applied amino acid and energy feeding of sows. Chapter 6. In: Farmer, C. (ed.) The gestating and lactating sow. Wageningen Academic Publishers, Wageningen, the Netherlands, pp. 117-145.

Vallet, J.L., Rempel, L.A., Miles, J.R. and Webel, S.K., 2014. Effect of essential fatty acid and zinc supplementation during pregnancy on birth intervals, neonatal piglet brain myelination, stillbirth, and preweaning mortality. Journal of Animal Science 92: 2422-2432.

Van Milgen, J., Noblet, J., Dubois, S. and Bernier, J.F., 1997. Dynamic aspects of oxygen consumption and carbon dioxide production in swine. British Journal of Nutrition 78: 397-410.

Wu, G. and Knabe, D.A., 1994. Free and protein-bound amino acids in sow's colostrum and milk. Journal of Nutrition 124: 415-424.

8. 初乳和常乳生产

H. Quesnel[1,2]*, C. Farmer[3] and P.K. Theil[4]

[1] INRA, UMR1348 PEGASE, 35590 Saint-Gilles, France; helene.quesnel@rennes.inra.fr

[2] Agrocampus Ouest, UMR1348 PEGASE, 35000 Rennes, France

[3] Agriculture and Agri-Food Canada, Dairy and Swine R & D Centre, Sherbrooke, QC, J1M 0C8, Canada

[4] Department of Animal Science, Aarhus University, Research Centre Foulum, Blichers Álle 20, Post box 50, 8830 Tjele, Denmark

摘要：初乳和常乳生产在确保仔猪生存和生长中起着至关重要的作用。母猪初乳产量是影响出生后几天内仔猪存活、仔猪健康和直至断奶后生长的一个限制因素，而常乳产量是仔猪生长速率的限制因素。在使用高产母猪品系的当前背景下，初乳和产奶量都更受限制。初乳和常乳在两个不同的哺乳期生理阶段产生，并且在分泌时间和组成上不同。与母猪常乳产量不同，母猪初乳产量不是由窝产仔数和哺乳强度决定的。人们对影响初乳产量的因素了解程度比常乳少。初乳生产是由激素控制的，催乳素和孕酮浓度分别对初乳产量有正面和负面影响。初乳产量随胎次而变化。最近的研究共同表明，妊娠晚期的母猪营养对于母猪的初乳产量是重要的，但是目前尚未知妊娠哪个阶段的对初乳生产最为关键。一方面受乳腺发育的潜在影响，另一方面，母猪代谢状态对初乳产量的潜在影响，显然需要更多的研究。产奶量随母猪遗传和胎次而变化，并受环境（噪声、环境温度）、母猪管理和猪舍环境的影响。此外，一些营养方面的问题对母猪产奶量也有重要影响，包括日粮中能量、氨基酸、生物活性成分、脂肪来源、饲喂曲线、母猪体况和母猪的身体动员。

关键词：初乳产量，乳生成，产奶量，母猪，仔猪

8.1 引言

在过去几十年中，提高母猪繁殖力和胴体质量一直是选育的主要目标。其结果是，在过去 20 年中，每胎繁殖力增加了 2～4 头仔猪，目前在许多国家每头母猪窝产仔数可达到 14～16 头。初乳和常乳生产在确保仔猪生存和直到断奶时生长

起到重要作用。选择多产性和瘦肉率的另一个结果是断奶前仔猪死亡率大量增加。活产仔猪的断奶前死亡率因国家而异,从巴西的约 9%,到法国、加拿大和丹麦的 14%(Baxter 和 Edwards,2013)。仔猪死亡率在出生后第 3 天最高,并且已经确定早期死亡主要是由于初乳的摄入量低(Edwards,2002;Le Dividich 等,2005)。母猪的踩压也被认为是哺乳期仔猪死亡的主要原因。当仔猪没有或者摄入少量初乳而较虚弱时,被踩压的风险可能增加。本章将介绍与母猪、仔猪、环境或母体营养对初乳和常乳产量有关的影响因素。

8.2 初乳、过渡乳和常乳的定义

初乳和常乳在分泌的时间和组成上不同。初乳是乳腺的第一种分泌物,其大部分在分娩前合成。与常乳相比,初乳具有高浓度的免疫球蛋白(Ig)(表 8.1),并且比常乳含有更低浓度的乳糖和脂质(表 8.2)。然而,乳腺分泌物随时间逐渐改变,因此可能可以更精细地区分开来。我们将初乳定义为分娩开始后 24 h 以内由仔猪摄入的乳腺分泌物,这与 Devillers 等(2004)所建议的相同;过渡乳定义为初乳之后到哺乳期第 4 天之间的乳腺分泌物,哺乳期第 10 天以后则定义为常乳。过渡乳富含脂质(表 8.1),而常乳的化学组成从哺乳的第 10 天开始基本上是恒定的(Csapó 等,1996;Klobasa 等,1987)。

表 8.1 母猪初乳和常乳中免疫球蛋白的含量(Loisel 等,2013)

	初乳		常乳	
开始分娩后时间	0 h	24 h	7 d	21 d
IgG (mg/mL)	51.9	10.4	—	—
IgA (mg/mL)	11.9	4.8	2.2	4.1

表 8.2 初乳、过渡乳和常乳中脂质、蛋白质、乳糖、干物质和能量的含量(Theil 等,2014b)

	初乳			过渡乳		常乳	
	早期	中期	后期				
产后时间	0 h	12 h	24 h	36 h	3 d	17 d	SEM
化学成分[1] (g/100 g)							
脂质	5.1[c]	5.3[c]	6.9[bc]	9.1[a]	9.8[a]	8.2[b]	0.5
蛋白质	17.7[a]	12.2[b]	8.6[c]	7.3[cd]	6.1[d]	4.7[e]	0.5
乳糖	3.5[d]	4.0[c]	4.4[bc]	4.6[b]	4.8[ab]	5.1[a]	0.1
干物质	27.3[a]	22.4[b]	20.6[b]	21.4[b]	21.2[b]	18.9[b]	0.6
能量 (kJ/100 g)[2]	260[d]	276[d]	346[c]	435[ab]	468[a]	409[b]	21

[1] 行中没有共同上标的值差异显著($P < 0.05$)。
[2] 由乳糖和脂肪含量计算出的能量(不包括蛋白质中的能量,因为蛋白质的作用在初乳(免疫作用)和常乳(生长作用)上不同,因此通常不会在很大程度上被氧化供能。

8.3 初乳生产

8.3.1 初乳对新生仔猪的作用

与许多哺乳动物的情况一样,出生时的仔猪突然暴露于冷环境中。因此,通过激活体温调节机制来维持身体的恒温是至关重要的。然而,与其他哺乳动物不同,新生仔猪没有可产热的褐色脂肪组织。此外,与大多数其他哺乳动物相比,仔猪的总脂质含量较低(小于2%;Seerley和Poole,1974)。因此,肝糖原和肌糖原就成了用于氧化供热而储备的主要营养物质。在没有初乳摄入的情况下,这些能量储备将在出生后12~17 h完全耗尽(Theil等,2011)。因此,通过提供用于温度调节的能量,初乳对于保证早期产后存活是必需的。初乳还为仔猪提供被动免疫。出生时仔猪缺乏免疫球蛋白,其免疫系统需要至少3~4周,即从出生到断奶的时间,才能完全发育(Rooke和Bland,2002)。初乳中存在的母体IgG可提供系统免疫,而初乳和常乳中存在的母体IgA,可保护肠黏膜免受病原体侵害,从而防止新生仔猪腹泻。初乳还含有免疫细胞和免疫调节因子,可在对病原体的反应中起作用,并且可以促进仔猪自身免疫系统的成熟(Salmon等,2009)。最后但并非最不重要的是,初乳还通过提供消化酶,刺激能量代谢和体温调节系统,帮助新生仔猪出生后的生理适应(Herpin等,2005)。此外,初乳富含有重要作用的生长因子,其主要用于刺激新生仔猪胃肠道的生长和发育(Xu等,2002)。

8.3.2 测量初乳产量的方法

量化母猪的初乳产量很有挑战性,因为没有直接的方法,所以采用诸如"称重-吮乳-称重"的方法(见测量猪产奶量的方法)。目前,大多数关于母猪初乳产量的研究,使用Devillers等(2004)开发的以下预测方程来量化单头仔猪的摄入量。方程如下:

$$CI = -217.4 + 0.217t + 1861019\,BW24/t + BWB\,(54.80 - 1861019/t)$$
$$(0.9985 - 3.7 \times 10^{-4} t_{FS} + 6.1 \times 10^{-7} t_{FS}^2)$$

式中,CI是单头仔猪初乳摄入量(g),t是第1次和第2次称重(其定义"初乳摄入持续时间")之间经过的时间(min),BW24是24 h时的体重(kg),BWB是初生重(kg),t_{FS}是出生和第一次哺乳之间的时间间隔(min)。

Devillers等(2004)的预测方程是使用人工饲喂仔猪(bottle-fed piglets)基础上建立的。然而,仔猪的预期初乳摄入量和母猪的初乳产量似乎被低估了,原因很可能是人工饲喂仔猪的身体运动比母猪饲喂下的仔猪低得多。量化母猪初乳摄入量的另一种方法是利用重水(DO)稀释技术,来测量仔猪初乳摄入量,和对应的母猪初乳产量。具体为,在出生后立即从仔猪收集初始血液样品(作为背景),然后向仔猪肌肉注射重水。注射后 1 h 内仔猪不能吮吸初乳,此时收集第二个血液样品,然后再使仔猪吮吸初乳。然后,在第一头仔猪出生后 24 h,从每头仔猪收集第三个血液样品,并且分析所有血液样品的重水含量。第二和第三血液样品之间的重水稀释比例用于量化来自初乳的水摄入量,收集初乳的化学分析值被用于将初乳水摄入量转化为初乳摄入量。该技术相当准确,但相当昂贵,并且需要大量的劳动力,因此不适用于大多数聚焦于初乳的生产研究。在最近的一项研究(Theil 等,2014a)中,发现通过重水稀释技术测量的初乳摄入量,比由 Devillers 等(2004)提出的方程式预测的初乳摄入量要高 43%。

8.3.3 影响初乳产量的因素

无论用于估算初乳产量的方法如何,母猪间分娩 24 h 内分泌的初乳总产量是高度变异的。当根据 Devillers 等(2007)提出的预测方程进行量化时,平均值为 3.3~3.7 kg,但实际个体产量范围却从 1 kg 到超过 6.0 kg(Devillers 等,2007;Quesnel,2011)。尽管在过去十年中对初乳生产的兴趣日益增加,但很少有研究提供了关于初乳产量及其影响因素的信息。因此,本文也将涉及可能影响哺乳早期仔猪生长的因素。事实上,从出生到哺乳期第 3、5 或 7 天的仔猪生长,都与出生后 24 h 内初乳摄入量有关。

母猪与仔猪

众所周知,母猪产奶量受到来自哺乳仔猪的刺激,其取决于窝产仔数和窝重。然而,仔猪的影响在初乳期阶段似乎并不那么重要。这首先表现为,在出生后第一个 24 h 期间用人工饲喂初乳的仔猪自由摄入量超过 450 g/kg 初生重,这是母猪饲喂仔猪平均摄入量的两倍(Devillers 等,2004)。该结果表明,母猪产生的初乳量通常比仔猪所能摄入的初乳量要少。其次,初乳在分娩后的头几个小时内可自由获得(DePassillé 和 Rushen,1989),并且直到产后 16~24 h,才需要仔猪经常地哺乳以保持乳头分泌(Atwood 等,1995;Theil 等,2006)。同样地,在出生时的窝产仔数或窝重,与哺乳早期初乳产量或仔猪生长之间,也没有发现相关性(Devillers 等,2007;Quesnel,2011)。另一方面,初乳产量与平均仔猪初生重呈正相关(Devillers 等,2007)。总之,这些结果表明初乳产量可能受到出生时仔猪的整体活力影响,并且在很大程度上取决于母猪生产初乳的能力。

母猪胎次

Devillers 等(2007)报道,2 胎和 3 胎母猪(4.3 kg)比初产母猪(3.4 kg)和老年母猪(3.6 kg)倾向于能产生更多的初乳。Decaluwé 等(2013)报道了 1~3 胎母猪比 4~7 胎母猪平均的初乳产量更高(3.7 和 2.8 kg)。此外另一研究结果为,1 胎和 2 胎母猪的初乳产量高于 3 胎和更高胎次的母猪(3.7 和 3.2 kg,Quesnel,未公开数据)。尽管 1 胎和 3 胎母猪也存在一些差异,但似乎结果一致显示,老年母猪

(至少4胎及4胎以上)产生的初乳比年轻母猪少。

乳腺发育

在讨论母猪初乳合成潜在影响因素时,忽略的一个方面是乳腺发育。如第4章所讨论的(Farmer和Hurley,2015),母猪产奶量取决于开始泌乳时乳腺中存在的泌乳细胞的数量(Head和Williams,1991),这也非常有可能适用于初乳产量的情况。当然,这还需要进一步研究证实。

母猪内分泌状态

对于母猪,催乳素的产前峰值对于乳生成是必需的(Farmer等,1998),这是由孕酮浓度的下降引起的(Taverne等,1982)。在初产母猪中,相对于分娩开始时的孕酮减少和催乳素增加的延迟,与初乳产量的急剧降低相关(Foisnet等,2010)。此外,最近发现在围产期催乳素和孕酮的相对浓度会影响初乳产量。具有较低孕酮浓度的初产母猪,具有更高催乳素浓度(高催乳素与孕酮比率)的趋势,在分娩前两天和分娩当天,会比具有低催乳素与孕酮比率的母猪产生更多的初乳(+ 0.6 kg)(Loisel,2014)。最后,当母猪在分娩后立即具有更高的孕酮循环浓度时,产仔后3或5天内的仔猪生长速度和存活率会降低(De Passillé等,1993;Quesnel等,2013)。值得注意的是,初产母猪的围产期催乳素浓度低于经产母猪(Quesnel等,2013)。因此,催乳素浓度似乎与同胎次母猪之间的初乳产量相关,但不适用于不同胎次之间的母猪。

母猪代谢状态

目前所知,妊娠晚期母猪的代谢状态与初乳生产之间仅部分研究结果具有相关性。有报道称,初乳产量与分娩前测量的尿素和肌酐浓度之间呈正相关(Loisel等,2014)。血浆尿素是蛋白质氧化的最终产物,肌酐可用作肌肉蛋白质动员的标志物。因此,更多的瘦肉组织和(或)更多的体蛋白动员,可能与初乳产量的增加相关。源自体蛋白质动员的氨基酸可以被乳腺用于蛋白质合成,这些蛋白质将用于乳腺发育、初乳免疫球蛋白和蛋白质的合成。它们也可能会用作生成葡萄糖的底物(Theil等,2012)。Decaluwé等(2013)报道了初乳产量和(异)丁酰肉碱[(iso)butyrylcarnitine]的循环浓度之间呈负相关,其可作为蛋白质分解代谢的指标。关于能量供应,初乳产量与分娩前一天测量的血浆游离脂肪酸浓度呈正相关(Loisel等,2014),这表明低(或负)能量平衡增强了初乳产量。与之相反,Decaluwé等(2013)发现,妊娠期最后5天脂肪组织的动员,会对初乳产量有负面影响。最后,Hansen等(2012a)观察到,在分娩后的第一个24 h期间的仔猪增重,与妊娠期最后1周的游离脂肪酸浓度或母猪能量平衡之间,没有相关性。这种差异可能与不同研究中的母猪体况、饲喂策略和日粮组成之间的差异有关。显然,母猪代谢状态在妊娠晚期对初乳产量的影响需要进一步研究。

母猪饲喂

针对母猪代谢状况,有关母猪营养如何影响初乳产量的了解不多。然而,最近的研究共同表明,妊娠晚期的母猪营养对于初乳生产有重要影响。实际上,从妊娠的第 108 天直到分娩,饲喂 1.3%CLA(顺式-9/反式-11 和反式-10/顺式-12 异构体的等量混合物)的母猪,倾向于产生较少的初乳(由 463 g/仔猪,减少到 409 g/仔猪;Krogh 等,2012)。在另一项研究(Flummer 和 Theil,2012)中,母猪饲喂常规哺乳日粮,并从妊娠第 108 天直到分娩补充 2.5 g/天的 Ca(HMB)(β-羟基-β-甲基丁酸钙),可通过初乳期的仔猪增重,推测产生了更多的初乳(HMB 组和对照组分别为 132 g 和 76 g,$P = 0.05$)。在整个妊娠期饲喂高纤维日粮可能有益于母猪的初乳生产,但是这也取决于日粮纤维来源。在 Theil 等(2014a)的一项研究中,母猪从交配到妊娠第 108 天,相比饲喂马铃薯渣或低纤维对照日粮(初乳摄入量分别为 393 和 414 g/仔猪;$P = 0.02$),当饲喂果胶渣或甜菜渣(初乳摄入量分别为 520 和 504 g/仔猪)时,仔猪的初乳摄入量(通过重水稀释技术测量)会更高。Quesnel 等(2009)报道了从妊娠第 26 天直到分娩,饲喂高纤维日粮比低纤维对照日粮(3.4 对 3.0 kg)的母猪初乳产量更高,但差异不显著。Loisel 等(2013)发现,与低纤维日粮相比,高纤维(来源于大豆皮、麦麸、带壳葵花籽粕和甜菜渣的混合物)没有显著差异。但是,当母猪饲喂高纤维日粮时,低出生重(<900 g)仔猪的初乳摄入量要高出近 60%。总的来说,这些数据表明,妊娠晚期的营养对于母猪的初乳产量有重要影响,但是目前不清楚妊娠的哪个阶段(最后 1 周,最后 3 周或最后 3 个月)对于初乳生产是最关键的。

8.4 产奶量

母猪产奶量是仔猪生长速率的限制因素(Harrell 等,1993)。事实上,向乳猪补充奶可明显增加它们的断奶重(Miller 等,2012)。仔猪断奶重反过来对断奶后的性能有重要影响,断奶体重和保育后期平均日增重存在线性关系(Cabrera 等,2010)。在当前使用高产母猪品系的背景下,母猪的产奶量更是成为限制生产的因素。因此,要制定出可能刺激母猪产奶量的管理策略,了解影响母猪产奶量的各种因素是很重要的。

8.4.1 泌乳模式

母猪泌乳可以分为四个阶段,即初乳阶段、上升阶段、稳定阶段和下降阶段。典型的母猪泌乳曲线如图 8.1 所示。在早期初乳阶段,乳腺分泌物可连续地提供给仔猪,并且如本章前面所述,每日初乳产量在母猪之间是非常不同的。相比之下,常乳不是连续提供的,而是只在射乳期放乳,每次持续 10~15 s(弗雷泽,1980)。在哺乳期的上升阶段,即哺乳期的第 2~10 天,哺乳频率倍增,每天授乳从 17 次增加到 35 次(Jensen 等,1991)。在哺乳期的第 1~3 周,每次哺乳所获得的奶量也从 29 g 增加到了 53 g(Campbell 和 Dunkin,1982)。哺乳期第 4 天的产奶量在 5~10 kg,平均约 8 kg(Toner 等,1996),大约是初乳总产量的两倍。上升阶段的持续时间也许会在 14~28 天变化,这取决于诸如品种、营养、胎次或用于估算猪产奶量的方法等因素(Elsley,1971;Harkins 等,1989)。根据最近的一项荟萃分析(meta-analysis),使用了 1980 年以来发表的众多研究,得出达到产奶高峰的平均时间为 18.7 天(Hansen 等,2012)。在上升阶段,产奶量会根据仔猪的需要进行调整,初生重较大的仔猪在每次哺乳时比其较轻的同窝仔猪吃到的奶更多(Campbell 和 Dunkin,1982)。尽管如此,产奶量仍限制了哺乳仔猪的生长速度。实际上,与未补充人工代乳品的猪相比,从哺乳期第 3 天直到断奶补充了人工代乳品的哺乳仔猪,增加了第 7、14 和 21 天的体重(Wolter 等,2002)。在现代养猪体系中,母猪通常没有达到哺乳期的下降阶段,因为它们都在稳定阶段就已断奶。

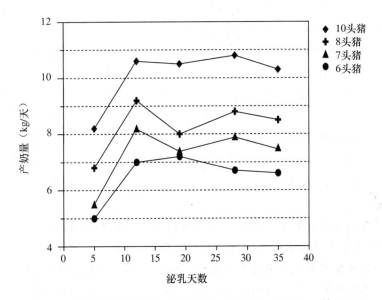

图 8.1 具有不同窝产仔数初产母猪的产奶量模式(Toner 等,1996)。

8.4.2 哺乳频率

在哺乳期,周期性哺乳在分娩开始后大约 10 h 开始(Lewis 和 Hurnik,1985),并逐渐增多(Algers 和 Uvnäs-Moberg,2007)。哺乳频率在决定产奶量中起着主要作用,并且在哺乳期的第 8~10 天达到最大值,之后降低(Puppe 和 Tuchscherer,2000)。有报道称,在哺乳早期(第 5 或 6 天)和哺乳高峰期(第 18 或 20 天),哺乳间隔分别为 36~40 min 和 39~48 min(Farmer 等,2001;Fisette 等,2004)。从乳腺中去除乳汁最为重要,以保持乳汁分泌(Theil 等,2005)。已有研究者提出,当乳成分在乳腺中积累时,会促使一种自分泌因子(译者注:一种乳清蛋白)的累积,该因子会抑制泌乳(Peaker 和 Wilde,1987)。然而,尽管已多次尝试,但这种蛋白质从未被纯化或测序。乳汁在被吮吸后 35 min 内,几乎可以重新注满乳腺(Spinka 等,1997),由此证明了将哺乳间隔尽可能保持在接近 35 min 的重要性。延长哺乳间隔仅略微增加了每次哺乳的产奶量(Spinka 等,1997)。

8.4.3 如何测量产奶量

有三种经常使用的量化母猪产奶量的方法,即称重—吮乳—称重法,重水稀释技术法和基于仔猪增重或窝增重的预测法。可以使用称重—吮乳—称重法直接测量产奶量。这意味着在每次哺乳结束后,对一系列连续的哺乳之前和之后的母猪,或整窝仔猪进行称重。出于实用和精确的原因,重复称量窝仔猪比称量母猪更常见。称重—吮乳—称重法适用于整个哺乳期,因为整窝仔猪在同时吮吸。即使这项技术可以直接量化产奶量,但可能低估了产奶量,因为一些仔猪在两次称重之间排尿和排便,因此必须考虑估算尿和/或粪便损失。此外,每小时抓放两次仔猪会扰乱母猪,并可能降低产奶量。据报道,通过称重—吮乳—称重法测量的产奶量比通过重水稀释技术测定的产奶量低估了约 20%(Hansen 等,2012 b)。

由于其精准度高,重水稀释技术被认为是测量仔猪奶摄入量和母猪产奶量的黄金方法(Theil 等,2007)。然而,其主要缺点是昂贵的成本和劳动密集。比起测定初乳摄入量,该技术更适合定量测定常乳,因为测量间隔可以延长到 3~4 天,这提高了试验精确度。该技术上文已介绍(如何测量初乳产量)。只要所有同窝仔猪的体增重已知,使用该技术的成本可以通过每窝只使用 3 头仔猪来量化乳摄入量而显著降低(Theil 等,2002)。

第三种方法是基于单头仔猪或整窝仔猪的活体增重来预测仔猪乳汁摄入量,或者,断奶时的体增重或体重可用作产奶量的指标,而不试图计算实际产奶水平。如果是从仔猪或窝增重计算出的产奶量,需要注意的是,随着哺乳期的推进,乳汁转化成仔猪体增重的效率在增加。然而,在 Theil 等(2002)的一项研究中发现,每克仔猪增重所需的奶量从第 3 天的 3.78 g/g,增加到第 10 天的 4.58 g/g,和第 17 天的 4.89 g/g。该研究还报道了仔猪的乳摄入量和活体增重之间呈曲线相关,这反映了在乳摄入量较低时,摄入量中有较高比例被用于身体维持,而在乳摄入量较高时,会沉积更多的体脂,这与低体增重有关(Noblet 和 Etienne,1987)。

最近,有研究者基于自 1980 年以来进行的母猪实验结果,进行了荟萃分析(meta-analysis),开发出用于计算母猪泌乳曲线的预测模型(Hansen 等,2012b)。该模型所需输入的数据为窝产仔数和窝增重(kg/天)。基于这些生产指标开发的体系可计算出从哺乳第 3 天到第 30 天的推定母猪的实际产奶量。该模型采纳了过去 30 年中母猪产奶量的公开数据,其中产奶量数据至少要在两个不同的哺乳阶段测量,并且还考虑到称重—吮乳—称重法比实际产奶量低估了 20%。该模型是免费提供的,可以作为电子表格下载。

8.4.4 影响产奶量的因素

遗传和胎次

已知母猪品种会影响产奶量(Grun 等,1993;Farmer 等,2001)。源自中国的母猪比典型的欧洲品种[如长白(Landrace),大白(Large White)]产奶更多,而后者比肉用品种[如杜洛克(Duroc),皮特兰(Pietrain)]产奶更多。母猪产奶量的遗传力是中等的,例如母女回归的遗传力为 0.27(York 和 Robison,1985),这说明了多年来对产奶量指标进行育种选择的积极影响。事实上,在 1971 年,Elsley(1971)曾报道母猪产奶量为 5.2 kg/天,而 Sauber 和 Stahly(1996)报道在 1996 年产奶量为 10.3 kg/天,King 和 Eason(1998)报道 1998 年的产奶量为 11.6 kg。之后,产奶量数值保持相当稳定,2011 年报道的值为 9.6 kg/天(De Oliveira Junior 等,2011)。有趣的是,母猪中乳头总数和功能性乳头(即没有内陷或瞎的乳头)的数量也可以有针对性地选择,分别具有 0.36~0.42 和 0.29 的遗传力(Chalkias 等,2013)。遗传和母猪产奶量之间的关系也可在另一项研究中说明,其通过在母猪基因组中插入牛基因,而将遗传修饰母猪的 α-乳清蛋白过表达(Noble 等,2002)。这些转基因母猪比对照母猪可多生产大约 50% 的 α-乳清蛋白,并且在哺乳期的第 3,6 和 9 天时,它们的产奶量高出 13%~23%。产奶量也受母猪胎次的影响,经产母猪比初产母猪产奶量大,而产奶量在 2~4 胎时最大,随后降低(Dourmad 等,2012)。Speer 和 Cox(1984)估算,第 2 胎母猪的产奶量比第 1 胎高 35%。功能性乳头的数量在第 1~7 胎次保持稳定,而在断奶时未使用的乳头数量在第 1~7 胎次,从 1.3 减少到 0.5(Caugant 等,2000)。

窝产仔数和哺乳强度

乳猪数量是母猪产奶量的主要决定因素。事实上,有研究发现产奶量与哺乳乳腺的数量成正比(Auldist 等,1998)。图 8.1 显示了不同窝产仔数的典型母猪生产曲线,表明随着窝产仔数增加,产奶量更高(Toner 等,1996)。然而,每头仔猪摄入的乳汁量随着窝产仔数的增加而减少,14 头仔猪时每头摄入 1.11 kg/天,而当窝产仔数为 6 时,每头摄入 1.63 kg/天(Auldist 等,1998)。仔猪在确定母猪产奶量中的作用在 Auldist 和 King(1995)的综述中有详细描述。不仅仔猪的数量,而且每头仔猪的大小也影响产奶量。Van der Steen 和 De Groot(1992)清楚地表示,初生重对仔猪的奶摄入量和断奶前生长速度有显著的影响。King 等(1997)也报道了仔猪体重和乳汁消耗量之间呈正相关关系,表明日龄较大和较重的仔猪,能够从哺乳母猪的乳腺中吸除更多的乳汁。哺乳强度本身似乎也在哺乳早期对产奶量起作用。Algers 和 Jensen(1991)报道了在哺乳的第 1~3 天乳头刺激的持续时间和强度与随后乳头产奶量的关系。然而,在后面的哺乳期并非这种情况,因为在哺乳期的第 11 和 18 天,排乳后按摩乳房的持续时间和随后的产奶量之间没有相关性(Thodberg 和 Sorensen,2006)。然而,从所有报道中不难发现,哺乳期仔猪对乳汁的需要对产奶量具有直接影响。这在 Auldist 和 King(1995)的研究中被很好地证明,其设计为母猪喂养 6 头仔猪,或者两窝仔猪,每窝 6 头仔猪每隔 30 min 交替地吮乳。

环境与管理

同一产房的母猪,会由于受其他母猪的听觉刺激而同步哺乳(Wechsler 和 Brodmann,1996)。当仔猪受到连续的嘈杂噪声(例如来自风扇)时,仔猪将减少吮乳刺激,从而降低产奶量(Algers 和 Jensen,1991)。有研究表明,通过回放母猪哺乳时哼哼声的录音,可以刺激哺乳(Stone 等,1974),并增加 14 日龄仔猪的体重(Cronin 等,2001),这种影响可一直持续到 27 日龄(Fisette 等,2004)。减小哺乳间隔时间以增加母猪产奶量的概念是有效的,但是在商业条件下实施它的手段尚未实现。其他环境因素,如热应激和光周期,也会影响母猪的泌乳性能。在许多研究中证明了高环境温度(范围为 27~32℃),会降低母猪产奶量(Farmer 和 Prunier,2002)。有迹象表明,在产房中延长光照时间可能对断奶前仔猪生长有益,但结果不一致。哺乳期对母猪给予 16 h 的光照而不是 8 h,可促进仔猪生长(Mabry 等,1982)或没有影响(Gooneratne 和 Thacker,1990;Greenberg 和 Mahone,1982)。初产母猪的管理会影响其后续哺乳期的生产力。在一项研究中,来自 2 胎母猪的仔猪可以吮吸在第 1 个哺乳期使用过和从未使用过的乳头,在第 56 天,曾经使用过的乳头具有更大的 56 天仔猪体重(Farmer 等,2012)。猪舍环境也可以影响产奶量。例如,更容易接触到乳房的分娩栏会使仔猪摄入更多乳汁(Pedersen 等,2011)。

营养、动员和体况

对哺乳母猪的大多数营养研究,得出饲料组成和饲喂水平对母猪产奶量没有影响的结论,这是令人惊讶的,因为有多达 50%~60% 的日粮营养物质会被分泌到猪乳中(Theil 等,2004)。大多数营养研究未能显示对产奶量产生任何影响的原因可能是:(1)不正当的饲养管理通过身体动员得到补偿;(2)产奶量难以量化;(3)产乳的遗传潜力因个体而异;(4)许多不同的因素影响产奶量,包括窝产仔数、哺乳阶段、饲喂水平、饲料组成、胎次、品种、体况、母猪代谢/内分泌状态和环境(温度、光照制度和猪舍环境)。已证明采食对产奶量有影响的营养研究,通常会对必需营养素或能量进行严格的限制。

严格限制日粮中赖氨酸(从 45 g/天降至 15 g/天)使母猪产奶量在哺乳第 8 天从 8.79 kg/天降至 7.87 kg/天,在哺乳第 18 天从 9.56 kg/天降至 7.42 kg/天(Kusina 等,1999)。尽管最小体动员所需的日赖氨酸估算为 45 g/天,但将后备母猪的赖氨酸供应量从 48 kg/天减少到 27 g/天,对其产奶量也没有影响(Touchette 等,1998)。当每千克泌乳饲料中含 5.5 g 的表观回肠可消化缬氨酸(译者注:原书为赖氨酸,经查原文应为缬氨酸)时,会显示产奶量降低(Paulicks等,2003)。除赖氨酸和缬氨酸以外的必需氨基酸需要量,还没有像生长猪一样进行大量试验,但Soltwedel等(2006)报道,赖氨酸和苏氨酸是饲喂基于玉米豆粕(译者注:原书为大豆soybean,经查原文应为豆粕soybean meal)基础日粮的哺乳母猪的第一和第二限制性氨基酸。与上述不同,日粮氨基酸的推荐水平通常是基于在猪乳中发现的氨基酸浓度(相对于赖氨酸)而建立,这意味着其前提是假设所有必需氨基酸(被消化的)用于乳合成的效率是相同的。当严重限制饲料供应,平均从5.3 kg/天降到2.6 kg/天时(即严格限制所有日粮营养物质,包括能量、蛋白质和必需氨基酸),也使母猪产奶量从哺乳第18天的10.1 kg/天降到第24天的6.6 kg/天(Verstegen等,1985)。然而,在同一研究中,在第7~14天(从8.3 kg/天降至7.0 kg/天)仅观察到产奶量的少量减少。这些研究强调,哺乳母猪的营养是重要的,但如果日粮供应不足,哺乳母猪会动用它们的体脂和肌肉来作为营养缓冲,以支持泌乳。这在King和Dunkin(1986)的研究中已很好地证明,其研究中共设6个饲养水平(1.5~4.8 kg/天)为母猪饲喂哺乳日粮。该研究显示,母猪会通过增加身体的营养物质动员来弥补饲料摄入不足。由于母猪能够补偿哺乳前3周日粮供应的不足,因此作为产奶量指标的仔猪生长仅在哺乳期的第4周时才受到损害。然而,必须记住,母猪哺乳期严重的身体动员已知会对后来的繁殖性能产生负面影响(Quesnel,2009),例如,损害出生时仔猪的均匀度(Wientjes等,2013),或者增加断奶后发情间隔。因此,避免母猪在哺乳期采食量过低,对母猪使用年限至关重要。

从实践的角度来看,了解如何能够改善哺乳母猪的食欲,以及如何能够提高已具有良好食欲和饲喂了适合哺乳日粮母猪的产奶量是很有趣的。Ramanau等(2004)提出补充L-肉碱可增加母猪产奶量,进而促进了仔猪生长。这是生物活性物质刺激泌乳的一个实例,因为母猪每天只补充了很少量的L-肉碱(250 mg/天)。还有一些关于日粮中常量化学成分变化的研究,其中脂肪已经受到一些关注。与8%鱼油或低脂肪对照日粮(补充3%动物脂肪)相比,一些脂源(8%动物脂肪或8%菜籽油)增加了在乳汁中分泌的能量。然而,向仔猪增加转移的能量,仅部分地由增加的产奶量和增加的乳脂含量获得,因为产奶量和乳组成都没有受显著影响(Lauridsen和Danielsen,2004)。然而,大多数日粮中添加脂肪的研究对产奶量没有影响(Theil等,2004;Tilton等,1999)。

另一个影响母猪产奶量的营养问题是饲喂曲线(Danielsen,2003)。如果从哺乳的第7天开始让母猪自由采食,其产奶量(8.3 kg/天)在数值上高于从哺乳第14天(8.1 kg/天)或从分娩(7.9 kg/天)开始,并且显著高于分娩前最后1周和整个哺乳期(7.4 kg/天)自由采食的母猪。这些发现表明,在分娩前后过度饲喂对母猪后续的泌乳是不利的,并且在分娩前后建立了产奶量的预期值。该数据还表明,高的饲料采食量有利于泌乳晚期的产奶,尽管与从哺乳期第7天自由采食相比,从第14天开始自由采食母猪的产奶量并没有显著降低。

体况是影响母猪产奶量的另一个重要指标,并且身体动员能力似乎比高采食量更重要(Beyer等,2007;Theil等,2012)。特别是在哺乳高峰期后,哺乳的持续性似乎与体况(如高水平的背膘厚)有关(Hansen等,2012a)。Kim等(2009)证明了哺乳母猪的理想氨基酸谱取决于哺乳期失重的程度,失重多的母猪比在整个哺乳期未失重的母猪需要更多的苏氨酸和亮氨酸(相对于赖氨酸)。这些结果表明,日粮中必需营养物质的比例和饲喂曲线是改善母猪产奶量的重要因素。

内分泌状态

产奶量受激素状态影响,并且人们曾使用各种激素处理以试图增加母猪产奶量。注射促甲状腺素释放因子(TRF)提高(Wung等,1977)或没有影响(Dubreuil等,1990)仔猪增重。在哺乳母猪日粮中添加促甲状腺素释放因子增加了仔猪体重,但也增加了断奶至发情间隔,使得这种做法不能完全满足母猪生产者(Cabell和Esbenshade,1990)。当观察外源性生长激素(GH)对母猪产奶量的影响时,早期研究报道产奶量增加了15%~22%(Harkins等,1989),然而,这些结果不能在以后的研究中重现(Cromwell等,1992;Toner等,1996)。生长激素释放因子的使用可增加循环生长激素浓度,但并未影响母猪产奶量、乳成分或仔猪生产性能(Farmer等,1992)。人们对另一种激素——催乳素在哺乳母猪上的应用特别感兴趣。实际上,在哺乳期间抑制催乳素会抑制母猪产奶(Farmer等,1998)。此外,自发性泌乳失败的母猪显示具有异常低的催乳素浓度(Whitacre和Threlfall,1981),随后在初产母猪(而不是经产母猪)中观察到,当在哺乳第1天接受单次注射猪催乳素,其窝增重增加了8%(Dusza等,1991)。然而,从哺乳期的第2~23天,每天向哺乳母猪注射猪催乳素,并未提高仔猪增重(Farmer等,1999)。这可能是由于以下事实:所有催乳素受体已经在对照动物中饱和,从而阻止了进一步增加催乳素浓度的任何有益效果。

8.5 结论

母猪产生的初乳和常乳的数量对于仔猪生存和/或性能至关重要,需要了解控制它们的因素。关于初乳产量,未来的研究应关注母猪营养和妊娠晚期代谢状况的影响以及乳腺发育的潜在影响。关于猪产奶量,未来的研究工作应集中于高产母猪的氨基酸需要、生物活性饲料原料的潜在影响,以及营养供应、身体动员和猪产奶量之间的复杂相互作用。

参考文献

Algers, B. and Jensen, P., 1991. Teat stimulation and milk production during early lactation in sows: effects of continuous noise. Canadian Journal of Animal Science 71: 51-60.

Algers, B. and Uvnäs-Moberg, K., 2007. Maternal behavior in pigs. Hormones and Behaviour 52: 78-85.

Atwood, C., Toussaint, J.K. and Hartmann, P.E., 1995. Assessment of mammary gland metabolism in the sow. II. Cellular metabolites in the mammary secretion and plasma during lactogenesis II. Journal of Dairy Research 62: 207-220.

Auldist, D.E. and King, R.H., 1995. Piglets' role in determining milk production in the sow. In: Hennessy, D.P. and Cranwell, P.D. (eds.) Manipulating pig production V., Australasian Pig Science Association, Werribee, Australia, pp. 114-126.

Auldist, D.E., Morrish, L., Eason, P. and King, R.H., 1998. The influence of litter size on milk production of sows. Animal Science 67: 333-337.

Baxter, E.M. and Edwards, S.A., 2013. Environmental factors behind piglet survival. In: Rodriguez-Martinez, H., Soede, N.M. and Flowers, W. (eds.) Control of pig reproduction IX. Nottingham University press, Society for Reproduction and Fertility, Nottingham, UK, pp. 129-143.

Beyer, M., Jentsch, W., Kuhla, S., Wittenburg, H. Kreienbring, F., Scholze, H., Rudolph, P. E. and Metges, C.C., 2007. Effects of dietary energy intake during gestation and lactation on milk yield and composition of first, second and fourth parity sows. Archives in Animal Nutrition 61: 452-468.

Cabell, S.B. and Esbenshade, K.L., 1990. Effect of feeding thyrotropin-releasing hormone to lactating sows. Journal of Animal Science 68: 4292-4302.

Cabrera, R.A., Boyd, R.D., Jungst, S.B., Wilson, E.R., Johnston, M.E., Vignes, J.L. and Odle, J., 2010. Impact of lactation length and piglet weaning weight on long-term growth and viability of progeny. Journal of Animal Science 88: 2265-2276.

Campbell, R.G. and Dunkin, A.C., 1982. The effect of birth weight on the esplusketimated milk intake, growth and body composition of sow-reared piglets. Animal Production 35: 193-197.

Caugant, A., Paboeuf, F., Quinio, P.Y. and Pellois, H., 2000. La perte des tétines des truies en cours de carrière. Rapport EDE-CA de Bretagne, Chambre d'Agriculture Bretagne, France.

Chalkias, H., Rydhmer, L. and Lundeheim, N., 2013. Genetic analysis of functional and non-functional teats in a population of Yorkshire pigs. Livestock Science 152: 127-134.

Cromwell, G.L., Stahly, T.S., Edgerton, L.A., Monegue, H.J., Burnell, T.W., Schenck, B.B. and Schricker, B.R., 1992. Recombinant porcine somatotropin for sows during late gestation and throughout lactation. Journal of Animal Science 70: 1404-1416.

Cronin, G.M., Leeson, E., Cronin, J.G. and Barnett, J.L., 2001. The effect of broadcasting sow suckling grunts in the lactation shed on piglet growth. Asian-Australasian Journal of Animal Science 14: 1019-1023.

Csapó, J., Martin, T.G., Csapó-Kiss, Z.S. and Házas, Z., 1996. Protein, fats, vitamin and mineral concentrations in porcine colostrum and milk from parturition to 60 days. International Dairy Journal 6: 881-902.

Danielsen, V., 2003. Feeding strategies for lactating sows. Grøn Viden, Husdyrbrug 33: 1-8. [in Danish]

De Oliveira Junior, G.M., Ferreira, A.S., Oliveira, R.F.M., Silva, B.A.N., De Figueiredo, E.M. and Santos, M., 2011. Behaviour and performance of lactating sows housed in different types of farrowing rooms during summer. Livestock Science 141: 194-201.

De Passillé, A.M.B. and Rushen, J., 1989. Using early suckling behavior and weight gain to identify piglets at risk. Canadian Journal of Animal Science 69: 535-544.

De Passillé, A.M.B., Rushen, J., Foxcroft, G.R., Aherne, F. and Schaefer, A., 1993. Performance of young pigs: relationship with periparturient progesterone, prolactin and insulin of sows. Journal of Animal Science 71: 179-184.

Decaluwé, R., Maes, D., Declerck, I., Cools, A., Wuyts, B., De Smet, S. and Janssens, G.P.J., 2013. Changes in back fat thickness during late gestation predict colostrum yield in sows. Animal 7: 1999-2007.

Devillers, N., Farmer, C., Le Dividich, J. and Prunier, A., 2007. Variability of colostrum yield and colostrum intake in pigs. Animal 1: 1033-1041.

Devillers, N., Van Milgen, J., Prunier, A. and Le Dividich, J., 2004. Estimation of colostrum intake in the neonatal pig. Animal Science 78: 305-313.

Dourmad, J.Y., Quiniou, N., Heugebaert, S., Paboeuf, F. and Ngo, T.T., 2012. Effect of parity and number of suckling piglets on milk production of sows. Proceedings of the EAAP-63rd annual meeting. Bratislava, Slovakia, 44 pp.

Dubreuil, P., Pelletier, G., Petitclerc, D., Lapierre, H., Couture, Y., Gaudreau, P., Morisset, J. and Brazeau, P., 1990. Influence of growth hormone-releasing factor and (or) thyrotropin-releasing factor on sows blood components, milk composition and piglet performance. Canadian Journal of Animal Science 70: 821-832.

Dusza, L., Sobczak, J., Jana, B., Murdza, A. and Bluj, W., 1991. Zastosowanie biolactinu-2 (pczyszczona prolaktyna swini) do stymulacji lactacji u loch. Medycyna Weter 47: 418-421.

Edwards, S.A., 2002. Perinatal mortality in the pig: environmental or physiological solutions? Livestock Production Science 78: 3-12.

Elsley, F.W.H., 1971. Nutrition and lactation in the sow. In: Falconer, I.R. (ed.) Lactation. Butterworths, London, UK, pp: 393-411

Farmer, C. and Hurley, W.L., 2015. Mammary development. Chapter 4. In: Farmer, C. (ed.) The gestating and lactating sow. Wageningen Academic Publishers, Wageningen, the Netherlands, pp. 73-94.

Farmer, C., Palin, M.F., Sorensen, M.T. and Robert, S., 2001. Lactational performance, nursing and maternal behavior of upton-meishan and large white sows. Canadian Journal of Animal Science 81: 487-493.

Farmer, C., Palin, M.F., Theil, P.K., Sorensen, M.T. and Devillers, N., 2012. Milk production in sows from a teat in second parity is influenced by whether it was suckled in first parity. Journal of Animal Science 90: 3743-3751.

Farmer, C., Petitclerc, D., Pelletier, G. and Brazeau, P., 1992. Lactation performance of sows injected with growth hormone-releasing factor during gestation and (or) lactation. Journal of Animal Science 70: 2636-2642.

Farmer, C. and Prunier, A., 2002. High ambient temperatures: how they affect sow lactation performance. Pig News and Information 23: 95N-102N.

Farmer, C., Robert, S. and Rushen, J., 1998. Bromocriptine given orally to periparturient or lactating sows inhibits milk production. Journal of Animal Science 76: 750-757.

Farmer, C., Sorensen, M.T., Robert, S. and Petitclerc, D., 1999. Administering exogenous porcine prolactin to lactating sows: milk yield, mammary gland composition, and endocrine and behavioral responses. Journal of Animal Science 77: 1851-1859.

Fisette, K., Laforest, J.P., Robert, S. and Farmer, C., 2004. Use of recorded nursing grunts during lactation in two breeds of sows. I. Effects on nursing behaviour and litter performance. Canadian Journal of Animal Science 84: 573-579.

Flummer, C. and Theil, P.K., 2012. Effect of β-hydroxy β-methyl butyrate supplementation of sows in late gestation and lactation on sow production of colostrum and milk and piglet performance. Journal of Animal Science 90(4): 372-374.

Foisnet, A., Farmer, C., David, C. and Quesnel, H., 2010. Relationships between colostrum production by primiparous sows and sow physiology around parturition. Journal of Animal Science 88: 1672-1683.

Fraser, D., 1980. A review of the behavioural mechanism of milk ejection of the domestic pig. Applied Animal Ethology 6: 247-255.

Gooneratne, A.D. and Thacker, P.A., 1990. Influence of an extended photoperiod on sow and litter performance. Livestock Production Science 24: 83-88.

Greenberg, L.G. and Mahone, J.P., 1982. Failure of a 16 h L:8 h D or an 8 h L:16 h D photoperiod to influence lactation or reproductive efficiency in sows. Canadian Journal of Animal Science 62: 141-145.

Grun, D., Reiner, G. and Dzapo, V., 1993. Investigations on breed differences in milk yield of swine. Part 1. Methodology of mechanical milking and milk yield. Reproduction in Domestic Animals 28: 14-21.

Hansen, A.V., Lauridsen, C., Sørensen, M.T., Bach Knudsen, K.E. and Theil, P.K., 2012a. Effects of nutrient supply, plasma metabolites and nutritional status of sows during transition on performance in the following lactation. Journal of Animal Science 90: 466-480.

Hansen, A.V., Strathe, A.B., Kebreab, E., France, J. and Theil, P.K., 2012b. Predicting milk yield and composition in lactating sows: a Bayesian approach. Journal of Animal Science 90: 2285-2298.

Harkins, M., Boyd, R.D. and Bauman, D.E., 1989. Effects of recombinant porcine somatotropin on lactational performance and metabolite patterns in sows and growth of nursing pigs. Journal of Animal Science 67: 1997-2008.

Harrell, R.J., Thomas, M.J. and Boyd, R.D., 1993. Limitations of sow milk yield on baby pig growth. In: Cornell University (ed.) Proceedings of the Cornell Nutrition Conference for Feed Manufacturers. October 19-21, 1993. Rochester, NY, USA, pp. 156-164.

Head, R.H. and Williams, I.H., 1991. Mammogenesis is influenced by pregnancy nutrition. In: Batterham, E.S. (ed.) Manipulating pig production III. Australasian Pig Science Association, Atwood, Australia, 33 pp.

Herpin, P., Louveau, I., Damon, M. and Le Dividich, J., 2005. Environmental and hormonal regulation of energy metabolism in early development of the pig. In: Burrin, D.G. and Mersmann, H. (eds) Biology of metabolism in growing animals. Elsevier Limited, Amsterdam, the Netherlands, pp. 353-374.

Jensen, P., Stangel, G. and Algers, B., 1991. Nursing and sucking behaviour of semi-naturally kept pigs during the first 10 days post partum. Applied Animal Behavior Science 31: 195-209.

Kim, S.W., Hurley, W.L., Wu, G. and Ji, F., 2009. Ideal amino acid balance for sows during gestation and lactation. Journal of Animal Science 87: E123-E132.

King, R.H. and Dunkin, A.C., 1986. The effect of nutrition on the reproductive performance of first-litter sows. 3. The response to graded increases in food intake during lactation. Animal Production 42: 119-126.

King, R.H. and Eason, P.J., 1998. The effect of bodyweight of sows on the response to dietary lysine during lactation. Journal of Animal Science 76(1): 162.

King, R.H., Mullan, B.P., Dunshea, F.R. and Dove, H., 1997. The influence of piglet body weight on milk production of sows. Livestock Production Science 47: 169-174.

Klobasa, F., Werhahn, E. and Butler, J.E., 1987. Composition of sow milk during lactation. Journal of Animal Science 64: 1458-1466.

Krogh, U., Flummer, C., Jensen, S.K. and Theil, P.K., 2012. Colostrum and milk production of sows is affected by dietary conjugated linoleic acid. Journal of Animal Science 90: 366-368.

Kusina, J., Pettigrew, J.E., Sower, A.F., White, M.E., Crooker, B.A. and Hathaway, M.R., 1999. Effect of protein intake during gestation and lactation in the lactational performance of primiparous sows. Journal of Animal Science 77: 931-941.

Lauridsen, C. and Danielsen, V., 2004. Lactational dietary fat levels and sources influence milk composition and performance of sows and their progeny. Livestock Production Science 91: 95-105.

Le Dividich, J., Rooke, J.A. and Herpin, P., 2005. Review: nutritional and immunological importance of colostrum for the newborn pig. Journal of Agricultural Science 143: 469-485.

Lewis, N.J. and Hurnik, F.J., 1985. The development of nursing behaviour in swine. Applied Animal Behavior Science 14: 225-232.

Loisel, F., 2014. Variabilité de la production de colostrum par la truie : rôle de la prolactine et de la progestérone et influence des fibres alimentaires. Ph.D. thesis. Agrocampus Ouest, France, 193 pp.

Loisel, F., Farmer, C., Ramaekers, P. and Quesnel, H., 2013. Effects of high fiber intake during late pregnancy on sow physiology, colostrum production and piglet performance. Journal of Animal Science 91: 5269-5279.

Loisel, F., Farmer, C., Ramaekers, P. and Quesnel, H., 2014. Colostrum yield and piglet growth during lactation are related to gilt metabolic and hepatic status prepartum. Journal of Animal Science 92: 2931-2941.

Mabry, J.W., Cunningham, F.L., Kraeling, R.R. and Rampacek, G.B., 1982. The effect of artificially extended photoperiod during lactation on maternal performance of the sow. Journal of Animal Science 54: 918-921.

Miller, Y.J., Collins, A.M., Smits, R.J., Thompson, P.C. and Holyoake, P.K., 2012. Providing supplemental milk to piglets preweaning improves the growth but not survival of gilt progeny compared with sow progeny. Journal of Animal Science 90: 5078-5085.

Noble, M.S., Rodriguez-Zas, S., Cook, J.B., Bleck, G.T., Hurley, W.L. and Wheeler, M.B., 2002. Lactational performance of first-parity transgenic gilts expressing bovine alpha-lactalbumin in their milk. Journal of Animal Science 80: 1090-1096.

Noblet, J. and Etienne, M., 1987. Body composition, metabolic rate and utilization of milk nutrients in suckling piglets. Reproduction Nutrition Development 27: 829-839.

Paulicks, B.R., Ott, H. and Roth-Maier, D.A., 2003. Performance of lactating sows in response to the dietary valine supply. Journal of Animal Physiology and Animal Nutrition 87: 389-396.

Peaker, M. and Wilde, C.J., 1987. Milk secretion: autocrine control. News in Physiological Science 2: 124-126.

Pedersen, M.L., Moustsen, V.A., Nielsen, M.B.F. and Kristensen, A.R., 2011. Improved udder access prolongs duration of milk letdown and increases piglet weigh gains. Livestock Science 140: 253-261.

Puppe, B. and Tuchscherer, A., 2000. The development of suckling frequency in pigs from birth to weaning of their piglets: a sociobiological approach. Animal Science 71: 273-279.

Quesnel, H., 2009. Nutritional and lactational effects on follicular development in the pig. In: Rodriguez-Martinez, H., Vallet, J.L. and Ziecik, A.J. (eds.) Control of pig reproduction VIII. Nottingham University Press, Nottingham, UK, pp. 121-134.

Quesnel, H., 2011. Colostrum production by sows: variability of colostrum yield and immunoglobulin G concentrations. Animal 5: 1546-1553.

Quesnel, H., Meunier-Salaün, M.C., Hamard, A., Guillemet, R., Etienne, M., Farmer, C., Dourmad, J.Y. and Père, M.C., 2009. Dietary fiber for pregnant sows: Influence on sow physiology and performance during lactation. Journal of Animal Science 87: 532-543.

Quesnel, H., Ramaekers, P., Van Hees, H. and Farmer, C., 2013. Short communication: relations between peripartum concentrations of prolactin and progesterone in sows and piglet growth in early lactation. Canadian Journal of Animal Science 93: 109-112.

Ramanau, A., Kluge, H., Spilke, J. and Eder, K., 2004. Supplementation of sows with L-carnitine during pregnancy and lactation improves growth of the piglets during the suckling period through increased milk production. Journal of Nutrition 134: 86-92.

Rooke, J.A. and Bland, I.M., 2002. The acquisition of passive immunity in the new-born piglet. Livestock Production Science 78: 13-23.

Salmon, H., Berri, M., Gerdts, V. and Meurens, F., 2009. Humoral and cellular factors of maternal immunity in swine. Developmental and Comparative Immunology 33: 384-393.

Sauber, T.E. and Stahly, T.S., 1996. Impact of dietary amino acid regimen on milk nutrient yield by sows differing in genetic capacity for lean tissue growth. Journal of Animal Science 74(1): 174.

Seerley, R.W. and Poole, D.R., 1974. Effect of prolonged fasting on carcass composition and blood fatty acids and glucose of neonatal swine. Journal of Nutrition 104: 210-217.

Soltwedel, K.T., Easter, R.A. and Pettigrew, J.E., 2006. Evaluation of the order of limitation of lysine, threonine, and valine, as determined by plasma urea nitrogen, in corn-soybean meal diets of lactating sows with high body weight loss. Journal of Animal Science 84: 1734-1741.

Speer, V.C. and Cox, D.F., 1984. Estimating milk yield. Journal of Animal Science 59: 1282-1285.

Spinka, M., Illmann, G., Algers, B. and Stetkova, Z., 1997. The role of nursing frequency in milk production in domestic pigs. Journal of Animal Science 75: 1223-1228.

Stone, C.C., Brown, M.S. and Waring, G.H., 1974. An ethological means to improve swine production. Journal of Animal Science 39: 137.

Taverne, M., Bevers, M. and Bradshaw, J.M.C., 1982. Plasma concentrations of prolactin, progesterone, relaxin and oestradiol-17β in sows treated with progesterone, bromocriptine or indomethacin during late pregnancy. Journal of Reproduction and Fertility 65: 85-96.

Theil, P.K., Cordero, G., Henckel, P., Puggaard, L., Oksbjerg, N. and Sørensen, M.T., 2011. Effects of gestation and transition diets, piglet birth weight, and fasting time on depletion of glycogen pools in liver and 3 muscles of newborn piglets. Journal of Animal Science 89: 1805-1816.

Theil, P.K., Flummer, C., Hurley, W.L., Kristensen, N.B., Labouriau, R.L. and Sørensen, M.T., 2014a. Mechanistic model to predict colostrum intake based on deuterium oxide dilution technique data and impact of gestation and lactation diets on piglet intake and sow yield of colostrum. Journal of Animal Science, in press. DOI: http://dx.doi.org/10.2527/jas2014-7841.

Theil, P.K., Jørgensen, H. and Jakobsen, K. 2004. Energy and protein metabolism in lactating sows fed two levels of dietary fat. Livestock Production Science 89: 265-276.

Theil, P.K., Jørgensen, H. and Jakobsen, K., 2002. Energy and protein metabolism in pregnant sows fed two levels of dietary protein. Journal of Animal Physiology and Animal Nutrition 86: 399-413.

Theil, P.K., Kristensen, N.B., Jørgensen, H., Labouriau, R. and Jakobsen, K., 2007. Milk intake and carbon dioxide production of piglets determined with the doubly labelled water technique. Animal 1: 881-888.

Theil, P.K., Labouriau R., Sejrsen, K., Thomsen, B. and Sørensen, M.T., 2005. Expression of genes involved in regulation of cell turnover during milk stasis and lactation rescue in sow mammary glands. Journal of Animal Science 83: 2349-2356.

Theil, P.K., Lauridsen, C. and Quesnel, H., 2014b. Neonatal piglet survival: impact of sow nutrition around parturition on foetal glycogen deposition, and production and composition of colostrum and transient milk. Animal 8: 1021-1030.

Theil, P.K., Nielsen, M.O., Sørensen, M.T. and Lauridsen, C., 2012. Lactation, milk and suckling. In: Bach Knudsen, K.E., Kjeldsen, N.J., Poulsen, H.D. and Jensen, B.B. (eds.) Nutritional physiology of pigs. Danish Pig Research Centre, Copenhagen, Denmark, pp. 1-47.

Theil, P.K., Sejrsen, K., Hurley, W.L., Labouriau, R., Thomsen, B. and Sørensen, M.T., 2006. Role of suckling in regulating cell turnover and onset and maintenance of lactation in individual mammary glands of sows. Journal of Animal Science 84: 1691-1698.

Thodberg, K. and Sorensen, M.T., 2006. Mammary development and milk production in the sow: effects of udder massage, genotype and feeding in late gestation. Livestock Science 101: 116-125.

Tilton, S.L., Miller, P.S., Lewis, A.J., Reese, D.E. and Ermer, P.M., 1999. Addition of fat to the diets of lactating sows: I. Effects on milk production and composition and carcass composition of the litter at weaning. Journal of Animal Science 77: 2491-2500.

Toner, M.S., King, R.H., Dunshea, F.R., Dove, H. and Atwood, C.S., 1996. The effect of exogenous somatotropin on lactation performance of first-litter sows. Journal of Animal Science 74: 167-172.

Touchette, K.J., Allee, G.L., Newcomb, M.D. and Boyd, R.D., 1998. The lysine requirement of lactating primiparous sows. Journal of Animal Science 76: 1091-1097.

Van der Steen, H. and De Groot, P.N., 1992. Direct and maternal breed effects on growth and milk intake of piglets: Meishan versus Dutch breeds. Livestock Production Science 30: 361-373.

Verstegen, M.W.A., Mesu, J., Van Kempen, G. and Geerse, C., 1985. Energy balances of lactating sows in relation to feeding level and stage of lactation. Journal of Animal Science 60: 731-740.

Wechsler, B. and Brodmann, N., 1996. The synchronization of nursing bouts in group-housed sows. Applied Animal Behavior Science 47: 191-199.

Whitacre, M.D. and Threlfall, W.R., 1981. Effects of ergocryptine on plasma prolactin, luteinizing hormone and progesterone in the periparturient sow. American Journal of Veterinay Research 42: 1538-1541.

Wientjes, J.G.M., Soede, N.M., Knol, E.F., Van den Brand, H. and Kemp, B., 2013. Piglet birth weight and litter uniformity: effects of weaning-to-pregnancy interval and body condition changes in sows of different parities and crossbred lines. Journal of Animal Science 91: 2099-2107.

Wolter, B.F., Ellis, M., Corrigan, B.P. and DeDecker, J.M., 2002. The effect of birth weight and feeding of supplemental milk replacer to piglets during lactation on preweaning and postweaning growth performance and carcass characteristics. Journal of Animal Science 80: 301-308.

Wung, S.C., Wu, H.P., Kou, Y.H., Shen, K.H., Koh, F.K. and Wan, W.C.M., 1977. Effect of thyrotropin-releasing hormone on serum thyroxine of lactating sows and the growth of their suckling young. Journal of Animal Science 45: 299-304.

Xu, R.J., Sangild, P.T., Zhang, Y.Q. and Zhang, S.H., 2002. Bioactive compounds in porcine colostrum and milk and their effects on intestinal development in neonatal pigs. In: Zabielski, R., Gregory, P.C. and Weström, B. (eds.) Biology of the intestine of growing animals. Elsevier Science, Amsterdam, the Netherlands, pp. 169-192.

York, D.L. and Robison, O.W., 1985. Genotypic and phenotypic parameters of milk production in primiparous Duroc sows. Journal of Animal Science 61: 825-833.

9. 母猪初乳和常乳的组成

W. L. Hurley

University of Illinois, Agricultural Education Program, 430 Animal Sciences Laboratory, 1207 W. Gregory, Urbana-Champaign, IL 61801, USA; wlhurley@illinois.edu

摘要：乳汁的成分包括蛋白质、脂质、碳水化合物、矿物质、维生素和细胞。这些成分的含量受多种因素的影响，其中所处哺乳阶段对其组成具有最显著的影响。与常乳相比，在分娩后最初的24 h里，来自母猪的乳房分泌物中其免疫球蛋白、一些微量矿物质和维生素，以及激素和生长因子的浓度更高，而乳糖浓度更低。母猪常乳中的脂肪浓度在哺乳的第2~4天短暂增加。常乳的组成在哺乳期第7~10天之后，与哺乳期的剩余阶段相对保持稳定。日粮可影响一些乳汁成分，包括脂肪、脂溶性维生素和一些矿物质的浓度，以及特定脂肪酸的比例。遗传、胎次、初乳和常乳产量和环境温度，已被证实会影响初乳和常乳的成分组成。本章总结了一些关于母猪初乳和常乳组成的文献，还提供了基于文献中报道的各组分浓度的基准平均值。

关键词：乳腺，蛋白质，脂肪，乳糖，免疫球蛋白

9.1 引言

在胎儿发育阶段，仔猪依赖于母猪为其提供所有营养物质、生长刺激因子和保护因子。一旦出生，仔猪继续依靠母猪，通过从乳腺分泌的流体完成这些物质输入。母猪的哺乳功能是母猪对胎儿期以后的仔猪生长和发育贡献的延续。初乳和常乳对新生仔猪的至关重要性已经被证实（Le Dividich 等，2005；Quesnel 等，2012；Theil 等，2014；也参见第8章；Quesnel 等，2015）。新生仔猪的快速发育与哺乳仔猪消耗的乳腺分泌物组成的快速变化相一致。

乳汁的组分包括碳水化合物、脂质、蛋白质、矿物质、维生素和细胞。这一表面上的乳汁成分列表，对这种异常复杂的液体来说是太过简单了。乳汁是一种介质，通过该介质，母猪将容易利用和容易消化的能量源，以脂质、氨基酸、矿物质、维生

素和一系列生物活性成分这几种形式,递送给新生仔猪。当考虑到其迅速合成、分泌、摄入和消化时,这种液体更显非凡。在乳汁喷射发生的几秒钟内,乳汁迅速从乳腺中吸除(Brooks 和 Burke,1998),并且一旦其进入仔猪的胃,就迅速形成凝乳。胃中的乳汁成分必须进入肠道,并且在 45min 内基本完成消化,以准备下一次哺乳的乳汁。初始底物在经历复杂细胞过程后,合成了乳糖、脂肪、蛋白质和其他乳成分,在合成后几个小时内,将很大程度上被消化并吸收到仔猪的血液中。

  本章重点关注母猪的初乳和常乳的组成,特别是在这些乳腺分泌物中发现的许多成分的含量。初乳和常乳中组分浓度的估算受多种因素的影响。与如何收集和分析乳腺分泌物样品相关的影响因素将在方法论部分概述。乳腺的生理状态是乳腺分泌物组成中的主要决定因素,这在初乳与常乳组成之间的差异中可以非常清楚地看到。乳腺分泌物的每个主要组成部分的章节都被组织起来,以说明在哺乳阶段的成分是如何变化的,以及其他因素如何影响其组成。对于母猪初乳和常乳组成的其他总结,读者可参考 Neuhaus(1961),Bowland(1966),Hartmann 和 Holmes(1989),以及 Darragh 和 Moughan(1998)。母猪乳组成与其他物种的比较可以在 Oftedal(1984),Oftedal 和 Iverson(1995)和 Park(2011)的综述中找到。

  本章中的表格为读者提供了在母猪初乳和常乳中发现的各组分的平均浓度,指出各组分报道值的范围,以及用于推导平均浓度的来源列表。这些列表是根据许多研究提出代表性的组分浓度值,这些研究在大多数情况下,提供了对哺乳期多个点的估算,或者提供了专门针对初乳成分的数据。这些表并不意味着对所有评估猪乳组成的研究进行了详尽的汇编。表 9.1 包含了用于计算表 9.2 中初乳(分娩至产后 24 h)和表 9.3 中常乳(产后 48 h 和之后)中平均组分含量的报道值的参考文献。

表 9.1 用于计算表 9.2 和 9.3 中母猪初乳和常乳主要成分含量估算值的参考资料。本表是一份不完整的参考文献报告,为哺乳期多天乳成分结果报道的汇总。

文献来源	第 1 采样时间和分娩情况	其他样品采样时间	报道的主成分[1]
Perrin, 1955	分娩时	3,6,9,12,15,18,21,24,27,30,33~48 h;3,4,5 天;2,3,4,5,6,7,8 周	ts, lac, fat, pro, ash, snf
Elliott 等, 1971	产后 12 h 内	7,14,21 天	ts, lac, fat, pro, ash, snf
Mahan 等, 1971	13 天	21,29 天	ts, lac, fat, pro, ash, snf
Fahmy, 1972	产后 3 h	14,28,25 天	ts, fat, pro, ash, ge
Coffey 等, 1982	24 h	72 h	fat
Boyd 等, 1982	产出第 1 头仔猪时	9,18 天	ts, fat, pro
White 等, 1984	15 天	22 天	ts, lac, fat, pro, ge

妊娠和哺乳母猪

续表 9.1

文献来源	第 1 采样时间和分娩情况	其他样品采样时间	报道的主成分[1]
Loudenslager 等,1986	分娩期间	2,21 天	fat
Klobasa 等,1987	仔猪吃奶前	6,12,18,24,48,72 h;5,7,14,21,28,35,42 天	ts, lac, fat, pro
Zou 等,1992	产出第 1 头仔猪 8 h 内	24 h;7,21 天	lac, fat, pro
Taugbøl 等,1993	分娩时	14 天	fat
Jackson 等,1995	产出第 1 头仔猪时	1,2,3,4,6,9,12,18,24,48,72,169 h	lac, fat, pro
Csapo 等,1996	产后立刻	12,24,48 h;3,5,10,20,45~60 天	ts, fat, pro
Dourmad 等,1998	3 天	7,15,22 天	ts, lac, fat, pro, ash, ge
Tilton 等,1999	18 天		ts, fat, pro, ash, ge
Alston-Mills 等,2000	3 天	21 天	lac, fat, pro
Kim and Mahan,2001	分娩时	2,3,4,5,6,7,10,14 天	pro
Noble 等,2002	在分娩的后 3 h	6,12,24,48 h;3,6,9,12,21 天	ts, lac, pro
Devillers 等,2004b	产出第 1 头仔猪时	6,12,24 h	ts, lac, fat, pro, ash, ge
Laws 等,2009	产出第 1 头仔猪 4 h 内	3,7,14,21 天	lac, fat, pro, ge
Farmer 等,2010	3 天	20 天	ts, lac, fat, pro
Foisnet 等,2010a	产出第 1 头仔猪时	3,6,24 h	ts, lac, fat, pro, ash, ge
Foisnet 等,2010b	产出第 1 头仔猪时	6,12,24,36,48 h	ts, lac, fat, pro, ash, ge
Leonard 等,2010	产出第 1 头仔猪 1 h 内	12 天	ts, fat, pro
De Quelen 等,2010	分娩前 12 h	3,7,21,32 天	fat
Ariza-Nieto 等,2011	初乳,未标明具体时间	7,14 天	fat, pro, ge
Foisnet 等,2011	产出第 1 头仔猪后	6,12,24,36 h	ts, lac, fat, pro, ash, ge
Krogh 等,2012	分娩时		ts, lac, fat, pro
Flummer and Theil,2012	产出第 1 头仔猪 3 h 内		ts, lac, fat, pro, ge
De Quelen 等,2013	分娩前 12 h	7,14,21,28 天	fat
Loisel 等,2013	产出第 1 头仔猪后	24 h;7,21 天	ts, lac, fat, pro, ash, ge

[1] ts=总固形物或干物质,lac=乳糖,fat=乳脂,pro=粗蛋白或总蛋白,ash=灰分,ge=总能量,snf=固体 非脂肪。

9.2 研究方法

9.2.1 乳腺分泌物的取样

所有报道初乳或常乳成分的研究都依赖于一些收集分泌物的方法。在分娩期间乳腺分泌物的收集，可以在不施用催产素刺激乳汁排出的情况下实现。内源催产素浓度在分娩期间以脉动形式升高（Gilbert 等，1994）。当对出生后约 6 h 的第一头仔猪收集初乳样品时，必需施加催产素（Jackson 等，1995）。催产素的施用量范围为 10～80 IU，其用量取决于施用途径，例如静脉内（10～20 IU；Devillers 等，2004b；Theil 等，2004），肌内（20～40 IU；Jackson 等，1995；Klobasa 等，1987）或乳房内注射（80 IU；Noble 等，2002）。通常，要将仔猪从母猪身边取走一段时间，以在乳汁样品收集之前积累乳汁（Klobasa 等，1987）。也有其他人使用挤奶机收集乳汁样品（Garst 等，1999）。在后一种情况下，母猪挤奶在一天中的时间会影响乳中脂肪、蛋白质和体细胞浓度（Garst 等，1999）。有关测量母猪初乳和常乳产量的更详细描述，请参见本书第 8 章初乳和常乳生产（Quesnel 等，2015）。乳汁通常从部分功能性乳腺，或从所有功能性乳腺收集样品，并且使用混合样品进行分析。乳脂、乳蛋白质和乳糖的浓度在母猪的前乳腺和后乳腺之间没有显著变化（Reynolds 和 Rook，1977）。

施用催产素可影响乳汁的成分，其中与手工刺激或静脉内注射 10 IU 催产素相比，静脉内注射 20 IU 催产素使得乳中的总固形物（干物质）、乳脂、乳糖和能量浓度较低（Hartog 等，1987）。与肌肉内注射催产素相比，静脉内注射催产素可导致较低的乳脂浓度，但给药途径不影响乳蛋白质浓度（Garst 等，1999）。乳成分随着乳汁排出期间除去的部分而变化，其中后乳（在仔猪完成哺乳后响应于催产素注射而除去的部分）具有较高浓度的总固形物、脂肪和能量，但与前乳部分相比，并未影响乳蛋白质和乳糖浓度（Atwood 和 Hartmann，1992）。

9.2.2 主要乳汁成分的分析

用于确定乳汁总组成的许多方法，来自美国官方分析化学家协会（Association of Official Analytical Chemists，AOAC）公布的官方分析方法。这里提供了用于乳成分分析方法的简要概述。总固形物或干物质，通常在乳汁样品干燥（也可以使用冷冻干燥）后，通过重量分析法测定。灰分含量是通过在马弗炉中 550℃下焚烧样品来测定。脂肪百分比通常使用巴布科克法（Babcock method）或格伯法（Gerber method）测定，两者都使用硫酸水解除脂质以外的有机组分。与一种溶剂萃取方法，即莫久尼耳法（Mojonnier method）相比，巴布科克法被注意到其测定的脂肪值略低（Fahmy，1972）。其他采用溶剂萃取的方法，如索氏提取法（Soxhlet apparatus）、瑞式法（Roese-Gottlieb method）、莫久尼耳法，它们采用乙醚或石油醚萃取，随后干燥并对萃取物称重。用于测定总乳蛋白质的方法包括凯氏定氮法（Kjeldahl method）、改进的劳瑞蛋白测定法（Lowry protein assay）（铜结合法，copper-binding assay）和改进蛋白质定量分析法（Bradford protein assay）（染料结合测定，dye-binding assay），以及最近的报道使用的自动化仪器直接测定氮元素含量。乳糖可通过还原糖法、酶水解法，或者通过从总固形物值中减去其他组分的差异法来测定。测定乳样品中乳糖的酶水解法，是通过 β-半乳糖苷酶将乳糖水解成 D-葡萄糖和 D-半乳糖，然后通过烟酰胺腺嘌呤二核苷酸氧化 D-半乳糖，然后测量还原的 NADH 的吸光度来进行的。总乳能量含量通常直接通过使用绝热弹式量热计的燃烧产热测出，或通过乳脂、总蛋白质和乳糖的百分比计算而得到（Klaver 等，1981）。最近的报道采用了傅立叶变换红外光谱分析（Fourier Transform Infared analysis）方法（Krogh 等，2012）。

9.3　物理化学性质

母猪初乳在分娩开始时的比重约 1.06 g/mL，反映了当时的高总蛋白质浓度。然后在第一天内下降，并且在大部分哺乳期间稳定在约 1.035 g/mL，然后在超过 6 周的延长哺乳期中略微升高（Fahmy，1972；Krakowski 等，2002；Sheffy 等，1952；Whittlestone，1952）。母猪初乳的 pH 比常乳的酸性更强。Kent 等（1998）报道，分娩前后初乳的 pH 为 5.7，在第 1 天升至 6.0，在第 9 天达到 6.9，这与其他一些报道一致（Coffey 等，1982；Miller 等，1971）。有其他报道称母猪常乳的 pH 范围更广，甚至会高于 7.0（DeRouchey 等，2003；Sheffy 等，1952；Whittlestone，1952）。乳汁的 pH 与水相和酪蛋白胶束相之间的钙平衡有关（Gaucheron，2005）。

描述母猪初乳或常乳其他物理化学性质的报道有限。母猪初乳的黏度在分娩时最高，然后下降，估算常乳黏度为 3.855 mPa·s（millipascal-seconds）（Whittlestone，1952）。在分娩的第 1 天，乳汁的电导率增加，然后缓慢下降，到约哺乳期的第 16 天，此时母猪常乳的平均电导率估算值为 3.248 ohm$^{-1} \times 10^3$（ohm 为欧姆）（Whittlestone，1952）。乳汁的凝固点通常用作判断是否向乳汁中掺水的手段。对于母猪常乳，凝固点估算值为 -0.563 ℃（Sheffy 等，1952）。

9.4　水和总固形物

水是乳腺分泌物的关键组分。它提供了一种介质，用于在乳汁合成和分泌期间混合其他组分，同时为仔猪提供水这种营养物质。作为乳汁其中一部分的水，其与乳腺分泌物中主要碳水化合物，即乳糖的合成和分泌密切相关。乳腺分泌物中的总固形物或干物质含量常被报道，水含量则少被提及。可通过在低于 100 ℃ 时蒸发来自乳样的水，使用重量分析法测定总固形物或干物质含量。因此，估算的总固形物含量包括了乳腺分泌物中的所有有机和无机组分。

9.4.1 哺乳阶段

根据母猪乳腺分泌物的总固形物含量(表9.2)可得出,水在分娩期间和分娩当时,占分泌物总质量的约73%。然后,水分含量在产后12 h增加至约80%,并且在整个哺乳期中保持在77%~81%(表9.3)。由于蛋白质,主要是免疫球蛋白的浓度快速下降,尽管同期乳糖含量有所增加,但仅能部分抵消蛋白质减少,因此导致在产后的最初1天中水分含量的增加(表9.2)。

表9.2 母猪初乳主要成分报道浓度的平均值和范围。

初乳	分娩后时间(h)					
	0[1]	3~4	6	12	18	24
总固形物(%)						
平均值[2]	26.7	28.1	23.8	20.1	18.4	20.1
范围[3]	24.0~30.2	26.7~28.9	21.8~26.6	18.4~21.6	17.7~19.4	17.2~23.4
研究数[4]	12	3	4	4	2	11
总蛋白质(%)						
平均值	16.6	16.7	13.8	9.6	9.4	7.7
范围	13.8~19.7	12.7~19.1	11.3~16.5	5.6~13.2	7.2~13.6	3.3~10.5
研究数	14	4	6	5	3	13
脂肪(%)						
平均值	6.4	6.1	5.9	5.9	6.4	8.0
范围	4.9~10.9	5.5~7.3	4.8~7.8	4.9~7.2	5.2~7.0	5.6~11.6
研究数	16	4	4	6	3	11
乳糖(%)						
平均值	2.8	2.7	3.0	3.6	4.1	3.9
范围	2.4~3.2	2.4~3.2	2.6~3.2	3.3~4.1	3.9~4.4	3.6~4.3
研究数	10	5	6	5	3	11
灰分(%)						
平均值	0.68		0.63	0.64		0.67
范围	0.54~0.70		0.61~0.68	0.63~0.66		0.61~0.68
研究数	7		4 (3~6 h)	4 (8~14 h)		8
能量(kJ/g)						
平均值	6.7		6.0			5.7
范围	5.5~8.3		5.2~6.7			4.6~6.4
研究数	7		2 (4~6 h)			8

[1] 定义为下列之一:立即分娩前,在第1头仔猪出生时,或在分娩期间。
[2] 来自分娩相应不同时间点乳腺分泌物研究的平均值。
[3] 来自确定平均值研究中值的范围。
[4] 确定平均值所包括的研究数。值是来自于多个动物(例如来自Perrin,1955)或多个实验(例如来自Jackson等,1995)的平均值,或来自于当研究中包括实验处理组时的对照组,或来自于给定的总组分浓度。用于计算平均值的参考文献列于表9.1。

表 9.3　母猪常乳中主要成分报道浓度的平均值和范围。

常乳	哺乳天数						
	2[1]	3	7	12～15	20～22	27～29	42～60
总固形物(%)							
平均值[2]	22.1	22.7	19.3	20.0	19.5	18.8	19.5
范围[3]	18.6～27.6	19.0～27.0	18.3～20.4	18.2～22.0	17.3～23.2	17.3～20.9	17.0～21.0
研究数[4]	5	6	6	4	8	4	6
总蛋白质(%)							
平均值	7.5	6.5	5.4	5.3	5.0	5.3	6.5
范围	5.4～10.4	4.6～9.9	3.6～6.4	4.7～7.1	3.6～6.0	5.36～5.4	5.4～8.2
研究数	7	10	9	12	9	3	6
脂肪(%)							
平均值	10.1	9.7	7.6	7.5	7.5	7.0	7.1
范围	6.5～12.9	5.4～13.0	4.5～8.8	5.3～10.8	5.0～11.5	5.2～9.8	5.3～8.8
研究数	5	9	11	10	16	6	6
乳糖(%)							
平均值	4.3	4.6	5.2	5.2	5.1	5.6	5.0
范围	4.0～4.5	3.8～5.3	4.7～5.6	5.1～6.3	4.0～5.8	4.9～6.0	4.3～5.7
研究数	5	8	8	7	13	3	5
灰分(%)							
平均值	0.75	0.79	0.81	0.90	0.86	0.89	1.02
范围	0.75～0.76	0.74～0.82	0.76～0.87	0.77～1.33	0.76～0.95	0.82～1.06	0.84～1.14
研究数	3	3	3	5	5	4	4
能量(kJ/g)							
平均值	6.5	6.0	5.4	4.9	5.0	4.4	
范围	—	5.5～6.5	5.1～5.6	4.6～5.3	4.2～5.1	4.2～4.7	
研究数	1	3	4	5	6	2	

[1] 定义为分娩后约 48h 或第 1 头仔猪出生后 > 24 h。
[2] 来自分娩相应不同时间点乳腺分泌物研究的平均值。
[3] 来自确定平均值研究中值的范围。
[4] 确定平均值所包括的研究数。值是来自于多个动物(例如来自 Perrin,1955)或多个实验(例如来自 Jackson 等,1995)的平均值,或来自于当研究中包括实验处理组时的对照组,或来自于给定的总组分浓度。用于计算平均值的参考文献列于表 9.1。

总固形物含量在分娩后的最初 4～6 h 内最高,然后在 12 h 后下降至乳腺分泌物质量的约 20%(表 9.2)。在第 2 天和第 3 天,总固形物短暂地增加至 22%～23%,反映了在那时发生的脂肪百分比的峰值。在剩余的哺乳期里,总固形物保持在约 19%。非脂固形物(solids-not-fat,SNF)为总固形物百分比和脂肪百分比之间的差值。非脂固形物含量可作为对脱脂乳组分的粗略估算,主要代表了蛋白质、乳糖和灰分。非脂固形物含量的报道在早期研究中很普遍(参见表9.1),然而,最近的研究或者分别报道每种主要的乳成分,或者关注于测量一种或两种特定乳成分,例如,脂肪或免疫球蛋白含量。

9.4.2 日粮和其他因素

已有报道称,日粮和环境因素对母猪乳中总固形物含量有影响。一些研究已观察到,用于补充日粮能量的不同来源会影响乳汁的总固形物含量(Van den Brand等,2000;White等,1984)。其他研究发现,在哺乳期间以玉米淀粉(Coffey等,1982)或动物脂肪形式来补充日粮能量,对乳总固形物含量没有影响(Coffey等,1982;Theil等,2004)。饲喂粗甘油(高至6.77%)给母猪,在哺乳的第18天有增加乳汁总固形物含量的趋势(Schieck等,2010)。妊娠期间的日粮纤维水平不影响初乳或常乳中总固形物含量(Loisel等,2013)。然而,在产前和哺乳期间的日粮中盐水平(0.1%对0.4%)影响了母猪的水摄入,但不影响哺乳第18天时的常乳总固形物含量(Seynaeve等,1996)。乳汁总固形物含量受环境温度的影响,总固形物百分比在20℃时比在29℃时低(Renaudeau和Noblet,2001)。

9.5 碳水化合物

乳糖是猪乳中的主要碳水化合物。它也是乳汁中的主要渗量(渗透压克分子),其在乳腺上皮细胞中合成,导致将水吸入分泌小泡(Peaker,1983)。L乳糖也被认为是乳汁中最稳定的组分,其浓度通常保持在一个狭窄的范围内。乳糖含量与脂肪或蛋白质相比,母猪的变异系数较低(Atwood和Hartmann,1992)。

9.5.1 哺乳阶段

在产后最初几个小时期间,初乳中的乳糖浓度比常乳低(表9.2和9.3)。乳糖的浓度在哺乳期的前2～3天逐渐增加。这个时期与乳生成的后期阶段一致(下面讨论)。这也是乳糖酶(在新生仔猪中水解乳糖的主要肠酶)活性增加的时期(第15章;LeHuërou-Luron和Ferret-Bernard,2015)。乳糖浓度可能会在延长的哺乳期的第7～8周时下降(表9.3),然而,该阶段的信息很有限。

相对于乳糖浓度(27～56 mg/mL;表9.2和表9.3),母猪乳腺分泌物中的葡萄糖浓度较低(18～135 μg/mL;Atwood和Hartmann,1995)。葡萄糖浓度从分娩时的低值增加到哺乳期第3天时的峰值,然后在第5天再次下降(Atwood和Hartmann,1995)。葡萄糖-6-磷酸在分娩后迅速增加,并且在第5天内持续升高,而葡萄糖-1-磷酸浓度在分娩后立即下降。半乳糖浓度在初乳中最高,在哺乳的第5天降低(Atwood和Hartmann,1995)。

猪乳还含有复杂的碳水化合物,包括29种不同的寡糖(Tao等,2010)。这些寡糖的浓度在初乳中最高,然后在哺乳期间下降,之后在第 24 天前又再次增加。此外,猪乳中寡糖的分布特征随着哺乳阶段而变化。

9.5.2 日粮

妊娠期日粮组成被认为可能会影响初乳和乳糖含量。然而,初乳或常乳中的乳糖浓度,不受以油脂形式为能量补充的妊娠日粮的影响(Jackson 等,1995;Laws 等,2009),也不受补充有共轭亚油酸的妊娠日粮的影响(Krogh 等,2012),或亮氨酸代谢物 β-羟基-β-甲基丁酸盐(Flummer 和 Theil,2012)的影响。高膳食纤维含量(23.4%)的妊娠日粮不影响初乳或乳糖浓度(Loisel 等,2013)。

在哺乳日粮中补充脂肪对乳糖含量没有影响(Coffey 等,1982;Lauridsen 和 Danielsen,2004;Theil 等,2004)。猪乳中的乳糖浓度不受哺乳日粮中的蛋白质水平、赖氨酸水平或支链氨基酸水平的影响(Dourmad 等,1998;King 等,1993;Richert 等,1997)。然而,猪乳的乳糖含量受能量来源的影响,饲喂淀粉会比牛油导致更高的乳糖百分比(Van den Brand 等,2000)。与对照组或高葡萄糖日粮相比,在哺乳期饲喂高果糖日粮的母猪在第 22 天的乳汁中的乳糖浓度较高(White 等,1984)。补充甘油的哺乳日粮可增加乳糖含量(Schieck 等,2010)。

有报道称,在哺乳日粮中补充叶酸,会使乳糖在哺乳的第 7 天降低(Wang 等,2011),然而,应当注意的是,该研究中的乳糖浓度比其他通常的报道值要低(表 9.3)。乳汁在哺乳期第 10~12 天的乳糖浓度不受日粮电解质平衡的影响(De Rouchey 等,2003)。

9.5.3 其他因素

Zou 等(1992)指出,与大约克母猪相比,第 1 次哺乳的梅山母猪初乳中的乳糖浓度较低。据报道,猪乳中的乳糖浓度在各品种之间没有差异(Alston-Mills 等,2000;Zou 等,1992),然而也有发现,与大约克母猪相比,具有 50% 梅山血缘的杂交母猪,其乳中的乳糖浓度较高(Farmer 等,2001)。母猪初乳中的乳糖浓度与初乳产量呈正相关(Foisnet 等,2010a)。类似地,哺乳期第 22 天的乳中乳糖含量与产奶量呈正相关($r=0.33$,$P<0.1$;White 等,1984)。猪乳的乳糖浓度不受胎次(Baas 等,1992;Goransson,1990;Klobasa 等,1987)、母猪体况(Klaver 等,1981)和环境温度(Renaudeau 和 Noblet,2001)的影响。

9.6 脂肪

乳腺分泌物中的脂肪含量被认为是最易变的组分。在考虑哺乳阶段、日粮和其他因素对初乳和常乳脂肪含量的影响时,这一点尤为明显。

9.6.1 哺乳阶段

在表9.1所列的各种研究中,报道了各哺乳阶段里,乳腺分泌物的平均脂肪含量或脂质浓度的分布情况(表9.2和表9.3)。在表9.2总结的各研究中,初乳在分娩时的脂肪含量为4.9%～10.9%,平均为6.4%。直到产后18 h,平均脂肪含量保持在5.9%～6.4%,然后在产后24 h增加至约8%。大多数研究发现脂肪含量的瞬时峰值,通常在产后24 h到第3天之间发生。尽管另有一些研究报道称,脂肪百分比甚至在第7天还在持续升高。已有报道称在第3天发现了高达13%的脂肪浓度(Csapo等,1996)。乳腺分泌物中脂肪的这种升高,恰巧同时发生在初乳阶段(分娩至产后24 h)和哺乳期的大约第4天之间的过渡乳阶段(Theil等,2014)。平均脂肪百分比从哺乳期的第7天到第6～8周都相对稳定在7.0～7.6(表9.3),这一阶段,正是常乳所处的时期(Theil等,2014)。

母猪乳汁中主要脂肪酸的平均浓度如表9.4所示。Csapo等(1996)报道了母

表9.4 母猪初乳和常乳中的氨基酸与常乳中脂肪酸的平均值和范围

	氨基酸(占总蛋白质%)[1]		脂肪酸(占总脂肪酸%)[2]	
	初乳(范围)	常乳(范围)		常乳(范围)
丙氨酸	4.6 (4.4～4.8)	3.4 (2.8～3.9)	C14:0	4.05 (2.3～6.4)
精氨酸	5.9 (5.6～6.1)	5.2 (4.6～5.8)	C16:0	29.3 (17.0～37.0)
天冬氨酸	8.6 (7.9～9.3)	8.1 (7.3～8.6)	C16:1	9.8 (7.4～13.8)
胱氨酸	1.8 (1.7～1.8)	1.5 (1.3～1.7)	C18:0	4.41 (2.6～6.0)
谷氨酸	17.8 (17.5～18.1)	22.0 (18.9～28.8)	C18:1[3]	32.07 (29.4～39.2)
甘氨酸	3.6 (3.1～4.0)	3.2 (2.3～3.6)	C18:2n-6	15.69 (8.9～25.9)
组氨酸	2.5 (2.1～3.3)	2.9 (2.3～3.9)	C18:3n-3	1.38 (0.6～2.9)
异亮氨酸	3.4 (2.4～3.9)	4.0 (2.9～4.4)	C20:4n-6	0.50 (0.1～0.9)
亮氨酸	9.7 (9.1～10.2)	8.8 (8.1～10.1)	C20:5n-3	0.38 (0.2～0.6)
赖氨酸	6.7 (6.3～7.3)	7.3 (7.0～7.9)	C22:5n-3	0.39 (0.2～0.7)
蛋氨酸	1.5 (1.2～1.7)	1.8 (1.4～2.0)	C22:6n-3	0.86 (0.2～2.1)
苯丙氨酸	4.4 (4.1～4.6)	3.9 (3.6～4.2)		
脯氨酸	9.9 (9.1～10.6)	11.9 (10.9～12.3)		
丝氨酸	6.8 (6.5～7.0)	5.3 (4.5～5.8)		
苏氨酸	5.9 (5.2～6.8)	4.1 (3.6～4.4)		
色氨酸	1.9 (1.6～2.2)	1.4 (1.3～1.6)		
酪氨酸	5.0 (4.0～6.1)	4.2 (3.9～4.9)		
缬氨酸	5.8 (5.0～6.3)	4.9 (3.9～5.5)		

[1] 对于氨基酸,初乳是以分娩后12 h内计,常乳以第17～28天计。用于计算平均值的参考文献:Bowland,1966;Csapo等,1996;Dourmad等,1998;Dunshea等,2005;Elliott等,1971;King,1998年。

[2] 对于脂肪酸,常乳以第7～20天计。用于计算平均值的参考文献:Csapo等,1996;Fritsche等,1993;Lauridsen和Danielsen,2004;Peng等,2010;Rooke等,1998;Taugbøl等,1993。

[3] 包括C18:1n-7和C18:1n-9;C18:1n-7约是总C18:1的5.5%。

猪初乳中脂肪酸的浓度。当在比较初乳(分娩时)和哺乳期第 20 天的脂肪酸比例时,在哺乳期 C16:0,C16:1 和 C18:3 比例会有所增加,而 C18:1 和 C18:2 比例会部分降低。母猪乳腺分泌物仅含有微量的短链和不超过 12 个碳(C12:0)的中链脂肪酸(Csapo 等,1996;Hartog 等,1987;Lauridsen 和 Danielsen,2004)。Csapo 等(1996)发现,不超过 0.25% 的总脂肪酸是由短链和 4~12 个碳(C4:0~C12:0)的中链脂肪酸组成。其他人报道了 C10:0 和 C12:0 的比例分别为总脂肪酸的 0.36% 和 0.49%(Lauridsen 和 Danielsen,2004)。

9.6.2 日粮

日粮对初乳或常乳成分的影响,主要是通过对乳脂含量的影响而实现。妊娠日粮预期可能会影响初乳组成,并可能对泌乳有后续影响。而且,只有在产前开始饲喂,哺乳日粮才会影响初乳组成。哺乳日粮将预期会影响常乳的组成。一些研究发现,补充日粮脂肪会增加母猪乳腺分泌物中的乳脂率(Boyd 等,1982;Pettigrew,1981;Shurson 和 Irvin,1992;Shurson 等,1986),而其他研究发现,在妊娠晚期和哺乳期补充日粮脂肪,对乳脂率没有显著影响(De Quelen 等,2010,2013;Farmer 等,2010;Jackson 等,1995;Lauridsen 和 Danielsen 等,2010;Miller 等,1971;Schieck 等,2010;Seerley 等,1981;Theil 等,2004;Tilton 等,1999)。与中链甘油三酯相比,母猪日粮中补充添加长链甘油三酯,可提高乳脂含量(Azain,1993)。在妊娠晚期日粮中补充共轭亚油酸,可提高初乳中脂肪百分比(Krogh 等,2012)。乳脂受日粮能量摄入量(Noblet 和 Etienne,1986;Schoenherr 等,1989)和日粮能量来源的影响(Coffey 等,1987;Van den Brand 等,2000)。另一方面,日粮粗蛋白质摄入量和日粮赖氨酸水平都不影响母猪乳脂(Dourmad 等,1998;Kusina 等,1999)。然而,乳脂浓度会受支链氨基酸的日粮摄入量影响(Richert 等,1997),并且在哺乳期的第 14 天,其也会受到来自日粮补充叶酸的影响(Wang 等,2011)。在产前两周对母猪限饲,可显著增加产后 15 h 收集的初乳中的脂肪百分比(Goransson,1990)。

许多研究已经证明,日粮的脂肪来源会影响脂肪酸在初乳和常乳中的比例(参见 Azain,1993;DeMan 和 Bowland,1963;Fritsche 等,1993;Lauridsen 和 Danielsen,2004;Peng 等,2010;Rooke 等,1998;Schmid 等,2008;Taugbøl 等,1993;Yao 等,2012)。Keenan 等(1970)报道了猪乳中磷脂的脂肪酸组成。本书第 16 章(Bontempo 和 Jiang,2015)阐述了各种日粮脂肪源对新生仔猪的影响。

9.6.3 其他因素

在对来源于欧洲品种的母猪进行比较中,Fahmy(1972)发现,常乳而非初乳中的脂肪百分比受品种影响。然而,后来的一项研究比较了杜洛克和长白两个品种,无论是每个品种内选择和未选择的品系,在乳脂含量方面没有显著差异(Shurson 和 Irvin,1992)。乳脂百分比受母猪胆固醇水平遗传选择的影响,低胆固醇母猪的乳脂百分比低于高胆固醇的母猪(Kandeh 等,1993)。中国梅山母猪的初乳和常乳的脂肪含量高于大约克母猪(Zou 等,1992)和包含几个欧洲品种血统的杂交母猪(Alston-Mills 等,2000)。梅山来源的母猪在哺乳期的第23天也比大约克有更多的乳脂含量(Farmer 等,2001)。

初乳脂肪百分比受初乳产量的显著影响,在低产奶量母猪的初乳中发现具有较高脂肪百分比(Foisnet 等,2010a)。类似地,母猪常乳的乳脂含量似乎也与产奶量呈负相关(White 等,1984),尽管其他研究没有发现母猪产奶量对脂肪百分比的影响(Garst 等,1999)。通过至少4个胎次的研究发现,胎次似乎不影响母猪初乳中的脂肪百分比(Mahan 和 Peters,2004)。后来的研究结果还表明,乳汁中的脂肪含量从第1胎到第4胎不断下降。Peters 和 Mahan(2008)观察到在1胎和6胎之间胎次对乳脂的二次效应,其中3胎和4胎具有最低乳脂含量。Goransson(1990)观察到第6胎母猪初乳中的脂肪百分比低于第1胎。然而,其他研究没有观察到不同胎次之间乳脂含量的显著差异(Baas 等,1992;Klobasa 等,1987)。增加仔猪数量与哺乳期第20天的乳脂含量减少有关(Baas 等,1992),但其他人没有观察到窝产仔数对脂肪百分比的影响(Klobasa 等,1987)。乳脂含量不受体蛋白质损失的影响(Clowes 等,2003)。

评估环境温度对母猪乳脂含量影响的研究有不同的结果,一些研究显示出在比较32℃和20°时,母猪的脂肪含量有降低的趋势(Schoenherr 等,1989),但当比较29℃与20℃(Renaudeau 和 Noblet,2001)或30℃与20℃(Prunier 等,1997)时,没有差异。在后一研究中,初乳的脂肪含量也不受环境温度的影响。母猪的乳脂含量不受季节的影响(Shurson 等,1986)。

9.7 蛋白质

乳腺分泌物中的蛋白质通常通过测定样品中总氮,然后乘以换算系数(例如 N‰×6.38)间接获得。因此,乳蛋白质通常基于样品的氮含量,称之为总蛋白质或粗蛋白质。乳腺分泌物含有其他非蛋白质形式的氮,例如游离氨基酸、肽、氨基糖和核苷酸。

9.7.1 哺乳阶段

母猪乳腺分泌物中总蛋白质的浓度在分娩时最高(表9.2)。在分娩后4h观察到与分娩时相似的蛋白浓度,然后在24h内下降超过50%。总蛋白质浓度的这些变化反映了免疫球蛋白浓度的变化(下面讨论)。乳的总蛋白质含量通常在

5.0%～6.5%(表9.3)。

9.7.2 其他因素

乳蛋白质含量一般不受日粮影响。Fahmy(1972)没有发现母猪品种对乳蛋白质含量的影响,然而,Shurson 和 Irvin(1992)观察到,在哺乳期的第 21 天,长白母猪的乳蛋白质百分比高于杜洛克母猪。Zou 等(1992)指出大约克母猪在第 7 天和第 21 天具有比梅山母猪更高的乳蛋白质含量。然而,在对来自梅山母猪和几个代表欧洲来源品种杂交母猪的比较中,没有显示出乳蛋白质含量的显著差异(Alston-Mills 等,2000),50%梅山血统母猪与大约克母猪在哺乳期的第 23 天,乳蛋白质含量相似(Farmer 等,2001)。猪乳的总蛋白质浓度不受胎次的影响(Baas 等,1992;Goransson,1990;Klobasa 等,1987)。

9.7.3 蛋白结合的氨基酸

表 9.4 总结了初乳和常乳的蛋白质组分中包含的氨基酸平均百分比。谷氨酸和脯氨酸分别占蛋白质结合氨基酸的 17%～22%和 10%～12%。支链氨基酸(异亮氨酸、亮氨酸和缬氨酸)总共占蛋白结合态氨基酸的 18%～19%。

9.7.4 主要乳蛋白质

在分娩期间或分娩后即刻,初乳中酪蛋白占总蛋白质比例的估算值为 9%～32%(Brent 等,1973;Csapo 等,1996)。在产后 24 h,免疫球蛋白浓度显著下降,酪蛋白占总蛋白质的比例增加至 30%～45%(Brent 等,1973;Csapo 等,1996)。在哺乳期的其余时间,大多数报道称,猪乳中酪蛋白含量占总蛋白质的比例通常在 50%～55%(Brent 等,1973;Csapo 等,1996;Mahan 等,1971;Richert 等,1997),但是,在哺乳晚期有人报道具有更高的估算比例(Mahan 等,1971)。初乳和常乳的重要蛋白质组分主要为乳清蛋白。

总乳清蛋白在母猪初乳中占总蛋白质的百分比在分娩时为 90%,此时免疫球蛋白为乳清蛋白的主要部分,然后在产后 24 h 降至约 70%(Csapo 等,1996;Klobasa 等,1987)。从哺乳期的第 10 天到第 60 天时,总乳清蛋白占真蛋白的百分比介于 47%～50%(Csapo 等,1996)。

β-乳球蛋白是猪乳中的主要乳清蛋白。β-乳球蛋白的浓度从初乳期到至少哺乳期的第 7 天都相对恒定,范围在 10～15 mg/mL(Hurley 和 Bryson,1999;Jackson,1990)。α-乳白蛋白浓度在初乳中最低,为 1.8～2.0 mg/mL,并且在哺乳期的前 7 天逐渐增加至约 3.3 mg/mL(Hurley 和 Bryson,1999;Jackson,1990)。猪乳中的其他乳清蛋白,乳清酸性蛋白或富半胱氨酸乳清蛋白,从初乳到哺乳第 7 天,其含量从约 0.3 mg/mL 增加到 0.9 mg/mL(Hurley 和 Bryson,1999;Jackson,1990)。乳铁蛋白浓度在分娩时的初乳中约 1.2 mg/mL,在哺乳期的第 3 天持续升高,然后在第 7 天下降至 0.3 mg/mL,随后浓度持续缓慢降低(Elliot 等,1984)。在分娩期母猪初乳中白蛋白的浓度为 19 mg/mL,然后在产后 12h 降至 8 mg/mL,随后在哺乳期的第 2 和第 3 周期间,进一步逐渐下降至 2.5~3.0 mg/mL 的稳定浓度(Klobasa 和 Butler,1987)。Zou 等(1992)通过蛋白质凝胶分析,注意到猪乳中有几种高分子质量蛋白质。虽然这些蛋白质尚未被鉴定,但它们似乎在猪的不同品种之间存在差异。

9.7.5 免疫球蛋白

初乳的主要蛋白质成分是免疫球蛋白,包括 IgG,IgA 和 IgM(Hurley 和 Theil,2013)。已有研究者鉴定了 IgG 同型的几个亚类(Butler 等,2009),但是,大多数针对初乳和常乳组成的研究是对分泌物中的总 IgG 进行定量。用于测量免疫球蛋白的方法较多,包括典型的免疫电泳法、放射免疫扩散法(Curtis 和 Bourne,1971;Frenyo 等,1981;Jensen 和 Pedersen,1979;Klobasa 等,1987)和近来的酶联免疫吸附试验法(ELISA;Foisnet 等,2010a,b,2011;Jackson 等,1995;Loisel 等,2013;Markowska-Daniel 和 Pomorska-Mol,2010;Quesnel 等,Rolinec 等,2012)。

免疫球蛋白的浓度在产后几个小时内的初乳中最高(表 9.5)。IgG 作为母猪初乳中的主要免疫球蛋白,至少在产后最初 6 h 保持高浓度。到 12 h 时,IgG 浓度与分娩时相比下降了近 50%,并且它们继续降低,至分娩后 24 h 和 48 h,分别降到分娩水平的约 16% 和 9%。IgA 和 IgM 浓度在产后第 1 天遵循类似 IgG 的下降模式,但它们的浓度下降速度比 IgG 更缓慢(表 9.5)。大约在哺乳期的第 3 天,IgA 成为乳汁中最主要的免疫球蛋白。

表 9.5 母猪初乳和常乳中免疫球蛋白报道浓度的平均值和范围。

	哺乳阶段						
	0 h	6 h	12 h	24 h	48 h	72 h	12~45 天
IgG (mg/mL)[1]							
平均值[2]	64.4	59.8	34.7	10.3	5.7	3.1	1.0
范围[3]	52~102	42~87	20~61	6~20	2~10	2~5	0.2~2.3
研究数[4]	12	7	5	11	6	6	4
IgA (mg/mL)							
平均值	13.1	11.4	9.3	5.0	3.8	4.1	4.0
范围	5.5~24	8~17	7~13	1.5~9.2	2.7~4.5	3.6~4.5	1.9~6.6
研究数	7	3	3	6	3	2	5

续表 9.5

	哺乳阶段						
	0 h	6 h	12 h	24 h	48 h	72 h	12～45 天
IgM (mg/mL)							
平均值	8.4	7.3	4.8	3.5	2.7	3.1	1.6
范围	1.3～10.7	1.4～5.9	1～9	1.8～4.5	1.8～4.5	1.7～4.5	0.9～2.4
研究数	5	3	3	4	3	2	3

[1] 总 IgG。
[2] 来自分娩相应不同时间点乳腺分泌物研究的平均值。
[3] 来自确定平均值研究中值的范围。
[4] 确定平均值所包括的研究数。用于估算平均值的研究包括：Curtis 和 Bourne，1971；Foisnet 等，2010a，b，2011；Frenyo 等，1981；Jackson 等，1995；Jensen 和 Pedersen，1979；Klobasa 等，1987；Loisel 等，2013；Markowska-Daniel 和 Pomorska-Mol，2010；Quesnel，2008；Rolinec 等，2012。

Klobasa 和 Butler(1987)发现，在乳房分泌物中免疫球蛋白浓度方面，母猪个体之间的变异性太高，无法识别除了哺乳阶段之外的显著差异。在不同胎次间的比较中没有观察到一致的差异，并且在比较母猪不同乳腺之间的初乳免疫球蛋白浓度时，也没有发现统计学上显著的差异。Quesnel 观察到分娩时初乳中的 IgG 浓度不受胎次影响，但是在产后 24 h，胎次大于 5 的母猪的 IgG 浓度大于初产母猪的 IgG 浓度(Quesnel，2011)。在产前两周对母猪进行限饲会降低初乳 IgA，但不降低 IgG 浓度(Goransson，1990)。

9.7.6 非蛋白氮

乳汁的非蛋白氮(non-protein nitrogen，NPN)组分包括游离氨基酸、核苷酸、氨基糖和其他含氮化合物。通常是通过用三氯乙酸沉淀使蛋白质分级，过滤上清液，并使用凯氏定氮法定量测定上清液中的氮，从而得到乳中的总非蛋白氮含量(Csapo 等，1996；Klobasa 等，1987)。初乳中非蛋白氮含量相对较低(64 mg N/100 g；Csapo 等，1996)。非蛋白氮含量在哺乳期会升高，报道范围在 68～158 mg N/100 g (Csapo 等，1996；Gurr，1981；Klobasa 等，1987；Mahan 等，1971；Perrin，1958；Sheffy 等，1952)。

Wu 和 Knabe(1994)测定了母猪初乳(收集产后 6～10 h 的)和常乳(至哺乳期的第 29 天)中游离氨基酸和蛋白结合态氨基酸的浓度。与在毫摩尔范围内的蛋白结合态氨基酸不同，大多数游离氨基酸存在于微摩尔浓度范围内。初乳中主要的游离氨基酸是组氨酸，其在随后的哺乳期间浓度降低，牛磺酸的浓度增加到哺乳的第 8 天，然后保持恒定。从初乳到哺乳的第 8 天，大多数其他游离氨基酸的浓度会增加。第 8 天和第 29 天之间游离氨基酸浓度的变化随不同氨基酸而变化(Wu 和 Knabe，1994)。游离非必需氨基酸的总浓度在初乳中最低，并在大部分哺乳期内增加，而游离必需氨基酸的总浓度在整个哺乳期保持恒定。尿素从初乳到哺乳期第3天浓度会增加,然后下降到与初乳相似的水平(Wu和Knabe,1994)。氨浓度在初乳中最高,然后下降到第8天,并在余后的哺乳期保持不变(Wu和Knabe,1994)。

母猪初乳中的核苷酸浓度在哺乳期会发生改变(Mateo等,2004)。例如,腺苷5′-单磷酸具有二次模式,在哺乳第7天达到峰值。然而,尿苷5′-单磷酸在分娩时最高,然后通过泌乳而下降(Atwood等,1995;Mateo等,2004)。多胺、精胺和亚精胺的浓度,从分娩到哺乳期的第1和第2周之间增加到峰值,然后随着哺乳进展而下降(Moytl等,1995)。对于那些多胺的浓度,可观察到相当大的个体差异。在母猪初乳和常乳中也鉴定出了许多含氮糖(Tao等,2010)。

9.8 能量

乳腺分泌物的总能是通过有机物质的燃烧和所释放的二氧化碳的量来估算。或者,一些研究估算总能是由猪乳中有机组分的含量来计算得到(Laws等,2009)。在产后和整个哺乳期,母猪乳腺分泌物中的总能的详细报道不如其他组分普遍(表9.2和表9.3)。初乳在分娩时的总能约为6.7 kJ/g,并且至少在哺乳的第3天内保持升高,但在之后的哺乳期降低。在分娩期初乳中估算的相对高水平的总能含量,部分地与初乳中高浓度的免疫球蛋白有关。然而,与其他乳蛋白质相比,免疫球蛋白对消化的抵抗力更强,并且与其他乳清蛋白相比,吸收的氨基酸比例更小(Danielsen等,2011;Yvon等,1993)。基于弹式量热法的总能估算,可能高估了仔猪从初乳中得到的能量。在哺乳期第2天和第3天观察到总能的瞬时增加,可能与在此阶段发生的脂肪含量达到峰值相关。

9.9 矿物质

乳腺分泌物的总无机组分通常被称为灰分。从表9.2和表9.3中总结的研究来看,分娩时初乳的灰分含量约为0.68%。灰分百分比在哺乳的第2天增加,之后继续逐渐增加直至约第2周,然后保持在约0.90%。表9.3中第42~60天的灰分百分比,是基于在延长的哺乳期期间该组分有限数量的研究报道。

表9.6总结了母猪初乳和常乳中各个矿物质浓度的报道。钙的浓度在初乳中相对较低,然后在整个哺乳期,至少直到哺乳期的第6或7周浓度升高(Coffey等,1982;Gueguen和Salmon-Legagneur,1959;Harmon等,1974;Miller等,1994;Perrin,1955)。大多数(但不是全部),钙结合在酪蛋白胶粒中。扩散性钙的浓度在初乳中最高,然后随着哺乳进展而下降(Kent等,1998),与酪蛋白含量的同时增加一致。磷酸盐浓度遵循与钙类似的模式,从初乳到常乳的含量有所增加(表9.6)。

钙的日粮水平不影响乳中钙浓度(Miller 等,1994),并且日粮中无机来源比有机来源的微量矿物质,可导致更高的乳钙浓度(Peters 等,2010)。

表 9.6 母猪初乳和常乳中常量和微量矿物质报道浓度的平均值和范围。

常量矿物质	初乳[1](mg/mL)		常乳(mg/mL)		参考文献[2]
	平均值	范围	平均值[3]	范围	
钙	0.80	0.48~1.52	2.00 (9~28 天)	1.51~2.54	1, 3, 4, 6, 7, 8, 9, 11, 13, 16, 17, 18, 19
磷	1.08	0.52~1.58	1.42 (9~28 天)	0.87~1.83	3, 4, 6, 7, 8, 9, 13, 16, 17, 18, 19
钾	1.29	1.10~1.62	0.89 (17~35 天)	0.36~1.57	3, 4, 6, 7, 11, 16, 17, 19
钠	0.83	0.68~1.00	0.42 (17~35 天)	0.33~0.54	3, 4, 6, 7, 11, 16, 17, 19
氯	0.94	0.93~0.96	0.69 (16~35 天)	0.60~1.06	11, 18, 19
柠檬酸盐	1.70		0.94 (9~35 天)	0.77~1.19	9, 11
镁	0.104	0.016~0.20	0.105 (9~28 天)	0.016~0.20	3, 4, 6, 7, 9, 16, 17, 18, 19
硫	1.00		0.27 (17~21 天)	0.04~0.51	17, 19
微量矿物质					
铝	3.5		2.1 (20~21 天)	0.8~3.7	3, 16, 17
硼	0.03		1.4 (20~21 天)	0.02~3.45	3, 16, 17
镉	0.04		0.04 (21 天)		3
铬	0.60		0.40 (20~21 天)	0.35~0.46	3, 16
铜	1.80	0.26~3.77	0.92 (17~21 天)	0.12~2.01	3, 4, 16, 17, 19
铁	2.84	1.7~5.4	1.96 (12~21 天)	1.27~4.6	2, 3, 4, 16, 19, 20, 24, 25
碘	135		48 (23~27 天)	14~73	22, 23
锰	0.26	0.06~0.45	0.15 (20~21 天)	0.06~0.36	3, 4, 16, 17
钼	0.04		0.06 (20~21 天)	0.02~0.10	3, 17
镍	0.42		0.31 (21 天)		3
铅	0.17		0.16 (21 天)		3
硒	0.13	0.02~0.24	0.05 (14~28 天)	0.02~0.14	10, 12, 14, 15, 16, 19, 21, 26, 27
锶	n/a		0.47 (21 天)		16
锌	15.1	9.2~16.1	6.8 (17~35 天)	5.1~8.3	3, 4, 5, 16, 17, 19, 20

[1] 定义为分娩 12 h 内。
[2] 参考文献:1,Alston-Mills 等,2000;2,Chaney 和 Barnhart,1963;3,Coffey 等,1982;4,Csapo 等,1996;5,Earle and Stevensen,1965;6,Elliott 等,1971;7,Fahmy,1972;8,Harmon 等,1974;9,Kent 等,1998;10,Kim 和 Mahan,2001;11,Konar 等,1971;12,Loudenslager 等,1986;13,Lyberg 等,2007;14,Mahan,2000;15,Mahan 等,1977;16,Mahan 和 Newton,1995;17,Park 等,1994;18,Perrin,1955;19,Peters 等,2010;20,Pond 等,1965;21,Quesnel 等,2008;22,Schone 等,1997;23,Schone 等,2001;24,Veum 等,1947;25,Veum 等,1965;26,Yoon 和 McMillan,2006;27,Zhun 等,2011。
[3] 取样母猪的哺乳天数。

初乳中钾和钠的浓度比常乳高。大多数乳成分的合成发生在乳腺上皮细胞的相对高钾和低钠的环境中。因此,乳腺分泌物反映了这种钾钠关系。研究发现氯也是在初乳中比在常乳中多。镁在初乳和常乳之间的浓度似乎没有显著差异。初乳中的硫含量高于常乳。柠檬酸盐作为一种阴离子,常与扩散性钙组分相关。柠檬酸盐浓度从初乳阶段到常乳阶段会下降(Holmes 和 Hartmann,1993;Kent 等,1998;Konar 等,1971),与乳生成的进程,和乳糖(作为乳中主要渗量)的增加一致(Holmes 和 Hartmann,1993;Peaker 和 Linzell,1975)。

母猪初乳和常乳中微量矿物质的浓度总结于表 9.6 中。特别值得注意的是,铜、铁、碘、锰和锌均是初乳比常乳具有更高的浓度。大多数研究发现,铁的日粮水平不影响乳腺分泌物中的铁浓度(Pond 和 Jones,1964;Pond 等,1965;Venn 等1947;Veum 等,1965),而其他研究观察到当母猪补充铁时,乳汁中铁浓度增加(Chaney 和 Barnhart,1963;Earle 和 Stevenson,1965)。铜在乳中的浓度不受日粮矿物质水平的影响,但受微量矿物质来源的影响(Peters 等,2010)。日粮中补充碘可增加猪乳中碘的浓度(Schone 等,1997,2001)。

硒与动物中的先天和获得性免疫应答相关联(Salman 等,2009)。初乳中的硒浓度高于常乳(表 9.6)。乳腺分泌物中硒浓度的下降发生在产后最初的 2~3 天(Kim 和 Mahan,2001;Loudenslager 等,1986;Quesnel 等,2008)。通过在日粮中补充矿物质可以增加乳腺分泌物中的硒浓度,并且在提高初乳和常乳中硒浓度方面,有机形式比无机形式更高效(Kim 和 Mahan,2001;Mahan,2000;Mahan 和 Peters,2004;Quesnel 等,2008;Yoon 和 McMillan,2006;Zhan 等,2011)。乳腺分泌物中的硒浓度,即使在日粮中补充无机硒的情况下,依然会随胎次增加至第 4 胎时而下降(Mahan 和 Peters,2004)。然而,补充有机硒似乎可以抵消母猪胎次对硒浓度的影响(Mahan 和 Peters,2004)。向哺乳母猪日粮中补充硒,可以增强乳中性粒细胞的杀菌活性(Wuryastuti 等,1993)。

9.10 维生素

母猪初乳和常乳中几种维生素的平均浓度总结于表 9.7 中。相对于常乳,母猪初乳中(在分娩后的 12 h 内)的维生素 A 浓度比较高。初乳中维生素 A 的估算值为 0.74 μg/mL(按照 1 IU=0.3 μg 视黄醇;Braude 等,1946)至 1.81 μg/mL(Elliott 等,1971)。维生素 A 的浓度相对于哺乳晚期,在哺乳的前 3 天的浓度保持升高,与这段时间乳腺分泌物的脂肪含量增加相对应。从第 5 天到第 28 天的乳汁中,维生素 A 浓度的估算值为 0.14~0.73 μg/mL。在超过 28 天的哺乳期中维生素 A 浓度的报道呈现出广泛的结果差异(Csapo 等,1996;Heidebrecth 等,1951)。初乳和常乳中维生素 A 的含量受维生素水平和日粮来源的影响。在妊娠期和哺乳期饲喂母猪黄玉米与白玉米,可增加乳汁中视黄醇浓度(Heying 等,2013)。在妊娠晚期和哺乳期日粮中补充鱼油,而不补充动物脂肪或其他各种油,可增加猪乳中维生素 A 的浓度(Lauridsen 和 Danielsen,2004)。另外,其他研究表明,与基础日粮相比,在妊娠晚期和哺乳期添加动物脂肪可增加初乳和常乳中的维生素 A 浓度(Coffey 等,1982)。Braude 等(1946,1947)报道,在夏季,猪乳中的维生素 A 浓度比冬季更高,这可能是由于夏季时母猪哺乳期的第 1 周后在草场上进行饲养,而冬季从产前 2 周到哺乳期,都是在室内饲养。哺乳母猪口服视黄醇可导致乳汁中的视黄醇浓度升高,并在给药后 7.5~10 h 达到峰值(Dever 等,2011)。

分娩第 1 天报道的初乳中维生素 E 浓度范围为 3.2~23.3 μg/mL,平均为 10 μg/mL。到第 2 天,维生素 E 浓度为 3.7~7.7 μg/mL(Csapo 等,1996;Lauridsen 和 Jensen,2005;Lauridsen 等,2002)。猪乳从哺乳的第 16~60 天,维生素 E 含量平均为 2.6 μg/mL(表 9.7)。大多数研究报道了哺乳阶段对乳腺分泌物中维生素 E 浓度的显著影响(Lauridsen 和 Danielsen,2004;Lauridsen 和 Jensen,2005;Lauridsen 等,2002;Mahan,1991;Mahan 等,2000;Malm 等,1976;Pinelli-Saavedra 等,2008)。在妊娠日粮和/或哺乳日粮中补充维生素 E,可增加初乳和常乳维生素 E 含量(Lauridsen 和 Jensen,2005;Mahan,1991;Mahan 等,2000;Malm 等,1976;Pinelli Saavedra 等,2008)。用椰子油补充妊娠晚期和哺乳期日粮,相比补充动物

表 9.7 母猪初乳和常乳中一些维生素报道浓度的平均值和范围   mg/mL

维生素	初乳[1]		常乳		参考文献[2]
	平均值	范围	平均值[3]	范围	
A	1.14	0.474~1.81	0.48 (4~60 天)	0.15~0.92	1,2,3,4,5,6,7,8,10,13,14
D	0.015		0.006 (21~60 天)	0.003~0.009	4,6
E	10.0	3.2~23.3	2.6 (16~60 天)	1.2~3.9	6,12,13,14,16,17,19
C	190	64~306	94 (15~60 天)	45~130	1,2,6,8,9,11
硫胺素(B_1)	0.83	0.50~1.45	0.74 (10~56 天)	0.68~0.80	1,2,8,18
核黄素(B_2)	2.64	0.45~6.50	1.28 (10~56 天)	0.46~2.1	1,2,8,18

[1] 定义为分娩 12 h 内。

[2] 参考文献:1,Bowland,1966;2,Bowland 等,1951;3,Braude 等,1947;4,Braude 等,1946;5,Coffey 等,1982;6,Csapo 等,1996;7,Dever 等,2011;8,Elliott 等,1971;9,Heidebrecht 等,1951;10,Heying 等,2013;11,Hidiroglou 和 Batra,1995;12,Lauridsen 和 Danielsen,2004;13,Lauridsen 等,2002;14,Lauridsen 和 Jensen,2005;15,Loudenslager 等,1986;16,Mahan,1991;17,Mahan 等,2000;18,Neuhaus,1961;19,Pinelli-Saavedra 等,2008。

[3] 包括了哺乳期大数的报道数据用以计算平均浓度。

脂肪或其他各种油,增加了猪乳中的维生素 E 浓度(Lauridsen 和 Danielsen,2004)。Mahan 等(2000)发现了胎次对维生素 E 具二次效应,最高浓度出现在第 2 胎和第 3 胎。有关猪乳中维生素 D(表 9.7)和维生素 K 的信息有限。维生素 D 浓度在分娩时的初乳中最高,约为 0.015 μg/mL,然后在哺乳后期约为 0.006 μg/mL(Bowland 等,1951;Csapo 等,1996)。维生素 K 的浓度范围为 0.089~0.101 μg/mL,但似乎不受哺乳阶段的影响(Csapo 等,1996)。

与常乳相比,维生素 C 浓度在初乳中含量最高,范围为 64~306 μg/mL,平均值为 190 μg/mL。常乳中维生素 C 的浓度平均为 94 μg/mL,范围为 45~130 μg/mL(表 9.7)。母猪初乳和常乳中 B 族维生素的信息也有限。初乳中的硫胺素(B_1)含量报道为 0.5~1.45 μg/mL,平均为 0.83 μg/mL,而常乳中平均为 0.74 μg/mL,范围为 0.7~0.8 μg/mL(表 9.6)。乳汁中的硫胺素含量受季节影响,在夏季,母猪哺乳期的大部分时间在草场上饲养时,其浓度较低(Braude 等,1947)。然而也有研究指出,与其他大多数维生素相反,硫胺素浓度在初乳中似乎不比在常乳中更大(Elliott 等,1971)。初乳中核黄素(维生素 B_2)报道的范围为 0.45~6.5 μg/mL,平均为 2.6 μg/mL(表 9.6)。常乳中核黄素浓度约为 1.28 μg/mL,范围为 0.4~8.2 μg/mL(表 9.6)。初乳中的核黄素浓度似乎高于常乳(Elliott 等,1971)。硫胺素和核黄素浓度都不受日粮蛋白质水平的影响(蛋白质水平高至 15%;Elliott 等,1971)。

在一篇总结了 1930—1961 年文献的综述中,Neuhaus(1961)报道了初乳中烟酸浓度约为 1.65 μg/mL,泛酸浓度为 1.3~6.8 μg/mL。常乳中的维生素 B_6(Coburn 等,1992;未指明哺乳阶段)包括吡哆醛,5′-磷酸吡哆醛和 5′-磷酸吡哆胺,它们的含量分别为 101,563 和 45 ng/mL,比初乳中报道的维生素 B_6 水平 25 ng/mL 更高(Neuhaus,1961)。初乳中的生物素浓度约为 53 ng/mL(Neuhaus,1961),在哺乳第 14 天的常乳中,其浓度为 24~68 ng/mL(Bryant 等,1985)。Bryant 等(1985)观察到,当在母猪日粮中补充生物素时,常乳中生物素浓度增加。

初乳中的叶酸(B)含量(从哺乳第 1 天数据获取的值)报道为 13~44 ng/mL(Barkow 等,2001;Ford 等,1975;O'Connor 等,1989)。然后叶酸浓度在哺乳第 7 天时下降 50%~60%(Ford 等,1975;Matte 和 Girard,1989)。在哺乳期的第 16~28 天,叶酸浓度为 2.3~13.4 ng/mL(Barkow 等,2001;Ford 等,1975;Matte 和 Girard,1989;O'Connor 等,1989)。常乳中叶酸浓度可以通过日粮补充(Barkow 等,2001),或每周肌肉注射叶酸(Matte 和 Girard,1989)来增加。初乳中的钴胺素(Cobalamin,维生素 B_{12})浓度约为 1.5 ng/mL(Ford 等,1975;Neuhaus,1961)。Ford 等(1975)也测定了哺乳的第 7~49 天猪乳中的钴胺素含量。钴胺素浓度在第 14 天最高(2.41 ng/mL),然后在余后的哺乳期保持在 1.40~1.64 ng/mL。

9.11 细胞

乳中的总体细胞浓度称为体细胞数(somatic cell count,SCC)。猪乳中体细胞数的报道估算值差异很大。一些研究者观察到在哺乳早期(通常被报道为哺乳期的第1或第2天),体细胞数介于 5.3×10^5/mL(每毫升细胞数)(Hurley 和 Grieve,1988)至 1.06×10^6/mL(Schollenberger 等,1986a)之间,最高可达 8×10^6/mL(Garst 等,1999)。Evans 等(1982)报道初乳体细胞数介于 $2\times10^5\sim5\times10^7$/mL(平均值=$1\times10^7$),Osterlundh 等(1998)报道中性粒细胞数为 7.9×10^6/mL。哺乳期观察到的体细胞数变化在不同的报道中也不同。Osterlundh 等(1998)发现中性粒细胞的浓度在哺乳期的第1~3天显著降低。Garst 等(1999)报道,体细胞数在哺乳期的第2~51天线性增加。Schollenberge 等(1986a)仅发现与初乳相比,体细胞数在哺乳期的第8~14天可以显著升高,其并未观察到体细胞数和哺乳期天数之间的相关性。其他研究者观察到,哺乳期第1~28天体细胞数没有显著变化(Hurley 和 Grieve,1988)。体细胞数在哺乳期第8天到第14天的报道范围从 2.9×10^5/mL(Hurley 和 Grieve,1988),到高达 8.5×10^6/mL(Garst 等,1999),也有报道为 $(1\sim2)\times10^6$/mL(Evan 等,1982;Schollenberger 等,1986a)。

母猪初乳和常乳中报道的体细胞数的差异可能来自几个原因。奶牛中的亚临床型乳腺炎通常在没有临床症状的情况下,通过观察体细胞数的相应增加来确定。对于母猪,临床型或者亚临床型乳腺炎的情况下,都预期会使体细胞数升高。在评估猪体细胞数的研究中,通常没有考虑亚临床型乳腺炎的存在。用于乳汁收集的催产素给药的途径也影响体细胞数,肌内注射比静脉内注射会导致更高的体细胞数(Garst 等,1999)。其他人报道,后乳,即在仔猪吮乳后通过催产素给药移除的乳汁,其总固形物和乳脂较高(Atwood 和 Hartmann,1992)。在牛奶中已报道过在除去的乳汁组分中体细胞数的类似变化(Paape 和 Tucker,1966)。因此,从母猪乳腺中若去除乳汁不完全,则可能导致低估体细胞数。此外,已使用过不同的方法来测定体细胞数,包括直接显微细胞计数(Evans 等,1982;Hurley 和 Grieve,1988),或通过主要用于牛奶分析的奶牛牛群改良实验室的仪器设备,来分析猪乳样品(Garst 等,1999)。乳中细胞通常是通过离心乳汁样品来收集,并评价所得沉淀细胞。然而,在已离心的乳脂或奶油组分中也发现了细胞(Phipps 和 Newbould,1966),因此在估算体细胞数时,应注意考虑到这一部分的影响。

乳细胞包括几种白细胞类型,包括中性粒细胞、巨噬细胞、淋巴细胞、嗜酸性粒细胞和上皮细胞(Lee 等,1983;Le Jan,1995;Schollenberger 等,1986b,c)。表9.8基于对所有5种细胞类型进行区分的研究,总结了母猪初乳和常乳中的细胞分类计数的平均值和范围。中性粒细胞是哺乳期所有阶段的主要细胞类型。在初乳中观察到的嗜中性粒细胞浓度最高,之后随着哺乳进程嗜中性粒细胞浓度下降。有报道称在哺乳后期,可观察到嗜中性粒细胞浓度更大的变异。巨噬细胞和淋巴细胞的平均浓度在整个哺乳期相对恒定,然而,在报道的细胞分类计数中存在相当大的变异性。嗜酸性粒细胞平均占乳腺分泌物中总细胞的不到2%。上皮细胞的比例在哺乳期显著变化,在初乳中观察到低百分比,在初乳阶段后快速增加。断奶会导致乳腺分泌物的90%~98%为嗜中性粒细胞(Lee等,1983)。

表9.8 母猪初乳和常乳中报道的细胞分类计数的平均值和范围[1]　　　　　　　　%

细胞类型	初乳[2]		常乳(3~7天)		常乳(14~21天)	
	平均值	范围	平均值	范围	平均值	范围
中性粒细胞	64.9	61.2~71.1	47.9	40.7~55.4	41.1	28.2~51.3
巨噬细胞	9.5	1.3~24.5	12.0	8.6~15.5	11.0	5.5~16.3
淋巴细胞	18.1	8.8~26.5	15.3	7.9~22.8	9.8	6.0~12.2
嗜酸性粒细胞	0.7	0.2~1.5	1.5	0.4~2.6	0.8	0.2~1.5
上皮细胞	6.0	0.4~19.6	23.2	6.1~36.8	37.3	31.3~48.0

[1] 用于计算平均细胞分类计数的参考文献:Evans等,1982;Lee等,1983;Schollenberger等,1986a;Wuryastuti等,1993。

[2] 在分娩2天内。

9.12 生物活性成分

初乳和常乳含有广泛的生物活性因子,包括免疫球蛋白和白细胞(如上所述)、酶、激素和生长因子等。在这里将简要讨论母猪初乳和常乳中的这些组分。

许多激素,生长因子和细胞因子已在不同物种的乳中鉴定(Baumrucker和Magliaro-Macrina,2011;Pakkanen和Aalto,1997)。许多这些生物活性成分被认为对新生仔猪具有作用。初乳中催乳素的浓度在分娩前最高,在产后最初24 h内迅速下降(Devillers等,2004a)。在初乳期之后,催乳素从约第5天开始至余后的哺乳期缓慢下降(Mulloy和Malven,1979)。猪乳中催乳素浓度受窝产仔数的影响,哺乳8头仔猪的母猪与哺乳10或12头仔猪的母猪相比,乳汁中催乳素浓度较低(Mulloy和Malven,1979)。至少在哺乳期的第13天,血清催乳素和乳汁中催乳素的浓度呈正相关(Mulloy和Malven,1979)。

猪乳中也发现有松弛素(Yan等,2006),作为激素前体分泌(Bagnell等,2009)。松弛素的浓度在初乳中最高,然后在哺乳期的最初4~6天下降(Yan等,2006)。这种乳源松弛素被认为是乳腺分泌假说(lactocrine hypothesis)中,母体对新生仔猪发育编程继续施加影响的一种途径(Bagnell等,2009)。

脱脂乳腺分泌物中瘦素的浓度,从分娩到哺乳期的第7天减少,然后直到哺乳期的第 22 天保持不变。然而,全乳中的瘦素浓度不受哺乳阶段的影响(Estienne 等,2000)。全乳和脱脂乳瘦素浓度都不显著受体况的影响,并且它们与背膘厚,或母猪血清瘦素浓度都不相关(Estienne 等,2000)。大约克(75%)×梅山(25%)杂交母猪的乳腺分泌物中瘦素平均浓度,显著大于梅山母猪(Mostyn 等,2006)。

母猪的乳腺产生雌二醇,这种激素似乎是由孕酮所控制(Staszkiewicz 等,2004)。母猪血浆中的雌二醇浓度在分娩前升高,并在产后最初 1 天快速下降(Devillers 等,2004a;Osterlundh 等,1998)。在分娩之前收集的全部初乳中雌二醇的浓度,比同期的血浆高 3~4 倍(Devillers 等,2004a)。在另一项使用无脂初乳的研究中,雌二醇的浓度约为血浆浓度的一半(Osterlundh 等,1998)。雌二醇浓度在产后第 1 天迅速下降,从分娩时约 1.5 ng/mL,到产后 24 h 降至 0.5 ng/mL(Devillers 等,2004a)。母猪全部初乳中的雌酮在分娩后立即测定时浓度最高(13.8 ng/mL),在最初的 24 h 内迅速下降,在产后 48 h 达到 0.5 ng/mL。整个母猪乳腺分泌物中的雌酮浓度不受腺体位置的影响(Farmer 等,1987)。母猪乳腺分泌物中孕酮的浓度遵循与雌激素类似的模式,浓度在初乳中最高,然后在产后的最初 1 天迅速下降(Devillers 等,2004a)。皮质醇也存在于母猪乳腺分泌物中,在分娩后不久达到其最高浓度,浓度为常乳的 2~2.5 倍(在无脂分泌物中测量;Osterlundh 等,1998)。

胰岛素和神经降压素的浓度在分娩时的初乳中升高,然后在哺乳期的第 3 天后下降,而铃蟾肽的浓度在相同时期保持相对恒定(Westrom 等,1987)。在哺乳早期,乳汁中甲状腺激素的浓度在初乳和常乳之间并无不同(Mostyn 等,2006)。有研究显示母猪乳腺分泌物中三碘甲状腺原氨酸浓度,而非甲状腺素浓度,会受品种的影响(Mostyn 等,2006)。

在猪乳中发现了几种生长因子和生长因子活性。胰岛素样生长因子-1(IGF-1)的浓度在产后第 1 天收集的初乳中最高(14~70 ng/mL),然后在产后第 2 天下降超过 50%,并继续下降直到哺乳期约第 10 天(范围为 3~14 ng/mL;Monaco 等,2005)。哺乳阶段对母猪乳腺分泌物中的 IGF-1 浓度有显著影响,而第 1 胎和第 2 胎之间没有显著差异(Monaco 等,2005)。Simmen 等(1990)也观察到哺乳阶段对乳腺分泌物中 IGF-1 浓度的影响以及品种效应。在猪乳中已鉴定出表皮生长因子样肽(Tan 等,1990)。通过放射性受体分析法测定的表皮生长因子浓度,在产后的最初 24 h 收集的乳腺分泌物中最高,在哺乳第 9 天下降 90%,然后直到至少第 27 天保持相对恒定(Jaeger 等,1987)。前列腺素样活性,从初乳中的高水平,到哺乳第 5 天时降低约 2.5 倍,并且至少直到哺乳第 20 天期间保持恒定(Maffeo 等,1987)。在母猪初乳和常乳中已经鉴定了许多细胞因子,包括 IL-4,IL-6,IL-10,IL-12,IFN-γ,TNF-α 和 TGF-β(Nguyen 等,2007)。鉴定的每种细胞因子都在初乳和早期常乳中具有最高浓度,之后随着继续哺乳而浓度降低,尽管各自下降的速率不同(Nguyen 等,2007)。

Chandan 等(1968)发现猪乳具有脂肪酶和核糖核酸酶活性,但没有溶菌酶活性。Krakowski 等(2002)报道,在分娩后立即测定,母猪初乳中的溶菌酶活性为 15~20 ng/mL。已在母猪初乳中鉴定出胰蛋白酶抑制因子活性(Jensen,1978)。母猪乳汁中的血浆铜蓝蛋白浓度,在泌乳第 3 天比第 33 天更高(Cerveza 等,2000)。

9.13 生理状态的影响

泌乳的开始,称为乳生成(生乳),已被描述发生在两个阶段(Hartmann,1973)。初始阶段包括乳腺细胞结构和酶的分化,为乳汁分泌做准备。该阶段与初乳形成一致。对于母猪,这个阶段的乳生成发生在妊娠的晚期阶段(Kensinger 等,1982)。第二阶段的乳生成,乳汁大量分泌,与分泌体积的快速增加相吻合,然而直到母猪产后 33~34 h,才可能完全开始(Krogh 等,2012;Theil 等,2014)。在分娩和乳生成阶段之间的生理协调中,可观察到自然变异性。例如,分娩开始后最初 24 h 内乳腺分泌物产量(根据仔猪体重变化确定)低的母猪,具有更高的总固形物、脂肪和总能含量,但其分娩时初乳中的乳糖含量,低于高初乳产量的母猪(Foisnet 等,2010a)。这提示在分娩时,与初乳产量高的母猪相比,低初乳产量的母猪还没有达到乳生成的高水平阶段。对于产后 24 h 乳腺分泌物的总固形物含量,初乳产量高、低母猪之间没有差异(Foisnet 等,2010a)。

在通过激素操控提早分娩或推迟分娩的研究中,也观察到分娩和乳生成时间的失准,在妊娠的第 114 天之前发生的诱导分娩,可以降低初乳分泌物中的脂肪含量,而初乳中的乳糖浓度不受诱导的早期分娩影响(Jackson 等,1995)。在妊娠第 113 天处理,比在第 114 天诱导分娩的母猪具有较高的乳糖含量,较低的蛋白质和灰分含量,并且在分娩时比未诱导的母猪具有更低的总固形物含量(Foisnet 等,2011)。另外,用孕激素延迟母猪分娩至第 116 天,不会影响分娩时或在分娩 24 h 时初乳的总固形物、乳糖、脂肪、蛋白质或灰分含量(Jackson 等,1995;Foisnet 等,2010b),然而总固形物和总能百分比在产后 48 h 降低(Foisnet 等,2010b)。

通过转基因修饰乳腺细胞功能从而改变生乳过程,也可以改变初乳组成。在表达牛 α-乳白蛋白基因的转基因初产哺乳母猪中,分娩 6 h 内初乳中的总固形物量低于非转基因母猪(Noble 等,2002)。来自转基因母猪的初乳中乳糖含量高于非转基因母猪,提示乳生成过程在转基因母猪中更优,这可以通过乳糖合成能力来体现。增加的乳糖合成将导致更多的水,从而稀释了其他初乳组分(Noble 等,2002)。

9.14 结论

母猪乳腺分泌物的组成最显著地受到哺乳阶段的影响。分娩后最初 24 h 的乳腺分泌物称为初乳。与常乳相比,初乳具有高浓度的蛋白质,特别是免疫球蛋白,一些微量矿物质(特别是铜、铁、碘和锌),一些维生素,激素和生长因子。乳糖在初乳中的浓度比常乳低。在哺乳期的第 2～4 天,乳脂浓度会有短期的升高。在哺乳期第 7～10 天后,乳汁的组成在哺乳期的其余部分相对稳定。日粮可影响一些猪乳成分,包括脂肪、脂溶性维生素和一些矿物质的浓度,以及某些特定脂肪酸的比例。猪乳的一些组分还受遗传、胎次、初乳和常乳产量以及环境温度的影响。

参考文献

Alston-Mills, B., Iverson, S.J. and Thompson, M.P., 2000. A comparison of the composition of milks from Meishan and crossbred pigs. Livestock Production Science 63: 85-91.

Ariza-Nieto, C., Bandrick, M., Baidoo, S.K., Anil, L., Molitor, T.W. and Hathway, M.R., 2011. Effect of dietary supplementation of oregano essential oils to sows on colostrum and milk composition, growth pattern and immune status of suckling pigs. Journal of Animal Science 89: 1079-1089.

Atwood, C.S. and Hartmann, P.E., 1992. Collection of fore and hind milk from the sow and the changes in milk composition during suckling. Journal of Dairy Research 59: 287-298.

Atwood, C.S., Toussaint, J.K. and Hartmann, P.E., 1995. Assessment of mammary gland metabolism in the sow. II. Cellular metabolites in the mammary secretion and plasma during lactogenesis II. Journal of Dairy Research 62: 207-220.

Azain, M.J., 1993. Effects of adding medium-chain triglycerides to sow diets during late gestation and early lactation on litter performance. Journal of Animal Science 71: 3011-3019.

Baas, T.J., Christian, L.L. and Rothschild, M.F., 1992. Heterosis and recombination effects in Hampshire and Landrace swine: I. Maternal traits. Journal of Animal Science 70: 89-98.

Bagnell, C.A., Steinetz, B.G. and Bartol, F.F., 2009. Milk-borne relaxin and the lactocrine hypothesis for maternal programming of neonatal tissues. Annals of the New York Academy of Science 1160: 152-157.

Barkow, B., Matte, J.J., Bohme, H. and Flachowsky, G., 2001. Influence of folic acid supplements on the carry-over of folates from sow to the piglet. British Journal of Nutrition 85: 179-184.

Baumrucker, C.R. and Magliaro-Macrina, A.L., 2011. Hormones in milk. Encyclopedia of Dairy Sciences (2nd ed.), pp. 765-771.

Bontempo, V. and Jiang, X.R., 2015. Feeding various fat sources to sows: effects on immune status and performance of sows and piglets. Chapter 16. In: Farmer, C. (ed.) The gestating and lactating sow. Wageningen Academic Publishers, Wageningen, the Netherlands, pp. 357-375.

Bowland J.P., Grummer, R.H., Phillips, P.H. and Bohstedt, G., 1951. Seasonal variation in the fat, vitamin A, and vitamin D content of sows colostrum and milk. Journal of Animal Science 10: 533-537.
Bowland, J.P., 1966. Swine milk composition – a summary. In: Bustad, L.K. and McClellan, R.O. (eds.) Swine in biomedical research. Pacific Northwest National, Richland, Washington, DC, USA, pp. 97-107.
Boyd, R.D., Moser, B.D., Peo, Jr., E.R., Lewis, A.J. and Johnson, R.K., 1982. Effect of tallow and choline chloride addition to the diet of sows on milk composition, milk yield and preweaning pig performance. Journal of Animal Science 54: 1-7.
Braude, R., Coates, M.E., Henry, K.M., Kon, S.K., Rowland, S.J., Thompson, S.Y. and Walker, D.M., 1947. A study of the composition of sow's milk. British Journal of Nutrition 1: 64-77.
Braude, R., Kon, S.K. and Thompson, S.Y., 1946. A note on certain vitamins of sow's colostrum. Journal of Dairy Research 14: 414-418.
Brent, B.E., Miller, E.R., Ullrey, D.E. and Kemp, K.E., 1973. Postpartum changes in nitrogenous constituents of sow milk. Journal of Animal Science 36: 73-78.
Brooks, P.H. and Burke, J., 1998. Behaviour of sows and piglets during lactation. In: Verstegen, M.W.A., Moughan, P.J. and Schrama, J.W. (eds.) The lactating sow. Wageningen Pers, Wageningen, the Netherlands, pp. 301-338.
Bryant, K.L., Kornegay, E.T., Knight, J.W., Webb, Jr., K.E. and Notter, D.R., 1985. Supplemental biotin for swine. II. Influence of supplementation to corn- and wheat-based diets on reproductive performance and various biochemical criteria of sows during four parities. Journal of Animal Science 60: 145-153.
Butler, J.E., Wertz, N., Deschacht, N. and Kacskovics, I., 2009. Porcine IgG: structure, genetics, and evolution. Immunogenetics 61: 209-230.
Cerveza, P.J., Mehrbod, F., Cotton, S.J., Lomeli, N., Linder, M.C., Fonda, E.G. and Wickler, S.J., 2000. Milk ceruloplasmin and its expression by mammary gland and liver in pigs. Archives of Biochemistry and Biophysics 373: 451-461.
Chandan, R.C., Parry, Jr., R.M. and Shahani, K.M., 1968. Lysozyme, lipase, and ribonuclease in milk of various species. Journal of Dairy Science 51: 606-607.
Chaney, C.H. and Barnhart, C.E., 1963. Effect of iron supplementation of sow rations on the prevention of baby pig anemia. Journal of Nutrition 81: 187-192.
Clowes, E.J., Aherne, F.X., Foxcroft, G.R. and Baracos, V.E., 2003. Selective protein loss in lactating sows is associated with reduced litter growth and ovarian function. Journal of Animal Science 81: 753-764.
Coburn, S.P., Mahuren, J.D., Pauly, T.A., Ericson, K.L. and Townsend, D.W., 1992. Alkaline phosphatase activity and pyridoxal phosphate concentrations in the milk of various species. Journal of Nutrition 122: 2348-2353.
Coffey, M.T., Seerley, R.W. and Mabry, J.W., 1982. The effect of source of supplemental dietary energy on sow milk yield, milk composition and litter performance. Journal of Animal Science 55: 1388-1394.
Coffey, M.T., Yates, J.A. and Combs, G.E., 1987. Effects of feeding sows fat or fructose during late gestation and lactation. Journal of Animal Science 65: 1249-1256.
Csapo, J., Martin, T.G., Csapo-Kiss, Z.S. and Hazas, Z., 1996. Protein, fats, vitamin and mineral concentrations in porcine colostrum and milk from parturition to 60 days. International Dairy Journal 6: 881-902.

Curtis, J. and Bourne, F.J., 1971. Immunoglobulin quantitation in sow serum, colostrum and milk and the serum of young pigs. Biochimica et Biophysica Acta 236: 319-332.

Danielsen, M., Pedersen, L.J. and Bendixen, E., 2011. An *in vivo* characterization of colostrum protein uptake in porcine during early lactation. Journal of Proteomics 74: 101-109.

Darragh, A.J. and Moughan, P.J., 1998. The composition of colostrum and milk. In: Verstegen, M.W.A., Moughan, P.J. and Schrama, J.W. (eds.) The lactating sow. Wageningen Pers, Wageningen, the Netherlands, pp. 3-21.

De Quelen, F., Boudry, G. and Mourot, J., 2010. Linseed oil in the maternal diet increases long chain-PUFA status of the foetus and the newborn during the suckling period in pigs. British Journal of Nutrition 104: 533-543.

De Quelen, F., Boudry, G. and Mourot, J., 2013. Effect of different contents of extruded linseed in the sow diet on piglet fatty acid composition and hepatic desaturase expression during the post-natal period. Animal 7: 1671-1680.

DeMan, J.M. and Bowland, J.P., 1963. Fatty acid composition of sow's colostrum, milk and body fat as determined by gas-liquid chromatography. Journal of Dairy Research 30: 339-343.

Den Hartog, L.A., Boer, H., Bosch, M.W., Klaassen, G.J. and Van der Steen, H.A.M., 1987. The effect of feeding level, stage of lactation and method of milk sampling on the composition of milk (fat) in sows. Journal of Animal Physiology and Animal Nutrition 58: 253-261.

DeRouchey, J.M., Hancock J.D., Hines, R.H., Cummings, K.R., Lee, D.J., Maloney, C.A., Dean, D.W., Park, J.S. and Cao, H., 2003. Effects of dietary electrolyte balance on the chemistry of blood and urine in lactating sows and sow litter performance. Journal of Animal Science 81: 3067-3074.

Dever, J.T., Surles, R.L., Davis, C.R. and Tanumihardjo, S.A., 2011. α-Retinol is distributed through serum retinol-binding protein-independent mechanisms in the lactating sow-nursing piglet dyad. Journal of Nutrition 141: 42-47.

Devillers, N., Farmer, C. Mounier, A.M., Le Dividich, J. and Prunier, A., 2004a. Hormones, IgG and lactose changes around parturition in plasma, and colostrum or saliva of multiparous sows. Reproduction Nutrition Development 44: 381-396.

Devillers, N., Van Milgen, J., Prunier, A. and Le Dividich, J., 2004b. Estimation of colostrum intake in the neonatal pig. Animal Science 78: 305-313.

Dourmad, J.Y., Noblet, J. and Etienne, M., 1998. Effect of protein and lysine supply on performance, nitrogen balance, and body composition changes of sows during lactation. Journal of Animal Science 76: 542-550.

Dunshea, F.R., Bauman, D.E., Nugent, E.A., Kerton, D.J., King, R.H. and McCauley, I., 2005. Hyperinsulinaemia, supplemental protein and branched-chain amino acids when combined can increase milk protein yield in lactating sows. British Journal of Nutrition 93: 325-332.

Earle, I.P. and Stevensen, J.W., 1965. Relation of dietary zinc to composition of sow colostrum and milk. Journal of Animal Science 24: 325-328.

Elliot, J.L., Senft, B., Erhardt, G. and Fraser, D., 1984. Isolation of lactoferrin and its concentration in sows' colostrum and milk during a 21-day lactation. Journal of Animal Science 59: 1080-1084.

Elliott, R.F., Vander Noot, G.W., Gilbreath, R.L. and Fisher, H., 1971. Effect of dietary protein level on composition changes in sow colostrum and milk. Journal of Animal Science 32: 1128-1137.

Estienne, M.J., Harper, A.F., Barb, C.R. and Azain, M.J., 2000. Concentrations of leptin in serum and milk collected from lactating sows differing in body condition. Domestic Animal Endocrinology 19: 275-280.

Evans, P.A., Newby, T.J., Stokes, C.R. and Bourne, F.J., 1982. A study of cells in the mammary secretions of sows. Veterinary Immunology and Immunopathology 3: 515-527.

Fahmy, M.H., 1972. Comparative study of colostrum and milk composition of seven breeds of swine. Canadian Journal of Animal Science 52: 621-627.

Farmer, C., Giguère, A. and Lessard, M., 2010. Dietary supplementation with different forms of flax in late gestation and lactation: effects on sow and litter performances, endocrinology, and immune response. Journal of Animal Science 88: 225-237.

Farmer, C., Houtz, S.K. and Hagen, D.R., 1987. Estrone concentration in sow milk during and after parturition. Journal of Animal Science 64: 1086-1089.

Farmer, C., Palin, M.F., Sorensen, M.T. and Robert, S., 2001. Lactational performance, nursing and maternal behavior of Upton-Meishan and large white sows. Canadian Journal of Animal Science 81: 487-493.

Flummer, C. and Theil, P.K., 2012. Effect of β-hydroxy β-methyl butyrate supplementation of sows in late gestation and lactation on sow production of colostrum and milk and piglet performance. Journal of Animal Science 90: 372-374.

Foisnet, A., Farmer, C., David, C. and Quesnel, H., 2010a. Relationships between colostrum production by primiparous sows and sow physiology around parturition. Journal of Animal Science 88: 1672-1683.

Foisnet, A., Farmer, C., David, C. and Quesnel, H., 2010b. Altrenogest treatment during late pregnancy did not reduce colostrum yield in primiparous sows. Journal of Animal Science 88: 1684-1693.

Foisnet, A., Farmer, C., David, C. and Quesnel, H., 2011. Farrowing induction induces transient alterations in prolactin concentrations and colostrum composition in primiparous sows. Journal of Animal Science 89: 3048-3059.

Ford, J.E., Scott, K.J., Sansom, B.F. and Taylor, P.J., 1975. Some observations on the possible significance of vitamin B12- and folate-binding proteins in milk. Absorption of [^{58}Co] cyanocobalamin by sucking piglets. British Journal of Nutrition 34: 469-492.

Frenyo, V.L., Peters, G., Antal, T. and Szabo, I., 1981. Changes in colostral and serum IgG content in swine in relation to time. Veterinary Research Communications 4: 275-282.

Fritsche, K.L., Huang, S.C. and Cassity, N.A., 1993. Enrichment of omega-3 fatty acids in suckling pigs by maternal dietary fish oil supplementation. Journal of Animal Science 71: 1841-1847.

Garst, A.S., Ball, S.F., Williams, B.L., Wood, C.M., Knight, J.W., Moll, H.D., Aardema, C.H. and Gwazdauskas, F.C., 1999. Influence of pig substitution on milk yield, litter weights, and milk composition of machine milked sows. Journal of Animal Science 77: 1624-1630.

Gaucheron, F., 2005. The minerals of milk. Reproduction Nutrition Development 45: 473-483.

Gilbert, C.L., Goode, J.A. and McGrath, T.J., 1994. Pulsatile secretion of oxytocin during parturition in the pig: temporal relationship with fetal expulsion. Journal of Physiology 475: 129-137.

Goransson, L., 1990. The effect on late pregnancy feed allowance on the composition of the sow's colostrum and milk. Acta Veterinaria Scandinavica 31: 109-115.

Gueguen, L. and Salmon-Legageur, E., 1959. La composition du lait de truie: variations des teneurs en quelques elements mineraux (P, Ca, K, Na, Mg). Comptes rendus hebdomadaires des séances de l'Académie des sciences 249: 784-786.

Gurr, M.I., 1981. Review of the progress in dairy science: human and artificial milks for infant feeding. Journal of Dairy Research 48: 519-554.

Harmon, B.G., Liu, C.T., Cornelius, S.G., Pettigrew, J.E., Baker, D.H. and Jensen, A.H., 1974. Efficacy of different phosphorus supplements for sows during gestation and lactation. Journal of Animal Science 39: 1117-1122.

Hartmann, P.E., 1973. Changes in the composition and yield of the mammary secretions of cows during the initation of lactation. Journal of Endocrinology 59: 231-247.

Hartmann, P.E. and Holmes, M.A., 1989. Sow lactation. In: Barnett, J.L. and Hennessy, D.P. (eds.) Manipulating pig production II. Australasian Pig Science Association, Werribee, Victoria, Australia, pp. 72-97.

Heidebrecht, A.A., MacVicar, R., Ross, O.R. and Whitehair, C.K., 1951. Composition of swine milk. I. Major constituents and carotene, vitamin A and vitamin C. Journal of Nutrition 44: 43-50.

Heying, E.K., Grahn, M., Pixley, K.V., Rocheford, T. and Tanumihardjo, S.A., 2013. High-provitamin A carotenoid (orange) maize increases hepatic vitamin A reserves of offspring in a vitamin A-depleted sow-piglet model during lactation. Journal of Nutrition 143: 1141-1146.

Hidiroglou, M. and Batra, T.R., 1995. Concentrations of vitamin C in milk of sows and plasma of piglets. Canadian Journal of Animal Science 75: 275-277.

Holmes, M.A. and Hartmann, P.E., 1993. Concentration of citrate in the mammary secretion of sows during lactogenesis II and established lactation. Journal of Dairy Research 60: 319-326.

Hurley, W.L. and Bryson, J.M., 1999. Enhancing sow productivity through an understanding of mammary gland biology and lactation physiology. Pig News and Information 20: 125N-130N.

Hurley, W.L. and Grieve, R.C.J., 1988. Total and differential cell counts and N-acetyl-β-D-glucosaminidase activity in sow milk during lactation. Veterinary Research Communications 12: 149-153.

Hurley, W.L. and Theil, P.K., 2013. Immunoglobulins in mammary secretions. In: McSweeney, P.L.H. and Fox, P.F. (eds.) Advanced dairy chemistry, volume 1A: proteins: basic aspects (4th Edition). Springer Science, New York, NY, USA, pp. 275-294.

Jackson, J.R., 1990. Effect of gestation length and supplemental dietary fat on colostrum and milk composition in swine. PhD Thesis, University of Illinois, Urbana-Champaign, IL, U.S.A., pp. 1-129.

Jackson, J.R., Hurley, W.L., Easter, R.A., Jensen, A.H. and Odle, J., 1995. Effects of induced or delayed parturition and supplemental dietary fat on colostrum and milk composition in sows. Journal of Animal Science 73: 1906-1913.

Jaeger, L.A., Lamar, C.H., Bottoms, G.D. and Cline, T.R., 1987. Growth-stimulating substances in porcine milk. American Journal of Veterinary Research 48: 1531-1533.

Jensen, P.T., 1978. Tryspin inhibitor in sow colostrum and its function. Annales De Recherches Veterinaires 9: 225-228.

Jensen, P.T. and Pedersen, K.B., 1979. Studies on immunoglobulins and trypsin inhibitor in colostrum and milk from sows and in serum of their piglets. Acta Veterinaria Scandinavica 20: 60-72.

Kandeh, M.M., Park, Y.W., Pond, W.G. and Young, L.D., 1993. Milk cholesterol concentration in sows selected for three generations for high or low serum cholesterol. Journal of Animal Science 71: 1100-1103.

Keenan, T.W., King, J.L. and Colenbrander, V.F., 1970. Comparative analysis of mammary tissue and milk lipids of the sow. Journal of Animal Science 30: 806-811.

Kensinger, R.S., Collier, R.J., Bazer, F.W., Ducsay, C.A. and Becker, H.N., 1982. Nucleic acid, metabolic and histological changes in gilt mammary tissue during pregnancy and lactogenesis. Journal of Animal Science 54: 1297-1308.

Kent, J.C., Arthur, P.G. and Hartmann, P.E., 1998. Citrate, calcium, phosphate and magnesium in sows' milk at initiation of lactation. Journal of Dairy Research 65: 55-68.

Kim, Y.Y. and Mahan, D.C., 2001. Prolonged feeding of high dietary levels of organic and inorganic selenium to gilts from 25 kg body weight through on parity. Journal of Animal Science 79: 956-966.

King, R.H., 1998. Dietary amino acids and milk production. In: Verstegen, M.W.A., Moughan, P.J. and Schrama, J.W. (eds.) The lactating sow. Wageningen Pers, Wageningen, The Netherlands, pp. 131-141.

King, R.H., Toner, M.S., Dove, H., Atwood, C.S. and Brown, W.G., 1993. The response of first-litter sows to dietary protein level during lactation. Journal of Animal Science 71: 2457-2463.

Klaver, J., Van Kempen, G.J.M., De Lange, P.G.B., Verstegen, M.W.A. and Boer, H., 1981. Milk composition and daily yield of different milk components as affected by sow condition and lactation/feeding regimen. Journal of Animal Science 52: 1091-1097.

Klobasa, F. and Butler, J.E., 1987. Absolute and relative concentrations of immunoglobulins G, M, and A, and albumin in lacteal secretion of sows of different lactation numbers. American Journal of Veterinary Research 48: 176-182.

Klobasa, F., Werhahn, E. and Butler, J.E., 1987. Composition of sow milk during lactation. Journal of Animal Science 64: 1458-1466.

Konar, A., Thomas, P.C. and Rook, J.A.F., 1971. The concentrations of some water-soluble constituents in the milks of cows, sows, ewes and goats. Journal of Dairy Research 38: 333-341.

Krakowski, L., Krzyzanowski, J., Wrona, Z., Kostro, K., and Siwicki, A.K., 2002. The influence of nonspecific immunostimulation of pregnant sows on the immunological value of colostrum. Veterinary Immunology and Immunopathology 87: 89-95.

Krogh, U., Flummer, C., Jensen, S.K. and Theil, P.K., 2012. Colostrum and milk production of sows is affected by dietary conjugated linoleic acid. Journal of Animal Science 90: 366-368.

Kusina, J., Pettigrew, J.E., Sower, A.F., White, M.E., Crooker, B.A. and Hathaway, M.R., 1999. Effect of protein intake during gestation and lactation on the lactational performance of primiparous sows. Journal of Animal Science 77: 931-941.

Lauridsen, C. and Danielsen, V., 2004. Lactational dietary fat levels and sources influence milk composition and performance of sows and their progeny. Livestock Production Science 91: 95-105.

Lauridsen, C., Engel, H., Jensen, S.K., Craig, A.M. and Traber, M.G., 2002. Lactating sows and suckling piglets preferentially incorporate RRR- over all-rac-α-tocopheryl into milk, plasma tissues. Journal of Nutrition 132: 1258-1264.

Lauridsen, C. and Jensen, S.K., 2005. Influence of supplementation of all-rac-α-tocopheryl acetate preweaning and vitamin C postweaning on α-tocopherol and immune responses of piglets. Journal of Animal Science 83: 1274-1286.

Laws, J., Amusquivar, E., Laws, A., Herrera, E., Lean, L.J., Dodds, P.F. and Clarke, L., 2009. Supplementation of sows diets with oil during gestation: sow body condition, milk yield and milk composition. Livestock Science 123: 88-96.

Le Dividich, J., Rooke, J.A. and Herpin, P., 2005. Nutrition and immunological importance of colostrum for the new-born pig. Journal of Agricultural Science 143: 469-485.

Le Huërou-Luron, I. and Ferret-Bernard, S., 2015. Development of gut and gut-associated lymphoid tissues in piglets. Role of maternal environment. Chapter 15. In: Farmer, C. (ed.) The gestating and lactating sow. Wageningen Academic Publishers, Wageningen, the Netherlands, pp. 335-355.

Le Jan, C., 1995. Epithelial cells in sow mammary secretions. Advances in Experimental Medicine and Biology 371A: 233-234.

Lee, C.S. McCauley, I.M. and Hartmann, P.E., 1983. Light and electron microscopy of cells in pig colostrum, milk and involution secretion. Acta Anatomica 116: 126-135.

Leonard, S.G., Sweeney, T., Babar, B., Lynch, B.P. and O'Doherty, J.V., 2010. Effect of maternal fish oil and seaweed extract supplementation on colostrum and milk composition, humoral immune response, and performance of suckled piglets. Journal of Animal Science 88: 2988-2997.

Loisel, F., Farmer, C., Ramackers, P. and Quesnel, H., 2013. Effects of high fiber intake during late pregnancy on sow physiology, colostrum production, and piglet performance. Journal of Animal Science 91: 5269-5279.

Loudenslager, M.J., Ku, P.K., Whetter, P.A., Ullrey, D.E., Whitehair, C.K., Stowe, H.D. and Miller, E.R., 1986. Importance of diet of dam and colostrum to the biological antioxidant status and parenteral iron tolerance of the pig. Journal of Animal Science 63: 1905-1914.

Lyberg, K., Andersson, H.K., Simonsson, A. and Lindberg, J.E., 2007. Influence of different phosphorus levels and phytase supplementation in gestation diets on sow performance. Journal of Animal Physiology and Animal Nutrition 91: 304-311.

Maffeo, G., Damasio, M., Balabio, R. and Jochle, W., 1987. Detection of prostaglandin-like substances in sow's milk. Zuchthygiene 22: 209-214.

Mahan, D.C., 1991. Assessment of the influence of dietary vitamin E on sows and offspring in three parities: reproductive performance, tissue tocopherol, and effects on progeny. Journal of Animal Science 69: 2904-2917.

Mahan, D.C., 2000. Effect of organic and inorganic selenium sources and levels on sow colostrum and milk selenium content. Journal of Animal Science 78: 100-105.

Mahan, D.C., Becker, D.E., Harmon, B.G. and Jensen, A.H., 1971. Effect of protein levels and *opaque-2* corn on sow milk composition. Journal of Animal Science 32: 482-496.

Mahan, D.C., Kim, Y.Y and Stuart, R.L., 2000. Effect of vitamin E sources (RRR- or all-*rac*-α-tocopheryl acetate) and levels on sow reproductive performance, serum, tissue, and milk α-tocopherol contents over a five-parity period, and the effects on the progeny. Journal of Animal Science 78: 110-119.

Mahan, D.C., Moxon, A.L. and Hubbard, M., 1977. Efficacy of inorganic selenium supplementation to sow diets on resulting carry-over to their progeny. Journal of Animal Science 45: 738-746.

Mahan, D.C. and Newton, E.A., 1995. Effect of initial breeding weight on macro- and micromineral composition over a three-parity period using a high-producing sow genotype. Journal of Animal Science 73: 151-158.

Mahan, D.C and Peters, J.C., 2004. Long-term effects of dietary organic and inorganic selenium sources and levels on reproducing sows and their progeny. Journal of Animal Science 82: 1343-1358.

Malm, A., Pond, W.G., Walker, Jr., E.F., Homan, M., Aydin, A. and Kirtland, D., 1976. Effect of polyunsaturated fatty acids and vitamin E level of the sow gestation diet on reproductive performance and on level of alpha tocopherol in colostrum, milk and dam and progeny blood serum. Journal of Animal Science 42: 393-399.

Markowska-Daniel, I. and Pomorska-Mol, M., 2010. Shifts in immunoglobulins levels in the porcine mammary secretions during whole lactation period. Bulletin of the Veterinary Institute in Pulway 54: 345-349.

Mateo, C.D., Peters, D.N. and Stein, H.H., 2004. Nucleotides in sow colostrum and milk at different stages of lactation. Journal of Animal Science 82: 1339-1342.

Matte, J.J. and Girard, C.L., 1989. Effects of intramuscular injections of folic acid during lactation on folates in serum and milk and performance of sows and piglets. Journal of Animal Science 67: 426-431.

Miller, G.M., Conrad, J.H. and Harrington, R.B., 1971. Effect of dietary unsaturated fatty acids and stage of lactation on milk composition and adipose tissue in swine. Journal of Animal Science 32: 79-83.

Miller, M.B., Hartsock, T.G., Erez, B., Douglass, L. and Alston-Mills, B., 1994. Effect of dietary calcium concentrations during gestation and lactation in the sow on milk composition and litter growth. Journal of Animal Science 72: 1315-1319.

Monaco, M.H., Gronlund, D.E., Bleck, G.T., Hurley, W.L., Wheeler, M.B. and Donovan, S.M., 2005. Mammary specific transgenic over-expression of insulin-like growth factor-I (IGF-I) increases pig milk IGF-I and IGF binding proteins, with no effect on milk composition or yield. Transgenic Research 14: 761-773.

Mostyn, A., Sebert, S., Litten, J.C., Perkins, K.S., Laws, J., Symonds, M.E. and Clarke, L., 2006. Influence of porcine genotype on the abundance of thyroid hormones and leptin in sow milk and its impact on growth, metabolism and expression of key adipose tissue genes in offspring. Journal of Endocrinology 190: 631-639.

Motyl, T., Ploszaj, T., Wojtasik, A., Kukulska, W. and Podgurniak, M., 1995. Polyamines in cow's and sow's milk. Comparative Biochemistry and Biophysics 111B: 427-433.

Mulloy, A.L. and Malven, P.V., 1979. Relationships between concentrations of porcine prolactin in blood serum and milk of lactating sows. Journal of Animal Science 48: 876-881.

Neuhaus, U., 1961. Die milchleistung der sau und die zusammensetzung und eigenschaften der sauenmilch. Zeitschrift für Tierzüchtung und Züchtungsbiologie 75: 160-191.

Nguyen, T.V., Yuan, L., Azevedo, S.P., Jeong, K.I., Gonzalez, A.M. and Saif, L.J., 2007. Transfer of maternal cytokines to suckling piglets: *in vivo* and *in vitro* models with implications for immunomodulation of neonatal immunity. Veterinary Immunology and Immunopathology 117: 236-248.

Noble, M.S., Rodriguez-Zas, S., Cook, J.B., Bleck, G.T., Hurley, W.L. and Wheeler, M.B., 2002. Lactational performance of first-parity transgenic gilts expressing bovine α-lactalbumin in their milks. Journal of Animal Science 80: 1090-1096.

Noblet, J. and Etienne, M., 1986. Effect of energy level in lactating sows on yield and composition of milk and nutrient balance of piglets. Journal of Animal Science 63: 1888-1896.

O'Connor, D.L., Picciano, M.F., Roos, M.A. and Easter, R.A., 1989. Iron and folate utilization in reproducing swine and their progeny. Journal of Nutrition 119: 1984-1991.

Oftedal, O.T., 1984. Milk composition, milk yield and energy output at peak lactation: a comparative review. Symposium of the Zoological Society of London 51: 33-85.

Oftedal, O.T. and Iversen, S.J., 1995. Comparative analysis of nonhuman milks. A. Phylogenetic variation in the gross composition of milks. In: Jensen, R.G. (ed.) Handbook of milk composition. Academic Press, Inc., San Diego, CA, USA, pp. 749-789.

Osterlundh, I., Holst, H. and Magnusson, U., 1998. Hormonal and immunological changes in blood and mammary secretion in the sow at parturition. Theriogenology 50: 465-477.

Paape, M.J. and Tucker, H.A., 1966. Somatic cell content variation in fraction-collected milk. Journal of Dairy Science 49: 265-267.

Pakkanen, R. and Aalto, J., 1997. Growth factors and antimicrobial factors in bovine colostrum. International Dairy Journal 7: 285-297.

Park, Y.W., 2011. Milks of other domesticated mammals (pigs, yaks, reindeer, etc). In: Fuquay, J.W., Fox. P.F. and McSweeney, P.L.H. (eds.) Encyclopedia of dairy sciences, volume 3 (2nd Edition). Academic Press, Inc., San Diego, CA, USA, pp. 530-537.

Park, Y.W., Kandeh, M., Chin, K.B., Pond, W.G. and Young, L.D., 1994. Concentrations of inorganic elements in milk of sows selected for high and low cholesterol. Journal of Animal Science 72: 1399-1402.

Peaker, M., 1983. Secretion of ions and water. In: Mepham, T.B. (ed.) Biochemistry of lactation. Elsevier Science Publishers, Amsterdam, the Netherlands, pp. 285-305.

Peaker, M. and Linzell, J.L., 1975. Citrate in milk: a harbinger of lactogenesis. Nature 253: 464.

Peng, Y., Ren, F., Yin, J.D., Fang, Q., Li, F.N. and Li, D.F., 2010. Transfer of conjugated linoleic acid from sows to their offspring and its impact on the fatty acid profiles of plasma, muscle, and subcutaneous fat in piglets. Journal of Animal Science 88: 1741-1751.

Perrin, D.R., 1955. The chemical composition of the colostrum and milk of the sow. Journal of Dairy Research 22: 103-107.

Perrin, D.R., 1958. The calorific value of milk of different species. Journal of Dairy Research 25: 215-220.

Peters, J.C. and Mahan, D.C., 2008. Effects of dietary organic and inorganic trace mineral levels on sow reproductive performances and daily mineral intakes over six parities. Journal of Animal Science 86: 2247-2260.

Peters, J.C., Mahan, D.C., Wiseman, T.G. and Fastinger, N.D., 2010. Effect of dietary organic and inorganic micromineral source and level on sow body, liver, colostrum, mature milk, and progeny mineral compositions over six parities. Journal of Animal Science 88: 626-637.

Pettigrew, Jr., J.E., 1981. Supplemental dietary fat for peripartal sows: a review. Journal of Animal Science 53: 107-117.

Phipps, L.W. and Newbould, F.H.S., 1966. Determination of leucocyte concentrations in cow's milk with a Coulter counter. Journal of Dairy Research 33: 51-64.

Pinelli-Saavedra, A., Calderon de la Barca, A.M., Hernandez, J., Valenzuela, R. and Scaife, J.R., 2008. Effect of supplementing sows' feed with α-tocopherol acetate and vitamin C on transfer of α-tocopherol to piglet tissues, colostrum, and milk: aspects of immune status of piglets. Research in Veterinary Science 85: 92-100.

Pond, W.G. and Jones, J.R., 1964. Effect of level of zinc in high calcium diets on pigs from weaning through one reproductive cycle and on subsequent growth of their offspring. Journal of Animal Science 23: 1057-1060.

Pond, W.G., Veum, T.L. and Lazar, V.A., 1965. Zinc and iron concentration of sows' milk. Journal of Animal Science 24: 668-670.

Prunier, A., Messias de Bragança, M. and Le Dividich, J., 1997. Influence of high ambient temperature on performance of reproductive sows. Livestock Production Science 52: 123-133.

Quesnel, H., 2011. Colostrum production by sows: variability of colostrum yield and immunoglobulin G concentrations. Animal 5: 1546-1553.

Quesnel, H., Farmer, C. and Devillers, N., 2012. Colostrum intake: influence on piglet performance and factors of variation. Livestock Science 146: 105-114.

Quesnel, H., Farmer, C. and Theil, P.K., 2015. Colostrum and milk production. Chapter 8. In: Farmer, C. (ed.) The gestating and lactating sow. Wageningen Academic Publishers, Wageningen, the Netherlands, pp. 173-192.

Quesnel, H., Renaudin, A., Le Floc'h, N., Jondreville, C., Père, M.C., Taylor-Pickard, J.A. and Le Dividich, J., 2008. Effect of organic and inorganic selenium sources in sow diets on colostrum production and piglet response to a poor sanitary environment after weaning. Animal 2: 859-866.

Renaudeau, D. and Noblet, J., 2001. Effects of exposure to high ambient temperature and dietary protein level on sow milk production and performance of piglets. Journal of Animal Science 79: 1540-1548.

Reynolds, L. and Rook, J.A.F., 1977. Intravenous infusion of glucose and insulin in relation to milk secretion in the sow. British Journal of Nutrition 37: 45-52.

Richert, B.T., Goodband, R.D., Tokach, M.D. and Nelssen, J.L., 1997. Increasing valine, isoleucine, and total branched-chain amino acids for lactating sows. Journal of Animal Science 75: 2117-2128.

Rolinec, M., Biro, D., Stastny, P., Galik, B., Simko, M. and Juracek, M., 2012. Immunoglobulins in colostrum of sows with porcine reproductive and respiratory syndrome – PPRS. Journal of Central European Agriculture 13: 303-311.

Rooke, J.A., Bland, I.M. and Edwards, S.A., 1998. Effect of feeding tuna oil or soyabean oil as supplements to sows in late pregnancy on piglet tissue composition and viability. British Journal of Nutrition 80: 273-280.

Salman, S., Khol-Parisini, A., Schafft, H., Lahrssen-Wiederholt, M., Hulan, H.W., Dinse, D. and Zentek, J., 2009. The role of dietary selenium in bovine mammary gland health and immune function. Animal Health Research Reviews 10: 21-34.

Schieck, S.J., Kerr, B.J., Baidoo, S.K., Shurson, G.C. and Johnston, L.J., 2010. Use of crude glycerol, a biodiesel coproduct, in diets for lactating sows. Journal of Animal Science 88: 2648-2656.

Schmid, A., Collomb, M., Bee, G., Butikofer, U., Wechsler, D., Eberhard, P. and Sieber, R., 2008. Effect of dietary alpine butter rich in conjugated linoleic acid on milk fat composition of lactating sows. British Journal of Nutrition 100: 54-60.

Schoenherr, W.D., Staley, T.S. and Cromwell, G.L., 1989. The effects of dietary fat or fiber addition on yield and composition of milk from sows housed in a warm or hot environment. Journal of Animal Science 67: 482-495.

Schollenberger, A., Degorski, A., Frymus, T and Schollenberger, A., 1986a. Cells of sow mammary secretions. I. Morphology and differential counts during lactation. Journal of Veterinary Medicine A 33: 31-38.

Schollenberger, A., Degorski, A., Frymus, T and Schollenberger, A., 1986b. Cells of sow mammary secretions. II. Some properties of phagocytic cells. Journal of Veterinary Medicine A 33: 39-46.

Schollenberger, A., Degorski, A., Frymus, T and Schollenberger, A., 1986c. Cells of sow mammary secretions. III. Characterization of lymphocyte populations. Journal of Veterinary Medicine A 33: 353-359.

Schone, F., Leiterer, M., Hartung, H., Jahreis, G. and Tischendorf, F., 2001. Rapeseed glucosinolates and iodine in sows affect the milk iodine concentration and the iodine status of piglets. British Journal of Nutrition 85: 659-670.

Schone, F., Leiterer, M., Jahreis, G. and Rudolph, B., 1997. Effect of rapeseed feedstuffs with different glucosinolate content and iodine administration on gestating and lactating sow. Journal of Veterinary Medicine A 44: 325-339.

Seerley, R.W., Snyder, R.A. and McCampbell, H.C., 1981. The influence of sow dietary lipids and choline on piglet survival, milk and carcass composition. Journal of Animal Science 52: 542-550.

Seynaeve, M., De Wilde, R., Janssens, G. and De Smet, B., 1996. The influence of dietary salt level on water consumption, farrowing, and reproductive performance of lactating sows. Journal of Animal Science 74: 1047-1055.

Sheffy, B.E., Shahani, K.M., Grummer, R.H., Phillips, P.H. and Sommer, H.H., 1952. Nitrogen constituents of sow's milk as affected by ration and stage of lactation. Journal of Nutrition 48: 103-114.

Shurson, G.C., Hogberg, M.G., DeFever, N., Radecki, S.V. and Miller, E.R., 1986. Effects of adding fat to the sow lactation diet on lactation and rebreeding performance. Journal of Animal Science 62: 672-680.

Shurson, G.C. and Irvin, K.M., 1992. Effects of genetic line and supplemental dietary fat on lactation performance of Duroc and Landrace sows. Journal of Animal Science 70: 2942-2949.

Simmen, F.A., Whang, K.Y., Simmen, R.C.M., Peterson, G.A., Bishop, M.D. and Irvin, K.M., 1990. Lactational variation and relationship to postnatal growth of insulin-like growth factor-I in mammary secretions from genetically diverse sows. Domestic Animal Endocrinology 7: 199-206.

Staszkiewicz, J., Franczak, A., Kotwica, G. and Koziorowski, M., 2004. Secretion of estradiol-17β by porcine mammary gland of ovariectomized steroid-treated sows. Animal Reproduction Science 81: 87-95.

Tan, T.J., Schober, D.A. and Simmen, F.A., 1990. Fibroblast mitogens in swine milk include an epidermal growth factor-related peptide. Regulatory Peptides 27: 61-74.

Tao, N., Ochonicky, K.L., German, B., Donovan, S.M. and Lebrilla, C.B., 2010. Structural determination and daily variations of porcine milk oligosaccharides. Journal of Agricultural and Food Chemistry 58: 4653-4659.

Taugbøl, O., Farmstad, T. and Saarem, K., 1993. Supplements of cod liver oil to lactating sows. Influence on milk fatty acid composition and growth performance of piglets. Journal of Veterinary Medicine A 40: 437-443.

Theil, P.K., Jørgensen, H and Jakobsen, K., 2004. Energy and protein metabolism in lactating sows fed two levels of dietary fat. Livestock Production Science 89: 265-276.

Theil, P.K., Lauridsen, C. and Quesnel, H., 2014. Neonatal piglet survival: impact of sow nutrition around parturition on foetal glycogen deposition, and production and composition of colostrum and transient milk. Animal 8: 1021-1030.

Tilton, S.L., Miller, P.S., Lewis, A.J., Reese, D.E. and Ermer, P.M., 1999. Addition of fat to the diets of lactating sows: I. Effects on milk production and composition and carcass composition of the litter at weaning. Journal of Animal Science 77: 2491-2500.

Van den Brand, H., Heetkamp, M.J.W., Soede, N.M., Schrama, J.W. and Kemp, B., 2000. Energy balance of lactating primiparous sows as affected by feeding level and dietary energy source. Journal of Animal Science 78: 1520-1528.

Venn, J.A.J, McCance, R.A. and Widdowson, E.M., 1947. Iron metabolism in piglet anaemia. Journal of Comparative Pathology and Therapeutics 57: 314-325.

Veum, T.T., Gallo, J.T., Pond, W.G., Van Vleck, L.D. and Loosli, J.K., 1965. Effect of ferrous fumarate in the lactation diet on sow milk iron, pig hemoglobin and weight gain. Journal of Animal Science 24: 1169-1173.

Wang, S.P., Yin, Y.L., Qian, Y., Li, L.L., Li, F.N., Tan, B.E., Tang, X.S. and Huang, R.L., 2011. Effects of folic acid on the performance of suckling piglets and sows during lactation. Journal of the Science of Food and Agriculture 91: 2371-2377.

Westrom, B.R., Ekman, R., Svendsen, L., Svendsen, J. and Karlsson, B.W., 1987. Levels of immunoreactive insulin, neurotensin, and bombesin in porcine colostrum and milk. Journal of Pediatric Gastroenterology and Nutrition 6: 460-465.

White, C.E., Head, H.H., Bachman, K.C. and Bazer, F.W., 1984. Yield and composition of milk and weight gain of nursing pigs from sows fed diets containing fructose or dextrose. Journal of Animal Science 59: 141-150.

Whittestone, W.G., 1952. The physical properties of sow's milk as a function of stage of lactation. Journal of Dairy Research 19: 330-334.

Wu, G. and Knabe, D.A., 1994. Free and protein-bound amino acids in sow's colostrum and milk. Journal of Nutrition 124: 415-424.

Wuryastuti, H., Stowe, H.D., Bull, R.W. and Miller, E.R., 1993. Effects of vitamin E and selenium on immune responses of peripheral blood, colostrum, and milk leukocytes of sows. Journal of Animal Science 71: 2464-2472.

Yan, W., Wiley, A.A., Bathgate, A.D., Frankshun, A.L., Lasano, S., Crean, B.D., Steinetz, B.G., Bagnell, C.A. and Bartol, F.F., 2006. Expression of LGR7 and LGR8 by neonatal porcine uterine tissues and transmission of milk-borne relaxin into the neonatal circulation by suckling. Endocrinology 147: 4303-4310.

Yao, W., Li, J., Wang, J.J., Zhou, W., Wang, Q., Zhu, R., Wang, F. and Thacker, P., 2012. Effects of dietary ratio of n-6 to n-3 polyunsaturated fatty acids on immunoglobulins, cytokines, fatty acid composition, and performance of lactating sows and suckling piglets. Journal of Animal Science and Biotechnology 3: 43-50.

Yoon, I. and McMillan, E., 2006. Comparative effects of organic and inorganic selenium on selenium transfer from sows to nursing pigs. Journal of Animal Science 84: 1729-1733.

Yvon, M., Levieux, D., Valluy, M.C., Pelissier, J.P. and Mirand, P.P., 1993. Colostrum protein digestion in newborn lambs. Journal of Nutrition 123: 586-596.

Zhan, X., Qie, Y., Wang, M., Li, X. and Zhao, R., 2011. Selenomethionine: an effective selenium source for sow to improve Se distribution, antioxidant status, and growth performance of pig offspring. Biological Trace Element Research 142: 481-491.

Zou, S., McLaren, D.G. and Hurley, W.L., 1992. Pig colostrum and milk composition: comparisons between Chinese Meishan and US breeds. Livestock Production Science 30: 115-127.

10. 母猪分娩期间和泌乳早期的圈舍、管理和环境

O. A. T. Peltoniemi and C. Oliviero

University of Helsinki, Dept. Production Animal Medicine, Paroninkuja 20, 04920 Saarentaus, Finland; olli.peltoniemi@helsinki.fi, claudio.oliviero@helsinki.fi

摘要：母猪分娩的成功取决于诸多方面，包括母性行为、产程、仔猪死亡率和初乳摄入量。母猪的分娩产程可以作为衡量分娩是否成功的一个简单指标。我们建议时间在300 min以内的分娩算做一次成功分娩。使用外源激素对分娩过程进行人为干预除了众所周知的好处外，其存在的危险也同样值得重视。用前列腺素诱导分娩可能导致未成熟的仔猪出生时出现先天性疾病。而过度使用或常规使用催产素会降低胎盘血流量，从而使子宫内胎儿出现缺氧的症状。母猪分娩时的疼痛管理对于哺乳的成功同样是一个值得高度重视的问题，无论何时当我们观察到母猪护仔出现异常行为时，疼痛都是我们应该考虑的一个潜在原因。饲喂方法被认为是超高产母猪在分娩期间繁殖管理的主要方面。新的饲喂方法如在分娩前在妊娠母猪的日粮中添加膳食纤维可以防止便秘、提高母猪饮水量、提高产奶量并改善仔猪生产性能。运用现代技术监督分娩可以减少大窝产仔相关损失。在育种计划中，新的母性特征部分如母猪行为、产程延长、初乳生产和仔猪质量参数，都可被用来进一步提高分娩和泌乳早期繁殖管理的成功率。

关键词：母猪，饲喂，疼痛，激素，便秘

10.1 引言

分娩在母猪的繁殖生产中是一个关键事件，并且它对仔猪生产有很大的经济影响。多年来在养猪生产中选育品种这种方法在提高每窝产仔数方面已经取得了良好的成果。然而，有资料显示产仔数与仔猪个体出生体重之间呈负相关（Kerr and Cameron, 1995; Roehe, 1999; Sorensen 等, 2000），仔猪出生时的死亡率随产仔数的增加而增加（Sorensen 等, 2000）。饲养计划的最终目标旨在增加母猪的产奶量，以便母猪在断奶前阶段能更好地支持所有仔猪的生长（Einarsson 和 Rojkittikhun, 1993）。通过减少哺乳时间和缩短断奶至发情间隔，使母猪的生产周期变得更快，以最大限度地提高每头母猪每年生产的仔猪数量。近年来养猪场的变

化越来越大,通常由于空间的功能分布,限制了每个母猪的可用空间。所有这些变化都有助于给母猪产仔增加环境压力。

10.2 分娩生理机制

分娩的生理过程十分复杂。事实上,分娩过程是由多种激素相互作用来进行调节的。在妊娠的最后阶段,激素如孕酮、黄体生成素、雌激素、皮质醇、催乳素、松弛素、前列腺素等就成为调节分娩这一生理过程的主要参与者(图10.1)。所有这些激素,受各种内部时钟和信号系统的调节,以非常精细的方式相互影响和相互调节(Anderson,2000)。

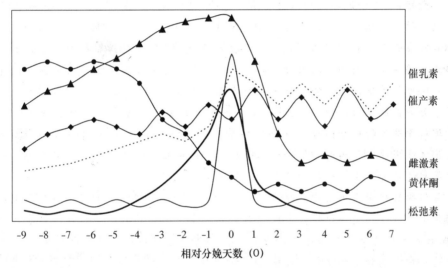

图10.1 母猪妊娠期间生殖激素浓度的示意图(Anderson,2000)。分娩前几天,激素浓度发生重大变化。

在分娩过程中行为表达同样非常重要。母猪的激素促使其表现出筑巢行为,这标志着母猪分娩活动的开始(Algers 和 Uvnäs-Moberg,2007)。然而,环境也可以对这种激素的模式产生重要但间接的影响。例如,限制性环境会影响筑巢行为,特别是在现代养猪生产系统中,母猪受到几个外部因素的压力(Damm 等,2003)。这种精细而微妙的激素过程显然需要非常明确的前提才能正确地进行。除了圈舍,营养和疾病都有可能直接或间接的影响激素的释放从而影响分娩过程。

在妊娠早期,在胚胎植入以及黄体开始发育活动后,恒定的更高浓度的孕酮开始主导激素模式(Meulen 等,1988)。这可以保持妊娠的顺利进行,而在妊娠后的近2/3阶段,黄体酮的血液循环浓度仍然很高。在分娩开始前24~48 h,孕酮浓度快速下降并对许多其他激素产生级联效应,而后其在妊娠期间保持相当稳定的浓度。而前列腺素开始达到最高浓度，催产素浓度有所增加并开始表现出高脉动活动（Gilbert 等，1994），催乳素浓度逐渐升高，雌激素水平逐渐升高，并且在快速上升达到巅峰后逐渐降低到基础水平（Anderson，2000；Ellendorff 等，1979；Kindahl 等，1982）（图10.1）。因此，可以看出产仔期大量激素变化的过程被局限在一个很短的时间内。同样的，在产仔期间皮质醇浓度也达到峰值，上升到2～3倍基底浓度后在24～36 h逐渐恢复到平衡状态。

10.3 行为和活动

在产仔过程中，上述强烈的激素活动不仅有助于诱导分娩过程，还可以诱发明显的行为变化。诸如呆立、刨地、来回走动等活动在分娩前24 h开始大幅增加（Hartsock 和 Barczewski，1997），并且母猪开始表现出一定的筑巢行为。孕酮减少和前列腺素增加引起的催乳素上升是触发筑巢行为的因素之一（Algers 和 Uvnäs-Moberg，2007）（Castrén 等，1994）。外部环境对筑巢行为的表达同样具有重要影响。事实上，适当筑巢材料的使用可以加快筑巢这一活动的进程（Damn 等，2000）。由于现代圈舍制度导致筑巢材料非常有限或者根本不存在并且母猪的活动范围也非常有限，这就促进了母猪在产床上产仔。在这些特殊情况下，内源激素活动所引发的筑巢行为就不能得到恰当的表达。在没有筑巢材料的情况下，受限母猪延长并且不能成功表现出筑巢行为（Damn 等，2003）。这种筑巢机会的缺乏使筑巢行为不能表达导致血浆皮质醇和促肾上腺皮质激素分泌开始增加（Jarvis 等，1997），这两种激素的增加量也可以用来衡量母猪所受应激程度。Gustafsson 等（1999）发现有过几次产仔经验的母猪即使在产仔箱没有垫草的情况下依然能够建立与野猪相同的巢。因此，无论有没有圈舍或者筑巢材料，这种与生俱来的行为都是即将分娩的一个明确标志。

10.4 成功分娩

分娩成功的前提是产仔母猪拥有适宜它表达筑巢行为的工具，如可利用和可自由移动的筑巢材料。此外，同样重要的是饲喂母猪的饲料要注意避免造成母猪的便秘和肥胖。并且，如果一次分娩的总时间少于5 h并且有超过90%的新生仔猪在72 h之内活着，那么这次分娩就可以被称作一次成功分娩。成功分娩的标准是仔猪成功接受母乳，并且在生产过程中母猪和仔猪均不产生并发症。要避免的典型并发症包括仔猪缺氧、母猪产道污染和随后的母猪产后炎症等。

10.5 分娩环境

现代化养猪场产房的四种基本类型为：母猪产床、母猪围栏、室内群养、室外散养。在欧洲和北美洲，母猪围栏式和母猪产仔箱式产房被广泛应用，而后者又在其中占主导地位（Hartsock 和 Barczewski，1997；Jensen 等，1997；Kemp 和 Soede，2012）。

10.5.1 产房：产床式和围栏式

通常产床由一个杆件系统组成，其可以将母猪限制在有限的空间内使母猪只能躺下或站立而不能转动和移动。并且产仔箱通常放置在水泥地面、漏缝地板或二者组合的地面上。由于母猪粗饲料和筑巢材料的使用受欧洲相关法规的限制只能描述为最低标准（91/630/EEC）。因此，不同国家间的差异会相当大，但在粗饲料方面的差异会很小甚至没有。在这些特殊的环境中，母猪很少有机会表现出适当的筑巢行为。在没有合适的筑巢材料的情况下，母猪会将它的注意力转向地板或产仔箱的围栏（Lawrence 等，1994）。在产床上的母猪通常比在围栏中的母猪更早开始分娩（Hansen 和 Curtis，1980）。应用产床是为了减少母猪将新生仔猪踩压而造成的损失（Edwards 和 Fraser，1997）。一些研究确实报道过产床式与围栏式产房相比可以更好地避免母猪将新生仔猪压死（Cronin 等，1996），然而，在其他方面的几项研究并没有发现两者间的差异（Aumaitre 和 Le Dividich，1984；Fraser，1990；Phillips 和 Fraser，1993）。最近，在丹麦的研究者提出了进一步需要向母猪提供筑巢材料这种方式来应对哺乳期母猪产房宽松问题的可行性（Hales 等，2014）。产仔围栏在尺寸和结构上有较大的不同，更大的产床可以允许母猪自由地移动和转身。在分娩前 24 h 产仔围栏中的母猪与产床上活动受限的母猪相比其开始更加频繁地移动和转身，而产床上的母猪的活动受限制（Hartsock 和 Barczewski，1997）。只要提供更多的空间，即使没有稻草，在分娩期间也能促进母性行为的表达。Vestergaard 和 Hansen（1984）发现在母猪产仔之前或者分娩中将其限制在产床上使母猪的分娩过程延长。若欠佳的建巢条件其直接影响或随后应激的间接影响都会造成分娩时间增长。因为围栏可以让母猪更好地表达筑巢行为，这在一定程度上对母猪和仔猪的健康和福利有益（Algers，1994）。

10.5.2 产房对与分娩相关激素变量的影响

有数据表明，在母猪产仔后 2~5 天，产床上的母猪分泌的唾液皮质醇浓度比围栏中母猪分泌的唾液皮质醇浓度更高（Oliviero 等，2008a）。在此期间产床组皮质醇一直保持较高浓度而围栏组皮质醇浓度则恢复到分娩前的水平（图 10.2）。而高浓度的唾液皮质醇则会导致母猪的分娩时间延长。如上所述，母猪若缺乏合适的表达筑巢行为环境(Jarvis 等，2002)其将会将压力水平提高到正常生理机制的阈值之上。并且产床上的母猪体内较低的催产素浓度似乎也可以解释这一现象(Oliviero 等，2008a)，因为催产素的分泌不足往往被看做是分娩过程延长的主要原因。

饲养在产床上的母猪产后催产素脉冲的平均浓度(仔猪产出后 6 min 内测定)往往低于圈养的母猪(Oliviero 等，2008a)。这些激素的浓度与产程的长短密切相关。此外,在不考虑产房不同的情况下,若产程超过 5 h 则与催产素浓度低密切相关。

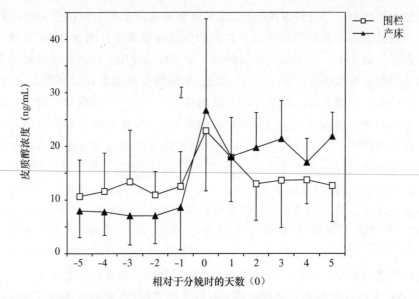

图 10.2　饲养在围栏圈和产床上母猪的平均唾液皮质醇浓度。结果为平均值±标准差 (Oliviero 等，2008a)。

10.5.3　产房对产程和仔猪死亡率的影响

产房和仔猪死亡率之间存在非常清晰的联系,因为它们均与产程相关(表 10.1)。很多研究发现在产床上产仔的母猪其平均产程会延长 90 min,并且在产床上产仔的母猪其产程持续超过 300 min。产程超过 300 min 的母猪与产程少于 200 min 的母猪相比其具有较高的死胎率。

在不考虑产房因素的情况下,仔猪的死亡率与产程长短有着较大的关联。产程超过 300 min 的母猪平均死胎头数为(1.4±1.5)头,而产程短于 300 min 的母猪平均死胎头数为 (0.5±0.9)头(Oliviero 等，2008a)。与围栏产的母猪相比,在产床上分娩的母猪产程更长,这可能与栏位本身以及缺乏足够的筑巢材料有关,这些因素都有可能干扰母猪的筑巢行为的自然表达,从而提高严仔母猪受到的应激水平(Lawrence 等，1994；Thodberg 等，1999)。如前所述，分娩时母猪体内高浓度的皮质醇证明了分娩本身会触发母猪的应激介导的反应。这种环境与母猪生理之间的相互作用同样也可能影响到母猪的健康状况，环境因素影响了母猪的筑巢行为并产生了一系列多米诺骨牌效应，因此，产程很可能影响母猪的健康。有数据表明产程≥4 h 的母猪与产程＜4 h 的母猪相比，前者在分娩后的第一天具有更高的发烧风险(Tummaruk 和 Sang-Gassanee，2013)。

表10.1　不同圈舍类型所对应的死胎数(Gu 等，2011；Oliviero 等，2008a，2010)。

圈舍	死胎数	平均产程(min)	窝
围栏圈	0.6±0.8	208±58	89
产仔箱	1.1±1.1	297±130	133

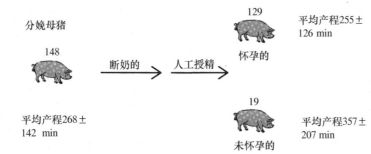

图10.3　初次受精未能妊娠的母猪上次产程为 100 min 长于成功妊娠母猪。

10.6　产程以及其对母猪繁殖能力的影响

分娩时环境与母猪生理机能之间复杂的相互作用可能会对母猪繁殖能力产生进一步的影响。最近的研究结果表明，产程较长的母猪(＞300 min)具有较高的返情率。虽然这一发现可为母猪繁殖管理提供有价值的信息。但现在还不能清晰地解释这一发现的原理。在半自然环境中饲养的家猪在产后 91~126 天开始缓慢而逐渐地断奶(Jensen 和 Recén，1989)。而商业养猪场中饲养的家猪哺乳期就相对较短，通常在 18~28 天，并且断奶到发情之间的间隔平均只有 5 天。因为分娩和随后人工授精间的间隔与自然条件下相比要短得多，这就很可能对分娩生理产生不利影响从而干扰断奶后母猪的发情。虽然子宫组织的恢复会在产后 3 周完成(Belstra 等，2005)，但分娩到随后人工授精间的间隔还是相对较短。卵泡发育、催产素活性减退或催产素受体调节等生理机制可能会因为环境因素而受到不良影响。然而，若想证实这一观点还需要对催产素和脂肪代谢展开进一步的研究。

10.6.1 人为干预分娩生理:在分娩过程中分娩的诱导和分娩过程中的催产素管理

前列腺素被广泛应用于诱导母猪分娩。若在早晨注射前列腺素,待产母猪预计在次日白天就会进行分娩。虽然前列腺素的使用很容易使母猪同步产仔,方便管理人员在白天进行管理,但是前列腺素的使用还是具有很多已知的风险。如仔猪还未足月就被诱导出生很容易导致出生的是未成熟的仔猪,这就对新生仔猪的饲养管理提出了更高的要求。众所周知,运用激素诱导母猪分娩可能使母猪的宫缩长时间中断(Smith,1982),并且会导致产后泌乳障碍综合征的发生(post-partum dysgalactia syndrome, PDS; Papadopoulos 等,2010)。此外,诱导分娩还会增加母猪患如卵巢脱落滞后这类先天性疾病的风险。因此,应接受兽医给的建议:对动物特别是后备母猪应该超过其预产期后再进行诱导分娩。如果后备母猪的妊娠期超过 117 天,建议进行诱导分娩。

催产素可以刺激乳汁分泌和子宫收缩。只需每隔 2~3 h 重复 4~5 次,每次使用 5~10 IU 剂量(IV)的催产素就可以观察到上述两种现象的出现(Martineau,2005)。两次注射间的时间间隔最少要 30 min。事实上,在大多数情况下 5 IU 的剂量就足以引起产奶量的下降,因此我们可以使用较低剂量来降低重复注射引起的副作用。注射合成的催产素可以非常有效地促进排乳和子宫收缩。最为常见的给药方式是肌肉注射,但静脉注射的疗效似乎更好。虽然催产素有效且安全,但反复地使用催产素还是会对母猪造成伤害。事实上,催产素的重复使用可能会降低种群的生长性能(Bilkei Papp,1994;Ravel 等,1996),并且导致乳汁中体细胞的数量增加(Garst 等,1999)。使用催产素的一个缺点是它会引起胎盘血管收缩,使胎儿缺氧(Rootwelt,2012)。因此,催产素不应该在常规条件下使用,而应该在推荐使用的条件下使用。兽医学上普遍认可的催产素的适应症为子宫原发性迟缓和 PDS。

10.6.2 分娩过程中的疼痛管理

分娩即使在猪身上也是一个十分痛苦的过程,疼痛甚至会降低母猪对仔猪的兴趣。在极端的情况下,疼痛甚至会引起母猪的攻击行为。分娩前,母猪的活动开始增加。分娩后,母猪侧卧可以被看做是母猪被动的行为(Malmkvist 等,2012;Oliviero 等,2008b;Wallenbeck 等,2008)。产后母猪可能会因为地板材料粗糙而引起皮肤损伤(Norring 等,2007)。此外,研究表明,在分娩时,非甾体类药物的使用可以减少母猪躺卧时间,因此在一定程度上保护了泌乳早期的母猪,使其不会出现皮肤损伤(Viitasaari 等,2013)。由于内毒素是造成 PDS 临床表现的主要成分,因此推荐使用可针对性减轻内毒素血症的药物来治疗 PDS。事实上,用非甾体类药物治疗受影响母猪,这对母猪的健康是有好处的。通常的治疗程序是在分娩的当天进行一次处理,有时在分娩次日进行第二次处理。已经提出的药物包括:氟尼辛(2 mg/kg)(Cerne等,1984)、托芬那酸(2~4 mg/kg)(Rose等,1996)和美洛昔康(0.4 mg/kg)(Hirsch等,2003)。因此,分娩的痛苦看起来可以适当地管理。根据我们了解,运用低剂量药物的现象比运用高剂量药物的现象更普遍。在泌乳前期,当母猪拒绝露出乳房为仔猪哺乳时,我们认为疼痛可能是导致这种现象的一个潜在原因。

10.7　泌乳早期的环境影响

　　自由放养母猪的母性行为通常在母猪临产前建立的隔离巢中进行(Jarvis等,2004)。在分娩前的几个小时母猪自然地表现出筑巢行为,例如搜寻、用鼻子拱地、用蹄子刨地等,显示出母猪渴望为其后代建立一个庇护所的意愿。然而,由于产床上缺乏空间和筑巢材料导致了分娩母猪的筑巢行为受到了限制。因此,提供与生物学相关的刺激物,如秸秆,就会对仔猪吃奶和母猪的哺乳行为产生一定的积极影响,如减少吃奶终止和频繁的争斗,以及延长仔猪吮吸乳头时间并且使仔猪更早地学会吮吸奶水行为,而这些因素都有利于仔猪早期吃奶(Herskin等,1999)。产床在一定程度上可以防止母猪和仔猪间的相互作用,并且在分娩期间产床上的空间还可以促进母猪母性行为的表达。

　　与拥有松散筑巢材料的围栏式饲养的母猪相比,在产床上饲喂的母猪由于缺乏合适的筑巢材料其催产素浓度会显著降低(Oliviero等,2008a)。最近的研究结果表明,分娩前充足的筑巢材料可以刺激母猪血液中催产素浓度在产前的3天到产后7天持续升高(Yun等,2013)。这表明在围产期母猪体内的催产素循环与丰富的筑巢材料引起的筑巢行为可能性二者之间存在着潜在的联系。然而,在这项研究中,饲养在宽松但筑巢材料有限的围栏中的母猪与饲养在产床上同样筑巢材料有限的母猪相比其催产素浓度并没有显著升高。这表明,筑巢材料增加与围栏式饲养的宽松空间相比,前者对催产素浓度的提高有着更大的作用(Yun等,2013)。

　　近年来的研究表明,在产床上生产的仔猪在早期泌乳期需要来自其额外的乳房刺激以获取乳汁(Yun等,2013)。这种长时间的乳房按摩可能会扰乱母猪,因为这在产床式产房中不可避免,很可能导致不良的动物福利,由于母猪长期处于被吮吸奶的压力之下而导致应付机制不足。事实上,Oliviero等(2008a)的报告指出,在产床上产仔的母猪可能很难拒绝护理仔猪,这就导致与在宽松围栏中饲养的母猪相比其唾液皮质醇浓度会显著升高(图10.2)。当给母猪提供充足的筑巢材料时，它们表现出更多的仔细的提前躺卧行为，并且为了照顾后代而表现出更好的母性行为。因此可以得出结论，在分娩前为母猪提供适当的空间和筑巢材料可以提高母猪的福利，为它们提供更好的表达自然行为的机会并且也可以减少哺乳期母猪潜在的压力。

10.8 分娩时的体况、脂肪代谢和肠道功能

10.8.1 身体状况与脂肪代谢

在分娩时，母猪的新陈代谢已经转变为分解代谢状态，这时它将身体储备的大量能量转化为母乳（Van den Brand 和 Kemp，2005）。血液非酯化脂肪酸（non-esterified fatty acids，NEFA）的水平上升可以明确地显示出与体重严重下降和低采食量二者相关的分解代谢状态（Messias de Branganca 和 Prunier，1999）。报告指出，在产仔前几天循环 NEFA 的浓度会迅速增加，在分娩的当天达到一个峰值（Le Cozler 等，1999；Oliviero 等，2009）。并且在母猪即将分娩时，它的身体会提前对即将到来的分娩做出反应开始产奶。另一方面，母猪的采食量和肠道功能开始降低，这样，外部能量的使用被分解代谢产生的能量所取代，这就可以将如肌肉蛋白和来自脂肪的游离脂肪酸等内部储存能量动员起来（Oliviero 等，2009）。在分解代谢阶段，饲料的能量含量似乎不如饲料本身的质量重要。另外，长期限制饲料（尽管数量适中）对母猪妊娠期间乳房发育和在妊娠末期乳腺基因的表达等方面均表现出不良影响（Farmer 等，2014）。在母猪临近分娩时，肠道蠕动减少常常会导致母猪出现轻度便秘（Kamphues 等，2000）。母猪的身体似乎具有可以主动降低其肠道活动的能力以用来支持其他的生理需求，如分娩。研究还表明，增加妊娠后期母猪的摄食能量会对母猪泌乳早期的采食量产生负面影响。这种采食量减少是因为妊娠后期能量摄入过多引起葡萄糖耐受性降低和胰岛素抵抗作用（Fangman 和 Carlson，2007）。

10.8.2 肠道活动和便秘

如前所述，母猪在妊娠末期常出现轻度的便秘症状并且在分娩时常经历便秘（Oliviero 等，2009）。此外，在这一阶段由于产奶引起的液体需求导致肠道内的吸水量开始增加（Mroz 等，1995）。提供少量纤维的饲料会加重便秘，从而增加吸收针对乳房的细菌毒素的风险（Smith，1985）。在其他研究中，与不便秘母猪相比便秘母猪出现乳房炎的概率较高（PDS）且对乳房健康的影响更为直接（Hermansson 等，1978；Persson，1996）。在妊娠后期，母猪饲养的一个常见做法就是增加饲料能量，减少饲料供给量。与标准妊娠饲料不同的是这种浓缩饲料通常所含的纤维

量较低。这种做法是为了确保母猪在妊娠后期获得足够的能量以满足即将到来的产奶的需要(Einarsson 和 Rojkittikhun,1993)。在生理上小肠蠕动较弱的时期,这种浓缩和低纤维素的结合饲料可能会导致更严重的便秘。大量的固体粪便可能会因为挤压而堵塞产道,从而使分娩更加困难(Cowart,2007)。目前还缺乏相关知识来证明这种严重便秘所引起的肠道疼痛是导致母猪福利下降的一个原因。在预产期的前五天,每天对猪的粪便进行定性评价,所得评分可用来监测母猪7~10天的肠活动。每天早晨清洁前,应通过目测对母猪粪便进行定性检查。其检查结果分为0~5的分值(图10.4)。当平均分为1.9~3.5时,肠道活动应该是正常的,当平均分为0~1.8时表明母猪发生不同程度的便秘,当评分低于0.9时病情严重。

	0	无粪便
	1	干颗粒,有一定形状的(未成规则形状的)
	2	介于干燥和正常之间,颗粒,有一定形状并成规则形状的
	3	正常的,软的,但坚硬的,且完全成规则形状的
	4	介于正常和湿润之间;仍然成一定形状的,但不坚硬的
	5	非常湿的粪便、无一定形状并且成液体的

图 10.4 粪便分数适用于评估一周内母猪的肠道活动(数据来源于 Oliviero 等,2009)。

当在妊娠后期提供高浓度的粗纤维(7%~10%)时,与低浓度粗纤维(<4%)的饲料相比,对母猪便秘的影响较小并且可以使母猪肠道活动恢复更快。最近的研究表明,在预产期前5天到分娩后5天之间饲喂7%粗纤维饲料的母猪的粪便平均分为2.1±1.3,而饲喂3%粗纤维饲料的母猪的粪便平均分为1.2±1.1(图10.5)。

在这项研究中,低纤维组22%的母猪表现出极其严重的便秘(连续5天以上没有产生粪便),而高纤维组中只有5%的母猪表现出这种情况(图10.6)。并且高纤维组的平均日耗水量(4.9~29.8 L)也同样高于低纤维组(3.3~20.2 L)

(图10.7)。当母猪在妊娠后期纤维素摄取不足时,在分娩时的轻度便秘会逐渐恶化成严重便秘,从而对母猪在产仔期间水的摄入和母猪福利产生不利影响。

产后的第1天和第5天,高纤维组仔猪重分别为(1.8±0.3)kg和(2.5±0.3)kg,低纤维组仔猪重分别(1.7±0.3)kg和(2.3±0.5)kg。并且高纤维组仔猪在1~5天的日增重也较大。这些结果意味着在母猪分娩仔猪前对其饲喂高纤维饲料会对母猪哺乳仔猪的表现产生积极影响。

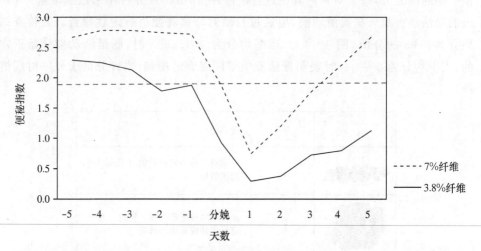

图10.5 饲喂含量不同的粗纤维的焦糖日粮的母猪肠道活动,表示为便秘指标。较低的值表示便秘状态较严重。正常的肠道活动被认为在虚线之上(Oliviero等,2009)。

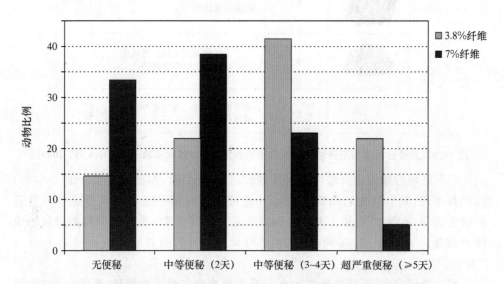

图10.6 饲喂7%纤维($n=40$)或3.8%的纤维($n=41$)日粮的母猪分娩前5天至分娩后5天期间,不同程度的便秘发生率。每个类别由不同数目的没有粪便的连续天数组成(Oliviero等,2009)。

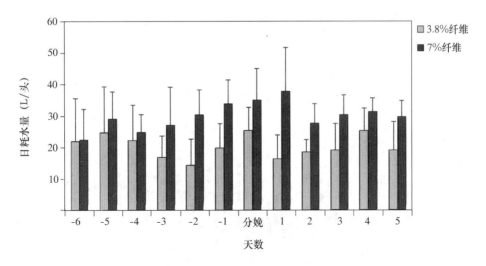

图 10.7 饲喂两种不同水平的粗纤维的母猪的平均日耗水量（Oliviero 等，2009）。

10.8.3 膘情和便秘对母猪分娩过程的影响

人们认为高背膘值会导致产程的延长（Oliviero 等，2010）。在图 10.8 中，C、D 两个区域内的所有母猪产仔时间均正常（＜300 min），而在 A 和 B 两个区域中的母猪其产仔时间均长于正常（＞300 min）。大部分较胖的母猪（背膘厚度＞17 mm）在 B 区，大部分较瘦的母猪（背膘厚度＜17 mm）在 C 区。

当观察粪便评分与分娩时间的关系时，平均便秘指数在 2±0.6（范围为 0.3～3），并且较低的便秘指数和生产时间呈负相关。图 10.9 说明了在 C、D 区的母猪产仔时间都较为正常（＜300 min），而在 A、B 区的母猪的产仔时间与正常产仔时间相比较长（＞300 min）。许多的便秘母猪（便秘指数评分＜1.9）在 A 区，而大部分不便秘的母猪（便秘指数评分＞1.9）在 D 区。

可以通过在孕期最后阶段增加饲料中纤维数量的方法来避免严重便秘病例的出现。这种日粮纤维能改善肠道活动，降低便秘程度。因此，高纤维饲料的使用似乎是一种有益的策略并且粗纤维在分娩母猪的粗饲料/底物中也同样起到很重要的作用。额外的粗饲料在支持母猪筑巢行为的同时还成为母猪的一个纤维来源可以减少母猪便秘。

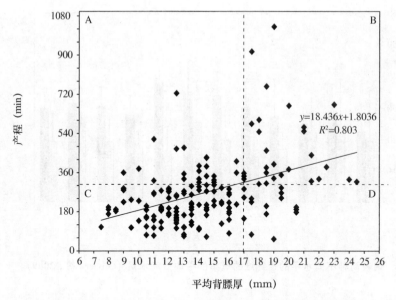

图 10.8 根据平均背膘厚和产程绘制的个体母猪。水平虚线区分长时间产程(> 300 min；A、B 区域)。垂直虚线区分较胖的母猪(B、D 区域)。实的回归线代表了平均背膘厚和产程之间的正相关关系 (Oliviero 等，2010)。

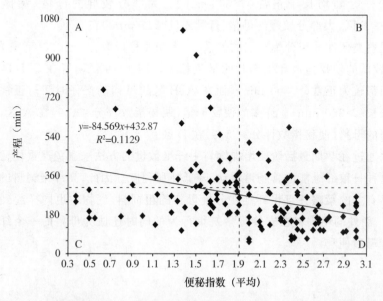

图 10.9 根据平均便秘指数(CI)和产程绘制的个体母猪。低 CI 值表示便秘母猪，而高 CI 值表示不便秘母猪。水平虚线区为较长产程(> 300 min；A、B 区域)；垂直虚线区为便秘母猪(A、C 区域)。实的回归线代表便秘指数与产程之间的负相关关系 (Oliviero 等，2010)。

10.9 分娩预测及管理技术

10.9.1 临产时的生理症状

除了行为标志,母猪在临产时还会表现出很多临床症状。一些研究表明,猪的体温从产前24～48 h开始直到产前12 h会逐渐上升到高于正常体温1～1.5℃。这个升高后的体温会一直维持到仔猪断奶(Elmore等,1979;King等,1972)。母猪的体温检测是一种值得信赖并且快速的帮助预测分娩日期的方法,并且其同样可以用来检测那些在分娩和泌乳早期即将到来的疾病如PDS。然而,传统的直肠温度检测并不实用。一种远程检测方法可以节省时间,减轻动物的应激。最近的研究表明猪在不同解剖部位的体表温度(眼睛、乳腺、耳朵后面、阴户和耳朵的内部)可以用红外摄像机(infrared camera,IRC)进行测量,并得出结论,红外热像仪可以通过对机体表面温度的常规测量来检测早期疾病(Schmidt等,2013;Traulsen等,2010)。在农场条件下运用IRC对体表温度进行单次测量时其不能提供重复的结果,但应用一个适当电脑程序就可以让从业者像使用监测工具一样更好地使用它。例如,运用IRC对母猪体温监测超过1周那么在分娩前它就会提供母猪的平均体温(Schmidt等,2013)。这样,无论在分娩早期、分娩过程中还是在哺乳早期检测到生理温度的异常升高都可以用来提醒养殖户母猪即将分娩或将要发病。

呼吸和心率在分娩前上升,到分娩后逐渐恢复正常。其他的众所周知的临产症状是乳房开始产奶,阴户出现肿胀。如前所述,筑巢行为也同样可以用来很好地预测分娩。在临产前的几个小时,母猪的活动会比正常水平增加三倍。这些症状的观察有助于预测母猪分娩,但是,在技术设备的帮助下,更精确的数据收集可以提供更准确的预测。

10.9.2 使用运动传感器预测分娩

使用光电传感器时需将它放置在产床上0.6 m高度处,使得母猪站起来时它同样可以对母猪进行检测。临产前24 h的母猪与临产前72～120 h的母猪相比其平均站立时间更长,这种情况的出现与分娩有关。与其他时间相比产前24 h的母猪躺下和站起的平均频率也开始增加(图10.10)。如果把重力传感器放在母猪下面的地板上,也能得到类似的结果。重力传感器在产前24 h记录到的峰值(活动)比任何其他检测的24 h都多。

运动传感器可以帮助预测分娩。这种传感器便宜并且方便安装在猪舍,它们可以为养殖户提供有用的信息以便养殖户们防止和应对产仔问题的发生。一些研究已经证实人类对产仔过程进行监督可以大大减小仔猪的死亡率(Andersen等,

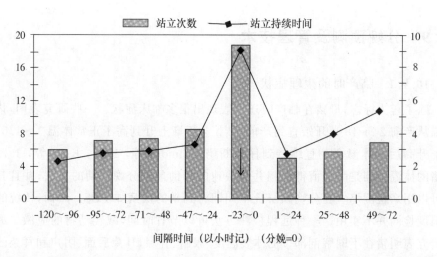

图 10.10 产前和产后母猪在不同时间间隔每个小时站立的平均持续时间和次数(黑色箭头)(Oliviero 等,2008b)。

2009;Holyoake 等,1995;White 等,1996)。母猪产仔时,提高预测的准确度可以提高在出生后一小时内经历低温饥饿等情况的仔猪的存活率。现今大多数母猪产仔时,在环境中得不到合适的协助,这就导致它们需要表达生理和行为的工具需求。这就要增加更多的投资来改善产房的质量,并且要减少由于缺乏监督而造成的仔猪死亡。

10.9.3 养猪业的影响

现今,自动化在畜牧业中扮演的角色越来越重要。在动物生产中,现代技术可以用来改善对母猪生理、行为和临床症状的监测,尽管如此,我们还是应尽量避免过度使用机器和计算机来管理人类与牲畜之间的关系。特别是考虑到近年来人们越来越喜欢集约化养殖所导致每个单元所饲养的动物数量越来越多。一个自动化系统,借助专业软件可以监测整个围产期母猪群的母猪,并在需要时发出警报是非常有用的。Oliviero 等(2008b)研究表明,若监测仪显示动物行为开始发生变换,则暗示着动物即将分娩。它也同样可以监测发生在围产期内的重要生理和临床症状。分娩后新生仔猪的体温是其在产后数小时后能否存活的关键因素(Le Dividich 和 Noblet,1983;Tuchscherer 等,2000)。关于远程设备的使用如运用热相机对母猪和仔猪的体温进行快速而准确的测量还需进一步的探究。

10.10 结论

现代技术在监视母猪分娩方面的应用结果可能鼓励养殖户从母猪和仔猪的视角来查看分娩。采用开放式产仔箱可以带来更显著的效益,表明在这类猪舍中死亡的仔猪较少,这说明动物福利与经济效益呈显著的正相关。在多数情况下,为了使母猪更好地表达筑巢行为,母猪有很大可能被允许在开放的产床上产仔,并且只需要注意在产仔后关闭产仔箱以防止仔猪被压死。在分娩前几天给母猪提供例如稻草之类的粗饲料和筑巢材料不仅可以使母猪更好地表达自己的筑巢行为,还为产房中母猪提供了缓解便秘的纤维素源。另外应仔细考虑和制定药物干预的分娩管理。使用激素制剂如前列腺素和催产素可能对母猪造成不良影响,在使用前需要加以考虑,因此,使用这些常规激素及其类似物并不是一个好的选择。疼痛管理是产仔管理中一个常常被遗忘的因素,因此其还有改进的余地。当观察到动物拒绝护理或出现异常行为时,疼痛应被看做是潜在的原因,并且应该推荐相应的止痛药。养殖户对产仔时母猪的便秘应多加注意,并认识到在妊娠后期通过增加纤维饲料来预防便秘的重要性。要仔细地观察畜群中便秘的迹象,并根据粪便质量评分对个体进行日常筛查。在产房中,对母猪粪便进行日常评价,特别是记录猪未排便天数,可以让养殖户和兽医更清晰地了解个别母猪的便秘状况,并及时治疗。在妊娠后期、分娩、哺乳期,母猪的饮水量同样非常重要。最近的研究结果显示,饲喂高纤维食物的母猪会喝更多的水,并在分娩后的第一周饲养出更重的仔猪,这就证明在这段时间水的充分利用是十分重要的。

参考文献

Algers, B., 1994. Health, behaviour and welfare of outdoor pigs. Pig News Information 15: 113-115.

Algers, B. and Uvnäs-Moberg, K., 2007. Maternal behavior in pigs. Hormones and Behaviour 52: 78-85.

Andersen, I., Haukvik, I. and Bøe, K., 2009. Drying and warming immediately after birth may reduce piglet mortality in loose-housed sows. Animal 3: 592-597.

Anderson, L.L., 2000. Reproductive cycle of pigs. In: Hafez, E.S.E. and Hafez, B. (eds.) Reproduction in farm animals – VII edition. Lippincott, Philadelphia, PN, USA, pp. 189-190.

Aumaitre, A. and Le Dividich, J., 1984. Improvement of piglet survival rate in relation to farrowing systems and conditions. Annales de Recherches Vétérinaires 15: 173-179.

Belstra, B., Flowers, W., Croom, W., De Groot, J. and See, M., 2005. Urinary excretion of collagen degradation markers by sows during postpartum uterine involution. Animal Reproduction Science 85:131-145.

Bilkei Papp, G., 1994. Perinatal losses general condition of sows III experiences obtained with prednisolone pre-treatment. Magy Allatorv Lap 49: 680-683.

Castrén, H., Algers, B., De Passillé, A.M., Rushen, J. and Uvnäs-Moberg, K., 1994. Nest-building in sows in relation to hormone release. Applied Animal Behaviour Science 40: 74-75.

Castrén, H., Algers, B., De Passillé, A.M., Rushen, J. and Uvnäs-Moberg, K., 1993. Early ejection, prolonged parturition and periparturient oxytocin release in the pig. Animal Production 57: 465-471.

Cerne, F., Jerkovic, I. and Debeljak, C., 1984. Influence of Finadyne on some clinical signs of MMA. In: Proceedings of the 8th International Pig Veterinary Society Congress. Ghent, Belgium, pp. 290.

Cowart, R.P., 2007. Parturition and dystocia in swine. In: Youngquist, R.S. and Threlfall, W.R., (eds.) Large animal theriogenology. Saunders Elsevier, St. Louis, MO, USA, pp. 778-784.

Cronin, G.M., Simpson, G.J. and Hemsworth, P.H., 1996. The effects of the gestation and farrowing environments on sow and piglet behaviour and piglet survival and growth in early lactation. Applied Animal Behaviour Science 46: 175-192.

Damm, B.I., Lisborg, L., Vestergaard, K.S. and Vanicek, J., 2003. Nest-building behavioural disturbances and heart rate in farrowing sows kept in crates and Schmid pens. Livestock Production Science 80: 175-187.

Damm, B.I., Vestergaard, K.S., Schrøder-Petersen, D.L. and Ladewig, J., 2000. The effects of branches on prepartum nest-building in gilts with access to straw. Applied Animal Behaviour Science 69: 113-124.

Edwards, S.A. and Fraser, D., 1997. Housing systems for farrowing and lactation. Pig Journal 39: 77-89.

Einarsson, S. and Rojkittikhun, T., 1993. Effects of nutrition on pregnant and lactating sows. Journal of Reproduction and Fertility Supplement 48: 229-239.

Ellendorff, F., Taverne, M., Elsaesser, F., Forsling, M., Parvizi, N., Naaktgeboren, C. and Smidt, D., 1979. Endocrinology of parturition in the pig. Animal Reproduction Science 2: 323-334.

Elmore, R.G., Martin, C.E., Riley, J.L. and Littledike, T., 1979. Body temperatures of farrowing swine. Journal of the American Veterinary Medical Association, 174: 620-622.

Fangman, T.J. and Carlson, M.S., 2007. Disease of the puerperal period.. In: Youngquist, R.S. and Threlfall W.R. (eds.) Large animal theriogenology 2. Saunders Elsevier, St. Louis, MO, USA, pp. 792-793.

Farmer, C., Palin, M.F. and Martel-Kennes, Y., 2014. Impact of diet deprivation and subsequent overallowance during gestation on mammary gland development and lactation performance. Journal of Animal Science 92: 141-151.

Fraser, D., 1990. Behavioural perspectives on piglet survival. Journal of Reproduction and Fertility 40: 355-370.

Garst, A.S., Ball, S.F., Williams, B.L., Wood, C.M., Knight, J.W., Moll, H.D., Aardema, C.H. and Gwazdauskas, F.C., 1999. Influence of pig substitution on milk yield, litter weights, and milk composition of machine milked sows. Journal of Animal Science 77: 1624-1630.

Gilbert, C.L., Goode, J.A. and MacGrath, T.J., 1994. Pulsatile secrection of oxytocin during parturition in the pig: temporal relationship with foetal expulsion. Journal of Physiology 475: 129-137.

Gu, Z., Gao, Y., Lin, B., Zhong, Z., Liu, Z., Wang, C. and Li, B., 2011. Impacts of freedom farrowing pen design on sow behaviours and performance. Preventive Veterinary Medicine 102: 296-303.

Gustafsson, M., Jensen, P., De Jonge, F.H., Illman, G. and Špinka, M., 1999. Maternal behaviour of domestic sows and crosses between domestic sows and wild boar. Applied Animal Behaviour Science 65: 29-42.

Hales, J., Moustsen, V.A., Nielsen, M.B. and Hansen, C.F., 2014. Higher preweaning mortality in free farrowing pens compared with farrowing crates in three commercial pig farms. Animal 8: 113-120.

Hansen, K.E. and Curtis, S.E., 1980. Prepartal activity of sows in stall or pen. Journal of Animal Science 51: 456-460.

Hartsock, T.G. and Barczewski, R.A., 1997. Prepartum behaviour in swine: effects of pen size. Journal of Animal Science 75: 2899-2903.

Hermansson, I., Einarsson, S., Larsson, K. and Backstrom, L., 1978. On the agalactia post partum in the sow. A clinical study. Nordisk Veterinaer Medicin 30: 465-473.

Herskin, M.S., Jensen, K.H. and Thodberg, K., 1999. Influence of environmental stimuli on nursing and suckling behaviour in domestic sows and piglets. Animal Science 68: 27-34.

Hirsch, A.C., Philipp, H. and Kleemann, R.. 2003. Investigation on the efficacy of meloxicam in sows with mastitis-metritis-agalactia syndrome. Journal of Veterinary Pharmacology and Therapeutics 26: 355-360.

Holyoake, P.K., Dial, G.D., Trigg, T. and King, V.L., 1995. Reducing pig mortality through supervision during the perinatal period. Journal of Animal Science 73: 3543-3551.

Jarvis J., Reed B.T., Lawrence A.B., Calvert S.K. and Stevenson J., 2004. Peri-natal environmental effects on maternal behaviour, pituitary and adrenal activation, and the progress of parturition in the primiparous sow. Animal Welfare 13: 171-181.

Jarvis, S., Lawrence, A.B., McLean, K.A., Deans, L.A., Chirnside, J. and Calvert, S.K., 1997.

Jarvis, S., Reed, B.T., Lawrence, A.B., Calvert, S.K. and Stevenson, J., 2002. Pituitary-adrenal activation in pre-parturient pigs (*Sus scrofa*) is associated with behavioural restriction due to lack of space rather than nesting substrate. Animal Welfare 11: 371-384.

Jensen, P., Broom, D.M., Csermely, D., Dijkhuizen, A.A., Hylkema, S., Madec, F., Stamataris, C. and Von Borell, E., 1997. The welfare of intensively kept pigs. Report of the Scientific Veterinary Committee of the European Union, pp. 31-34.

Jensen, P. and Recén, B., 1989. When to wean – observations from free-ranging domestic pigs. Applied Animal Behaviour Science 23: 49-60.

Kamphues, J., Tabeling, R. and Schwier, S., 2000. Effects of different feeding and housing conditions on dry matter content and consistency of faeces in sows. Deutsche tierärztliche Wochenschrift 107: 380.

Kelley, K.W. and Curtis S.E., 1978. Effects of heat stress on rectal temperature, respiratory rate and sitting and standing activity rates in peripartal sows and gilts. Journal of Animal Science 46: 356.

Kemp, B. and Soede, N.M., 2012. Reproductive issues in welfare-friendly housing systems in pig husbandry: a review. Reproduction in Domestic Animal 47(5): 51-57.

Kerr, J. C. and Cameron, N.D., 1995. Reproductive performance of pigs selected for components of efficient lean growth. Animal Science 60: 281-290.

Kindahl, H., Alonso, R., Cort, N. and Einarsson, S., 1982. Release of prostaglandin F2α during parturition in the sow. Zentralblatt für Veterinärmedizin 29: 504-510.

King, G.J., Willoughby, R.A. and Hacker, R.R., 1972. Fluctuations in rectal temperature of swine at parturition. Canadian Veterinary Journal 13: 72-74.

Lawrence, A.B., Petherick, J.C., McLean, K.A., Deans, L.A., Chirnside, J., Vaughan, A., Clutton, E. and Terlouw, E.M.C., 1994. The effect of environment on behaviour, plasma cortisol and prolactin in parturient sows. Applied Animal Behaviour Science 39: 313-330.

Le Cozler, Y., Beaumal, V., Neil, M., David, C. and Dourmad, J.Y., 1999. Changes in the concentrations of glucose, non-esterified fatty acids, urea, insulin, cortisol and some mineral elements in the plasma of the primiparous sow before, during and after induced parturition. Reproduction Nutrition Development 39: 161-169.

Le Dividich, J. and Noblet, J., 1983. Thermoregulation and energy metabolism in the neonatal pig. Annales de recherches vétérinaires 14: 375-381.

Malmkvist, J., Pedersen, L.J., Kammersgaard, T.S. and Jørgensen, E., 2012. Influence of thermal environment on sows around farrowing and during the lactation period. Journal of Animal Science 90: 3186-3199.

Martineau, G.P., 2005. Postpartum dysgalactia syndrome and mastitis in sows. In: Kahn C.M. (ed.) Reproduction. The Merck Veterinary Manual (9th ed.). Merck Co, Inc, Whitehouse Station, NJ, USA, pp: 1134-1137.

Messias de Branganca, M. and Prunier, A., 1999. Effects of low feed intake and hot environment on plasma profiles of glucose, nonesterified fatty acids, insulin, glucagon, and IGF-1 in lactating sows. Domestic Animal Endocrinology 16: 89-101.

Meulen, J., Helmond, F.A., Oudenaarden, C.P.J. and Van der Meulen, J., 1988. Effect of flushing of blastocysts on days 10-13 on the life-span of the corpora lutea in the pig. Journal of Reproduction and Fertility 84: 157-162.

Mroz, Z., Jongbloed, A.W., Lenis, N.P. and Vreman, K., 1995. Water in pig nutrition: physiology, allowances and environmental implications. Nutrition Research Reviews 8: 137-164.

Norring, M., Valros, A., Munksgaard, L., Puumala, M., Kaustell, K.O. and Saloniemi, H., 2007. The development of skin, claw and teat lesions in sows and piglets in farrowing crates with two concrete flooring materials. Acta Agriculturae Scandinavica, Section A – Animal Science 56: 148-154.

Oliviero, C., Heinonen, M., Valros, A. and Peltoniemi, O.A.T., 2010. Environmental and sow-related factors affecting the duration of farrowing. Animal Reproduction Science 119: 85-91.

Oliviero, C., Heinonen, M., Valros. A., Hälli, O. and Peltoniemi, O.A.T., 2008a. Effect of the environment on the physiology of the sow during late pregnancy, farrowing and early lactation. Animal Reproduction Science 105: 365-377.

Oliviero, C., Kokkonen, T., Heinonen, M., Sankari, S. and Peltoniemi, O.A.T., 2009. Feeding sows a high-fibre diet around farrowing and early lactation: impact on intestinal activity, energy balance-related parameters and litter performance. Research in Veterinary Science 86: 314-319.

Oliviero, C., Kothe, S., Heinonen, M., Valros, A. and Peltoniemi. O., 2013. Prolonged duration of farrowing is associated with subsequent decreased fertility in sows. Theriogenology 79: 1095-1099.

Oliviero, C., Pastell, M., Heinonen, M., Heikkonen, J., Valros, A., Ahokas, J., Vainio, O. and Peltoniemi, O.A.T., 2008b. Using movement sensors to detect the onset of farrowing. Biosystems Engineering 100: 281-285.

Osterlundh, I., Holst, H. and Magnusson, U., 1998. Hormonal and immunological changes in blood and mammary secretion in the sow at parturition. Theriogenology 50: 465-477.

Papadopoulos, G., Vanderhaege, C., Janssens, G., Dewulf, D. and Maes, D., 2010. Risk factor associated with *post partum* dysgalactia syndrome in sows. The Veterinary Journal 184: 167-171.

Persson, A., 1996. Lactational disorders in sows, with special emphasis on mastitis. In: Proceeding of the 47th meeting of the European association for animal production. Lillehammer, Norway, pp. 26-29.

Phillips, P.A. and Fraser, D., 1993. Developments in farrowing housing for sows and litters. Pigs News Info. 14: 51-55.

Ravel, A., D'Allaire, S. and Bigras Poulin, M., 1996. Influence of management, housing and personality of the stockperson on preweaning performances on independent and integrated swine farms in Québec. Preventive Veterinary Medicine 29: 37-57.

Roehe, R., 1999. Genetic determination of individual birthweight and its association with sows' productivity traits using Bayesian analysis. Journal of Animal Science 77: 330-343.

Rootwelt, V., 2012. Piglet stillbirth and neonatal death. PhD Thesis, Norwegian School of Veterinary Medicine, Oslo, Norway.

Rose, M., Schnurrbusch, U. and Heinrotzi, H.. 1996. The use of cequinome in the treatment of pig respiratory disease and MMA syndrome. Proceedings of the International Pig Veterinary Society Congress 14: 317-317.

Schmidt, M., Lahrmann, K.H. and Ammon, C., 2013. Assessment of body temperature in sows by two infrared thermography methods at various body surface locations. Journal of Swine Health Production 21: 203-209.

Smith, B.B., 1985. Pathogenesis and therapeutic management of lactation failure in periparturient sows. Compendium Continuing Education Practice 7: 523.

Smith, W.C., 1982. The induction of parturition in sows using prostaglandin F2alpha. New Zealand Veterinary Journal 30: 34-37.

Sorensen, D., Vernersen, A. and Andersen, S., 2000. Bayesian analysis of response to selection: a case study using litter size in Danish Yorkshire pigs. Genetics 156: 283-295.

The effect of environment on behavioural activity, ACTH, beta-endorphin and cortisol in prefarrowing gilts. Animal Science 65: 465-472.

Thodberg, K., Jensen, K.H., Herskin, M.S. and Jørgensen, E., 1999. Influence of environmental stimuli on nest-building and farrowing behaviour in domestic sows. Applied Animal Behaviour Science 63: 131-144.

Traulsen, I., Naunin, K., Müller, K. and Krieter, J., 2010. Application of infrared thermography to measure body temperature of sows. Züchtungskunde 82: 437-446.

Tuchscherer, M., Puppe, B., Tuchscherer, A. and Tiemann, U., 2000. Early identification of neonates at risk: traits of newborn piglets with respect to survival. Theriogenology 54: 371-388.

Tummaruk, P. and Sang-Gassanee, K., 2013. Effect of farrowing duration, parity number and the type of anti-inflammatory drug on postparturient disorders in sows: a clinical study. Tropical Animal Health and Production 45: 1071-1077.

Van den Brand, H. and Kemp, B., 2005. Dietary fat and reproduction in the post partum sow. In: Ashworth, C.J. and Kraeling, R.R., (eds.) Control of pig reproduction VII, Nottingham University Press, Nottingham, UK, pp. 177-189.

Vestergaard, K. and Hansen, L.L., 1984. Tethered versus loose sows: ethological observations and measures of productivity. I. Ethological observations during pregnancy and farrowing. Annales de Recherches Vétérinaires 15: 245-256.

Viitasaari, E., Hänninen, L., Heinonen, M., Raekallio, M., Orro, T., Peltoniemi, O. and Valros, A., 2013. Effects of post-partum administration of ketoprofen on sow health and piglet growth. Veterinary Journal 198: 153-157.

Wallenbeck, A., Rydhmer, L. and Thodberg, K., 2008. Maternal behaviour and performance in first-parity outdoor sows. Livestock Science 116: 216-222.

White, K.R., Anderson, D.M. and Bate, L.A., 1996. Increasing piglet survival through an improved farrowing management protocol. Canadian Journal of Animal Science 76: 491-495.

Yun, J., Swan, K.M., Vienola, K., Farmer, C., Oliviero, C., Peltoniemi, O. and Valros, A., 2013. Nest-building in sows: effects of farrowing housing on hormonal modulation of maternal characteristics. Applied Animal Behaviour Science 148: 77-84.

11. 仔猪死亡：原因和预防

S. A. Edwards[1]* and E. M. Baxter[2]

[1] Newcastle University, School of Agriculture, Food and Rural Development, Agriculture Building, Newcastle upon Tyne, NE1 7RU, United Kingdom; sandra.edwards@ncl.ac.uk

[2] Animal Behaviour and Welfare, Animal and Veterinary Sciences Group, ScotlanD's Rural College (SRUC), West Mains Road, Edinburgh, EH9 3JG, United Kingdom

摘要：根据猪的进化生物学，猪更倾向于生产更多的后代，在当前农场条件下，这种趋势由于人们根据遗传选择强度增加繁殖力而进一步加剧。仔猪从出生到断奶的死亡率通常达到 16%～20%，引起死亡的主要原因是死胎、踩压和饥饿。然而，这些主要的原因掩盖了真正的死亡诱因，如母猪和仔猪的生物学特性、新生仔猪环境等方面的交互风险因素。子宫内部的营养竞争可能导致死胎或弱仔猪、仔猪初生重低和产后生存机会的降低。产前应激可以影响仔猪的活力和体温调节能力，导致其无法从分娩过程中窒息应激、接踵而至的体温过低和对于初乳激烈竞争的众多挑战中幸存。降低仔猪死亡率需要遗传、营养、管理和饲养员等方面的综合技术措施。抗病性、适应性等育种目标的选择，母猪最佳营养优化，以及通过环境条件的调控，避免母猪产前应激都会提高初生健仔猪数，并且刺激良好的母性行为。产仔过程中的监督和帮助，包括饲养员的协助，可促进所有仔猪及时吃上充足的初乳，提供保温箱来防止仔猪体温过低，以及仔猪固定奶头来确保适当大小和均匀的整窝仔猪，这些在提高仔猪成活率方面都非常关键。随着社会对养猪生产的道德和福利问题的持续关注，除了不减轻对猪农的经济压力来保持竞争外，在充分表达母猪行为学需求的条件下，实现一个高的每年每头母猪断奶仔猪数，将继续构成挑战。

关键词：新生仔猪，存活，产仔猪，行为，管理

11.1 引言

猪是一种多胎动物，它的自然生物学特征是生产更多的子代。这种进化策略

被认为是一种"父母代乐观"的表现形式,由于每胎分娩大量仔猪,使得每窝中一个或者更多仔猪死亡对猪的繁育影响较小(Forbes 和 Mock,1998；Mock 和 Forbes,1995)。虽然这种策略使得母猪在哺乳期能有效利用资源,但是它不能提前预测,因此,在适合的季节繁育更多的后代,需要在出生前对任何个体仔猪的投入要低,那些弱仔猪在这个时期应该尽早淘汰,或者至少同窝仔猪内做出调整,这可以通过母猪在同窝内资源的不均衡分配来达到,包括早出生和晚出生的仔猪,如果不进行均窝调整,将导致同窝仔猪间为获得有限的营养而激烈竞争,引起弱仔猪死亡风险升高。在产前环境中,母猪有限的资源是用于胎盘发育所需要的子宫空间,30%～50%排的卵未能在妊娠过程中存活(Geisert 和 Schmitt,2002；Geisert 等,1991；Pope,1994)。母猪胚胎发育是不同步的,致使有些胎儿发育比其他较快,引起雌激素(17β-雌二醇)释放,产生潜在的不利于其他胎儿的子宫环境,阻碍同窝其他胎儿发育迟缓,最终退化。随后,如果对于子宫而言,着床的胎儿数量太大,限制胎盘发育将会导致养分供给减少、胎儿死亡或者仔猪出生后的成活率降低(Foxcroft 等,2009)。在出生后的环境中,有限的资源是母乳,仔猪出生时利用发育好的锋利的獠牙对抗同窝仔猪以守卫乳头。当同窝仔猪数量超过母猪功能性乳头数时,对于能力不足的仔猪的影响是致命的。

当前社会发展要求现代畜牧业朝着高效节约型方向发展,因此实施了超高产品系育种计划,这些品系可生产超量后代。在驯化过程中,选择性育种窝产仔数增加已经超过100%。丹麦在这方面已经取得了最显著的进步；在13年间,从每窝总产仔数4头发展到每窝平均16.6头仔猪,而欧洲达到每窝平均12.7头仔猪(BPEX,2013；Rutherford 等,2013)。然而,由于对生存和生产的其他性状的负面影响,相关的死亡率被许多人认为高得令人无法接受(Baxter 等,2013；Rutherford 等,2013)。此外,其他重要经济性状强烈的选择强度,如瘦肉组织快速生长减少胴体脂肪含量,已经导致仔猪在出生时的生理成熟降低,造成初生仔猪的活力和体温调节能力下降(Herpin 等,1993)。因此,在农场条件下降低仔猪死亡率是一个持续的挑战。

为了开发有效的环境和生物学方法来应对这一挑战,了解不同类型的仔猪死亡原因和易发生这些情况的相关危险因素是非常重要的。

11.2 死亡原因

目前总死亡率(出生+难产死亡)相当于每窝16%～20%(BPEX,2013)。已经有许多与仔猪死亡率相关的综合评论文章报告了仔猪断奶前死亡的原因(Dyck 和 Swierstra,1987；Edwards,2002；Edwards 等,1994；English 和 Edwards,1996；English 和 Morrison,1984；English 和 Smith,1975；Lay 等,2002；

Marchant 等,2000；Mellor 和 Stafford,2004；Svendsen,1992)。这些结果有很大的一致性,特别是关于死亡的主要原因和影响因素,认为出生后的前 72 h 是仔猪成活的决定性的时期。Dyck 和 Swierstra(1987)研究确定了仔猪死亡的八个具体原因,但主要的三个,分别是死胎、母猪的踩压和饥饿。然而,在分析死亡原因时,很难发现单一的因素。仔猪死亡率的综合性意味着母猪、仔猪和环境都是相互起作用的因素。如果准确判断这些复杂的相互作用造成的死亡原因,需要对仔猪个体进行长期的研究和仔细解剖。许多调查数据已经表明,误诊在农场的死因诊断报告中是常见的。例如,死胎占总出生仔猪的 4%~8%(English 和 Edwards,1996)或约占总死亡率的 30%~40%,踩压是长久以来仔猪死亡的主要诱因。然而,踩压和饥饿不是相互排斥的条件；它可能是一个接着一个,并且两者都受体温过低的影响(Curtis,1970)。直接或间接导致的体温过低被认为是造成死亡的一个主要原因,高于踩压、饥饿、疾病或弱仔的原因(Curtis,1970)。由于不能连续监测直肠温度,所以母猪踩压是有记录以来最容易辨认出的死因,而忽视了低温造成的影响。因此,农场主的记录往往误导仔猪死亡原因(Christensen 和 Svensmark,1997；Vaillancourt 等,1990)而低估了真正的危险因素。为了降低仔猪死亡率,这些易感因素及其复杂的相互作用(图 11.1)是需要解决的。

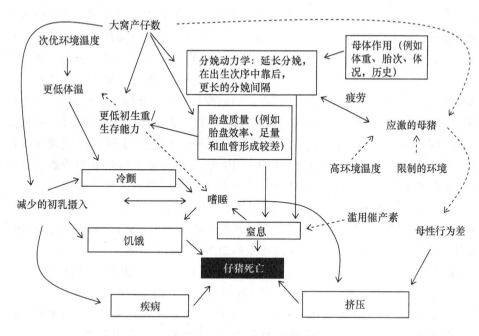

图 11.1 仔猪死亡的诱发因素。

11.2.1 母猪的影响

母猪在胎儿和产后生存中起着至关重要的作用,它不仅在子宫内和哺乳期间提供重要的营养,而且显示良好的母性。另外,母猪的基因型和胎次,以及妊娠和哺乳期的身体状况,也都会影响仔猪的存活。

子宫内对仔猪性能的影响

胎儿的生存取决于子宫的血流量、适当分配母体营养和胎盘转运的效果。在胚胎延伸期,子宫内的死亡率与在拥挤的子宫环境中着床后的损失相关(Foxcroft 等,2006)。胎盘发育受到子宫挤压的损害,胎盘功能不全是胎儿子宫内生长限制/延迟(intrauterine growth restriction/retardation,IUGR)的主要原因。这样的产前损伤可能导致死胎或产生弱仔和出生后的仔猪成活率低。品种间的差异说明了最佳产前环境的重要性;梅山母猪更适合满足胎儿的需求,比西方同类母猪更有成效地生产更多的可育仔猪,同时生产出更少的死胎(Canario 等,2006)。它们的成功主要是通过增加胎盘效率显著增加胎儿大小与胎盘大小的比例,克服子宫能力的限制(Biensen 等,1998),这包括增加胎盘的血管密度:子宫内膜的连接以响应胎儿增加的需求(Biensen 等,1998;Wilson 等,1998)。尽管梅山仔猪体型较小,但它们在出生时比现代西方高产品系的仔猪更具生理成熟性,具有较高的胴体脂肪和较大的脂肪组织细胞比例(Herpin 等,1993)。相比之下,在现代高产品系中增加窝产仔数的选择已经实现了大量后代的预期生产,但是付出了生存能力降低的代价。再加上较小的仔猪患病率增加造成同窝仔猪内体重变异增大;两者都是已知的仔猪死亡的风险因素(Foxcroft 等,2006)。仔猪的发育程度及其大小对生存产生重大影响。最近涉及生长迟缓的形态学指标研究已经强调了把仔猪体型作为生存预测因子的重要性(Baxter 等,2008),其可以诱发同样体重仔猪的不同生存结果。Baxter 等(2008)发现,与存活的同窝仔猪相比死胎比较长而瘦,具有低出生庞代络指数(ponderal index,PI,评价体格营养状况的一个参考指标)和体重指数(body mass index,BMI)(指示 IUGR);与初生重相比 PI 和 BMI 更能预测仔猪是否存活。他们还将 IUGR(即 BMI 和 PI)的总体测量值与较低的胎盘总数和总体胎盘密度相关联,胎盘性状不仅是母体和胎儿之间营养转移的关键而且是出生时仔猪活力的决定因素。Rootwelt 等(2012)不仅重复了 IUGR 的总测量值与仔猪存活相关联的结果,而且还将胎盘区域与仔猪活力相关联,从而强调了胎盘质量不仅对产前存活有影响,而且也对出生后的生存能力有影响。

除了 PI 和 BMI 之外,仔猪的其他身体特征也与产后仔猪的生存有关(Amdi 等,2013;Hales 等,2013)。Hales 等(2013)已经开发了一个 IUGR 评分,基于 Chevaux 等(2010)对于具有不同程度 IUGR 的仔猪头部形态的研究。根据某些特征的存在或不存在,他们将仔猪可观察的 IUGR 评分划分为正常 IUGR、轻度 IUGR 到严重 IUGR。这些特点是:(1)陡峭的海豚般的前额;(2)凸出的眼睛;(3)垂直于嘴的皱纹。如果存在 2 或 3 个特征,则给出严重的 IUGR 评分。如果存在 1 个特征,则给出轻度 IUGR 评分。如果没有一个特征,仔猪被认为是正常的。

Hales等(2013)研究表明出生后早期死亡与较低的BMI、更大的生长限制水平和更大的窝仔猪数相关联。Amdi等(2013年)进一步发现,分类为正常的仔猪在产后0～12 h初乳摄入量较大,产后12～24 h有更高的初生重、顶臀长度、BMI和PI,并且有比严重IUGR的仔猪获得更高生命力的趋势。IUGR的仔猪还可以显示差异器官发育,包括"脑保护效应",作为对胎盘功能不全的胎儿适应性反应的一部分(Roza等,2008)。差异器官发育可以影响认知和免疫功能(Wu等,2006)、对生殖能力有长期的损害(Da Silva-Buttkus等,2003)并且影响后期应激反应的下丘脑-垂体-肾上腺轴功能(Kranendonk等,2006)。因此,仔猪出生时的身体状况影响其产后存活、生理、行为生存能力以及如何对出生后的挑战作出反应,而胎盘的质量决定了出生时身体状况。

分娩动力学:持续较长分娩时间对仔猪成活的影响

分娩时间延长、最后1/3分娩、脐带过早破裂、母猪行为和生理特征、胎次和血红蛋白浓度小于9 g/100 mL均为引起死胎的主要影响因素(Baxter等,2008;Fahmy和Friend,1981;Randall,1972a,b;Van Dijk等,2005;Van Rens和Van der Lende,2004;Zaleski和Hacker,1993)。这些因素导致窒息死亡(Alonso-Spilsbury等,2005;Mota-Rojas等,2005;Van Dijk等,2005;Van der Lende等,2001),或造成产后生存机会很低的缺氧仔猪。这些发病诱因不是独立的(图11.1);例如延长的分娩时间受到窝产仔数增加(Andersen等,2011;Van Rens和Van der Lende,2004)、母猪与疲劳相关的应激(Vallet等,2002;Van Kempen,2007)、分娩环境的制约(Oliviero等,2006)和较高的环境温度(Vanderhaeghe等,2010)的影响。仔猪大小也可以影响分娩时间和死胎率。初生重与死胎率之间存在二次线性关系;尽管由于不同的原因,非常小的仔猪和"巨大的"仔猪同样面临危险,较重的仔猪通常由于分娩困难而比体重小的同窝仔猪更容易缺氧(Trujillo-Ortega等,2007)。

妊娠期间母猪分娩的动力和许多其他发育过程可能受到妊娠期及以后时期母猪饲养的影响。妊娠期间过度饲喂影响母猪的分娩能力而间接损害仔猪;如果母猪过胖,仔猪太大,则分娩间隔时间会增加,脐带闭塞和死胎的可能性也会增加(Leenhouwers等,2001)。相反,如果母猪在妊娠期间喂养不足,则对胎盘大小、仔猪生长和仔猪出生后使用的体内储备能量有不利影响(Wu等,2004)。妊娠期营养不良也可导致乳腺发育异常(Head和Williams,1992;Kim等,1999),哺乳期产奶量不足可能会影响后代未来的繁殖能力(Kerr和Cameron,1995;O'Dowd等,1997),甚至有可能是两代间的影响(Gluckman和Hanson,2004),因为当母猪还是胎儿的时候,它自身的代谢功能可能已经被编程。第4章(Farmer和Hurley,2015)和第8章更详细地讨论了妊娠后期的饲养管理。

高胎次母猪会产生更多的死胎(Leenhouwers等,2003;Randall和Penny,

1970),研究发现,产生死胎的母猪通常会在后来的仔猪窝中持续产出死胎,从而使这个性状具有遗传性(Roehe 等,2010)。

初乳的可用性

新生仔猪早期摄入初乳对其生存至关重要。仔猪不仅是依赖于初乳的摄入和代谢产热来抵御寒冷(在后面的部分讨论)(Herpin 等,1994),而且也是直接从摄入的初乳中摄取母体免疫球蛋白获得被动免疫保护的唯一途径(Rooke 和 Bland,2002)。哺乳动物产奶量和初乳组成分别在第 8 章(Quesnel 等,2015)和第 9 章(Hurley,2015)中详细讨论,因此这里不再进一步讨论,只是再次强调越来越多的证据表明,高产基因型母猪的初乳转移受母猪产奶量而不是仔猪获取的限制(Devillers 等,2011)。

在分娩后的前几个小时,初乳可以自由获取(De Passillé 和 Rushen,1989)。然后,大约每 20 min 发生一次周期性的下降。这个保护性的发展在第 13 章(Špinka 和 Illmann,2015)中有更详细的介绍,但很明显,如果母猪在这段时间内没有显示适当的母性,那么后代的初乳摄入量就会受到严重影响。

母性行为

对于一头仔猪在生命早期摄取初乳,它必须能够安全和容易地找到母猪乳头。因此,母猪必须舒适地保持不动;侧躺,露出乳房并发出有节奏的哼哼声,提醒仔猪吸吮。焦躁不安的母性行为,特别是在分娩期间,不仅会阻碍早产仔猪吃初乳,而且会延长分娩时间(其后果已被讨论),并增加踩压的风险。

一般来说,踩压是大多数新生仔猪死亡的最终原因,十分重要,也是低体温-饥饿-踩压综合性原因的一部分。猪产生踩压的危险行为包括躺卧前对仔猪引导行为的缺乏、从躺卧位置赶出仔猪、姿势变化的频率和特点以及对被压仔猪的反应缓慢(Andersen 等,2005;Marchant 等,2001;Thodberg 等,2002)。对被压仔猪反应缓慢受到母猪身体状况的影响,这可能与胎次、本身反应能力及分娩环境相关。较老的母猪腿部可能会更虚弱,阻碍姿势变化(Damm 等,2005;Pedersen 等,2006),而产仔箱可以有效地限制母猪对踩压仔猪的反应。

母猪的性格也影响仔猪的死亡率(Marchant 等,2001)。关于母性能力有个体差异,认为是"踩压者"的母性行为不同于"非踩压者"(Andersen 等,2005;Jarvis 等,2005)。母猪个体行为模式的一致性以及群体内的高度变化表明了选择"非踩压者"母猪的可能性(Grandison 等,2003)。研究发现,母猪对人类的恐慌和紧张与仔猪踩压有关(Lensink 等,2009),可能是通过增加了对干扰的反应。当每窝产仔数量超过功能性乳头数量时,同窝仔猪之间乳头竞争的发生率增加,提高了对哺乳行为的干扰,这可能导致母猪的不安和大量母猪终止哺乳(13 章;Špinka 和 Illmann,2015)。踩压不是唯一可以直接导致仔猪死亡的母性行为——以母猪攻击或驱赶的形式进行的"假母性"行为可能导致创伤或致命的伤害。这种攻击在后备母猪中更为普遍(Chen 等,2007;Harris 等,2003;Vangen 等,2005),可能是因为后备母猪对新生仔猪的恐新反应。更多攻击的报道是当母猪被限制在限位栏中并且不能逃避新生仔猪的注意时(Lawrence 等,1994),而这种行为在群养系统中很少见。在分娩期间,有攻击性的母猪更为不安,并对其仔猪过度反应(Ahlstrom 等,2002)。因此,即使仔猪逃避直接攻击,也可能会被踩压,并且导致仔猪吃乳推迟。此外,如果母猪过度躺卧或性情凶猛,将导致其仔猪非致命伤害,仔猪将处于感染和后期死亡的风险中。

11.2.2 仔猪的因素

本节将详细介绍仔猪出生的身体状况和行为损害的后果,特别是与寒冷、初乳摄入减少和踩压的联系。

大小和活力

Roehe 和 Kalm(2000)报道,初生重低于 1 kg 仔猪的断奶前死亡率为 40%,初生重在 1~1.2 kg 的为 15%,出生体重超过 1.6 kg 时仅为 7%。然而,如前所述,考虑发育状况而不仅仅是大小,并且区分胎龄小(small for gestational age,SGA)、低初生重仔猪和承受 IUGR 的仔猪仍然是重要的。虽然定义有所不同,但是在出生时体重小于同窝 1/10 的仔猪也显示正常的异速生长,通常被归类为 SGA。另一方面,不成比例的仔猪(表明它们尚未达到子宫内生长潜力)被归类为 IUGR(Bauer 等,1998;Hales 等,2013)。区分是重要的,因为 SGA 仔猪可能比 IUGR 仔猪具有更多潜在的恢复能力,因为 IUGR 仔猪有其他异常,倾向于低生存能力。Baxter 等(2008 年)证实,生理成熟和有活力的小型仔猪可以在危机四伏的围产期生存。基于新生仔猪使用与计算机登记系统相连接的人造乳头的力量和持久度,采用复杂的活力测量,结果表明,一只小而有活力的仔猪可以与其有较大差异的仔猪一样生存。活力或精力描述了仔猪积极生存的行为,这在出生时可能变化很大(Herpin 等,1996;Zaleski 和 Hacker,1993)。发现功能性乳头和吸吮初乳使那些更快找到乳房的仔猪生存力更强(Baxter 等,2008;Herpin 等,2002;Tuchscherer 等,2000)。随后,仔猪将争取获得并保持拥有优选的乳头,如果不能在该乳头上进行最佳的按摩和哺乳行为,则初乳和产奶量以及仔猪的存活可能受损。未能确立固定乳头的仔猪生长速度更慢(De Passillé 等,1988),当它们设法获得一个乳头时,无法获得最有效的乳头,导致饥饿或更低的母乳摄入量。DePassillé 和 Rushen(1989)发现,出生较早、较重的仔猪获得了更多的乳头选择,确定乳头更加快速,吸乳更频繁,最终比有活力但行动少的同窝仔猪具有明显的优势。新生仔猪对环境的竞争性在第 13 章(Špinka 和 Illmann,2015)中有更详细的讨论。

出生时的生理成熟程度不仅影响仔猪早期摄入初乳的行为能力,而且未成熟的器官发育可能影响仔猪消化和吸收其获得初乳的能力。在出生之后,肠闭合开始之前(约 48 h)(Cranwell,1995),有一段有限的时间肠道可以渗透大分子,例如可以获得被动免疫的免疫球蛋白。在此期间，仔猪重要的是获得和消化最大量的初乳，因为肠闭合过程是由初乳摄入刺激的。在仔猪体循环中，有两个主要的病原菌侵入机会。第一个是在产后最初 24 h 内，受到可以延缓肠道闭合的初乳摄入的影响。第二个是在母猪的抗体浓度下降和仔猪从被动到主动免疫之间过渡的时间（Gaskins 和 Kelley,1995），这在早期的产后生活中受到低初乳摄入的损害。

性别

有证据表明，母仔猪从出生到断奶的幸存机会比公仔猪更大（Baxter 等，2012；Hales 等，2013；Lay 等，2002）。Baxter 等（2012）表明，尽管公仔猪出生体重比母仔猪重，但与母仔猪相比，公仔猪较差的体温调节能力，在 24 h 时直肠温度有明显降低。此外，来自雄性较多窝的仔猪显示，体温调节能力降低，吸收初乳较慢，并且有更多的因疾病相关原因而死亡的可能。这些结果表明，雄性偏多窝的仔猪死亡率反映了内在的、与性别有关的因果易感因素。但当竞争资源有利于更大、社会等级更高的个体时，这种可能会被掩盖，如果雄性只能获得稀少的资源，那么雄性可能不能够去竞争。

体温调节能力

新生仔猪皮肤几乎无毛，没有褐色脂肪组织以促进代谢产热（Herpin 等，2002）。由于它生理能力极限的代谢运转，因此具有差的体温调节能力（Herpin 等，2002；Mellor 和 Stafford, 2004）。新生仔猪产生热量的能力对其生存至关重要，并且依赖于各种器官和过程的协调功能。当仔猪处于低温条件时，仔猪由于寒冷的环境或较低的产热能力而导致过多热量流失，这些是至关重要的。如果低体温变得不可逆转，那么仔猪将被冻死，或由于低温间接地导致死亡，例如增加踩压的可能性（Curtis, 1970）。较小的仔猪面临更大的低体温风险，因为每单位体重的热量损失与身体尺寸成反比（Herpin 等，2002）。通过连续测量记录仔猪出生后早期的直肠温度，证明初生重对产后 36 h 内仔猪温度变化的影响（图 11.2）。

当环境温度下降到 15～20℃，所有仔猪进入子宫外环境时都会感到寒冷。初生仔猪的代谢能力及其行为能力会影响体温下降的程度和持续时间。为应对寒冷，产热反应的启动和持续性（即代谢率的持续增加）取决于初乳的摄入和代谢（Herpin 等，1994）。

出生后早期的行为

易发生踩压的仔猪行为常常是由于缺氧、饥饿或体温过低等生理挑战而引起的，这可能会增加嗜睡。仔猪在找到温暖的乳房、获得宝贵的初乳和建立固定乳头的先天需求之间存在一个平衡，以免被母猪踩压的风险。Weary 等（1996）得出结论，踩压是新生仔猪面临营养挑战的部分结果；一个体重较轻的仔猪在母猪躺卧或站立的危险区域下花费更多的时间。如果仔猪的能量储备很低，那么它会因为太弱小而不能逃避一只移动的母猪。由于出生时的热损失减少主要是通过行为调整

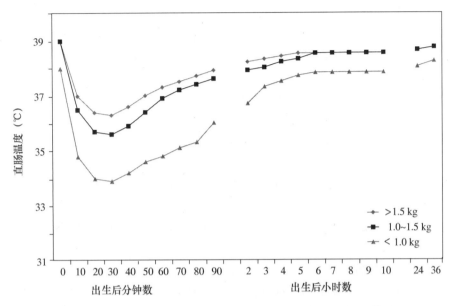

图 11.2　产后 36 h 仔猪出生体重与直肠温度分布之间的关系（Pattison 等，1990）。

来实现的,包括挤在一起和姿势的适应(Hrupka 等,2000),仔猪选择的躺卧的位置可能对其生存至关重要。对于面临严峻生理挑战的低生存能力的仔猪来说,远离乳房和远离同窝仔猪同样危险,因为尽管踩压的风险降低,但是饥饿和体温过低的风险要大得多。环境因素增添了这些风险。

11.2.3　环境因素

体温调节

新生仔猪较低的临界温度约为 34℃ (Mount,1968),但在子宫外环境中不能达到。将环境加热至 34℃ 会对母猪的温度舒适(Mount,1968)产生负面影响,母猪的温度舒适范围为 12～22℃ (Black 等,1993),这与新生仔猪的温度舒适范围明显不同。因此,仔猪产后即会经历传导、对流、辐射和蒸发的快速热损失(Curtis,1970)。这些热损失过程不仅受到已经讨论的身体和行为因素的影响,也受到环境的影响。传导热损失的早期研究(Mount,1967；Curtis,1970 年修订)显示,与水泥地板接触的仔猪比用垫料(即 2.5 cm 厚的秸秆)绝缘的仔猪多损失 40% 以上的热量。对"自然"分娩环境的调查显示,母猪对于极端气候的筑巢行为是缓冲热损失的途径（Algers 和 Jensen,1990；Baxter 等,2009)。辐射热损失受温度梯度、周围表面的面积和仔猪与这些表面之间的距离的影响,并且不会因环境温度升高而过热。最后,蒸发热损失的问题尤为严重,因为它是将胎盘液作为蒸汽蒸发消耗热量。所有的仔猪都要经历这种快速冷却,因为母猪和其他动物的母体一样,不会通过舔新生仔猪来除去胎盘液体。由于较高的相对体表面积、空气流速的增加(例如通气系统的通风，Curtis，1972）和低环境空气压力，低初生重仔猪会增加相对蒸发损失。即使胎盘液已经消散，呼吸期间仍然会发生黏膜表面的蒸发热损失。

竞争

第二大环境挑战来自社会环境。如前所述，特别是在产仔数非常大的窝里，对于新生仔猪确定自己的乳头来说，同窝仔猪的竞争是很激烈的，并且在放乳时对于乳头激烈的竞争中，仔猪可能会由于同窝仔猪的牙齿和/或地板表面的摩擦而受伤（Drake 等，2008；Fraser，1975；Mouttotou 和 Green，1999）。这些竞争效应意味着，除了有效管理猪群（例如通过交叉喂养）外，不管其内在精力如何，低初生重的仔猪都有挨饿的风险。因此，窝内初生重的变化可能比个体出生体重作为生存的决定因素更为重要（Milligan 等，2002；Roehe，1999；Quiniou 等，2002）。这个因素对仔猪存活的关键作用并不新鲜（English 和 Smith，1975），但仍需要有效的解决。

饲养员的影响

在产房内有一些明显的区域，饲养员可能对仔猪死亡率产生影响。例如，如果不能在初乳摄入或维持体温调节的情况下帮助弱仔猪，并且在母猪拥有比功能性乳头更多的仔猪时没有干预，则不可避免地导致仔猪死亡。此外，在卫生和疾病治疗方面的不良管理将会增加新生仔猪致命疾病和感染的风险。然而，应该认识到从母猪饲养开始，繁殖周期所有阶段的护理标准可以间接影响仔猪的生存。多年来，Hemsworth 和同事们已经清楚地发现不同的猪场员工对于猪的态度和随后他们对猪的行为，会影响猪的恐惧程度，增加的恐惧最终导致对生产和繁殖性能的影响（Hemsworth 等，1995）。这些影响的意义可以很容易地被忽略，因为因果关系在时间上分离，特别是在大单位，并且工作人员可能专门从事生产周期的某些部分，这可能加剧这种脱节（Rutherford 等，2012；Baxter 等，2013）。在产房中，人类行为和猪的恐惧水平共同影响仔猪的死亡率。例如，当母猪恐惧水平和神经紧张度高时，人类的存在可能造成死胎（Hemsworth 等，1999）以及由于踩压和进攻造成死亡的危险因素（Hemsworth 等，1995；Lensink 等，2009）。此外，Marchant-Forde（2002）在后备母猪妊娠期间进行的人为方法测试，将它们归类为行为"害羞-胆大"分类，结果表明，极端害羞的后备母猪更可能粗鲁地对待它们的后代。因此，恐惧症的增加可能是导致仔猪侵略性的重要风险因素（English 等，1977）。Janczak 等（2003）表明，母猪的焦虑可能导致其母性不良。事实上，他们发现，恐惧行为的程度与后备母猪在大约两个月大的时候的焦虑程度以及之后的母性品质是相关联的，这反映在仔猪死亡率上。

妊娠期间的紧张经历不仅影响母猪的行为，而且对后代发育有长期不利影响，包括跨代影响。Rutherford 等（2014 年）研究发现，初产母猪在妊娠中期遭受社会综合压力（Jarvis 等先前证明的产前压力模型，2006）。神经生物学研究表明，焦虑母猪的雌性后代更容易发生焦虑,而那些继续成为母猪的雌性表现出母性异常的行为。它们对仔猪也更有反应性,与对照母猪相比,它们花费更多的时间看着自己的仔猪;这些差异对仔猪存活有负面影响。

11.3 预防死亡

仔猪死亡率可以通过动物遗传选择策略的变化、分娩环境的变化或管理变化来解决。随着母猪产仔能力进一步增加,所有这些措施都需要优化。

11.3.1 动物的选择

除了出生数量外,调整包括育种目标中与新生仔猪生存相关性状的选择标准,是可持续发展的战略,已经在提高仔猪存活率方面取得了成功(Roehe 等,2009,2010)。以在增加窝产仔数量方面取得成功而闻名的丹麦猪业,认识到产仔增加的同时也伴随着最初发生的死亡率的显著增加(全部断奶死亡率增加了5%)。

在2004年,将选择标准从总产仔数改变为"第5天活仔猪数"(live piglets at day 5, LP 5)(Su 等,2007)和死亡数,虽然仍然很高,但随着每窝增加2.3头断奶仔猪数,结果趋于稳定。随着农场主对高产母猪的饲养经验的增加,也反映了较多仔猪的成功管理,但也表明 LP 5 选择法可能不会导致很多受影响的仔猪(例如病理生长迟缓的个体)用于产仔数的选择。减少窝内的变化,尤其是出生体重,是许多作者讨论的另一个重要的育种目标(Damgaard 等,2003;Huby 等,2003;Knol 等,2002a,b;Rydhmer,2000)。选择改善胎盘效率是另一种改善仔猪成活率的潜在策略(Van Rens 等,2005)。此外,改良母性的育种(Baxter 等,2011a;Gade 等,2007;Grandison,2005)以及规定更强壮的仔猪遗传特征都具有降低死亡率的潜力。

11.3.2 环境设计

先前减少仔猪死亡率的研究几乎全部集中在对分娩环境的改变。如增加对出生地点和产仔区域的宏观和微观气候的控制(Morrison 等,1983;Schmid,1994),当然,引入产仔箱(Robertson 等,1966)是将工具运用到提高仔猪成活率的实例。虽然产仔箱对仔猪生存有显著的益处(Edwards 和 Fraser,1997),但是它对母猪福利施加了物理限制,阻止了重要的特定行为的表现,例如筑巢。Lawrence 等(1994)发现,限位的母猪的皮质醇水平升高,可能是因为其无法进行自然行为并导致在限制性环境中增加了伤害的可能因素的结果(Jarvis 等,1998,2004;Lawrence 等,1994;McLean 等,1998)。通过减少临产母猪的压力来改善母性和分娩进程可间接协助仔猪。通过确保母猪适当的状况,最大限度地减少热应激,并允许筑巢行为和减少挫折,可以缩短分娩持续时间和缺氧风险(Damm 等,2003,2005;Jarvis 等,2001,2002;Thodberg 等,1999)。在选择实验(Herskin

等，1998)中，产床和垫料来改善母性，是通过减少姿势变化的数量，从而降低踩压风险。提供垫料不仅可以改善母性，而且还为新生仔猪提供了微环境。Morrison等(1983)的早期实验证明了科学的设备使用可以减少出生时体温降低的问题，在出生部位和附近的乳房使用补充辐射加热，仔猪将度过产后早期的生活。类似这种设备在产仔箱系统中很容易实现，该系统可以确定母猪的位置，但具挑战性的是在松散的产仔栏中，出生地点无法可靠预测。已经证明在宽松的产仔栏中，通过在仔猪卧区使用地板加热，可显著降低死亡率(Malmkvist 等，2006)。在使用宽松产仔栏的情况下，结合仔猪保护功能，例如倾斜的墙壁，来协助受约束而躺卧的母猪并促进仔猪逃跑，也是成功的关键特征。对新分娩系统的设计和对现有系统的调整，应考虑基于了解猪行为的选择，以优化母猪和仔猪的福利(Baxter 等，2011b)。

11.3.3 饲养员的干预

妊娠期的管理

在妊娠期间优化管理以降低母猪及其发育后代所经历的社会和营养压力是减少仔猪死亡率的重要方面，并在第 3 章(Spoolder 和 Vermeer，2015)、5 章(Meunier-Salaün 和 Bolhuis，2015)和 12 章(Prunier 和 Tallet，2015)进一步讨论。此外，营养干预措施也可能会增强仔猪的生存能力。能提高胚胎质量和随后的初生重及均匀性的营养干预措施包括在育种之前在母猪日粮中添加可发酵成分(Van den Brand 等，2009)和胎盘发育时的必需氨基酸(Wu 等，2004)，而在妊娠晚期补充必需脂肪酸可以增加仔猪活力(Rooke 等，2001，见 16 章；Bontempo 和 Jiang，2015)。Campos 等(2012)发表了最近对这些仔猪福利的评论，而 Meunier-Salaün 等(2001)和 De Leeuw 等(2008)讨论了营养干预对母猪福利的影响。也可以采用行为干预措施；在分娩前每天短时间对母猪积极的处理可以减少恐惧（Andersen 等，2006)，对母体行为和仔猪生存具有潜在的益处，如第 12 章(Prunier 和 Tallet，2015)中更详细的讨论。

围产期管理：诱导、监督与协助

对分娩时间和出生间隔很长的干预监督，可以大大减少死胎的发生率(English 和 Edwards，1996)，而出生后马上仔细照顾弱仔猪可以减少产后死亡率(Andersen 等，2009；Christison 等，1997；Holyoake 等，1995)。为了实现监督的这些好处，有必要确保分娩时饲养员的存在。药物诱导分娩可用于同步分娩，并可以促进可能降低仔猪死亡率的监督和干预措施(Cerne 和 Jochle，1981)。然而，如果由于不正确的做法而导致分娩并发症和死胎的增加，这也可能适得其反(Mota-Rojas 等，2002)。诱导的时机至关重要，因为晚期胎儿发育和成熟是生存的诱发因素(Randall，1972b；Van der Lende 等，2001)。在分娩前的日子里，胎儿经历生长速度(Biensen 等，1998)和发育的增加，其中包括子宫外生命的最后生理准备，特别是肺的成熟。因此，过早诱导出生可能导致新生仔猪受损。普遍的共识是不要在妊娠113天之前引起分娩(Kirkden等,2013年),并避免对后备母猪的干预,因为其预产期往往不准确。

另一种常见的药物干预是催产素的使用,旨在帮助在疲劳状况下母猪的分娩进程。然而,这也应该谨慎使用,因为它可以通过加强子宫肌肉收缩,阻止母猪和胎猪之间的气体交换以及增加脐带闭塞的机会而增加胎儿窒息(Alonso-Splisbury 等,2004,2005;Mota-Rojas等,2002,2006)。这些研究中催产素的使用导致仔猪心动过缓和胎粪染色增加,伴随严重酸中毒和生存受损。因此,滥用旨在改善分娩的药物,是引起死胎、死亡率增加和活仔猪在某种程度上遭受缺氧的重要风险因素。

产后管理:初乳摄入量和仔猪哺育

为了确保所有仔猪在出生后不久就能吃到重要的初乳,饲养员可以通过将仔猪放在乳头上并协助哺乳来加快进程。提供协助将提高低活力仔猪的存活率。例如,Muns等(2014)研究表明,在出生4 h内,对体重小于1.35 kg的仔猪提供母猪初乳的口服补充,在产后4天会增加IgG的水平。当窝产仔数大于功能乳头数时,当存在更多的低活力,生长迟缓的仔猪时,这种护理干预措施特别重要。

其他干预措施包括分批哺乳和交叉哺育,以实现同窝仔猪均衡或标准化(即相似尺寸的仔猪)。使用看护母猪系统和使用人工饲养系统也可以帮助提高存活率(参见Baxter等,2013),对管理策略和大窝仔猪的福利影响的全面检查)。分批哺乳是一种用于初产母猪分娩期间培养机会有限时的技术。它包括将同窝仔猪分成两组,并允许每个组在指定的非竞争时间到乳房摄取初乳。这可能需要密集劳动和注意时间管理,来确保这一制度下的仔猪能够定期和正确地轮换。因此,如果有机会达到,这是首选。另外,如果操作正确,交叉哺育可以提高仔猪的生存(Cecchinato等,2008;English等,1977),并且可以减少对一大窝中本可能有剩余的仔猪或不能与自己同窝较大仔猪争夺功能性乳头的低初生重的仔猪进行管理干预的需要。然而,有各种福利问题与一些培育实践有关。这些担忧与出生后的培养时间和过度哺育问题(Baxter等,2013)有关。移动太早(即6 h以前)可能使仔猪无法获得初乳,而移动太迟(即24~48 h后)可能导致更大的争斗(Horrell,1982)、更多的中断哺乳和寄养母猪更大可能的拒绝(Price等,1994)。一些农场管理者会反复交叉培育仔猪,把它们在母猪之间移动,以实现更均匀的断奶重量。然而,这种做法对母猪和仔猪都是有非常大的伤害,因此产生反作用,不断交叉培育的仔猪不能定期喂奶,面部损伤增加,断奶重量的表现没有改善(Robert和Martineau,2001)。

使用奶妈母猪来解决大窝仔猪的挑战,现在已经在丹麦和荷兰等国家普遍存在,在这些国家,超高产育种计划已经使过剩仔猪的生产保持一致。然而,这些系统尚未在其他国家广泛使用。涉及使用奶妈母猪的管理过程主要有两种:即一步法和两步法。一步法管理选择的奶妈母猪至少21天断奶,然后哺育来自新分娩母猪至少12 h时龄的剩余仔猪。奶妈母猪喂养第二窝至少21日龄断奶之后,她再返回到空怀母猪舍继续服务。两步法管理有时被称为"阶梯培养",涉及两个哺乳母猪的使用。确定在28日龄(或至少21日龄)自己仔猪断奶的一头中等母猪(中期母猪),然后确定第二步奶妈母猪的仔猪为4～7日龄。这些仔猪都被中等母猪哺育。然后第二步母猪哺育过剩的、体重大的、新分娩的仔猪(详见Baxter等,2013)。这种策略需要勤奋的饲养员管理,并且对于仔猪和母猪的健康和福利都有风险(Baxter等,2013),特别是在表现不佳的情况下。

目前,荷兰、美国和德国日益广泛使用人造饲养系统来处理过剩的仔猪。例如,仔猪哺育平台(Rescue Deck)系统是一个设计特别的辅助哺育平台,主要用于哺育产房上有过剩或低活力的仔猪。哺育平台使用漏缝地板、具有充足的热源和光照,并有人造乳汁、水和用于较大日龄仔猪的饲喂系统。小猪通常在3～20日龄时在这里哺育,并且这个系统确实会"拯救"可能死亡的仔猪(Van Dijk,2012)。然而,关于人工饲养系统在福利和"救助"仔猪的长期生存前景方面的优缺点的科学证据很少,如果采用这种做法,则需要进一步调查。尽管讨论的管理干预旨在促进仔猪的生存,但显然它们也可能对仔猪(和母猪)的福利产生负面影响,特别是在管理不善的情况下。

11.4 结论

仔猪死亡率是一个病因复杂的问题,易受品种自然生物学的影响,并受农场的经济选择压力而加剧。与动物特性、周围环境和人类行为相关的风险因素相互作用,进而影响分娩和早期生长。虽然有史以来的重点是把控制分娩时的物理环境作为改善仔猪成活率的手段,但是由于超高产母猪系的发展引起的近期挑战需要更多的方法。造成死亡的事件目前正在分娩之前发生,因此需要通过增加有利于仔猪生存的遗传性状的结合、增强胚胎发育和新生仔猪活力的营养干预以及熟练的饲养员在仔猪出生和初乳摄入时帮助弱仔猪和外来仔猪的方式来处理。

随着社会越来越关注猪生产中的伦理和福利问题,但是为了保持竞争力而没有减轻对养猪者的财政压力,所以在允许充分表达母猪行为需要的环境中,实现每头母猪每年大量断奶仔猪的要求将继续面临挑战。无论如何,持续降低仔猪死亡率符合经济和社会目标,因此必须成为未来研究和开发的重点课题。

参考文献

Ahlstrom, S., Jarvis, S. and Lawrence, A.B., 2002. Savaging gilts are more restless and more responsive to piglets during the expulsive phase of parturition. Applied Animal Behaviour Science 76: 83-91.

Algers, B. and Jensen, P., 1990. Thermal microclimate in winter farrowing nests of free-ranging domestic pigs. Livestock Production Science 25: 177-181.

Alonso-Spilsbury, M., Mota-Rojas, D., Martinez-Burnes, J., Arch, E., Lopez Mayagoitia, A., Ramirez-Necoechea, R., Olmos, A. and Trujillo, M.E., 2004. Use of oxytocin in penned sows and its effect on fetal intra-partum asphyxia. Animal Reproduction Science 84: 157-167.

Alonso-Spilsbury, M., Mota-Rojas, D., Villanueva-Garcia, D., Martinez-Burnes, J., Orozco, H., Ramirez-Necoechea, R., Mayagoitia, A.L. and Trujillo, M.E., 2005. Perinatal asphyxia pathophysiology in pig and human: a review. Animal Reproduction Science 90: 1-30.

Amdi, C., Krogh, U., Flummer, C., Oksbjerg, N., Hansen, C.F. and Theil, P.K., 2013. Intrauterine growth restricted piglets defined by their head shape ingest insufficient amounts of colostrum. Journal of Animal Science 91: 5605-5613.

Andersen, I., Nævdal, E. and Böe, K., 2011. Maternal investment, sibling competition, and offspring survival with increasing litter size and parity in pigs (Sus scrofa). Behavioral Ecology and Sociobiology: 1-9.

Andersen, I.L., Berg, S. and Boe, K.E., 2005. Crushing of piglets by the mother sow (Sus scrofa) – purely accidental or a poor mother? Applied Animal Behaviour Science 93: 229-243.

Andersen, I.L., Berg, S., Boe, K.E. and Edwards, S., 2006. Positive handling in late pregnancy and the consequences for maternal behaviour and production in sows. Applied Animal Behaviour Science 99: 64-76.

Andersen, I.L., Haukvik, I.A. and Bøe, K.E. 2009. Drying and warming immediately after birth may reduce piglet mortality in loose-housed sows. Animal 3: 592-597.

Bauer, R., Walter, B., Gaser, E., Rosel, T., Kluge, H. and Zwiener, U., 1998. Cardiovascular function and brain metabolites in normal weight and intrauterine growth restricted newborn piglets – Effect of mild hypoxia. Experimental Toxicology Pathology 50: 294-300.

Baxter, E.M., Jarvis, S., D'Eath, R.B., Ross, D.W., Robson, S.K., Farish, M., Nevison, I.M., Lawrence, A.B. and Edwards, S.A., 2008. Investigating the behavioural and physiological indicators of neonatal survival in pigs. Theriogenology 69: 773-783.

Baxter, E.M., Jarvis, S., Palarea-Albaladejo, J. and Edwards, S.A., 2012. The Weaker Sex? The propensity for male-biased piglet mortality. PLoS ONE 7: e30318.

Baxter, E.M., Jarvis, S., Sherwood, L., Farish, M., Roehe, R., Lawrence, A.B. and Edwards, S.A., 2011a. Genetic and environmental effects on piglet survival and maternal behaviour of the farrowing sow. Applied Animal Behaviour Science 130: 28-41.

Baxter, E.M., Jarvis, S., Sherwood, L., Robson, S.K., Ormandy, E., Farish, M., Smurthwaite, K.M., Roehe, R., Lawrence, A.B. and Edwards, S.A., 2009. Indicators of piglet survival in an outdoor farrowing system. Livestock Science 124: 266-276.

Baxter, E.M., Lawrence, A.B. and Edwards, S.A., 2011b. Alternative farrowing systems: design criteria for farrowing systems based on the biological needs of sows and piglets. Animal 5: 580-600.

Baxter, E.M., Rutherford, K.M.D., D'Eath, R.B., Arnott, G., Turner, S.P., Sandoe, P., Moustsen, V.A., Thorup, F., Edwards, S.A. and Lawrence, A.B., 2013. The welfare implications of large litter size in the domestic pig II: management factors. Animal Welfare 22: 219-238.

Biensen, N.J., Wilson, M.E. and Ford, S.P., 1998. The impact of either a Meishan or Yorkshire uterus on Meishan or Yorkshire fetal and placental development to days 70, 90, and 110 of gestation. Journal of Animal Science 76: 2169-2176.

Black, J.L., Mullan, B.P., Lorschy, M.L. and Giles, L.R., 1993. Lactation in the sow during heat stress. Livestock Production Science 35(1-2): 153-170.

Bontempo, V. and Jiang, X.R., 2015. Feeding various fat sources to sows: effects on immune status and performance of sows and piglets. Chapter 16. In: Farmer, C. (ed.) The gestating and lactating sow. Wageningen Academic Publishers, Wageningen, the Netherlands, pp. 357-375.

British Pig Executive (BPEX), 2013. BPEX/AHDB 2012 Pig cost of production in selected countries. BPEX, Agriculture and Horticulture Development Board (AHDB), Stoneleigh, UK.

Campos, P.H.R.F., Silva, B.A.N., Donzele, J.L., Oliveira, R.F.M. and Knol, E.F., 2012. Effects of sow nutrition during gestation on within-litter birth weight variation: a review. Animal 6: 797-806.

Canario, L., Cantoni, E., Le Bihan, E., Caritez, J.C., Billon, Y., Bidanel, J.P. and Foulley, J.L., 2006. Between-breed variability of stillbirth and its relationship with sow and piglet characteristics. Journal of Animal Science 84: 3185-3196.

Cecchinato, A., Bonfatti, V., Gallo, L. and Carnier, P., 2008. Survival analysis of preweaning piglet survival in a dry-cured ham-producing crossbred line. Journal of Animal Science 86: 2486-2495.

Cerne, F. and Jochle, W., 1981. Clinical evaluations of a new prostaglandin analog in pigs: 1. Control of parturition and of the MMA-syndrome. Theriogenology 16: 459-467.

Chen, C., Gilbert, C.L., Yang, G., Guo, Y., Segonds-Pichon, A., Ma, J., Evans, G., Brenig, B., Sargent, C., Affara, N. and Huang, L., 2007. Maternal infanticide in sows: Incidence and behavioural comparisons between savaging and non-savaging sows at parturition. Applied Animal Behaviour Science 109: 238-248.

Chevaux, E., Sacy, A., Le Treut, Y. and Martineau, G., 2010. Intrauterine Growth Retardation (IUGR): morphological and behavioural description. In: IPVS (ed.) Proceedings of the 21st IPVS Congress. July 18-21, 2010. Vancouver Canada, p. 209.

Christensen, J. and Svensmark, B., 1997. Evaluation of producer-recorded causes of preweaning mortality in Danish sow herds. Preventive Veterinary Medicine 32: 155-164.

Christison, G.I., Wenger, I.I. and Follensbee, M.E., 1997. Teat seeking success of newborn piglets after drying or warming. Canadian Journal of Animal Science 77: 317-319.

Cranwell, P.D., 1995. Development of the neonatal gut. In: Varley, M.A. (ed.) The neonatal pig development and survival. CAB International, Wallingford, UK, pp. 99-154.

Curtis, S.E., 1970. Environmental-thermoregulatory interactions and neonatal piglet survival. Journal of Animal Science 31: 576-587.

Curtis, S.E., 1972. Air environment and animal performance. Journal of Animal Science 35: 628-634.

Da Silva-Buttkus, P., Van den Hurk, R., Te Velde, E.R. and Taverne, M.A., 2003. Ovarian development in intrauterine growth retarded and normally developed piglets originating from the same litter. Reproduction 126: 249-258.

Damgaard, L.H., Rydhmer, L., Lovendahl, P. and Grandinson, K., 2003. Genetic parameters for within-litter variation in piglet birth weight and change in within-litter variation during suckling. Journal of Animal Science 81: 604-610.

Damm, B.I., Pedersen, L.J., Heiskanen, T. and Nielsen, N.P., 2005. Long-stemmed straw as an additional nesting material in modified Schmid pens in a commercial breeding unit: effects on sow behaviour, and on piglet mortality and growth. Applied Animal Behaviour Science 92: 45-60.

Damm, B.I., Pedersen, L.J., Marchant-Forde, J.N. and Gilbert, C.L., 2003. Does feed-back from a nest affect periparturient behaviour, heart rate and circulatory cortisol and oxytocin in gilts? Applied Animal Behaviour Science 83: 55-76.

De Leeuw, J.A., Bolhuis, J.E., Bosch, G. and Gerrits, W.J.J., 2008. Effects of dietary fibre on behaviour and satiety in pigs. Proceedings of the Nutritional Society 67: 334-342.

De Passillé, A.M.B. and Rushen, J., 1989. Suckling and teat disputes by neonatal piglets. Applied Animal Behaviour Science 22: 23-38.

De Passillé, A.M.B., Rushen, J. and Harstock, T.G., 1988. Ontogeny of teat fidelity in pigs and its relation to competition at suckling. Canadian Journal of Animal Science 68: 325-338.

Devillers, N., Le Dividich, J. and Prunier, A., 2011. Influence of colostrum intake on piglet survival and immunity. Animal 5: 1605-1612.

Drake, A., Fraser, D. and Weary, D.M., 2008. Parent-offspring resource allocation in domestic pigs. Behavioral Ecology and Sociobiology 62: 309-319.

Dyck, G.W. and Swierstra, E.E., 1987. Causes of piglet death from birth to weaning. Canadian Journal of Animal Science 67: 543-547.

Easicare, 1995. Pig management yearbook (7th edition). Easicare Computers, Driffield, UK.

Edwards, S.A., 2002. Perinatal mortality in the pig: environmental or physiological solutions? Livestock Production Science 78: 3-12.

Edwards, S.A. and Fraser, D., 1997. Housing systems for farrowing and lactation. The Pig Journal 39: 77-89.

Edwards, S.A., Smith, W.J., Fordyce, C. and Macmenemy, F., 1994. An analysis of the causes of piglet mortality in a breeding herd kept outdoors. Veterinary Record 135: 324-327.

English, P.R. and Edwards, S.A., 1996. Management of the nursing sow and her litter. In: Taverner, M.R. and Dunkin, A.C. (eds.) Pig production: C10 (World Animal Science). Elsevier Health Sciences, Oxford, UK, pp. 113-140.

English, P.R. and Morrison, V., 1984. Causes and prevention of piglet mortality. Pig News and Information 5: 369-375.

English, P.R. and Smith, W.J., 1975. Some causes of death in neonatal piglets. Veterinary Annual 15: 95-104.

English, P.R., Smith, W.J. and MacLean, A., 1977. The sow – improving her efficiency. Farming Press Limited, Ipswich, UK.

Fahmy, M.H. and Friend, D.W., 1981. Factors influencing, and repeatability of the duration of farrowing in Yorkshire sows. Canadian Journal of Animal Science 61: 17-22.

Farmer, C. and Hurley, W.L., 2015. Mammary development. Chapter 4. In: Farmer, C. (ed.) The gestating and lactating sow. Wageningen Academic Publishers, Wageningen, the Netherlands, pp. 73-94.

Forbes, L.S. and Mock, D.W., 1998. Parental optimism and progeny choice: when is screening for offspring quality affordable? Journal of Theoretical Biology 192: 3-14.

Foxcroft, G.R., Dixon, W.T., Dyck, M.K., Novak, S., Harding, J.C.S. and Almeida, F.C.R.L., 2009. Prenatal programming of postnatal development in the pig. In: Rodriguez-Martinez, H., Vallet, J.L. and Ziecik, A.J. (eds.) Control of pig reproduction VIII. Nottingham University Press, Nottingham, UK, pp.213-231.

Foxcroft, G.R., Dixon, W.T., Novak, S., Putman, C.T., Town, S.C. and Vinsky, M.D.A., 2006. The biological basis for prenatal programming of postnatal performance in pigs. Journal of Animal Science 84: E105-E112.

Fraser, D., 1975. Teat order of suckling pigs. 2. Fighting during suckling and effects of clipping eye teeth. Journal of Agricultural Science 84: 393-399.

Gade, S., Bennewitz, J., Kirchner, K., Looft, H., Knap, P.W., Thaller, G. and Kalm, E., 2007. Genetic parameters for maternal behaviour traits in sows. Livestock Science 114: 31-41.

Gaskins, H.R. and Kelley, K.W., 1995. Immunology and neonatal mortality. In: Varley, M.A. (ed.) The neonatal pig develoment and survival. CAB International, Wallingford, UK, pp. 39-56.

Geisert, R.D., Morgan, G.L., Zavy, M.T., Blair, R.M., Gries, L.K., Cox, A. and Yellin, T., 1991. Effect of asynchronous transfer and estrogen administration on survival and development of porcine embryos. Journal of Reproduction and Fertility 93: 475-481.

Geisert, R.D. and Schmitt, R.A.M., 2002. Early embryonic survival in the pig: can it be improved? Journal of Animal Science 80: E54-E65.

Gluckmann, P.D. and Hanson, M.A., 2004. Maternal constraints of fetal growth and its consequence. Seminars in Neonatology 9: 419-425.

Grandinson, K., 2005. Genetic background of maternal behaviour and its relation to offspring survival. Livestock Production Science 93: 43-50.

Grandinson, K., Rydhmer, L., Strandberg, E. and Thodberg, K., 2003. Genetic analysis of on-farm tests of maternal behaviour in sows. Livestock Production Science 83: 141-151.

Hales, J., Moustsen, V.A., Nielsen, M.B.F. and Hansen, C.F., 2013. Individual physical characteristics of neonatal piglets affect preweaning survival of piglets born in a noncrated system. Journal of Animal Science 91: 4991-5003.

Harris, M.J., Li, Y.Z. and Gonyou, H.W., 2003. Savaging behaviour in gilts and sows. Canadian Journal of Animal Science 83: 819-821.

Head, R.H. and Williams, I.H., 1992 Mammogenesis is influenced by pregnancy nutrition. In: Proceedings of the Third Biennial Conference of the Australasian Pig Science Association (APSA). November 24-27, 1991. Albury, Australia, p. 33.

Hemsworth, P.H., Coleman, G.J., Cronin, G.M. and Spicer., E.M., 1995. Human care and the neonatal pig. In: Varley, M.A. (ed.) The neonatal pig. Development and survival. CAB International, Wallingford, UK, pp. 313-331.

Hemsworth, P.H., Pedersen, V., Cox, M., Cronin, G.M. and Coleman, G.J., 1999. A note on the relationship between the behavioural response of lactating sows to humans and survival of piglets. Applied Animal Behaviour Science 65: 43-52.

Herpin, P., Damon, M. and Le Dividich, J., 2002. Development of thermoregulation and neonatal survival in pigs. Livestock Production Science 78: 25-45.

Herpin, P. and Le Dividich, J., 1995. Thermoregulation and the Environment. In: Varley, M.A. (ed.) The neonatal pig. Development and survival. CAB International, Wallingford, UK, pp. 57-95.

Herpin, P., Le Dividich, J. and Amaral, N., 1993. Effect of selection for lean tissue-growth on body-composition and physiological-state of the pig at birth. Journal of Animal Science 71: 2645-2653.

Herpin, P., Le Dividich, J., Berthon, D. and Hulin, J.C., 1994. Assessment of thermoregulatory and postprandial thermogenesis over the first 24 hours after birth in pigs. Experimental Physiology 79: 1011-1019.

Herpin, P., Le Dividich, J., Hulin, J.C., Fillaut, M., DeMarco, F. and Bertin, R., 1996. Effect of the level of asphyxia during delivery on viability at birth and early postnatal vitality of newborn pigs. Journal of Animal Science 74: 2067-2075.

Herskin, M.S., Jensen, K.H. and Thodberg, K., 1998. Influence of environmental stimuli on maternal behaviour related to bonding, reactivity and crushing of piglets in domestic sows. Applied Animal Behaviour Science 58: 241-254.

Holyoake, P.K., Dial, G.D., Trigg, T. and King, V.L., 1995. Reducing pig mortality through supervision during the perinatal period. Journal of Animal Science 73: 3543-3551.

Horrell, R.I., 1982. Immediate behavioural consequences of fostering 1-week-old piglets. The Journal of Agricultural Science 99: 329-336.

Hrupka, B.J., Leibbrandt, V.D., Crenshaw, T.D. and Benevenga, N.J., 2000. Effect of sensory stimuli on huddling behavior of pigs. Journal of Animal Science 78: 592-596.

Huby, M., Canario, L., Tribout, T., Caritez, J.C., Billon, Y., Gogué, J. and Bidanel, J.P., 2003. Genetic correlations between litter size and weights, piglet weight variability and piglet survival from birth to weaning in large white pigs. In: Proceedings of the 54th Annual Meeting of the European Association for Animal Production Roma, Italy. Wageningen Pers, Wageningen, the Netherlands, pp. 362.

Hurley, W.L., 2015. Composition of sow colostrum and milk. Chapter 9. In: Farmer, C. (ed.) The gestating and lactating sow. Wageningen Academic Publishers, Wageningen, the Netherlands, pp. 193-229.

Janczak, A.M., Pedersen, L.J. and Bakken, M., 2003. Aggression, fearfulness and coping styles in female pigs. Applied Animal Behaviour Science 81: 13-28.

Jarvis, S., Calvert, S.K., Stevenson, J., Van Leeuwen, N. and Lawrence, A.B., 2002. Pituitary-adrenal activation in pre-parturient pigs (*Sus scrofa*) is associated with behavioural restriction due to lack of space rather than nesting substrate. Animal Welfare 11: 371-384.

Jarvis, S., D'Eath, R.B. and Fujita, K., 2005. Consistency of piglet crushing by sows. Animal Welfare 14: 43-51.

Jarvis, S., Lawrence, A.B., Mclean, K.A., Chirnside, J., Deans, L.A. and Calvert, S.K., 1998. The effect of environment on plasma cortisol and beta-endorphin in the parturient pig and the involvement of endogenous opioids. Animal Reproduction Science 52: 139-151.

Jarvis, S., Moinard, C., Robson, S.K., Baxter, E., Ormandy, E., Douglas, A.J., Seckl, J.R., Russell, J.A. and Lawrence, A.B., 2006. Programming the offspring of the pig by prenatal social stress: neuroendocrine activity and behaviour. Hormones and Behavior 49: 68-80.

Jarvis, S., Reed, B.T., Lawrence, A.B., Calvert, S.K. and Stevenson, J., 2004. Peri-natal environmental effects on maternal behaviour, pituitary and adrenal activation, and the progress of parturition in the primiparous sow. Animal Welfare 13: 171-181.

Jarvis, S., Van der Vegt, B.J., Lawrence, A.B., Mclean, K.A., Deans, L.A., Chirnside, J. and Calvert, S.K., 2001. The effect of parity and environmental restriction on behavioural and physiological responses of pre-parturient pigs. Applied Animal Behaviour Science 71: 203-216.

Kerr, J.C. and Cameron, N.D., 1995. Reproductive-performance of pigs selected for components of efficient lean growth. Animal Science 60: 281-290.

Kim, S.W., Hurley, W.L., Han, I.K., Stein, H.H. and Easter, R.A., 1999. Effect of nutrient intake on mammary gland growth in lactating sows. Journal of Animal Science 77: 3304-3315.

Kirkden, R.D., Broom, D.M. and Andersen, I.L., 2013. Piglet mortality: the impact of induction of farrowing using prostaglandins and oxytocin. Animal Reproduction Science 138: 14-24.

Knol, E.F., Ducro, B.J., Van Arendonk, J.A.M. and Van der Lende, T., 2002a. Direct, maternal and nurse sow genetic effects on farrowing-, pre-weaning- and total piglet survival. Livestock Production Science 73: 153-164.

Knol, E.F., Leenhouwers, J.I. and Van der Lende, T., 2002b. Genetic aspects of piglet survival. Livestock Production Science 78: 47-55.

Kranendonk, G., Hopster, H., Fillerup, M., Ekkel, E.D., Mulder, E.J., Wiegant, V.M. and Taverne, M.A., 2006. Lower birth weight and attenuated adrenocortical response to ACTH in offspring from sows that orally received cortisol during gestation. Domestic Animal Endocrinology 30: 218-238.

Lawrence, A.B., Petherick, J.C., Mclean, K.A., Deans, L.A., Chirnside, J., Vaughan, A., Clutton, E. and Terlouw, E.M.C., 1994. The effect of environment on behavior, plasma-cortisol and prolactin in parturient sows. Applied Animal Behaviour Science 39: 313-330.

Lay, Jr., D.C., Matteri, R.L., Carroll, J.A., Fangman, T.J. and Safranski, T.J., 2002. Preweaning survival in swine. Journal of Animal Science 80: E74-E86.

Leenhouwers, J.I., De Almeida, C.A., Knol, E.F. and Van der Lende, T., 2001. Progress of farrowing and early postnatal pig behavior in relation to genetic merit for pig survival. Journal of Animal Science 79: 1416-1422.

Leenhouwers, J.I., Wissink, P., Van der Lende, T., Paridaans, H. and Knol, E.F., 2003. Stillbirth in the pig in relation to genetic merit for farrowing survival. Journal of Animal Science 81: 2419-2424.

Lensink, B.J., Leruste, H., LeRoux, T. and Bizeray-Filoche, D., 2009. Relationship between the behaviour of sows at 6 months old and the behaviour and performance at farrowing. Animal 3: 128-134.

Malmkvist, J., Pedersen, L.J., Damgaard, B.M., Thodberg, K., Jørgensen, E. and Labouriau, R., 2006. Does floor heating around parturition affect the vitality of piglets born to loose housed sows? Applied Animal Behaviour Science 99: 88-105.

Marchant Forde, J.N., 2002. Piglet- and stockperson-directed sow aggression after farrowing and the relationship with a pre-farrowing, human approach test. Applied Animal Behaviour Science 75: 115-132.

Marchant, J.N., Broom, D.M. and Corning, S., 2001. The influence of sow behaviour on piglet mortality due to crushing in an open farrowing system. Animal Science 72: 19-28.

Marchant, J.N., Rudd, A.R., Mendl, M.T., Broom, D.M., Meredith, M.J., Corning, S. and Simmins, P.H., 2000. Timing and causes of piglet mortality in alternative and conventional farrowing systems. Veterinary Record 147: 209-214.

Mclean, K.A., Lawrence, A.B., Petherick, J.C., Deans, L., Chirnside, J., Vaughan, A., Nielsen, B.L. and Webb, R., 1998. Investigation of the relationship between farrowing environment, sex steroid concentrations and maternal aggression in gilts. Animal Reproduction Science 50: 95-109.

Mellor, D.J. and Stafford, K.J., 2004. Animal welfare implications of neonatal mortality and morbidity in farm animals. Veterinary Journal 168: 118-133.

Meunier-Salaün, M.C. and Bolhuis, J.E., 2015. High-Fibre feeding in gestation. Chapter 5. In: Farmer, C. (ed.) The gestating and lactating sow. Wageningen Academic Publishers, Wageningen, the Netherlands, pp. 95-116.

Meunier-Salaun, M.C., Edwards, S.A. and Robert, S., 2001. Effect of dietary fibre on the behaviour and health of the restricted fed sow. Animal Feed Science and Technology 90: 53-69.

Milligan, B.N., Fraser, D. and Kramer, D.L., 2002. Within-litter birth weight variation in the domestic pig and its relation to pre-weaning survival, weight gain, and variation in weaning weights. Livestock Production Science 76: 181-191.

Mock, D.W. and Forbes, L.S., 1995. The evolution of parental optimism. Trends in Ecology & Evolution 10: 130-134.

Morrison, V., English, P.R. and Lodge, G.A., 1983. The effect of alternative creep heating arrangements at two house temperatures on piglet lying behaviour and mortality in the neonatal period. Animal Production 36: 530-531.

Mota-Rojas, D., Martinez-Burnes, J., Trujillo-Ortega, M.E., Alonso-Spilsbury, M., Ramirez-Necoechea, R. and Lopez, A., 2002. Effect of oxytocin treatment in sows on umbilical cord morphology, meconium staining, and neonatal mortality of piglets. American Journal of Veterinary Research 63: 1571-1574.

Mota-Rojas, D., Rosales, A.M., Trujillo, M.E., Orozco, H., Ramirez, R. and Alonso-Spilsbury, M., 2005. The effects of vetrabutin chlorhydrate and oxytocin on stillbirth rate and asphyxia in swine. Theriogenology 64: 1889-1897.

Mota-Rojas, D., Trujillo, M.E., Martinez, J., Rosales, A.M., Orozco, H., Ramirez, R., Sumano, H. and Alonso-Spilsbury, M., 2006. Comparative routes of oxytocin administration in crated farrowing sows and its effects on fetal and postnatal asphyxia. Animal Reproduction Science 92: 123-143.

Mount, L.E., 1967. The heat loss from new-born pigs to the floor. Research in Veterinary Science 8: 175-186.

Mount, L.E., 1968. The climatic physiology of the pig. Edward Arnold, London, UK.

Mouttotou, N. and Green, L.E., 1999 Incidence of foot and skin lesions in nursing piglets and their association with behavioural activities. Veterinary Record 145: 160-165.

Muns, R., Silva, C., Manteca, X. and Gasa, J., 2014. Effect of cross-fostering and oral supplementation with colostrums on performance of newborn piglets. Journal of Animal Science 92: 1193-1199.

O'Dowd, S., Hoste, S., Mercer, J.T., Fowler, V.R. and Edwards, S.A., 1997. Nutritional modification of body composition and the consequences for reproductive performance and longevity in genetically lean sows. Livestock Production Science 52: 155-165.

Oliviero, C., Heinonen, M., Valros, A., Halli, O. and Peltoniemi, O.A.T., 2006. Duration of farrowing is longer in sows housed in farrowing crates than in pens. Reproduction in Domestic Animals 41: 367.

Pattison R.J.; English P.R.; Macpherson O.; Roden J.A.; Birnie M., 1990. Hypothermia and its attempted control in newborn piglets. Animal Production 50(3): 568.

Pedersen, L.J., Jorgensen, E., Heiskanen, T. and Damm, B.I., 2006. Early piglet mortality in loose-housed sows related to sow and piglet behaviour and to the progress of parturition. Applied Animal Behaviour Science 96: 215-232.

Pope, W.F., 1994. Embryonic mortality in swine. In: Zavy, M.T. (ed.) Embryonic mortality in domestic species. CRC Press, Boca Raton, FL, USA, pp. 53-78.

Price, E.O., Hutson, G.D., Price, M.I. and Borgwardt, R., 1994. Fostering in swine as affected by age of offspring. Journal of Animal Science 72: 1697-1701.

Prunier, A. and Tallet, C., 2015. Endocrine and behavioural responses of sows to human interactions and consequences on reproductive performance. Chapter 12. In: Farmer, C. (ed.) The gestating and lactating sow. Wageningen Academic Publishers, Wageningen, the Netherlands, pp. 279-295.

Quesnel, H., Farmer, C. and Theil, P.K., 2015. Colostrum and milk production. Chapter 8. In: Farmer, C. (ed.) The gestating and lactating sow. Wageningen Academic Publishers, Wageningen, the Netherlands, pp. 173-192.

Quiniou, N., Dagorn, J. and Gaudre, D., 2002. Variation of piglets' birth weight and consequences on subsequent performance. Livestock Production Science 78: 63-70.

Randall, G.C., 1972a. Observations on parturition in the sow. I. Factors associated with the delivery of the piglets and their subsequent behaviour. Veterinary Record 90: 178-182.

Randall, G.C., 1972b. Observations on parturition in the sow. II. Factors influencing stillbirth and perinatal mortality. Veterinary Record 90: 183-186.

Randall, G.C.B. and Penny, R.H.G., 1970. Stillbirth in the pig: an analysis of the breeding records of five herds. British Veterinary Journal 126: 593-603.

Robert, S. and Martineau, G.P., 2001. Effects of repeated cross-fosterings on preweaning behavior and growth performance of piglets and on maternal behavior of sows. Journal of Animal Science 79: 88-93.

Robertson, J.B., Laird, R., Forsyth, J.K.S., Thomson, J.M. and Walker-Love, T., 1966. A comparison of two indoor farrowing systems for sows. Animal Production 8: 71.

Roehe, R., 1999. Genetic determination of individual birth weight and its association with sow productivity using Bayesian analyses. Journal of Animal Science 77: 330-343.

Roehe, R. and Kalm, E., 2000. Estimation of genetic and environmental risk factors associated with pre-weaning mortality in piglets using generalized linear mixed models. Animal Science 70: 227-240.

Roehe, R., Shrestha, N.P., Mekkawy, W., Baxter, E.M., Knap, P.W., Smurthwaite, K.M., Jarvis, S., Lawrence, A.B. and Edwards, S.A., 2009. Genetic analyses of piglet survival and individual birth weight on first generation data of a selection experiment for piglet survival under outdoor conditions. Livestock Science 121: 173-181.

Roehe, R., Shrestha, N.P., Mekkawy, W., Baxter, E.M., Knap, P.W., Smurthwaite, K.M., Jarvis, S., Lawrence, A.B. and Edwards, S.A., 2010. Genetic parameters of piglet survival and birth weight from a two-generation crossbreeding experiment under outdoor conditions designed to disentangle direct and maternal effects. Journal of Animal Science 88: 1276-1285.

Rooke, J.A. and Bland, I.M., 2002. The acquisition of passive immunity in the new-born piglet. Livestock Production Science 78: 13-23.

Rooke, J.A., Sinclair, A.G., Edwards, S.A., Cordoba, R., Pkiyach, S., Penny, P.C., Penny, P., Finch, A.M. and Horgan, G.W., 2001 The effect of feeding salmon oil to sows throughout pregnancy on pre-weaning mortality of piglets. Animal Science 73: 489-500.

Rootwelt, V., Reksen, O., Farstad, W. and Framstad, T., 2012. Associations between intrapartum death and piglet, placental, and umbilical characteristics. Journal of Animal Science 90: 4289-4296.

Roza S.J., Steegers, E.A., Verburg, B.O., Jaddoe, V.W., Moll, H.A., Hofman, A., Verhulst, F.C. and Tiemeier, H., 2008. What is spared by fetal brain-sparing? Fetal circulatory redistribution and behavioral problems in the general population. American Journal of Epidemiology. 68: 1145-1152.

Van Dijk, A.J., 2012. A rescue package for piglets in need. In Pig Progress. Available at: http://tinyurl.com/kc9cbuc.

Van Dijk, A.J., Van Rens, B.T.T.M., Van der Lende, T. and Taverne, M.A.M., 2005. Factors affecting duration of the expulsive stage of parturition and piglet birth intervals in sows with uncomplicated, spontaneous farrowings. Theriogenology 64: 1573-1590.

Van Kempen, T., 2007. Sports supplements to facilitate parturition and reduce perinatal mortality. In P., Garnsworthy, and J. Wiseman (eds). Recent Advances in Animal Nutrition. Nottingham University Press, Nottingham, UK.

Van Rens, B.T.T.M., de Koning, G., Bergsma, R. and Van der Lende, T., 2005. Preweaning piglet mortality in relation to placental efficiency. Journal of Animal Science 83: 144-151.

Van Rens, B.T.T.M. and Van der Lende, T., 2004. Parturition in gilts: duration of farrowing, birth intervals and placenta expulsion in relation to maternal, piglet and placental traits. Theriogenology 62: 331-352.

Vanderhaeghe, C., Dewulf, J., Ribbens, S., De Kruif, A. and Maes, D., 2010 A cross-sectional study to collect risk factors associated with stillbirths in pig herds. Animal Reproduction Science 118: 62-68.

Vangen, O., Holm, B., Valros, A., Lund, M.S. and Rydhmer, L., 2005. Genetic variation in sows' maternal behaviour, recorded under field conditions. Livestock Production Science 93: 63-71.

Weary, D.M., Pajor, E.A., Thompson, B.K. and Fraser, D., 1996. Risky behaviour by piglets: a trade off between feeding and risk of mortality by maternal crushing? Animal Behaviour 51: 619-624.

Wilson, M.E., Biensen, N.J., Youngs, C.R. and Ford, S.P., 1998. Development of Meishan and Yorkshire littermate conceptuses in either a Meishan or Yorkshire uterine environment to day 90 of gestation and to term. Biology of Reproduction 58: 905-910.

Wu, G., Bazer, F.W., Cudd, T.A., Meininger, C.J. and Spencer, T.E., 2004. Maternal nutrition and fetal development. The Journal of Nutrition 134: 2169-2172.

Wu, G., Bazer, F.W., Wallace, J.M. and Spencer, T.E., 2006. Board-invited review: Intrauterine growth retardation: implications for the animal sciences. Journal of Animal Science. 84: 2316-2337.

Zaleski, H.M. and Hacker, R.R., 1993. Comparison of viability scoring and blood-gas analysis as measures of piglet viability. Canadian Journal of Animal Science 73: 649-653.

12. 母猪与人类相互作用对母猪内分泌、行为以及繁殖性能的影响

A. Prunier* and C. Tallet

INRA, UMR1348 PEGASE, 35590 Saint-Gilles, France; Agrocampus Rennes, UMR1348 PEGASE, 35000 Rennes, France; armelle.prunier@rennes.inra.fr

摘要：农场工作人员与能繁母猪及其仔猪接触的时间尽管相对较短（每个繁殖周期每头母猪约 4 h），但也非常重要，因为它可能会影响动物的性能和福利。在此期间，许多人—猪的相互作用发生，特别是在出生、繁殖和分娩时。从动物角度看，这些相互作用可以被认定为积极、无影响或消极的。在本章中，总结了关于母猪及其仔猪的所有类型管理的文献。首先，涵盖了造成动物疼痛的养殖管理方法可能对哺乳仔猪及其与人类的关系产生的直接和长期影响，以及这些方法对母猪的影响。其次，讨论与母猪相互作用的结果，从而区别在转群/运输期间的操作与繁殖、分娩管理和其他管理相关的操作。评估了母猪和仔猪对人类不同操作类型所产生的反应的后果以及这种操作对繁殖性能的可能影响。最后，克服猪管理系统固有的负面相互作用，开发新实践方法，特别是使用动物表达自己的需求或状态的技术。这篇综述清楚地表明了人与动物相互作用对猪和人类的相互影响，不仅对性能有影响，而且还对猪的福利和工作中的人类满意度产生了影响。

关键词：福利、猪、人类-动物关系、应激

12.1 引言

2009 年，法国传统农场处理一头繁殖母猪的平均工作时数约为 14 h（Roguet 等，2011）。这个时间包括断奶后专门用于猪的工作以及与动物无接触的工作（对猪圈的清洁和消毒，在批次之间维护和修理设备，商业任务），但不包括用于农场饲料生产和处理粪污的时间。可以估计，在传统的养猪场中，一个饲养员在一个周期内在每头繁殖母猪身上每天花费不到 4 h，包括繁殖（查情和人工授精不到 1 h）、分娩管理和用于哺乳一窝仔猪日常工作（约 1 h），饲养、健康监控，从一个猪圈或猪棚转移到另一个（Roguet 等，2011）。然而，当猪场员工正在对一头母猪进行工作时，相邻的母猪也可能受到猪场员工行为和他/她正在工作的母猪的反应的影响。猪

场员工与繁殖母猪及其仔猪接触的时间非常重要,因为它可能影响动物的表现和福利(Hemsworth,2003;Kirkden 等,2013;Rushen 等,1999)。

从猪的角度来看,人与猪的相互作用可以表现为积极、无影响或消极(Hemsworth,2003;Rushen 等,1999)。这种分类取决于相互作用的性质以及动物感知的方式。当相互作用包含害怕刺激的因素时,例如大群、大噪音、喊叫或猪场员工的突然动作,或者当其引起痛苦时,例如饲养习惯中固有的某些做法(例如注射或断尾),可以认为是对动物负向的或有害的(Hemsworth,2003;Rushen 等,1999)。如果不包含这些痛苦或恐惧的挑衅性因素,那么一些人-猪的相互作用就可以被认为是无影响的。例如:清洁卫生和材料或对动物的监控。这些工作给猪提供了熟悉人们存在的机会,从而减少它们的恐惧。如果工作是愉悦的元素,如抚摸、轻声说话,或当猪场员工与积极元素相关联时,分发饲料等,这样的人与动物的相互作用是正面的(Boivin 等,2003;Sommavilla 等,2011)。最后,将母猪从一个圈或者猪棚转移到另一个的一些工作可能是积极的、无影响的或消极的,这取决于猪场员工处理动物的方式。可以轻轻地(使用柔和的声音,友好的拍打等),无影响的或粗鲁的(使用喊叫、电动刺棒等)。在现代传统的猪场中,由于饲料和其他工作的自动化以及广泛使用最小化清洁任务的漏缝地板,无影响和积极相互作用的场合越来越少。因此,动物与猪场员工的直接经历越来越偏向于消极的相互作用。这种后果可能会更加明显,因为猪似乎将对一个工作者的厌恶习惯扩大到所有人(Hemsworth 等,1994,1996b)。此外,观察到积极相互作用中的少量消极相互作用和消极相互作用一样,能够使猪产生对人类的恐惧(Hemsworth 等,1987)。甚至恶意的极端处理(电刺棒短暂的电击),这些数据表明,偶尔的负面经历可能会对猪感知人的方式产生重大影响。此外,也可能导致对人产生应激(Boivin 等,2003)。

本章的目的是总结关于繁殖母猪的人-动物相互作用的研究。根据母猪利用年限期间发生的这种相互作用的影响,评估对其行为、生理和性能方面的影响。最后,通过猪场员工更好地理解猪发出来的表达它们需求的信号,改善各种应激情况。

12.2 哺乳仔猪的处理

许多饲养管理技术常常应用于幼小仔猪。如公猪的去势手术、断尾、剪牙、铁剂注射、打耳缺、打耳标或两性耳刺。而且,仔猪经常在出生时处理,特别是在分娩管理的农场,目的是降低新生仔猪死亡率。所有这些程序可能会对仔猪产生影响,也可能对母猪产生影响,因为母猪对处理它们的后代有相应反应。

12.2.1 造成动物疼痛的操作对哺乳仔猪的影响

仔猪出生后不久,就进行处理,从头部和口部移除胎衣和黏液,用毛巾或"棉布"进行干燥,将其放置在母猪乳房旁或保温灯下。尽管干燥是否有积极效果依然存在疑问,但这种处理通常提高了仔猪的存活率(Andersen 等,2009;Holyoake 等,1995;Kirkden 等,2013)。这些操作,特别是干燥,可能对仔猪是有害的,因此可能影响其后续行为。据我们所知,在猪方面,还没有数据可以证实这一假设,但是关于小马驹的研究数据表明,出生后长期的有害处理(卧位的限制和维护,暴露于新颖的触觉刺激),改变了它们成年时对该应激因子的反应方式(Durier 等,2012)。

现代猪场中,除了在受法律限制的一些国家,初生仔猪上下犬齿和切牙切除(总共 8 颗牙齿)非常普遍(Fredriksen 等,2009),它通常在出生后几天内进行;还有其他常规做法,如注射铁剂、断尾且有时进行阉割。剪牙是用钳子夹住牙齿或用旋转砂轮磨削进行的。断尾也是非常普遍的,断尾是用解剖刀、剪刀/线切割机或用烫铁灼烧进行的。通过断尾移除的尾巴的比例是可变的:从尾巴的尾尖到尾部的 3/4 或更多。耳刺、耳缺(用切口器进行 V 字切割)或耳标(使用涂抹器的塑料标签)在繁殖母猪场中很常见,以便识别它们。补铁管理在常规农场是系统管理的。可以通过口服途径或肌肉注射进行。雄性猪的手术去势非常普遍,尽管其应用在不同国家之间有很大差异(Fredriksen 等,2009),至少在欧洲国家是高度不统一的。通常在产后前几天或几周内进行。一些养猪生产者在仔猪出生当天或第二天进行阉割,同时进行断尾、铁剂注射和剪牙。阴囊的一个或两个切口用锋利的解剖刀或剪刀实现。每个睾丸都从周围的组织中解剖出来,通过剪切或拉/撕裂精索将其去除。为了开展以上操作,需要抓住仔猪,迫使它保持不动。这种处理本身对于仔猪而言是紧张的,并引起喊叫和防御行为(Marchant-Forde 等,2009;Torrey 等,2009;Weary 等,1998a)。

在进行这些饲养程序期间,通过记录行为数据和内分泌数据表明,它们是痛苦的,手术去势诱发最严重的疼痛(Hay 等,2003;Marchant-Forde 等,2009;Noonan 等,1994;Prunier 等,2005;Sutherland 等,2008;Torrey 等,2009;White 等,1995)。在欧洲国家,养猪生产者已开始使用药物,去势前使用止痛注射剂(美洛昔康),在阉割前减轻疼痛,或使用二氧化碳或异氟烷进行全身麻醉或利多卡因的局部麻醉(Anonymous,2010;Fredriksen 等,2009)。然而,即使这些方法减轻疼痛,它们也并非完全有效,动物仍然会感到一些不适甚至疼痛(Prunier 等,2006;Von Borell 等,2009)。此外,其他的管理程序通常是在没有任何治疗疼痛缓解的情况下进行的(Fredriksen 等,2009)。由于这些令猪只感到厌恶的操作都是通过人的行为实现造成的,这促进猪对人类的恐惧。采用这些操作,对于为了繁殖而饲养的雌性仔猪来说是特别重要的。据我们所知,现在还没有对这些新生仔猪的痛苦处理产生的长期影响进行评价。

12.2.2 非痛苦的相互作用对哺乳仔猪的影响

除了饲养操作的影响之外,在日常活动(清洁、饲喂、动物检查)期间,人接触猪只的质量可能会短期和长期地影响仔猪的行为。例如,哺乳仔猪定期遭受消极的人类互动(以侵略性的语调,转移仔猪,展示姿势威胁)则表现为,在断奶当天,仔猪对前期消极处理的人与仔猪遭受无影响的互动(很少注意母猪和仔猪,声音柔和)相比具有更多的逃避反应(Sommavilla 等,2011)。然而,当出现一个陌生人观察仔猪时,两组之间的逃避反应没有差异。断奶后的几天,与接受无影响相互作用的仔猪相比,遭受令其厌恶的人类相互作用的仔猪更具侵略性,经常休息时间更少。然而,两组之间的增长率和采食量没有差异。

很少有人试图评价那些处理仔猪的负面方法的长期影响。然而,有些数据表明,在年龄较小时猪对人类的恐惧可能对后期繁殖产生负面影响。事实上,对人类高度恐惧的母猪在8岁时的受胎率低于同龄的更少恐惧的猪(Janczak 等,2003)。在这个试验中,大多数动物与人类的相互作用从出生时是消极的(粗暴处理)。这表明早期的行为可能有长期的影响结果。

人的处理也可能是积极的,但对仔猪的积极处理的影响研究很少。据我们所知,只有 Hemsworth 和他的合作者在这个方面进行过研究。这可能是由于难以处理非常年幼的动物,以及由于母亲的存在而导致的干扰。然而,产后第一天/几周被认为是一个有利于发展具有长期影响的社会联系的敏感时期(Bateson,1979)。事实上,在没有这种接触的情况下,在出生的前3~8周期间,给仔猪提供抚摸和爱抚,与没有这些接触的仔猪相比,更能引起仔猪对人的积极效应(Hemsworth 和 Barnett,1992;Hemsworth 等,1986b)。10~18 周龄,处理的仔猪比非处理动物更快地接近人类,与人类相互作用更多。此外,积极处理的公猪与未处理的公猪相比表现出更多的性行为,射精时间长6~7个月。

12.2.3 施用于仔猪的疼痛操作方式对母猪的影响

在农场员工对仔猪实施疼痛(非福利)操作时,动物通常表现出防御性行为和喊叫(Hay 等,2003;Marchant-Forde 等,2009;Noonan 等,1994;Prunier 等,2005;Sutherland 等,2008;Torrey 等,2009;White 等,1995)。这些喊叫可以具有特定的特征,如手术去势后所表现出的那种叫声(Puppe 等,2005),仔猪向母猪发出了信号,因此引起它的反应(Weary 和 Fraser,1995)。由于这些负面行为都是由于人的操作,它们可以促成母猪对人类的厌恶反应。在短期,母猪会因猪场员工对仔猪的处理产生侵略性反应。长期来看,即使缺少科学数据来证实这一假设,母猪也可能对人类变得更加不安、怀疑和侵略。

12.3 后备母猪和经产母猪的处理

繁殖母猪目前正在接受人类的各种处理:注射疫苗或药物治疗,从一个猪舍或生产单元转移到另一个,查情,人工授精和分娩期间的仔猪接生等。这些手段是消极的,因为它们诱导疼痛,甚至强度低的时候,也可能增加心理压力,例如,当需要限制在猪舍的角落(大群饲养的动物接受注射)时。疼痛必然与一些处理方法相关联,例如注射疫苗接种,而其他处理,如转移猪,可以是无影响的、厌恶的或甚至是积极的,取决于猪场员工如何执行。因此,在本章中,我们将分别评估从一个猪舍/单元转移到另一个,以及其他管理实践的影响。我们还将评估可以通过改善人畜关系的具体目标进行的积极处理的影响。

12.3.1 后备母猪和经产母猪转群

在断奶后转移到产床或配种舍是母猪生命中多次发生的常规管理程序。母猪对转群情况的感觉取决于现实条件。实际上,这种转移猪可以分组或单独进行、粗鲁地或平静地进行、短距离或长距离。以组来转猪具有防止动物免受因分离造成应激的优点。猪场人员可以在转移过程中使用许多工具,以便转群过程的顺利进行。一些工具,如电子刺棒,猪对此非常厌恶(Gonyou 等,1986;Hemsworth 和 Barnett,1991;Hemsworth 等,1996a)。Fowler(2008)提出使用不那么厌恶的棍棒或赶猪挡板。猪场人员也可以简单地用自己的身体引导动物:他们可以自己站在动物身后,大声说话,用手推它们。这些方法的有效性并不明确,其使用取决于猪场员工的习惯。然而,猪显示对人的姿势行为特别敏感(Hemsworth 等,1986c;Miura 等,1996;Nawroth 等,2014),因此人的位置和姿势可能影响转猪的容易性。猪对人的声音的反应尚不清楚,但猪对高调或大的声音有应激(Eguchi 等,2007;Weeks,2008),因此大声或大喊可能会对转猪产生负面影响。猪对人类触觉也很敏感(Hemsworth 和 Coleman,2011),并且可能受到被赶猪挡板或用手推的影响。这些接触可能不利于转移猪只,因为它显示出转移速度与猪场员工推的次数呈负相关(Lensink 等,2009a)。然而,推的后果是造成更大的事故,而不是转猪困难的原因。除了人的行为,转猪路线的设计也可能会影响转移的容易性(Grandin,2010)。

据我们所知,在猪舍常规转移期间对应激水平下猪的生理和行为测量的数据不可用。然而,众所周知,使用电子刺棒(Gonyou 等,1986;Hemsworth 和 Barnett,1991;Hemsworth 等,1996a)增加了皮质醇的释放,并且任何粗鲁的处理方式(喊、踢等)均可能使动物产生应激。

12.3.2 其他管理工作的影响

母猪被重复进行各处理。这些人-动物的相互作用可能是使动物感到厌恶的,有两个原因:它们本身是痛苦的(例如用针注射)或者处理伴随着呼喊、糟杂声、击打、踢。当这些相互作用本身并不痛苦时,例如查情或者用超声进行妊娠诊断时,它们可能根据猪场员工的行为而被认为无影响或积极。

当母猪在户外时,戴上鼻环是很常见的,以防止母猪破坏牧场。这是一个有效的方法,但这种方法却减少了动物的福利(Horrell 等,2001)。鼻子打环的过程很可能是痛苦的，动物会再次将人类的处理与疼痛联系起来，但缺少科学数据来证实这一假设。

母猪通常通过注射接种疫苗，这可能是令它们厌恶的。实际上，在后备母猪或哺乳期母猪的肘部皮下注射盐水显示增加血液中皮质醇浓度，显示了一个应激反应（Robert 等，1989）。注射程序也可能增加对人的恐惧。事实上，对生长猪日常肌肉注射盐水 3 周后，与没有接受任何注射的猪相比对人的躲避更多（Hemsworth 等，1996a）。例如，与对照猪相比，经常被注射的猪，人与猪安全距离增加，头部朝向人次数减少，与人相互作用的时间增加。

在当前农场，大多数母猪繁殖都是通过人工授精。广泛使用两种技术，即子宫颈或子宫内输精法，超过子宫颈几厘米的深部输精法。在这两种情况下，将导管固定在子宫颈中，其中增厚的头部是柔性材料。应该轻轻地完成，以避免受伤。然而，在拔管或回流精液中导管尖端上存在血液是常见的（Sbardella 等，2014）。在子宫内授精的情况下尤其如此，显示用这种技术授精的母猪超过 20% 的存在流血现象。据我们所知，没有评估与导管进入相关的创伤是否疼痛，但可以怀疑它引起疼痛。

12.3.3　积极处理的影响

许多饲养员通过给予它们额外的接触和在母猪群中花费时间，付出特别的努力使后备母猪适应当前的环境。他们希望与母猪建立友好关系，使其更容易处理，分娩期间对人的躲避反应较少，从而促进对仔猪的观察和照顾。母猪是群居的。例如，Dellmeier 和 Friend(1991)提出，人类模拟母猪的社会问候（即母猪与母猪的问候）可以减少攻击性的反应。这种被动的声音交流，可使母猪对人产生接触兴趣，而不是敌意，这将有助于母猪对人产生积极接触行为。

在妊娠的最后几天，包括柔和的触觉、视觉和听觉的积极处理可以有效地迅速增加母猪对处理者的积极配合。身体的一些部位比其他部位更敏感，即使这还没有经过科学测试。当饲喂母猪时，采取这种积极的处理将更加有效。Dellmeier 和 Friend(1991)以大量饲养的母猪为例，声称母猪可以轻而易举地通过抚摸和轻挠或者轻拍进行训练。事实证明，每天接触 1 min，7 天之后足以使接近处理者的母猪比例从 33% 提高到 83%（English 等，1999）。随着这种积极的接触，后备母猪的接近人的行为反应也迅速增加，但水平仍然较低（59%），低于更有经验的母猪（English 等，1999）。妊娠后期的 10 天内，除了 1 min 的触觉接触外，还可提供饲料奖励（坚果），以提高母猪对人的信心评分（Andersen 等，2006）。

积极性处理对母亲行为的影响尚不清楚。但研究表明，通常对哺乳行为、姿势变化或仔猪死亡率没有影响（Andersen 等，2006；Enlighs 等，1999）。然而，Enlish 等(1999)发现，母猪的积极处理减少了出生间隔时间，而这取决于胎次。Andersen 等(2006)在最新一项研究中发现，积极的操作处理在妊娠结束时对母猪产程没有影响。然而,与对照组的母猪(母猪没有额外的积极接触),在妊娠结束时具有较高自信评分的母猪往往具有较短的产程。在积极处理的母猪组中,观察到类似的差异,但是较不明显,可能是由于积极的处理过程弥补了母猪的自信不足的现象。因此,人的积极行为(自发或大量的积极处理后)可能有助于分娩过程。

12.3.4 不同胎次母猪对处理反应的差别

可能由于母猪对外界事物有个学习过程,因此,胎次可能影响母猪对人的处理的反应。然而,相关研究较少,关于胎次的影响没有普遍的规则。在某些情况下,高胎次的母猪对人类的反应较少。例如,对于人类的恐惧似乎随胎次而下降,后备母猪相对第三胎次的母猪对人类的友好接触较少(即表示对人类的接触意愿较小)(English 等,1999)。初生母猪对于人对其仔猪的处理(Held 等,2006)或反复播放的由人类处理的仔猪尖叫声音(Hutson 等,1992)也比经产母猪反应更快。它们可能会了解到,这种处理情况没有任何负面后果,从而随着时间的推移反应变得更少。研究证明,这些学习过程也显示在 90 kg 猪中,训练要转移的猪减少了应激反应并提高了转移猪过程的速度(Lewis 等,2008)。

研究发现,在其他情况下,表现为胎次越高,躲避反应越强,转入产房更快和更容易(Lensink 等,2009a)。更高的躲避反应表明,更老的母猪对人类的恐惧更大。更高的转移速度也可能与更大的恐惧有关,因为害怕的动物可能会更快地移动,以逃避身后的猪场员工。这一现象得到了转移速度与躲避反应之间的正相关的支持(Lensink 等,2009a)。观察到障碍数量与躲避反应之间的负相关性同样支持在高胎次母猪中观察到的较低数量的障碍数是由于更加恐惧的表现。Hemsworth 等(1999)的研究结果也发现,更高胎次的母猪对人类躲避反应的增加,但与 Grandinson 等(2003)和 Vangen 等(2005)结果相反。在某一特定年龄,胎次对恐惧的积极或消极影响可能取决于到那个年龄的相互作用的结果。如果人类与对猪的消极和积极的相互作用之间的平衡明显偏向消极的相互作用,那么猪对人类的恐惧越来越大。

12.4 人与动物相互作用对繁殖性能的影响

普遍认为,尽管不同猪场具有相似的猪遗传背景、类似的圈舍以及农场管理(例如饲喂,分批次分娩等),但一些猪场员工能够实现比其他猪场员工使猪具有更好的繁殖性能。有些猪场虽然具有非常相似管理特征但母猪繁殖性能有所不同(Hemsworth 等,1981b;Ravel 等,1996)。这种差异通常归因于猪场员工的质量,包括技术技能和猪场员工与动物之间存在的相互作用质量(Kirkden 等,2013)。

人-动物相互作用的质量应该影响动物的恐惧程度,从而影响动物的应激水平。在繁殖方面,高度的恐惧会刺激肾上腺,从而抑制下丘脑-垂体-卵巢轴(Hem sworth,2003)。在农场水平(Coleman 等,2000;Hemsworth 等,1981a,1989,1999)或动物水平(Janczak 等,2003;Lensink 等,2009a,b)评估了恐惧程度与猪的繁殖性能之间的关系。在农场层面,观察到妊娠期间(Hemsworth 等,1981a)或哺乳期间(Hemsworth 等,1999)测量的恐惧指标与繁殖性能间存在显著关系。此外,已经证明在常规工作(例如,清洁、发情检测、饲喂)期间观察到的非常负面的相互作用(例如,用力的打击)的数量与妊娠期间的恐惧性行为指标正相关并且与繁殖性能呈负相关。然而,恐惧与较少负面的相互作用(例如"正常"的打击和踢声)之间的相关性较弱,与正相互作用(例如,拍拍或抚摸)之间的相关性非常低(Hemsworth 等,1996a)。尽管在执行的负面相互作用的百分比和母猪的恐惧程度方面取得了成功,但试图通过一些针对猪场员工的培训计划来改善人猪相互作用的质量并不能提高猪的繁殖性能(Coleman 等,2000)。同样,在一个更经典的实验研究中,妊娠的最后 11～13 天的积极处理(抚养、食物奖励和温和的谈话)并没有改变母亲在哺乳期的行为或其繁殖性能(死胎数、产活仔数和压死仔猪数;Andersen 等,2006)。

通过分析恐惧和繁殖性能之间的关系时,孕 8 周龄母猪对人类的恐惧程度与分娩时产仔间隔时间存在显著的正相关。母猪对人类的恐惧可能与其胆小或易应激的性情有关。Janczak 等(2003)还报道了母猪恐惧程度与死胎数和新生儿死亡率呈正相关关系的趋势。同样,Lensink 等(2009b)研究发现,在 6 月孕龄或妊娠后期的母猪中,在第一次分娩时,被压死的仔猪数量往往会随着对人类的恐惧程度增加而增加。一些研究还显示,在妊娠结束时,猪对人的躲避反应与早期哺乳期的压死仔猪水平之间存在显著相关性(Lensink 等,2009a)。然而,恐惧程度也随母猪胎次的增加而增加,以至于恐惧与胎次的影响无法分离。事实上,压死仔猪发生率随着胎次的增加而增加是由于增加的体重和母猪跛行引起的(Prunier 等,2014;Weary 等,1998b)。

试验研究证明,高水平的恐惧刺激肾上腺轴,反过来又抑制下丘脑-垂体-卵巢轴。人类通过不同形式对生长猪重复处理(每头猪 3～12 周,每周 3～5 天,每天 0.5～5 min),在基础情况下或对应激者或促肾上腺皮质激素(ACTH)注射的皮质醇对生长猪进行肾上腺的发育(以皮层的表面积或腺体的重量计算)以及血浆中皮质醇浓度(游离或总体)进行测量(gonyou 等,1986;Hemsworth 和 Barnett,1991;Hemsworth 等,1997;Hemsworth 等,1998;Hemsworth 等,1986;Hemsworth 和 Barnett,1991;Hemsworth 等,1981a,1986a,1987,1996a)。结果显示,当猪进行包括电击在内的恶性处理时肾上腺被长期激活。然而,当最小的相互作用与积极相互作用(由于触觉刺激如摩擦或抚摸)或负向相互作用(由于每日肌肉注射)相比时,对肾上腺轴无显著影响(Gonyou 等,1986;Hemsworth 等,1986a,1987,1996a)。在妊娠早期,令人厌恶的处理的后备猪,其游离和总的皮质醇浓度比在积极处理的后备母猪中更大(Pedersen 等,1998)。而对与人类接触最少的后备母猪中,皮质醇的浓度通常更接近于积极处理的后备母猪而不是令其厌恶处理的。然而,来自肾上腺轴的激素对生殖轴的各种成分的负面影响尚未明确(Turner 等,2005;Von Borell 等,2007)。因此,人类相互作用的质量对后备母猪和母猪繁殖性能的影响应该是在人类施加的重复的令其厌恶的相互作用不是极端的情况下进行研究。

可以认为,猪对人类的高度恐惧使许多养殖工作复杂化,并使其效率降低。例如,如果母猪的恐惧程度很高,它们会在分娩过程中对人类的接近感到紧张,这会促使分娩产程更长或压死猪发生率升高从而增加仔猪的死亡率。Berger 等(1997)发现在分娩时农场主查看母猪的比例越高,母猪群体中的死亡率越高。这可能被解释为人类存在的负面影响干扰了分娩过程。还可能是,具有较高死亡率的猪群的农场主付出更多努力以减少高死亡率,因此在分娩期间实施更多的行为以帮助母猪。如果两个可能都是真实的,那么将会建立一个负面的恶性循环。

最后,可能猪场员工对猪的行为与猪场员工的技术技能是正相关的(图 12.1)。发现这些猪场员工的行为与他们对猪的一般理念和态度有关(Hemsworth,2003)。还证明了人的特征(例如自我保证、自律或情绪)与母猪繁殖表现之间的重要关系(Ravel 等,1996)。除了这些个人特征外,工作的组织管理、住房和设备方面也会帮助或者使猪场员工对动物的工作复杂化从而影响人与动物的相互作用的质量,以及执行任务的效率(图 12.1)。因此,与猪的相互作用的质量和由猪场员工进行的工作的质量可能是高度相关的。据我们所知,目前还没有科学数据来证实这一情况。

12.5 更好地了解动物的表达信号,改善人与动物之间的相互作用

一些文献显示,人类对猪的信号敏感,并尝试解释其意义。例如,在引起痛苦的操作(例如手术去势)期间,仔猪发出的喊叫可以被解释为指示具有负面的强烈的情绪状态(Tallet 等,2010)。有趣的是,猪饲养员似乎比刚毕业的学生和更有经验的病理学家把这些喊叫解释得更加恰当。反复听到这些负面声音可能会导致一种习惯(Talling 等,1998),但是猪场员工也可能采取一种心理防御来保护它们或分离它们,如兽医学生对动物的痛苦表现的那样(Paul 和 Podbersek,2000)。

动物的肢体语言在评估其情感方面也很重要。Welmesfelder 和她的同事们开发了一种使用视频评估动物的方法,即所谓的"定性评估"方法来评估情绪(Wemelsfelder,2007)。已经证明了其评估许多物种(包括猪)情绪的效率(Wemelsfelder 等,2000,2003,2012)。实际上,使用这种方法的结果与行为的定量观察相关联,例如,动物可以被分类为紧张、极为紧张或恐惧。此外,该评估是高度

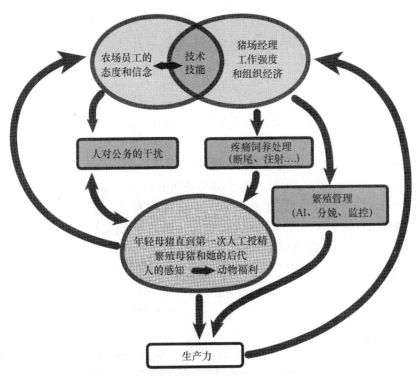

图12.1 动物、农场管理和猪场员工之间的相互作用的示意图,动物福利的可能后果,猪场员工的工作满意度和母猪的繁殖表现。

可重复的,似乎不依赖于人类与动物之前的经验(Wemelsfelder等,2012)。

因此,猪场员工可以感知和解释猪释放的信号,只要他们不会对猪产生冷漠的态度,这可以作为他们适应与动物接触的基础。如果猪场员工面对动物表达诸如恐惧或痛苦的负面状态,就应该用这种对猪行为的理解来开发实际的解决方案。目的是减少与不良福利相关的表现。这样的适应应该会使猪受益,也可以使工作人员受益。事实上,人类对处于积极状态的猪很敏感(Fiorelli等,2012;Tallet等,2010;Wemelsfelder等,2012)。通过猪场员工处理猪只,满足对猪的需求,更好的了解可以导致猪的负面反应较少,从而有助于增加工人的满意度,因为动物的行为与工作满意度之间存在直接联系(Hemsworth和Coleman,2011)。

12.6 结论

母猪和人之间的相互作用发生在母猪的整个生命周期中,并且可能影响母猪随后对人类的反应,特别是它们的恐惧反应。这些相互作用在母猪生命期间不均匀分布,而是在特殊时刻发生频繁:出生后不久,繁殖(查情,人工授精)以及分娩前后。母猪与人的相互作用的质量可能会通过不同类型的机制影响动物的繁殖表现，包括在母猪中建立习惯性应激，这对繁殖性能有害，猪场员工在有效执行养殖任务出现困难时，以及母猪的紧张可能会在分娩后不久引起压死仔猪。这些相互作用的质量不仅影响了母猪和仔猪的福利和表现，而且影响了猪场员工对自己工作的评价。再者，可能涉及各种类型的机制，包括人与动物之间情感的传递，以及猪场员工满意地来执行任务或从动物获得更好的繁殖表现。因此，提高相互作用的质量将有利于人类和动物，可以激活各种方法以改善这些相互作用。正如Hemsworth及其同事们所表明的那样，通过学习计划来提高猪场员工的态度是非常有希望的（Coleman等，2000）。改善猪场建筑和设备也应有助于更好的人－动物相互作用。在将猪从一个猪舍或生产单元转移到另一个时尤其重要（Hemsworth，2007）。要完成任务的组织工作和关于猪的管理决定也是重要的，因为它们会影响工作量，从而影响到猪场员工能够轻松工作。

参考文献

Andersen, I.L., Berg, S., Boe, K.E. and Edwards, S., 2006. Positive handling in late pregnancy and the consequences for maternal behaviour and production in sows. Applied Animal Behaviour Science 99: 64-76.

Andersen, I.L., Haukvik, I.A. and Boe, K.E., 2009. Drying and warming immediately after birth may reduce piglet mortality in loose-housed sows. Animal 3: 592-597.

Anonymous, 2010. European declaration on alternatives to surgical castration of pigs on the invitation of the European Commission and the belgium presidency. Available at: www.alternativepig.eu/partnership/declaration.html.

Bateson, P., 1979. How do sensitive periods arise and what are they for ? Animal Behaviour 27: 470-486.

Berger, F., Dagorn, J., Le Denmat, M., Quillien, J.P., Vaudelet, J.C. and Signoret, J.P., 1997. Perinatal losses in outdoor pig breeding. A survey of factors influencing piglet mortality. Annales De Zootechnie 46: 321-329.

Boivin, X., Lensink, B.J., Tallet, C. and Veissier, I., 2003. Stockmanship and farm animal welfare. Animal Welfare 12: 479-492.

Coleman, G.J., Hemsworth, P.H., Hay, M. and Cox, M., 2000. Modifying stockperson attitudes and behaviour towards pigs at a large commercial farm. Applied Animal Behaviour Science 66: 11-20.

Dellmeier, G.R. and Friend, T.H., 1991. Behaviour and extensive management of domestic sows (*Sus scrofa*) and litters. Applied Animal Behaviour Science 29: 327-341.

Durier, V., Henry, S., Sankey, C., Sizun, J. and Hausberger, M., 2012. Locomotor inhibition in adult horses faced to stressors: a single postpartum experience may be enough! Frontiers in Psychology 3: 442.

Eguchi, Y., Uetake, K. and Tanaka, T., 2007. Study of technical development using olfactory stimuli and auditory stimuli for behavior control of wild boar. Journal of Azabu University 15/16: 190-193.

English, P.R., Grant, S.A., McPherson, O. and Edwards, S.A., 1999. Evaluation of the effects of the positive 'befriending' of sows and gilts ('pleasant' treatment) prior to parturition and in early lactation on sow behaviour, the process of parturition and piglet survival. In: Russel, A.J.F., Morgan, C.A., Savory, C.J., Appleby, M.C. and Lawrence, T.L.J. (eds.) Farm animal welfare – who writes the rules? Proceedings of an international symposium organized by The British Society of Animal Science, Edinburgh, UK, 1999. Occasional Publication – British Society of Animal Science, pp. 132-136.

Fiorelli, C., Mouret, S. and Porcher, J., 2012. Rationalities for working with animals: producing, living together and self-fulfilling. INRA Productions Animales 25: 181-192.

Fowler, M., 2008. Restraint and handling of wild and domestic animals. Third edition. Wiley-Blackwell, Hoboken, NJ, USA, 470 pp.

Fredriksen, B., Font i Furnols, M., Lundström, K., Migdal, W., Prunier, A., Tuyttens, F.A.M. and Bonneau, M., 2009. Practice on castration of piglets in Europe. Animal 3: 1480-1487.

Gonyou, H.W., Hemsworth, P.H. and Barnett, J.L., 1986. Effects of frequent interactions with humans on growing pigs. Applied Animal Behaviour Science 16: 269-278.

Grandin, T., 2010. How to improve livestock handling and reduce stress. In: Grandin, T. (ed.) Improving animal welfare – a practical approach. Cambridge University Press, Cambridge, UK, pp. 64-87.

Grandinson, K., Rydhmer, L., Strandberg, E. and Thodberg, K., 2003. Genetic analysis of on-farm tests of maternal behaviour in sows. Livestock Production Science 83: 141-151.

Hay, M., Vulin, A., Génin, S., Sales, P. and Prunier, A., 2003. Assessment of pain induced by castration in piglets: behavioral and physiological responses over the subsequent 5 days. Applied Animal Behaviour Science 82: 201-218.

Held, S., Mason, G. and Mendl, M., 2006. Maternal responsiveness of outdoor sows from first to fourth parities. Applied Animal Behaviour Science 98: 216-233.

Hemsworth, P.H., 2003. Human-animal interactions in livestock production. Applied Animal Behaviour Science 81: 185-198.

Hemsworth, P.H., 2007. Behavioural principles of pig handling. In: Grandin, T. (ed.) Livestock handling and transport. CABI, Wallingford, UK, pp. 214-227.

Hemsworth, P.H. and Barnett, J.L., 1991. The effects of aversively handling pigs, either individually or in groups, on their behaviour, growth and corticosteroids. Applied Animal Behaviour Science 30: 61-72.

Hemsworth, P.H. and Barnett, J.L., 1992. The effects of early contact with humans on the subsequent level of fear of humans in pigs. Applied Animal Behaviour Science 35: 83-90.

Hemsworth, P.H., Barnett, J.L. and Campbell, R.G., 1996a. A study of the relative aversiveness of a new daily injection procedure for pigs. Applied Animal Behaviour Science 49: 389-401.

Hemsworth, P.H., Barnett, J.L. and Hansen, C., 1981a. The influence of handling by humans on the behavior, growth and corticosteroids in the juvenile female pig. Hormones and Behavior 15: 396-403.

Hemsworth, P.H., Barnett, J.L. and Hansen, C., 1986a. The influence of handling by humans on the behavior, reproduction and corticosteroids of male and feamle pigs. Applied Animal Behaviour Science 15: 303-314.

Hemsworth, P.H., Barnett, J.L. and Hansen, C., 1987. The influence of inconsistent handling by humans on the behaviour, growth and corticosteroids of young pigs. Applied Animal Behaviour Science 17: 245-252.

Hemsworth, P.H., Barnett, J.L., Coleman, G.J. and Hansen, C., 1989. A study of the relationships between the attitudinal and behavioural profiles of stockpersons and the level of fear of human and reproductive performance of commercial pigs. Applied Animal Behaviour Science 23: 301-314.

Hemsworth, P.H., Barnett, J.L., Hansen, C. and Gonyou, H.W., 1986b. The influence of early contact with humans on subsequent behavioural response of pigs to humans. Applied Animal Behaviour Science 15: 55-63.

Hemsworth, P.H., Brand, A. and Willems, P., 1981b. The behavioural response of sows to the presence of human beings and its relation to productivity. Livestock Production Science 8: 67-74.

Hemsworth, P.H. and Coleman, G.J., 2011. Changing stockperson attitudes and behaviour. In: Hemsworth, P.H. and Coleman, G.J. (eds.) Human-livestock interactions: the stockperson and the productivity and welfare of intensively farmed animals. CABI, Wallingford, UK, pp. 135-152.

Hemsworth, P.H., Coleman, G.J. and Barnett, J.L., 1994. Improving the attitude and behaviour of stockpersons towards pigs and the consequences on the behaviour and reproductive performance of commercial pigs. Applied Animal Behaviour Science 39: 349-362.

Hemsworth, P.H., Gonyou, H.W. and Dziuk, P.J., 1986c. Human communication with pigs: the behavioural response of pigs to specific human signals. Applied Animal Behaviour Science 15: 45-54.

Hemsworth, P.H., Pedersen, V., Cox, M., Cronin, G.M. and Coleman, G.J., 1999. A note on the relationship between the behavioural response of lactating sows to humans and the survival of their piglets. Applied Animal Behaviour Science 65: 43-52.

Hemsworth, P.H., Price, E.O. and Borgwardt, R., 1996b. Behavioural responses of domestic pigs and cattle to humans and novel stimuli. Applied Animal Behaviour Science 50: 43-56.

Holyoake, P.K., Dial, G.D., Trigg, T. and King, V.L., 1995. Reducing pig mortality through supervision during the perinatal period. Journal of Animal Science 73: 3543-3551.

Horrell, R.I., A'Ness, P.J.A., Edwards, S.A. and Eddison, J.C., 2001. The use of nose-rings in pigs: consequences for rooting, other functional activities, and welfare. Animal Welfare 10: 3-22.

Hutson, G.D., Argent, M.F., Dickenson, L.G. and Luxford, B.G., 1992. Influence of parity and time since parturition on responsiveness of sows to a piglet distress call. Applied Animal Behaviour Science 34: 303-313.

Janczak, A.M., Pedersen, L.J., Rydhmer, L. and Bakken, M., 2003. Relation between early fear- and anxiety-related behaviour and maternal ability in sows. Applied Animal Behaviour Science 82: 121-135.

Kirkden, R.D., Broom, D.M. and Andersen, I.L., 2013. Piglet mortality: management solutions. Journal of Animal Science 91: 3361-3389.

Lensink, B.J., Leruste, H., De Bretagne, T. and Bizeray-Filoche, D., 2009a. Sow behaviour towards humans during standard management procedures and their relationship to piglet survival. Applied Animal Behaviour Science 119: 151-157.

Lensink, B.J., Leruste, H., Le Roux, T. and Bizeray-Filoche, D., 2009b. Relationship between the behaviour of sows at 6 months old and the behaviour and performance at farrowing. Animal 3: 128-134.

Lewis, C.R.G., Hulbert, L.E. and McGlone, J.J., 2008. Novelty causes elevated heart rate and immune changes in pigs exposed to handling, alleys, and ramps. Livestock Science 116: 338-341.

Marchant-Forde, J.N., Lay, Jr., D.C., McMunn, K.A., Cheng, H.W., Pajor, E.A. and Marchant-Forde, R.M., 2009. Postnatal piglet husbandry practices and well-being: the effects of alternative techniques delivered separately. Journal of Animal Science 87: 1479-1492.

Miura, A., Tanida, H., Tanaka, T. and Yoshimoto, T., 1996. The influence of human posture and movement on the approach and escape behaviour of weanling pigs. Applied Animal Behaviour Science 49: 247-256.

Nawroth, C., Ebersbach, M. and Von Borell, E., 2014. Juvenile domestic pigs (*Sus scrofa domestica*) use human-given cues in an object choice task. Animal Cognition 17: 701-713.

Noonan, G.J., Rand, J.S., Priest, J., Ainscow, J. and Blackshaw, J.K., 1994. Behavioural observations of piglets undergoing tail docking, teeth clipping and ear notching. Applied Animal Behaviour Science 39: 203-213.

Paul, E.S. and Podberscek, A.L., 2000. Veterinary education and student's attitude towards animal welfare. Veterinary Records 149: 269-272.

Pedersen, V., Barnett, J.L., Hemsworth, P.H., Newman, E.A. and Schirmer, B., 1998. The effects of handling on behavioural and physiological responses to housing in tether-stalls among pregnant pigs. Animal Welfare 7: 137-150.

Porcher, J., 2011. The relationship between workers and animals in the pork industry: a shared suffering. Journal of Agricultural and Environmental Ethics 24: 3-17.

Prunier, A., Bonneau, M., Von Borell, E.H., Cinotti, S., Gunn, M., Frediksen, B., Giersing, M., Morton, D.B., Tuyttens, F.A.M. and Velarde, A., 2006. A review of the welfare consequences of surgical castration in piglets and the evaluation of non-surgical methods. Animal Welfare 15: 277-289.

Prunier, A., Lubac, S., Mejer, H., Roepstorff, A. and Edwards, S.A., 2014. Health, welfare and production problems in organic suckling piglets. Organic Agriculture 4: 107-121.

Prunier, A., Mounier, A.M. and Hay, M., 2005. Effects of castration, tooth resection, or tail docking on plasma metabolites and stress hormones in young pigs. Journal of Animal Science 83: 216-222.

Puppe, B., Schon, P.C., Tuchscherer, A. and Manteuffel, G., 2005. Castration-induced vocalisation in domestic piglets, *Sus scrofa*: complex and specific alterations of the vocal quality. Applied Animal Behaviour Science 95: 67-78.

Ravel, A., D'Allaire, S. and Bigras-Poulin, M., 1996. Influence of management, housing and personality of the stockperson on preweaning performances on independent and integrated swine farms in Quebec. Preventive Veterinary Medicine 29: 37-57.

Robert, S., De Passillé, A.M.B., St-Pierre, N., Dubreuil, P., Pelletier, G., Petitclerc, D. and Brazeau, P., 1989. Effect of the stress of injections on the serum concentration of cortisol, prolactin, and growth-hormone in gilts and lactating sows. Canadian Journal of Animal Science 69: 663-672.

Roguet, C., Renaud, H. and Duflot, B., 2011. Productivité du travail en elevage porcin: comparaison européenne et facteurs de variation. Journees de la Recherche Porcine en France 43: 251-252.

Rushen, J., Taylor, A.A. and De Passillé, A.M., 1999. Domestic animals' fear of humans and its effect on their welfare. Applied Animal Behaviour Science 65: 285-303.

Sbardella, P.E., Ulguim, R.R., Fontana, D.L., Ferrari, C.V., Bernardi, M.L., Wentz, I. and Bortolozzo, F.P., 2014. The post-cervical insemination does not impair the reproductive performance of primiparous sows. Reproduction in Domestic Animals 49: 59-64.

Sommavilla, R., Hoetzel, M.J. and Dalla Costa, O.A., 2011. Piglets' weaning behavioural response is influenced by quality of human-animal interactions during suckling. Animal 5: 1426-1431.

Sutherland, M.A., Bryer, P.J., Krebs, N. and McGlone, J.J., 2008. Tail docking in pigs: acute physiological and behavioural responses. Animal 2: 292-297.

Tallet, C., Spinka, M., Maruscakova, I. and Simecek, P., 2010. Human perception of vocalizations of domestic piglets and modulation by experience with domestic pigs (*Sus scrofa*). Journal of Comparative Psychology 124: 81-91.

Talling, J.C., Waran, N.K., Wathes, C.M. and Lines, J.A., 1998. Sound avoidance by domestic pigs depends upon characteristics of the signal. Applied Animal Behaviour Science 58: 255-266.

Torrey, S., Devillers, N., Lessard, M., Farmer, C. and Widowski, T., 2009. Effect of age on the behavioral and physiological responses of piglets to tail docking and ear notching. Journal of Animal Science 87: 1778-1786.

Turner, A.I., Hemsworth, P.H. and Tilbrook, A.J., 2005. Susceptibility of reproduction in female pigs to impairment by stress or elevation of cortisol. Domestic Animal Endocrinology 29: 398-410.

Vangen, O., Holm, B., Valros, A., Lund, M.S. and Rydhmer, L., 2005. Genetic variation in sows' maternal behaviour, recorded under field conditions. Livestock Production Science 93: 63-71.

Von Borell, E., Baumgartner, J., Giersing, M., Jäggin, N., Prunier, A., Tuyttens, F.A.M. and Edwards, S.A., 2009. Animal welfare implications of surgical castration and its alternatives in pigs. Animal 3: 1488-1496.

Von Borell, E., Dobson, H. and Prunier, A., 2007. Stress, behaviour and reproductive performance in female cattle and pigs. Hormones and Behavior 52: 130-138.

Weary, D.M., Braithwaite, L.A. and Fraser, D., 1998a. Vocal response to pain in piglets. Applied Animal Behaviour Science 56: 161-172.

Weary, D.M. and Fraser, D., 1995. Signalling need: costly signals and animal welfare assessment. Applied Animal Behaviour Science 44: 159-169.

Weary, D.M., Phillips, P.A., Pajor, E.A., Fraser, D. and Thompson, B.K., 1998b. Crushing of piglets by sows: effects of litter features, pen features and sow behaviour. Applied Animal Behaviour Science 61: 103-111.

Weeks, C., 2008. A review of welfare in cattle, sheep and pig lairages, with emphasis on stocking rates, ventilation and noise. Animal Welfare 17: 275-284.

Wemelsfelder, F., 2007. How animals communicate quality of life: the qualitative assessment of behaviour. Animal Welfare 16: 25-31.

Wemelsfelder, F., Batchelor, C., Jarvis, S., Farish, M. and Calvert, S., 2003. The relationship between qualitative and quantitative assessments of pig behaviour. In: Zootecniche, F.I.Z.e. (ed.) Proceedings of the 37th International Congress of the International Society for Applied Ethology, Brescia, Italy, p. 42.

Wemelsfelder, F., Hunter, A.E., Paul, E.S. and Lawrence, A.B., 2012. Assessing pig body language: agreement and consistency between pig farmers, veterinarians, and animal activists. Journal of Animal Science 90: 3652-3665.

Wemelsfelder, F., Hunter, E.A., Mendl, M.T. and Lawrence, A.B., 2000. The spontaneous qualitative assessment of behavioural expressions in pigs: first explorations of a novel methodology for integrative animal welfare measurement. Applied Animal Behaviour Science 67: 193-215.

White, R.G., Deshazer, J.A., Tressler, C.J., Borcher, G.M., Davey, S., Waninge, A., Parkhurst, A.M., Milanuk, M.J. and Clemens, E.T., 1995. Vocalization and physiological response of pigs during castration with or without a local anesthetic. Journal of Animal Science 73: 381-386.

13. 哺乳行为

M. Špinka* and G. Illmann

Institute of Animal Science, Department of Ethology, Pratelstvi 815, 10400 Prague Uhrineves, Czech Republic; spinka. marek@vuzv. cz

摘要：家猪在初乳阶段哺乳和后期哺乳之间的哺乳行为有所不同。在初乳生产阶段,母猪被动地暴露她的乳房,从而使新生仔猪有机会通过从不同的奶头自由地吮吸初乳。猪的哺乳行为特别复杂,因为具体的特征包括哺乳同步性,非营养性哺乳,寄养的哺乳潜力,哺乳开始和终止的个体发育变化以及母猪—窝仔猪与仔猪—仔猪合作与冲突之间的平衡。听觉、嗅觉和触觉交流在哺乳相互作用中发挥关键作用。母乳生产和放乳的生理控制机制决定了哺乳行为的许多方面,但另一方面,哺乳行为对哺乳母猪的生理机能,包括激素水平有很强的反馈。哺乳行为的功能,基于进化理论启发的问题"行为对动物的生存/繁殖带来哪些利益?"可为哺乳行为的不同特征之间的联系提供重要的见解。

关键词：哺乳,乳汁摄入,发声,母性行为,断奶

13.1 引言

家猪的哺乳行为是哺乳动物描述的最复杂的哺乳相互作用(Fraser,1980; Špinka 等,2002)。事实上,猪的哺乳相互作用包含了几个独特的特征,包括高频率哺乳、非常简单的放乳、常规的非营养性哺乳以及母猪和其仔猪之间丰富的触觉和声音交流。本章总结了哺乳期间母猪与仔猪之间相互作用的机制和功能,重点介绍哺乳期间哺乳行为发生的变化。

在分娩后的第一天发生的哺乳相互作用在行为和生理上不同于后期的哺乳行为。早期哺乳相互作用也具有提供初乳的特定功能,开始在仔猪之间建立乳头秩序的过程,以及确保个体仔猪的存活与否。因此,这个短而有活力的早期阶段在本章的第一节分开对待。第二部分致力于在第一个产后日结束时,直到第 3～6 周哺乳期(人为强制)结束哺乳期间,现代饲养的母猪保持全面的建立哺乳模式。然而,这种哺乳模式从第二周开始逐渐变化,因为母猪倾向于将哺乳投入限制在窝仔中,这种"逐渐断奶"的过程是本章第三部分的主题。

13.2 分娩后哺乳

13.2.1 临产时母猪的活动变化

母猪在分娩时和产后早期不是特别活跃。在明显的产前筑巢活动之后(Jensen,1986；Thodberg 等,1999),随着分娩临近,母猪平静下来。在分娩期间,母猪通常在侧卧位不动,相当安静(Jarvis 等,1999；Jensen,1986；Thodberg 等,1999),尽管在分娩的初始阶段,姿势变化可能比在后期阶段更频繁(Jarvis 等,1999；Thodberg 等,1999)。通过持续的侧卧,母猪将乳房暴露给仔猪以进行"奶头尝试",如下所述。有时候,母猪就会通过鼻子对前面刚出生的几头仔猪增加闻的频率(Jensen,1986)。产后 10 h 后,这种鼻子和鼻子接触的母体反应性活动增加(Pedersen 等,2003)。在这样的接触过程中,当仔猪接近她的鼻子时,母猪通过嗅觉探索来响应(Illmann 等,2001；Jensen,1988)。

13.2.2 早期初乳可用性和放乳

在分娩期间,初乳(早期奶水富含能量和免疫球蛋白)几乎连续利用,它是由浓度高的催产素刺激产生(Devillers 等,2007；Farmer 和 Quesnel,2009)。并且,在分娩期间,初乳已单独排乳(Castrén 等,1993；dePassillé 和 Rushen,1989；Fraser 和 Rushen,1992；Rushen 和 Fraser,1989)。逐渐地,由初始释放很少或没有释放的周期,逐渐变为越来越多的初乳在由间歇的催产素刺激引起的周期性排放中变得可用。出生后 5 h,这些奶水的排放是短暂和频繁的,发生时间不等,每小时约 3 次(Castrénet 等,1989b；Fraser,1984)。然后,它们的频率下降,且分娩开始后 8~11 h,奶水只有在以每小时约 1.5 次的稳定速率发生排乳时才能使用。这些放乳是自发发生,不需要仔猪按摩。产后 12~16 h,仔猪开始加进乳头按摩(用鼻子按摩乳头附近)进入吮吸乳汁行为,并且放乳行为逐渐依赖于仔猪对乳房的触觉刺激。

仔猪奶头尝试逐渐转换到同步吮乳

新生仔猪一般在出生后一两分钟内通过与躺卧母猪的垂直面接触,寻找到乳房而吃到乳汁,一旦仔猪吮吸到乳头,它会沿着乳房尝试向不同的乳头移动,咬或推同窝仔猪,以获得其他奶头(dePassillé 和 Rushen,1989)。这种奶头尝试行为持续约产后 8 h(dePassillé 等,1988a；Graves,1984)。在此期间,每头仔猪平均吸食 7 个不同的乳头,而对先前的乳头没有偏好。在前 8 h 吸乳更多的仔猪具有较大的免疫球蛋白水平(dePassilléet 等,1988b)。在奶头尝试期间,仔猪具有各自的吮乳节奏,因为它们在出生后 2~3 h 活跃,然后睡着,并且稍后恢复活动(Castrén 等,1989b)。个体仔猪的奶头尝试在前 12 h 逐渐转变为同步吸吮乳汁(dePassillé 和 Rushen,1989)。在分娩期间,仔猪吸吮乳汁的数量和体重增加方面有一定程度正相关(dePassillé 和 Rushen,1989；Fraser 和 Rushen,1992；Rushen 和 Fraser,1989)。然而,这些早期的吮乳发生时间较短,并且主要是少数仔猪。产后 10 h,乳汁的获得主要发生在周期性吮乳期间；然而，超过50%的仔猪没有这些吮乳行为（dePassillé 和 Rushen,1989）。在产后第一天,85%的仔猪都有了这种吮乳行为，在第三天时,平均每个吮乳行为和100%吮乳达到了同步。

13.2.3　母猪早期哺乳声音

产后母猪躺在侧面哺乳位置时会发出深层有节奏的哼叫声。发出这种哼叫声最高峰从分娩开始,在产后5 h内,峰值变得越来越频繁,与放乳行为有关。然而,当这种哼叫声频率低时,仔猪也会吮吸奶头（Castrénet 等,1989b）,所以在这一时期,看起来这种哼叫声可能主要吸引仔猪到乳房处,而不是排乳信号。由于哺乳周期特征在前12 h已经确立,独特的哼叫声高峰与仔猪的同步吮乳相关,这种哼叫声主要发生在哺乳行为期间。这种行为模式在产后第二天就已经确立（Castrén 等,1989b；dePassillé 和 Rushen,1989）。

仔猪乳头竞争

仔猪对乳头的竞争是初生仔猪行为的主要方面。仔猪是早熟的,能够在出生后几分钟内进行积极的竞争（dePassillé 等,1988a；Rushen 和 Fraser,1989）。初生仔猪甚至具有特异性的定向排列的犬牙和门牙以进行撕咬竞争（Fraser 和 Thompson,1991）。早出生的仔猪比晚出生的同窝仔猪能够尝试并赢得更多的乳头（dePassillé 和 Rushen,1989）。虽然出生体重小,出生晚以及只吸乳后乳头的仔猪倾向于获得较少的初乳,但差异很小,大多数仔猪获得与正常血清免疫球蛋白浓度一致的初乳量（Fraser 和 Rushen,1992）。出生顺序对仔猪获得的免疫球蛋白G水平几乎没有影响,除非产程超过4~5 h,否则晚产仔猪可能处于危险之中（Devillers 等,2011）。在奶头尝试期间,随着尝试乳头的变化,仔猪在每一个具有功能的乳头上互相争斗。争斗的最高频率发生在产后3 h。在这些奶头尝试争斗中,嘴里有奶头的仔猪赢得了2/3的争斗（dePassillé 和 Rushen,1989）。随着不同仔猪逐渐限制它们吸吮奶头的数量,这个"家庭法院"的优势发展成持久的乳头所有权。到第一天,5%~50%的仔猪建立了乳头固定行为,而其他仔猪在乳头变换之前与奶头的所有权斗争（dePassilléet 等,1988a；Puppe 和 Tuchscherer,1999）。随着争斗减少,乳头秩序稳定性增加,直到第四天达到85%~95%的水平。乳头所有权意味着仔猪在连续的哺乳期间固定吮吸同一个乳头,并且激烈地保护其免受同胞仔猪的占领（dePassillé 等,1988a；Puppe and Tuchscherer,1999）。

未能较早获得乳头的仔猪要付双重成本；它们耗费力量,试图夺取其他仔猪的乳头,它们经常获取较少的初乳,而初乳是作为最初的能源和免疫力的重要因素（Farmer 和 Quesnel,2009）。竞争并不是普遍无法健康成长及导致饥饿死亡的原因（Andersen 等,2011；Drake 等,2008）。窝产仔数在决定乳头竞争后果的严重性方面很重要。虽然不清楚兄弟姐妹之间的争斗性是否随着窝仔数量的增加而增加（D'Eath 和 Lawrence,2004；dePassillé 和 Rushen,1989；Scheel 等,1977）,但在大窝中,竞争较为严重（Andersen 等,2011）。大窝的仔猪具有较低的出生重,每增加一头小猪出生,每头仔猪摄取的初乳量降低约 10%,因为总初乳产量在不同窝之间是不变的(Devillers 等,2007)。仔猪的出生重是另一个重要因素,因为较重的仔猪在奶头上获得更多的竞争力获胜(Scheel 等,1977),增加更多的重量(Milligan 等,2001),具有较低的死亡率(Tuchscherer 等,2000)。由于压死或饥饿引起的仔猪的死亡率主要发生在产后的前几天内,主要影响大窝中的低体重仔猪。这通常与无法确保接触到乳头以及得不到足够的初乳和早期奶水的摄入等因素有关。

13.3 正常哺乳期间的哺乳行为

本章的第二部分描述了在初乳期之后,即从循环哺乳建立到断奶时的母猪哺乳行为的特征和结构。

13.3.1 典型的猪哺乳行为构成

猪哺乳通常由五个主要阶段组成(图 13.1):哺乳开始、排乳前乳头按摩、排乳、排乳后乳头按摩和哺乳终止(Fraser,1980;Schön 等,1999;Špinkaet 等,2002)。

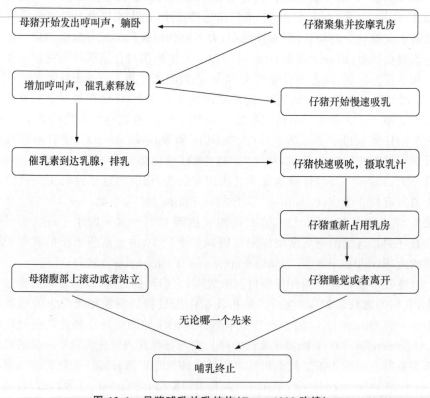

图 13.1　母猪哺乳放乳结构(Faser,1998 改编)。

哺乳开始有两种方式。第一，母猪可以自发地侧卧位躺卧在哺乳位置，从而广泛暴露其乳房，并有节奏地开始发出哼叫声。在这些刺激的反应中，仔猪在乳房处聚集并开始按摩奶头。第二，个体仔猪可以接近母猪，在母猪嘴附近发出声音，并开始用鼻子按摩奶头。这种行为吸引更多的仔猪加入刺激，直到母猪在哺乳位置躺好并开始发出哼叫声。接下来的阶段是预按摩，在此期间，所有或大多数仔猪强烈地摩擦"它们"各自的奶头附近，偶尔吸吮它们。一些仔猪可以通过接近母猪的鼻子而中断按摩，在那里它们发出短频率的"吵闹声"，同时通过鼻对鼻嗅与母亲接触。

在乳头刺激1~3 min后，垂体后部（神经垂体）释放出催产素，并通过血液流到乳腺时触发了排乳行为（Ellendorff等，1982）。猪的排乳持续时间平均只有20 s。在生理上，催产素的释放使得包围腺泡的肌上皮细胞收缩，并且将乳汁排到乳管中，从那里仔猪可以通过吮吸到口中，此时所有腺体同时出现排乳。由于母猪没有乳池，所以仔猪获得乳汁必须通过乳房压力的增加以及仔猪口和舌头的积极吮吸。排乳可以通过不同的仔猪行为来识别：随着奶水开始流动，仔猪停止按摩其鼻子附近的乳头，将乳头紧紧地含在嘴里，从而通过快速用嘴吮吸自乳管的奶水。

在短暂的排乳之后，仔猪又恢复乳房按摩，尽管按摩的次数逐渐下降，但可能还要继续几分钟，这是后期按摩阶段。哺乳终止是由仔猪活动行为引起的，当它们在母猪乳房附近入睡或在其他地方休息时，或者在母猪周边玩耍或者站立时，仔猪这些行为使母猪哺乳行为停止了。

在排乳前按摩期间，母猪发出有节律的哼叫声，表现为每两秒钟发出一次咕噜声。放乳前大约20 s，哼叫声速率突然增加到每秒大约两哼叫声的一个峰值，然后减少（Illmann等，1999）。

13.3.2 哺乳频率

母猪平均每50 min哺乳一次，每次哺乳时间间隔变化较大。昼夜哺乳节奏几乎没有差异。哺乳节奏与奶水释放的生理学关系以及母猪与窝仔之间的相互作用有关。每次排乳后，有大约20 min的不应期，在此期间不可能通过奶头刺激引发下一次放乳，无论其频率程度如何。母猪体内抑制系统十分重要，因为静脉内注射催产素能够在自然排乳的1 min内可引起乳腺收缩（Ellendorff等，1982）。影响哺乳频率的另一个因素是仔猪活动-休息循环行为。在摄取奶汁和排乳按摩后，仔猪大部分变得不活跃或在乳房附近或最舒适的地方睡着，通常在几十分钟后，它们再次在乳房附近开始活动并开始一个新的哺乳行为。

哺乳发起是哺乳频率的决定性因素。在分娩后的前几天，母猪启动了80%~100%的哺乳。在哺乳的第二周，这一比例下降到50%以下，到4周龄时，母猪不到10%（Algers和Jensen，1985；Castrénet等，1989b；Jensen等，1991；Šilerová等，2006）。

随着哺乳行为的进行,哺乳期的哺乳频率逐渐变化,但在前4~6周内变化相对较小。从最初每24 h的24~26次哺乳,直到哺乳第二周,稍微增加每天有28~30次哺乳,其后缓慢下降。从第二周开始,母猪试图以几种方式限制仔猪的吮乳活动,即通过花费更多的时间趴着休息,或者当仔猪通过乳房活动要求获得乳头时,拒绝放下身体(Valros等,2002)。然而,即使在4~5周的断奶日龄,母猪仍然每天哺乳20~24次(Valros等,2002)。在迥然不同的猪舍系统和环境中,从漏缝地板栏位到限位和群养以及半限位栏位,哺乳前四周的哺乳频率似乎遵循相同的模式(Jensen和Recén,1989;Newberry和Wood-Gush,1985;Pajor等,2002;Šilerová等,2013;Wallenbeck等,2008)。然而,在母猪可以离开仔猪并进入"无仔猪"区域的饲养系统中,哺乳频率至少降低10%(Arey和Sancha,1996;Pajor等,2002)并且个体间母猪哺乳行为的差异性较大,有些母猪甚至过早断奶(Bøe,1993b)。当哺乳允许在半自然环境中无人干扰继续时,哺乳频率继续下降,直到仔猪10~16周龄时断奶(Jensen,1988;Jensen和Recén,1989;Newberry和Wood-Gush,1985)。

哺乳期母猪的哺乳频率和产奶量之间存在复杂的关系。每个母猪通常每天都会增加或减少其哺乳频率,而不会有明显的外部因素(Illmann和Madlafousek,1995;Špinka等,1997)。这些自发的日常哺乳频率变化与母猪奶产量的变化相平衡(Špinka等,1997)。也就是说,如果一头母猪逐渐增加哺乳次数,那么仔猪在第二天就会获得更多的乳汁。这是因为在上一次放乳后35 min内,乳腺几乎完全重新填充了新生产的奶水(Špinka等,1997),而进一步的奶水积累则渐渐地减慢。如果仔猪成功吸吮更加频繁,它们会得到更多次放乳,因此,可获得更大的总奶水摄入量。相反地,大于常规的间隔会导致较少量的放乳和较低的奶水摄取量。哺乳频率的降低也可能导致催乳素浓度降低和改变生长激素分泌模式(Rushen等,1993),而更频繁的哺乳可能导致胰岛素浓度改变,从而导致一个分解代谢更快的状态(Špinka等,1999)。因此,对哺乳频率的个体调整可以看作是母猪和仔猪之间的相互作用影响,以决定转移的奶量(Špinka等,2011),从而科学使用母猪身体储备。

除了每个母猪本身哺乳频率的波动之外,哺乳频率在不同母猪之间也存在一定程度的差异(Špinka等,2002;Valros等,2002),尤其是哺乳后期(Wallenbeck等,2008)。在哺乳的第一周,哺乳频率的母猪之间差异与奶水产量似乎并不相关(Špinka等,1997),但其后在哺乳期,由于个体行为特征、环境因素(运动自由)(Bøe,1993a)或基因型(梅山猪对欧洲大白猪,Sinclair等,1998),母猪的哺乳期频率有所不同(Valros等,2002),哺乳行为越频繁的母猪越能获得更大的窝增重。

13.3.3 乳头按摩

每个哺乳行为循环,从哺乳期开始,当窝仔猪在乳房聚集并开始乳头按摩或吮吸行为,直到母猪或仔猪的哺乳终止时间持续数分钟。除了短暂的放乳外,哺乳的整个持续时间的仔猪活动称为乳头按摩。在乳头按摩期间,仔猪轮流吮吸乳头以及用嘴强烈摩擦乳头周围。这种触觉刺激的强度通常在乳头周围的干净皮肤和其他地方的脏体表面之间形成鲜明对比。

哺乳期乳头按摩阶段不会调动奶水产生,问题是,这个显而易见耗时行为的功能是什么？但排乳前按摩是触发排乳所必需的。从乳头刺激开始直到母猪发出哼叫声的速率峰值(表示为排乳)的平均时间从产后第1天的约2 min降至第7天的1 min(Algers等,1990),然后保持稳定。参与预按摩的仔猪比例越大,触发放乳的时间就越短。由于预按摩伴随着母猪的大声哼叫声,似乎其功能是允许整窝仔猪聚集在乳房并且为了使所有仔猪能找到她的乳头。

排乳后按摩的功能不太清楚。排乳后乳头按摩的持续时间和强度有很大差异。在每窝水平上,排乳后按摩比排乳前按摩时间长得多,但是个别仔猪的排乳后按摩时间较少(Jensen,1988)。Algers和Jensen(1985)提出了所谓的餐厅假说,也就是,奶头按摩中可以通过仔猪调节对乳头刺激使其产生奶水。实验证据表明,在前期哺乳中体重增加较少的仔猪和整个窝仔猪进行了更长时间或更强烈的排乳后按摩(Špinka和Algers,1995),从而表明增加了哺乳后乳头按摩是由于奶水供应不足。然而,这种强化乳头按摩导致产奶量增加的证据并不能使人信服(Illmann等,1998;Jensen等,1998;Špinka and Algers,1995)。由于缺乏更好的支持替代理论,仔猪通过排乳后乳头按摩的触觉行为表明它们对母乳的需求,并且母猪根据这种信号修改奶水生产仍然被认可。

在生理上,由仔猪进行的乳头按摩次数与几种激素的释放有关,包括催乳素、生长激素抑制素、生长激素、胰高血糖素和血管活性多肽(Algers等,1991;Rushen等,1993;Špinka等,1999)。由于这些激素可以影响母性行为或改善母猪的消化效率(Algers等,1991;Uvnäs-Moberg等,1984),因此,乳头按摩显然可以有其他影响,不仅仅影响奶水产量。

在哺乳的前几天,母猪一旦很想哺乳便允许仔猪按摩乳房,平均来说,排乳后按摩持续3 min或更长时间。随着哺乳期的进展,母猪终止越来越多的哺乳次数,而到第4周时,65%～100%的哺乳行为终止(Damm等,2003;Valros等,2002)。因此,随着泌乳进展,排乳后按摩的持续时间被缩短(Valros等,2002),并且到第4周时,排乳后按摩的持续时间下降到90 s以下(Jensen和Recén,1989)。

13.3.4 非营养性哺乳

在猪的哺乳行为中,由于非营养性哺乳行为的存在,哺乳行为与奶水摄入之间的关系变得复杂。非营养性哺乳的开始方式与正常的营养哺乳相同,即仔猪刺激乳头和母猪调整哺乳位置并发出典型的哼叫声(Illmann等,1999)。然而,在非营养性哺乳期间,尽管乳头刺激,母猪并不会增加哼叫声的速率(Illmann等,1999),

妊娠和哺乳母猪

没有催产素被释放(Ellendorff 等,1982),也没有奶水被排出,因此,没有奶水摄入(Fraser,1977;Illmann 等,1999)。在行为上,非营养性哺乳可以通过缺乏哼叫声峰值来区分,甚至更可靠地通过缺少通常在吮吸奶过程中出现的仔猪快速的嘴巴移动进行判断(图 13.2)。

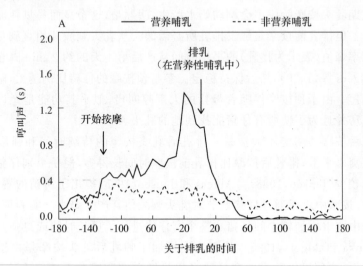

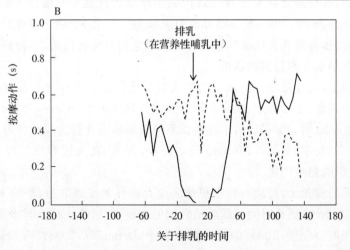

图 13.2 母猪哼叫频率(A)和仔猪进行的乳头按摩动作频率(B)在一个营养和一个非营养哺乳中(由 Illmann 等的研究改编,1999)。

非营养性哺乳并非罕见的例外,不能被认为是生理异常或病理性障碍。非营养性哺乳行为占所有哺乳事件的 5%~30%,并且发生所有类型的饲养模式中的平均频率相似,从集约化的(Fraser,1977;Illmann 和 Madlafousek,1995;Illmann 等,1999;Puppe 和 Tuchscherer,2000;Valros 等,2002;Whatson 和 Bertram,

13.哺乳行为

1980)到半开放(Castrén 等,1989a;Newberry 和 Wood-Gush,1985)以及散养型的(Horrell,1997)。一旦奶水释放的循环模式已经确立(Castrénet 等,1989a;Illmann 和 Madlafousek,1995),非营养性哺乳开始出现在产后 24 h 内,并且在整个泌乳过程中发生,也可能在第 2 周一些高峰期发生(Valros 等,2002)。

非营养性哺乳延长了哺乳期的排乳间隔时间(Castrén 等,1989a;Fraser,1977;Špinka 等,2011)。因为奶水在延长间隔的时期乳腺中供应的乳汁只是少量的增加(Špinka 等,1997),非营养性哺乳造成的结果是仔猪的奶水摄取总体下降。因此,在一些哺乳事件中没有排乳可能是因为仔猪高频率地吮吸行为引起母猪下调产奶量的策略(Špinka 等,2011)。这一假设得到了以下事实的支持:如果哺乳行为在最后一次排出乳汁后不久就又出现,而且哺乳行为是由仔猪而不是母猪发起的(Castrénet 等,1989a),则非营养性哺乳行为的可能性就会强烈增加。这种策略似乎是基于生理学的,即与上述中心不应期相关,而不是认知导向的。试验证实,在母猪中,哺乳的动机与瞬时释放奶水的能力分开的。事实上,Fraser(1975)表明,当母猪与它的窝仔猪分离时,并通过手按摩刺激出现排乳催产素反射(无需移出奶水),此后不久,母猪与它的窝仔猪重聚,母猪则愿意哺乳,但哺乳总是没有排乳,尽管它的窝仔猪是饥饿的,它的乳腺中也充满了奶水。

虽然非营养性哺乳行为没有给仔猪提供任何奶水,但它们在母猪的激素状态下仍然可以发挥作用,因为催乳素和生长激素在非营养性哺乳之后增加,尽管浓度仍然低于营养哺乳后(Rushen 等,1993)。

13.3.5 同步哺乳和寄养哺乳

当哺乳母猪大群饲养时,它们的哺乳变得高度同步。例如,在 3~4 头母猪组中,1 min 内发生的哺乳比例总计达到 81%(Maletínská 和 Špinka,2001)。声音接触足以使圈养在单体限位栏中的母猪发生同步(Šilerová 等,2013;Špinka 等,2004;)。已经研究了这种可能性,以便通过重复播放哺乳发声来使用这种敏感性并引起更频繁的哺乳(从而增加仔猪生长)。虽然一些研究发现了积极的影响(Cronin 等,2001),但大多数没有发现(Farmer 等,2004;Špinka 等,2004;Widowski 等,1984),所以该方法尚未实际应用。

为什么母猪之间存在这样一种强烈的倾向来同步它们的哺乳?一种可能性是,当哺乳母猪在小群饲养时,哺乳同步的目的是保持对异体哺乳的限制,即寄养哺乳。在自由放养和大群饲养的母猪中,仔猪通过异体哺乳永久切换到另一头母猪,或通过随机性哺乳在其他母猪上哺乳,同时留在自己的母亲定期哺乳。当哺乳紧密同步时,异体哺乳的机会减少了,因为随机性的异体哺乳仔猪不能同时吸吮自己的母亲和另一只母猪的奶水(Illmann 等,2005;Maletínská 和 Špinka,2001)。

13.4 哺乳过程中哺乳/吮乳行为的功能

哺乳行为的功能观点源于进化论。它是基于这样的假设：动物行为已经通过自然选择形成，从而使动物做出相应行为是为了获得好处如食物、安全性、信息等，从而有助于动物的生存和繁殖。将这种观点应用于驯养动物的行为并非没有争议，因为动物在其自然环境中与野生祖先通常是有很大的不同。例如，家猪与野猪不同之处在于它具有更大的窝产仔数，通常较温顺（Gustafsson 等，1999），自由采食，其身体和社会环境由人类确定和提供。然而，基本功能问题是"这种行为对动物有什么好处？"在过去几十年中对家畜的行为研究证明是有成果的（Drake 等，2008；Fraser 等，1995；Puppe 和 Tuchscherer，2000；Špinka 等，2011）。

哺乳为年轻动物提供奶水。母猪-仔猪、仔猪-仔猪和母猪-母猪的相互作用的丰富结构和活力意味着哺乳不仅仅是将母乳从母猪乳腺转移到后代的胃中。哺乳相互作用共同确定提供多少奶水，如何将这些奶水分配给个体仔猪，以及如何防止/调节对无母亲的仔猪的奶水供应。这些机制可以看作是个体行为者（即母猪和仔猪）的策略之间相互作用的结果。从母猪的角度来看，最好的策略是支持尽可能多的后代的存活和成长。然而，如果一些仔猪活力不好，或者母猪可用的资源不能支持所有后代，则可能为了有利于繁殖成功早点放弃一些后代，并能为剩余的健康年幼仔猪提供更多的奶水。此外，如果母猪在哺乳期间使用了大量的身体储备，或被体重较大的仔猪吸食，那么它未来的繁殖可能会受到影响。例如，断奶至发情间隔可能会延长，下一窝窝产仔数会减少（Bierhals 等，2012；Schenkel 等，2010；Ten Napel 等，1995）。因此，在哺乳互动中，有一些特征显示了母亲和窝仔之间的合作，以确保有效的奶水转移，但也可能会反映父母子女冲突关于要养育多少后代以及为它们投入多少的特征，以便未来的繁殖不会受到影响。

从仔猪的角度来看，与其他仔猪合作刺激母猪产奶是很好的。同时，获得更多的奶水生产是一个有利的策略，因为它将有助于加快增长。这为同窝兄弟姐妹对奶水分配创造了竞争的空间。此外，母猪的有益策略可能是将母乳平均分配给有活力的后代，因此，母猪可能通过自己的行为来抑制同窝仔猪竞争。

13.4.1 初乳生产和窝产仔数

在产后早期乳头尝试期间，哺乳相互作用的最重要功能是为仔猪提供初乳。分娩和产后的母猪通常使其乳房保持侧卧不动，从而使仔猪能够自由地对所有的乳头进行尝试。仔猪在奶头尝试期间争斗很多，一个仔猪赢得争斗的比例显著影响仔猪在产后第一天吮吸奶水，但对其吮吸奶头的数量没有影响（dePassillé 和 Rushen，1989；dePassillé 等，1988a）。因此，似乎母猪的乳房被动暴露为仔猪提供了一个公平的竞争环境，大多数仔猪可以在其中获得适当的初乳摄入量。然而，随着仔猪的数量接近或甚至超过功能性乳头的数量,母猪为所有仔猪提供可靠的初乳来源的能力可能会达到极限。影响初乳产量的因素见第8章(Quesnel等,2015)。众所周知,初乳的总产量与产仔数无关,在不同的产仔数之间是不变的(Devillers等,2007;Quesnel,2011),但每只仔猪的初乳摄入量随窝产仔数的增加而减少。由于母猪产奶量差异很大,据估计,约35%的母猪不能产生足够的初乳来充分供应所有仔猪(Quesnel等,2012)。此外,仔猪出生体重在窝内差异较大的母猪将会产生较少的初乳(Quesnel,2011)。因此,大窝中的出生体重低的仔猪处于不能获得至少200 g初乳的风险,这是足够的免疫球蛋白水平和最小生长所需的量。Andersen等(2011)提出,大窝中的新生儿死亡率是"适应性的",即这是一种面向后代质量与其质量或活力之间权衡的母猪的策略。根据这个"窝减少"假设,母亲被动的功能是允许同伴自由竞争的过程,导致丢弃最弱的后代,以便为存活的大多数仔猪提供更多的资源。虽然这种观点可能是有争议的,但它突出表明这个事实,即在出生时对大窝的激烈选择导致更高的仔猪死亡率,因此,断奶仔猪数量的净增长越来越小(Milligan等,2002)。

仔猪死亡主要发生在出生后1~3天,因此在哺乳期的其他时间里,仔猪的数量几乎是稳定的。这对奶水生产具有重要的意义,因为未吮吸的腺体在第一周内迅速回落(Kim等,2001),总产奶量受到影响(参见第4章关于乳腺发育的更多细节(Farmer和Hurley,2015))。确定的窝产仔数在很大程度上决定了接下来的产奶量,这一点由哺乳期第3周的母猪体内蛋白质的动员随着产仔数量呈线性增加来证明。在较小的窝中,一些仔猪能够吸到两个奶头(Illmann等,2007),但不知道这是否增加了奶水产量。

13.4.2 奶水生产和哺乳频率

母猪允许的哺乳次数可以直接影响产奶量。仔猪如何刺激更大的奶产量,已经提出了两种机制。第一个是在哺乳的排乳后阶段的乳头按摩。证据表明,最近几次哺乳期间的排乳后按摩的持续时间对于后面哺乳期间的奶水产量具有短期的积极影响(Jensen等,1998;Špinkaand Algers,1995)。然而,在哺乳数日甚至数周后进行更长时间的排乳后按摩不会增加乳汁产量(Špinka 和 Algers,1995;Torrey 和 Widowski,2007)。第二个和更好的效果是通过由仔猪发起更多频率的哺乳,从而增加了排乳的频率,更频繁的乳腺排空增加了奶水生产。从功能的角度来看,通过更频繁的哺乳启动增加奶水产量,来自更加饥饿的仔猪的反馈回路与鸟的父母与其雏鸟之间的供应-乞讨-供应周期类似,并且可以被理解为诚实的需求信号。但如果哺乳变得太频繁,可能引起非营养性哺乳,这可能会成为母猪对小猪哺乳的成本(Špinka等,2011)。因此,窝仔猪对奶水的需要是以哺乳起始的频率和较长时间乳头按摩的倾向来表达的。母猪可以通过允许营养哺乳和允许长时间的排乳后按摩阶段来同意它;或者母猪可以通过不释放奶水和/或通过抑制排乳按摩来抵抗需求。这种反馈循环可能有助于奶水生产的稳定性,使奶水产量被调整到由母猪负担得起仔猪需要的水平。

13.4.3 奶水分配到乳腺/仔猪

仔猪通过它们在乳房附近的行为,可以单独影响母猪产出的奶水产量。更强壮的仔猪能够从较弱的同窝兄弟姐妹那里夺取乳头,导致奶水分布非常不平等的危险。然而,无论是非常不平等地给仔猪分配奶水(Milligan 等,2002),还是通过激烈的兄弟姐妹竞争来浪费能量,这对母猪的繁殖都不是有利的。母猪的哺乳行为和生理似乎是通过自然选择来塑造兄弟姐妹之间的竞争。首先,大声哼叫声的预按摩阶段使所有的仔猪有机会在乳房上聚集并找到其乳头,并且增加哼叫声的频率以对所有仔猪宣布放乳开始。其次,所有奶头的奶水可用性的短时间同步使更强壮的仔猪也没有机会篡夺多于一个奶头的奶水生产。因此,排乳与母猪的哼叫声信号相结合确保了所有同窝的大部分仔猪的奶水摄入量免受干扰(Drake 等,2008;Fraser,1980)。

在哺乳阶段,仔猪竞争可分为三个阶段。第一阶段为乳头尝试阶段(大约前12 h),在此阶段,仔猪主要竞争吮吸的乳头数量和初乳摄取量。在分娩开始后约12 h,短暂和快速的排乳机制已经建立,仔猪早期竞争产奶乳头,而不是竞争哺乳奶头的数量。第二阶段(大约前三天),小猪努力拥有自己的奶头并且获得前期高度可靠的奶头,从更频繁的吮乳和更快的早期生长中获益(dePassillé 等,1988a)。在以往的哺乳期使用的乳头平均比以前未使用过的乳头产生更多的奶水(Farmer 等,2012)。前乳头往往更受仔猪喜欢,比后乳头产生的奶水要多一些,但这更多是由于它们被体重较大的仔猪更加一致和更加密集地使用,而不是由于固有的更高的生产力(Fraser 和 Rushen,1992;Orihuela 和 Solano,1995)。在第一次竞争阶段,以及竞争程度较低的第二次竞争阶段,竞争阶段的同窝同胞竞争强烈,有助于出生体重较低仔猪的不成比例的死亡率,尤其是在出生体重差异较大的窝中(Milligan 等,2002)。一旦固定了奶头顺序,到第四天,直接竞争就会消失,间接竞争就会发生。在此,如果通过更多次的奶水吸取行为或更强烈的奶头按摩,能够在奶头中提高奶水产量,仔猪可能会以同窝同胞为代价来获得额外的母乳。在同一窝中,仔猪的乳汁摄入量差异很大,从而体重差异也很大,但只有约20%这种差异可以通过乳头可用性和奶头位置的差异来解释(Fraser 和 Thompson,1986)。事实上,体重的内在差异系数不会从出生到断奶期间增加(Milligan 等,2002)。这表明仔猪对乳头的竞争不会加剧同窝仔猪的体重的初始差异。

13.4.4 确保母猪-仔猪关系,异体哺乳调节

虽然在大多数现代养猪生产环境中,哺乳母猪均为单独饲养,因此,不会遇到外来仔猪,但这不是猪哺乳行为进化的环境。在天然野猪群中,同时有几头母猪哺乳,因此可能进行异体哺乳(非后代哺乳)。如果几个母猪圈养在室内猪栏或户外

散养,可能会定期发生异体哺乳(Kim 等,2001;Maletínská 和 Špinka,2001;Newberry 和 Wood-Gush,1985)。从功能的角度来看,至少在某些情况下,非后代仔猪在一定程度上可能适应母亲哺乳(Roulin,2003;Roulin 和 Heeb,1999),例如,在野猪群体中异体仔猪哺乳通常是一头遗传相关的母猪的后代(Kaminski 等,2005)。然而,非后代的仔猪由于与寄养母亲的遗传相关性低,要求更加频繁的吮乳来适应母猪。因此,母猪将其自己的仔猪与外来仔猪区分开来,对于调节后者能够获取她的奶水供应是有益的。

早期母猪-仔猪的相互作用可以迅速使仔猪与母猪建立母性关系。通过嗅闻仔猪,母猪在产后 24 h 内迅速记录了个体仔猪的嗅觉身份特征。这可以在分娩期间的短暂活动期间或者当仔猪在乳头尝试或循环吮乳期间到达躺卧母猪的鼻子时,或者之后当母猪恢复哺乳期间的活动时(Maletínská 等,2002),通过个体气味识别自己的仔猪,这可以使母猪鉴别外来仔猪并且通过目标的攻击行为劝阻异体哺乳(Olsen 等,1998)。当进行仔猪的交叉寄养时,需要考虑这个过程。同样,通过母猪提早暴露它的气味,仔猪也可以了解它们母亲的嗅觉特征(Horrell 和 Hodgson,1992)。

此外,通过声道可以进行相互识别。不同母猪哼叫声的绝对频率和单个哼叫声频率组成均有所不同(Schönet 等,1999;Špinkaet 等,2002)。这些特征可以作为不同母猪的声学特征。因此,即使几头母猪由于哺乳同步而同时发出哼叫声,仔猪也能够快速找到它们的母亲。通过及时到达乳房,自己的仔猪能够击败异体吃奶仔猪并吃到奶。同一窝内的仔猪在其呼声的声学质量上具有相似性(Illmann 等,2002)和母猪使用这些线索优先照顾其自己的仔猪,例如,当仔猪与母猪分开时。

哺乳行为中的另外两个特征也可能具有降低异体哺乳发生率的作用:即哺乳同步和非营养性哺乳。哺乳同步、非营养性哺乳的发生和异体哺乳之间存在有趣的互动。大多数哺乳是同步的,在这些哺乳中,同时发生的排乳阻碍了异体哺乳。有趣的是,当哺乳不同步时,非营养性哺乳的比例增大(Illmann 等,2005),表明在非同步化哺乳期间哺乳同步和不释放乳汁的事实是哺乳母猪使用的两个补充策略,以减少异体哺乳。

13.4.5 哺乳期逐渐断奶

研究发现,广义上的逐渐减少母乳投入的断奶过程已经在出生后数天内开始,并且在哺乳期间继续(Drake 等,2008;Puppe 和 Tuchscherer,2000)。这个过程被解释为母猪-后代仔猪冲突的一个例子,也就是说后代仔猪想从母猪身上获得比母猪所能给的更多资源,对当前后代的过度投入可能会危及母猪未来的繁殖性能。哺乳期哺乳行为确实有突出的变化。在早期哺乳阶段,几乎所有的哺乳行为都是由母猪发起的,几乎没有一个由它终止。因此,母猪积极促进奶水转移,并提供给仔猪无限的机会,使其在出生后的前几天按摩奶头。然而,在哺乳的第 2 周,只有一半的哺乳是由母猪发起的,而在第 4 周,这个比例下降到接近零。同时,第 4 周,母猪终止的哺乳比例增加到50%～100%。因此,在哺乳期早期和后期几乎绝对逆转角色。然而,随着哺乳期的进行,哺乳行为其他方面的发展并不表明明显的后代-母猪冲突或断奶过程。哺乳期的第1周至第4周的哺乳频率仅下降约10%(Puppe和Tuchscherer,2000;Valros等,2002)。非营养哺乳行为是母猪产奶量与其仔猪之间冲突的迹象,非营养哺乳行为似乎于2～3周达到高峰,然后下降,而不是像预期那样随着哺乳期的进行母猪-仔猪冲突逐渐加剧时持续不断地增加。因此,营养哺乳的数量与产奶量是最为密切相关的行为测量方法,但很少记录,可能是在哺乳期前4周几乎是稳定的(Valros等,2002)。重要的是,商业种母猪的日产奶量增加直到第三周上半周,然后保持稳定至哺乳期的第30天(Hansen等,2012)。

研究证明,在允许母猪更大自由活动的圈舍系统中,逐渐断奶过程进展得更快(Drake等,2008)。实际上,哺乳期的母猪安置在"远离"的猪圈中,在那里它们可以离开仔猪并留在无猪仔区域,比正常猪圈饲养的母猪的哺乳频率低约10%(Drake等,2008;Pajor等,2002)。但是,防止仔猪接近母猪是不人道的限制。来自其他类型圈舍的数据并不能显示出单独的猪圈,群养猪圈和户外开放饲养的母猪之间的4～6周的哺乳频率具有明显差异(Newberry和Wood-Gush,1985;Šilerová等,2006)。

总之,在哺乳的前几周,哺乳开始和终止发生了突飞猛进的变化。然而,哺乳频率和哺乳期奶水产量表明,在4～6周之前,仔猪与母猪分离之前,奶水转运并没有被自然断奶过程中所减弱,并且在各种圈舍系统中没有实质的不同。

13.5 结论

猪的哺乳是复杂的,因为具体的特征包括短期的排乳、仔猪的长期乳房按摩、同步哺乳、非营养性哺乳及哺乳开始和终止的个体发生的变化。为了充分了解猪的哺乳,必须结合对生理学基础知识的理解,彻底阐明母猪和仔猪的行为,包括听觉、嗅觉和触觉的交流。基于进化理论的哺乳行为的功能观点有助于突出母猪与其后代之间的合作和冲突的潜力。

参考文献

Algers, B. and Jensen, P., 1985. Communication during suckling in the domestic pig. Effects of continuous noise. Applied Animal Behaviour Science 14: 49-61.

Algers, B., Madej, A., Rojanasthien, S. and Uvnäs-Moberg, K., 1991. Quantitative relationships between suckling-induced teat stimulation and the release of prolactin, gastrin, somatostatin, insulin, glucagon and vasoactive intestinal polypeptide in sows. Veterinary Research Communications 15: 395-407.

Algers, B., Rojanasthien, S. and Uvnäs-Moberg, K., 1990. The relationship between teat stimulation, oxytocin release and grunting rate in the sow during nursing. Applied Animal Behaviour Science 26: 267-276.

Andersen, I.L., Naevdal, E. and Bøe, K.E., 2011. Maternal investment, sibling competition, and offspring survival with increasing litter size and parity in pigs (Sus scrofa). Behavioral ecology and sociobiology 65: 1159-1167.

Bierhals, T., Magnabosco, D., Ribeiro, R.R., Perin, J., da Cruz, R.A., Bernardi, M.L., Wentz, I. and Bortolozzo, F.P., 2012. Influence of pig weight classification at cross-fostering on the performance of the primiparous sow and the adopted litter. Livestock Science 146: 115-122.

Bøe, K., 1993a. Maternal behaviour of lactating sows in a loose-housing system. Applied Animal Behaviour Science 35: 327-338.

Bøe, K., 1993b. Weaning of Pigs: behavioural strategies of the sow and piglets and the effect of age at abrupt weaning and the post-weaning environment on the behaviour of pigs. Agricultural University of Norway, Oslo, Norway.

Castrén, H., Algers, B. and Jensen, P., 1989a. Occurrence of unsuccessful sucklings in newborn piglets in a semi-natural environment. Applied Animal Behaviour Science 23: 61-73.

Castrén, H., Algers, B., de Passillé, A.M., Rushen, J. and Uvnäs-Moberg, K., 1993. Early milk ejection, prolonged parturition and periparturient oxytocin release in the pig. Animal Production 57: 465-471.

Castrén, H., Algers, B., Jensen, P. and Saloniemi, H., 1989b. Suckling behavior and milk consumption in newborn piglets as a response to sow grunting. Applied Animal Behaviour Science 24: 227-238.

Cronin, G.M., Leeson, E., Cronin, J.G. and Barnett, J.L., 2001. The effect of broadcasting sow suckling grunts in the lactation shed on piglet growth. Asian-Australasian Journal of Animal Sciences 14: 1019-1023.

Damm, B.I., Pedersen, L.J., Jessen, L.B., Thamsborg, S.M., Mejer, H. and Ersboll, A.K., 2003. The gradual weaning process in outdoor sows and piglets in relation to nematode infections. Applied Animal Behaviour Science 82: 101-120.

de Passillé, A.M.B. and Rushen, J., 1989. Suckling and teat disputes by neonatal piglets. Applied Animal Behaviour Science 22: 23-38.

de Passillé, A.M.B., Rushen, J. and Hartsock, T.G., 1988a. Ontogeny of teat fidelity in pigs and its relation to competition at suckling. Canadian Journal of Animal Science 68: 325-338.

de Passillé, A.M.B., Rushen, J. and Pelletier, G., 1988b. Sucking behavior and serum immunoglobulin levels in neonatal piglets. Animal Production 47: 447-456.

D'Eath, R.B. and Lawrence, A.B., 2004. Early life predictors of the development of aggressive behaviour in the domestic pig. Animal Behaviour 67: 501-509.

Devillers, N., Farmer, C., Le Dividich, J. and Prunier, A., 2007. Variability of colostrum yield and colostrum intake in pigs. Animal 1: 1033-1041.

Devillers, N., Le Dividich, J. and Prunier, A., 2011. Influence of colostrum intake on piglet survival and immunity. Animal 5: 1605-1612.

Drake, A., Fraser, D. and Weary, D.M., 2008. Parent-offspring resource allocation in domestic pigs. Behavioral Ecology and Sociobiology 62: 309-319.

Ellendorff, F., Forsling, M.L. and Poulain, D.A., 1982. The milk ejection reflex in the pig. Journal of Physiology 333: 577-594.

Farmer, C., Fisette, K., Robert, S., Quesnel, H. and Laforest, J.P., 2004. Use of recorded nursing grunts during lactation in two breeds of sows. II. Effects on sow performance and mammary development. Canadian Journal of Animal Science 84: 581-587.

Farmer, C. and Hurley, W.L., 2015. Mammary development. Chapter 4. In: Farmer, C. (ed.) The gestating and lactating sow. Wageningen Academic Publishers, Wageningen, the Netherlands, pp. 73-94.

Farmer, C., Palin, M.F., Theil, P.K., Sorensen, M.T. and Devillers, N., 2012. Milk production in sows from a teat in second parity is influenced by whether it was suckled in first parity. Journal of Animal Science 90: 3743-3751.

Farmer, C. and Quesnel, H., 2009. Nutritional, hormonal, and environmental effects on colostrum in sows. Journal of Animal Science 87: 56-65.

Fraser, D., 1975. The nursing and suckling behaviour of pigs. III. Behaviour when milk ejection is elicited by manual stimulation of the udder. British Veterinary Journal 131: 417-425.

Fraser, D., 1977. Some behavioural aspects of milk ejection failure by sows. British Veterinary Journal 133: 126-133.

Fraser, D., 1980. A review of the behavioural mechanism of milk ejection of the domestic pig. Applied Animal Ethology 6: 247-255.

Fraser, D., 1984. Some factors influencing the availability of colostrum to piglets. Animal Production 39: 115-123.

Fraser, D., Kramer, D.L., Pajor, E.A. and Weary, D.M., 1995. Conflict and cooperation: sociobiological principles and the behaviour of pigs. Applied Animal Behaviour Science 44: 139-157.

Fraser, D. and Rushen, J., 1992. Colostrum intake by newborn piglets. Canadian Journal of Animal Science 72: 1-13.

Fraser, D. and Thompson, B.H., 1986. Variation in piglets weights – relationship to sucklign behavior, parity number and farrowing crate design. Canadian Journal of Animal Science 66: 31-46.

Fraser, D. and Thompson, B.K., 1991. Armed sibling rivalry among suckling piglets. Behavioural Ecology and Sociobiology 29: 9-15.

Graves, H.B., 1984. Behavior and ecology of wild and feral swine (*Sus scrofa*). Journal of Animal Science 58: 482-492.

Gustafsson, M., Jensen, P., De Jonge, F., Illmann, G. and Špinka, M., 1999. Maternal behaviour of domestic sows and crosses between domestic sows and wild boar. Applied Animal Behaviour Science 65: 29-42.

Hansen, A.V., Strathe, A.B., Kebreab, E., France, J. and Theil, P.K., 2012. Predicting milk yield and composition in lactating sows: a Bayesian approach. Journal of Animal Science 90: 2285-2298.

Horrell, I., 1997. The characterisation of suckling in wild boar. Applied Animal Behaviour Science 53: 271-277.

Horrell, I. and Hodgson, J., 1992. The basis of sow-piglet identification. 2. Cues used by piglets to identity their dam and home pen. Applied Animal Behaviour Science 33: 329-343.

Illmann, G. and Madlafousek, J., 1995. Occurrence and characteristics of unsuccessful nursings in minipigs during the first week of life. Applied Animal Behaviour Science 44: 9-18.

Illmann, G., Pokorná, Z. and Špinka, M., 2005. Nursing synchronization and milk ejection failure as maternal strategies to reduce allosuckling in pair-housed sows (*Sus scrofa domestica*). Ethology 111: 652-668.

Illmann, G., Pokorná, Z. and Špinka, V., 2007. Allosuckling in domestic pigs: teat acquisition strategy and consequences. Applied Animal Behaviour Science 106: 26-38.

Illmann, G., Schrader, L., Špinka, M. and Šustr, P., 2002. Acoustical mother-offspring recognition in pigs (*Sus scrofa domestica*). Behaviour 139: 487-505.

Illmann, G., Špinka, M. and De Jonge, F., 2001. Vocalizations around the time of milk ejection in domestic piglets: a reliable indicator of their condition? Behaviour 138: 431-451.

Illmann, G., Špinka, M. and Štětková, 1998. Influence of massage during simulated non-nutritive nursing on piglets milk intake and weight gain. Applied Animal Behaviour Science 55: 279-289.

Illmann, G., Špinka, M. and Štětková, Z., 1999. Predictability of nursings without milk ejection in domestic pigs. Applied Animal Behaviour Science 61: 303-311.

Jarvis, S., Mclean, K.A., Calvert, S.K., Deans, L.A., Chirnside, J. and Lawrence, A.B., 1999. The responsiveness of sows to their piglets in relation to the length of parturition and the involvement of endogenous opioids. Applied Animal Behaviour Science 63: 195-207.

Jensen, P., 1986. Observations on the maternal behaviour of free-ranging domestic pigs. Applied Animal Behaviour Science 16: 131-142.

Jensen, P., 1988. Maternal behaviour and mother-young interactions during lactation in free-ranging domestic pigs. Applied Animal Behaviour Science 20: 297-308.

Jensen, P., Gustafsson, M. and Augustsson, H., 1998. Teat massage after milk ingestion in domestic piglets: an example of honest begging? Animal Behaviour 55: 779-786.

Jensen, P. and Recén, B., 1989. When to wean – observations from free-ranging domestic pigs. Applied Animal Behaviour Science 23: 49-60.

Jensen, P., Stangel, G. and Algers, B., 1991. Nursing and sucking behaviour of semi-naturallye kept pigs during the first 10 days post partum. Applied Animal Behaviour Science 31: 195-209.

Kaminski, G., Brandt, S., Baubet, E. and Baudoin, C., 2005. Life-history patterns in female wild boars (*Sus scrofa*): mother-daughter postweaning associations. Canadian Journal of Zoology-Revue Canadienne De Zoologie 83: 474-480.

Kim, S.W., Easter, R.A. and Hurley, W.L., 2001. The regression of unsuckled mammary glands during lactation in sows: the influence of lactation stage, dietary nutrients, and litter size. Journal of Animal Science 79: 2659-2668.

Maletínská, J., Špinka, M., Víchová, J. and Stěhulová, I., 2002. Individual recognition of piglets by sows in the early post-partum period. Behaviour 139: 975-991.

Maletínská, M. and Špinka, M., 2001. Cross-sucking and nursing synchronization in group housed lactating sows. Applied Animal Behaviour Science 75: 17-32.

Milligan, B.N., Fraser, D. and Kramer, D.L., 2001. Birth weight variation in the domestic pig: effects on offspring survival, weight gain and suckling behaviour. Applied Animal Behaviour Science 73: 179-191.

Milligan, B.N., Fraser, D. and Kramer, D.L., 2002. Within-litter birth weight variation in the domestic pig and its relation to pre-weaning survival, weight gain, and variation in weaning weights. Livestock Production Science 76: 181-191.

Newberry, R.C. and Wood-Gush, G.M., 1985. The suckling behaviour of domestic pigs in a semi-natural environment. Behaviour 95: 11-25.

Olsen, A.N.W., Dybkjaer, L. and Vestergaard, K.S., 1998. Cross-suckling and associated behaviour in piglets and sows. Applied Animal Behaviour Science 61; 13-24.

Orihuela, A. and Solano, J.J., 1995. Managing "teat order" in suckling pigs (*Sus scrofa domestica*). Applied Animal Behaviour Science 46: 125-130.

Pajor, E.A., Weary, D.M., Caceres, C., Fraser, D. and Kramer, D.L., 2002. Alternative housing for sows and litters part 3. effects of piglet diet quality and sow-controlled housing on performance and behaviour. Applied Animal Behaviour Science 76: 267-277.

Pedersen, L.J., Damm, B.I., Marchant-Forde, J.N. and Jensen, K.H., 2003. Effects of feed-back from the nest on maternal responsiveness and postural changes in primiparous sows during the first 24 h after farrowing onset. Applied Animal Behaviour Science 83: 109-124.

Puppe, B. and Tuchscherer, A., 1999. Developmental and territorial aspects of suckling behaviour in the domestic pig (*Sus scrofa f. domestica*). Journal of Zoology London 249: 307-313.

Puppe, B. and Tuchscherer, A., 2000. The development of suckling frequency in pigs from birth to weaning of their piglets: a sociobiological approach. Animal Science 71: 273-279.

Quesnel, H., 2011. Colostrum production by sows: variability of colostrum yield and immunoglobulin G concentrations. Animal 5: 1546-1553.

Quesnel, H., Farmer, C. and Devillers, N., 2012. Colostrum intake: Influence on piglet performance and factors of variation. Livestock Science 146: 105-114.

Quesnel, H., Farmer, C. and Theil, P.K., 2015. Colostrum and milk production. Chapter 8. In: Farmer, C. (ed.) The gestating and lactating sow. Wageningen Academic Publishers, Wageningen, the Netherlands, pp. 173-192.

Roulin, A., 2003. The neuroendocrine function of allosuckling. Ethology 109: 185-195.

Roulin, A. and Heeb, P., 1999. The immunological function of allosuckling. Ecology Letters 2: 319-324.

Rushen, J., Foxcroft, G. and de Passillé, A.M., 1993. Nursing-induced changes in pain sensitivity, prolactin, and somatotropin in the pig. Physiology and Behaviour 53: 265-270.

Rushen, J. and Fraser, D., 1989. Nutritive and nonnutritive sucking and the temporal organization of the suckling behavior of domestic piglets. Developmental Psychobiology 22: 789-801.

Scheel, D.E., Graves, H.B. and Sherritt, G.W., 1977. Nursing order, social dominance and growth in swine. Journal of Animal Science 45: 219-229.

Schenkel, A.C., Bernardi, M.L., Bortolozzo, F.P. and Wentz, I., 2010. Body reserve mobilization during lactation in first parity sows and its effect on second litter size. Livestock Science 132: 165-172.

Schön, P.C., Puppe, B., Gromyko, T. and Manteuffel, G., 1999. Common features and individual differences in nurse grunting of domestic pigs (*Sus scrofa*): a multi-parametric analysis. Behaviour 136: 49-66.

Šilerová, J., Špinka, M. and Neuhauserová, K., 2013. Nursing behaviour in lactating sows kept in isolation, in acoustic and visual contact. Applied Animal Behaviour Science 143: 40-45.

Šilerová, J., Špinka, M., Šárová, R., Slámová, K. and Algers, B., 2006. A note on differences in nursing behaviour on pig farms employing individual and group housing of lactating sows. Applied Animal Behaviour Science 101: 167-176.

Sinclair, A.G., Edwards, S.A., Hoste, S. and Mccartney, A., 1998. Evaluation of the influence of maternal and piglet breed differences on behaviour and production of Meishan synthetic and European White breeds during lactation. Animal Science 66: 423-430.

Špinka, M. and Algers, B., 1995. Functional view on udder massage after milk let-down in pigs. Applied Animal Behaviour Science 43: 197-212.

Špinka, M., Gonyou, H.W., Li, Y.Z.Z. and Bate, L.A., 2004. Nursing synchronisation in lactating sows as affected by activity, distance between the sows and playback of nursing vocalisations. Applied Animal Behaviour Science 88: 13-26.

Špinka, M., Illmann, G., Algers, B. and Štětková, Z., 1997. The role of nursing frequency in milk production in domestic pig. Journal of Animal Science 75: 1223-1228.

Špinka, M., Illmann, G., Haman, J., Šimeček, P. and Šilerová, J., 2011. Milk ejection solicitations and non-nutritive nursings: an honest signaling system of need in domestic pigs? Behavioral Ecology and Sociobiology 65: 1447-1457.

Špinka, M., Illmann, G., Štětková, Z., Krejčí, P., Tománek, M., Sedlák, L. and Lidický, J., 1999. Prolactin and insulin in lactating sows as dependent on nursing frequency. Domestic Animal Endocrinology 17: 53-64.

Špinka, M., Stěhulová, I., Zacharová, J., Maletínská, J. and Illmann, G., 2002. Nursing behaviour and nursing vocalisations in domestic sows: repeatability and relationship with maternal investment. Behaviour 139: 1077-1097.

Ten Napel, J., De Vrires, A.G., Buiting, G.A.J., Luiting, P., Merks, J.M.W. and Brascamp, E.W., 1995. Genetics of the interval from weaning to estrus in first-litter sows: Distribution of data, direct response of selection, and heritability. Journal of Animal Science 73: 2193-2203.

Thodberg, K., Jensen, K.H., Herskin, M.S. and Jorgensen, E., 1999. Influence of environmental stimuli on nest building and farrowing behaviour in domestic sows. Applied Animal Behaviour Science 63: 131-144.

Torrey, S. and Widowski, T.M., 2007. Relationship between growth and non-nutritive massage in suckling pigs. Applied Animal Behaviour Science 107: 32-44.

Tuchscherer, M., Puppe, B., Tuchscherer, A. and Tiemann, U., 2000. Early identification of neonates at risk: traits of newborn piglets with respect to survival. Theriogenology 54: 371-388.

Uvnäs-Moberg, K., Eriksson, M., Blomquist, L., Kunavongkrit, A. and Einarsson, S., 1984. Influence of suckling and feeding on insulin, gastrin, somatstatin and VIP levels in peripheral venous blood of lactating sows. Acta Physiologica Scandinavica 121: 31-38.

Valros, A.E., Rundgren, M., Špinka, M., Saloniemi, H., Rydhmer, L. and Algers, B., 2002. Nursing behaviour of sows during 5 weeks lactation and effects on piglet growth. Applied Animal Behaviour Science 76: 93-104.

Wallenbeck, A., Rydhmer, L. and Thodberg, K., 2008. Maternal behaviour and perfonnance in first-parity outdoor sows. Livestock Science 116: 216-222.

Whatson, T.S. and Bertram, J.M., 1980. A comparison of incomplete nursing in the sow in two environments. Animal Production 30: 105-114.

Widowski, T.M., Curtis, S.E. and McFarlane, J.M., 1984. Can piglets be induced by environmental stimuli to nurse? Journal of Animal Science 59: 149-150.

14. 乳腺血流和营养摄取

C. Farmer[1*], N.L. Trottier[2] and J.Y. Dourmad[3]

[1] Agriculture and Agri-Food Canada, Dairy and Swine R & D Centre, 2000 College St., Sherbrooke, QC, J1M 0C8, Canada; chantal.farmer@agr.gc.ca

[2] Michigan State University, Department of Animal Science, 2209 Anthony Hall, East Lansing, MI 48824, USA

[3] INRA-Agrocampus Ouest, UMR1348 PEGASE, 35590 Saint-Gilles, France

摘要:母乳是哺乳仔猪营养的主要来源,考虑到我们目前母猪基因型的大窝产仔数,必须最大程度地使乳腺营养利用达到最大化。可用于乳腺组织的营养物质的数量取决于血液中营养物质的浓度和其流向哺乳期腺体的速度。可以通过测量乳房动静脉差异来估计乳房的营养可用性,并且可以直接或通过间接计算来测量乳腺血流量。对于所有这些检测,需要进行乳腺静脉和动脉血液取样,导管必须插入乳腺静脉。流向乳房的血液受多种因素的影响,如窝产仔数、饲喂时间、姿势行为、血管活性物质和环境温度;然而,乳腺血流中最重要的效应因子是产奶。葡萄糖占母猪乳腺组织吸收的总碳量的 40%~60%,其乳腺摄取似乎由葡萄糖转运蛋白介导。乳腺使用的其他高能前体是甘油三酸酯、磷脂、乙酸酯、丙酸酯和乳酸。通过对母猪乳腺氨基酸的吸收进行的广泛的研究可知,其受到日粮、哺乳期和奶水需求的影响。最近的数据还显示,氨基酸摄取受细胞内通道氨基酸转运蛋白的控制,而这些蛋白又受日粮和生理状态的影响。乳腺摄取激素的数据是矛盾的;然而,内分泌参与乳汁合成过程的调节是通过乳腺组织中特定激素受体的存在来证明的。

关键词:日粮营养,哺乳期,乳腺,乳腺摄取,母猪

14.1 引言

母猪产奶量是哺乳仔猪生长的主要决定因素,反过来,断奶仔猪的体重对断奶后的增重有重要的影响(Klindt,2003)。然而,由于发育不全的乳腺组织,哺乳期间的自由采食量低,以及妊娠晚期和哺乳期间存在延长的分解代谢状态的原因,初产母猪不能产生最佳的产奶量(参见本书的第 4 章,Farmer 和 Hurley,2015;第 7 章,Theil,2015)。因此,必须寻找提高母猪产奶量的方法,更好地理解控制产奶所

14. 乳腺血流和营养摄取

涉及的机制，能够帮助实现这一目标。可通过增加血液中的营养物质浓度和血液流向乳腺腺体来增强可用于乳腺组织的营养成分(Renaudeau 等，2002)。因此，本章重点是乳腺血液流和猪的乳腺组织摄取营养和激素。将描述直接和间接估计血液流量的方法，并将提供与本章相关的当前知识的更新。将覆盖主要营养素如葡萄糖和氨基酸以及其他能量前体的乳腺摄取，重点是与葡萄糖和氨基酸转运蛋白有关的新发现。最后，总结了有关乳腺摄取激素的相关信息，并讨论了母猪乳腺摄取的特定激素对营养摄取的潜在影响。

14.2 血流

14.2.1 乳腺循环系统的解剖学

如图 14.1 所示，母猪乳腺的动脉、静脉和淋巴循环通过从腋窝纵向延伸到腹股沟区域的网络(腹部中线)的两侧提供。与反刍动物只有一条动脉(阴部动脉)向乳房两侧供血(Barone，1996)相比，母猪乳房的两侧有几条动脉供应(Trottier 等，1995a)。阴部外动脉通过腹股沟下延，其分为侧前、中前和内前分支供给后乳腺(Ghoshal，1975)。前乳腺主要由源自内部胸动脉的腹壁下动脉提供(Ghoshal，1975；Trottier 等，1995a)。离开乳腺的血液使用两种不同的途径(Turner，1952)；前腺体通过两个平行于乳腺系统两侧的皮下腹静脉排出，到达胸腔静脉(Trottier 等，1995a)。离开腹股沟区域腺后部的血液通过相同的皮下腹部静脉进行尾静脉排出，从而排入外部的静脉(Turner，1952)。与迄今为止研究的产多胎的种类相

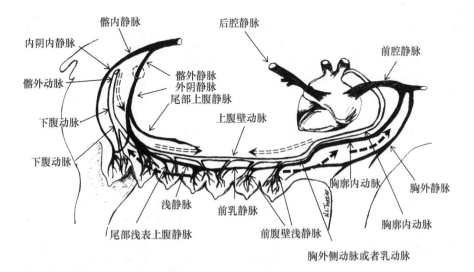

图 14.1 哺乳母猪乳腺循环系统的图解。椭圆形线代表腹股沟。单个虚线箭头表示静脉血流方向，双虚线箭头表示动脉血流方向。动脉空心表示，静脉实心表示。绘图不按实际比例。

比，由乳腺实质产生的浅表静脉和从上腹静脉分支的静脉通过静脉吻合连接在每对腺体的右侧和左侧之间(Lignereux 等,1996)。

14.2.2 乳腺血流量估计

需要乳房静脉和动脉血液采样来测量血液流动并估计乳腺的摄取量。哺乳期母猪的乳腺血流量可进行直接和间接的估计。通过间接方法获得的所有猪乳腺血流值(Guan 等,2002,2004a,b；Linzell 等,1969a；Nielsen 等,2002a,b；Trottier 等,1997)都是基于 Fick 的原始扩散原理(Fick,1870)。Kety 和 Schmidt(1945)根据 Fick 的原理,使用一氧化二氮作为惰性标记来测量脑血流量。Linzell 等(1969a)使用 Kety-Schmidt 方法(Kety 和 Schmidt,1945)和 DO 作为惰性标记来估计单个乳腺的血流量。然后将该值用于估计整个乳房的血流量(表 14.1)。在同一研究中,Linzell 等(1969a)也首次测量了营养物浓度的动静脉差异通过采集乳腺静脉和动脉血的样品通过外阴部动脉和排入单个功能性腺体的乳腺静脉导管技术(图 14.1)。在 5 h 期间测量腺体的产奶量,并在研究结束时对总乳腺组织称重,以便估计每 100 g 乳腺组织的产奶量和血流量。为了与其他研究进行比较,本章中整个乳房的产乳量和血流量是基于乳房重量为 770 g(表 14.1)。

表 14.1 根据提供最低营养需求的饲料喂养的母猪的各种研究数据估计,产生 1 kg 牛奶所需的流经整个乳腺的乳腺血浆流、产奶量和血浆量(Farmer 等,2008b)。

研究	方法	窝产仔数	血浆流 (L/min)	产奶量 (L/天)	血浆 (L/L 奶)
Guan 等(2002)	Fick (phe + tyr)[1]	12	4.9	12.1	581
Guan 等(2004a)	Fick (lysine)	11	4.5	11.4	560
Guan 等(2004b)	Fick (lysine)	11	3.6	10.7	490
Linzell 等(1969a)[2]	Fick (^{3}HO)	6	1.9[3]	6.2	441[3]
Nielsen 等(2002b)	Fick (lysine)	8	3.2	10	550
Nielsen 等(2002a)	Fick (lysine)	12	5.5	7.4	1050
Nielsen 等(2002a)	Fick (methionine)	12	4.4	7.4	858
Nielsen 等(2002a)	Fick (calcium)	12	2.8	7.4	548
Renaudeau 等(2002)	Flow probe	12	3.6[3]	11	471[3]
Renaudeau 等(2002)	Flow probe	12	3.9[4]	11	511
Trottier 等(1997)	Fick (lysine)	11	3.0	8.1	541

[1] 苯丙氨酸 + 酪氨酸。

[2] 对于进行血液流量测量的腺体的重量为 770 g,数据来源于报道的每 100 g 组织的乳产量为 43 mL/h 或 134 mL/天,每 100 g 组织的血流量为 41 mL/min。血液流过该腺体重新计算为 L/(腺体·天),其中 41 mL/(min·100 g)组织代表 316 mL/(min·腺体)或 455 L/(天·腺体)。功能腺体的数量是基于哺乳仔猪的数量,以计算通过所有功能腺体的乳腺血流量。通过将 4.62 kg 的总乳腺组织重量除以 770 g 的单个腺体重量(如上估计)来验证该值。因此,总乳房血流量估计为 2728 L/天(455 L/(天·腺体)×6 个腺体),总乳产量为 6.19 L/天(1032 mL/天×6 腺体)。估计每个腺体的产奶量为 1032 mL/天,其中每天 134 mL/100 g 组织×770 g/腺体。

[3] 数据为每升奶的血流和血流升数。

[4] 估计从报告的血液与奶的比例为 511 以及平均产奶量为 11 L/天。

当应用 Fick 的原理来测量哺乳期母猪的血液流量时,可以使用任何动脉部位进行外部惰性标记物的动脉输注和血液取样的导管放置点。然而,Fick 方法对于确定血液流量的长期变化(即每天)而言更为可行,而不是确定在短期内发生的变化(即分钟)。最近的研究(Guan 等,2002,2004a,b;Manjarín 等,2012b;Nielsen 等,2002a,b;Trottier 等,1997)应用 Fick 的原理、通过使用一个内部标记来估计乳腺血流量。内部标记的优点是不需要外部标记物的动脉输注。在这些研究中,使用赖氨酸或苯丙氨酸+酪氨酸,其主要前提是赖氨酸和苯丙氨酸被猪乳腺忽略不计地分解代谢或利用,除了它们掺入乳蛋白中的部分(Guan 等,2002,2004a)。然而,这种假设可能导致乳腺血浆流的低估。例如,Lapierre 等(2009)报道了赖氨酸可通过泌乳奶牛非乳蛋白合成途径被利用和氧化。此外,NRC(2012)最近估计,在母猪中,日粮赖氨酸用于乳蛋白合成利用率达到 67%;虽然这个值似乎被高估了,但 Lapierre 等在奶牛中的研究结果(2009)与 NRC(2012)估计结果表明,猪乳腺的赖氨酸利用可能比以往提出的更重要。

使用 Fick 原理的研究报告结果显示大多数血浆流量值范围为 1.9~4.9 L/min,相应的血浆:奶(体积:体积)和产奶量分别为 441~581 和 6.2~12.1 L/天(表 14.1)。血浆:当赖氨酸和甲硫氨酸分别用作内部标记物时,也报道了浆:奶比值高达 1050 和 858(Nielsen 等,2002a,表 14.1)。考虑到报道的产仔数为 12 时平均每日产奶量较低,为 7.4 kg(Nielsen 等,2002a),这些值便产生了问题。在同一研究中,当使用钙作为标记物时,平均血浆流量为 2.8 L/min,观测值在报道范围内。因为没有其他研究用钙估计血浆流量,因此钙是否是有效的标记还未可知。与氨基酸不同,钙不被代谢,因此它具有作为"惰性"标记定义的优点。另一方面,考虑到 Ca 的动静脉差异和提取率较低,其作为内标的有效性是有问题的(Nielsen 等,2002a)。

Renaudeau 等(2002)通过在右外部的阴部外动脉周围植入超声波血流探针,直接测量通过哺乳母猪的乳腺血液流量。平均乳房血流量为 910±283 μL/min,相当于 3.6 L/min 或 4984 L/天。这些值与以往引用的使用 Fick 原理从文献获得的值相比较(表 14.1)。这种方法具有提供了评估乳房血流中的短期变化(即分钟)的优点。

14.2.3 调节乳腺血流量

目前,对母猪乳腺和各个乳腺的血液流量调节情况了解甚少。似乎在调节血液流量方面发挥重要作用的一个主要因素是奶水移出。Renaudeau 等(2002)报道,母猪静脉内给予催产素后,乳腺血流量明显减少,开始护理后 12 min 恢复到基础值。此后,在挤奶开始后 22 min,血流量增加并达到最大值。有人提出,哺乳后乳腺血流的增加可能是为了应对哺乳行为或奶量的下降。Olsson 等(2003)显示,静脉注射催乳素导致哺乳期山羊的乳腺血流短期增加。还表明，在天然排乳期间或在母猪中静脉内注射催产素之后，乳腺内压力增加，并且乳头之间的排乳时间可能受到血流的影响（Kent等，2003）。在断奶时，哺乳和乳腺血流之间的联系也非常明显，因此在断奶后8 h和16 h血流量分别降低了40%和60%（Renaudeau等，2002）。Nielsen等报道了产仔数与乳腺血浆流量之间呈正相关（2002b）；随着仔猪数量从3增加到13，乳腺血浆流量从大约2000 L/天线性增加到5000 L/天以上，显示了功能腺体数量和总需求量的增加。母猪的乳腺血流量的微小变化也可见于体位变化和相对于饲喂的时间。Renaudeau等（2002年）报告说，母猪站立与躺卧相比，乳腺的血流量下降了6%，采食后达到峰值高达7.7%。还探索了环境温度对乳腺血流的影响（Renaudeau等，2003）。母猪暴露于环境温度28℃时不影响产奶量，乳腺血流量增加约8%。作者认为，这种小的增加是由于血液流向皮下毛细血管以消散热量所致。

一氧化氮也称为内皮衍生因子，刺激血管平滑肌松弛，导致血管舒张和血流增加。Lacasse等（1996）证实，哺乳期山羊中一氧化氮供体的乳内输注迅速增加了乳腺血流量。然而，乳腺血流量和乳产量之间的因果关系仍有待证明。

在泌乳母猪的分离血管试验中研究了乳静脉和乳房前动脉对潜在的其他血管活性物质的反应。去甲肾上腺素、5-羟色胺、前列腺素、前列环素、组胺和钾对这两种血管都具有血管收缩作用，而乙酰胆碱对动脉和静脉具有松弛作用，β-腺苷仅对乳静脉具有松弛作用（Busk等，1999）。在啮齿动物和人类中，肾上腺素能神经纤维引起乳腺中的血管收缩，从而减少通过乳腺的血液流动（Erikson等，1996）。Franke-Radowiecka和Wasowicz（2002）表明，在猪中，大多数肾上腺素能和乙酰胆碱酯酶阳性神经纤维定位于乳头和乳腺的皮下组织中。Franke-Radowiecka和Wasowicz（2002）报道，胆碱性神经支配远远低于猪乳腺中的肾上腺素性神经支配。对母猪的头端乳腺的神经供应不同于腹股沟腺体。前乳腺从胸部神经接受神经支配，而腹股沟乳腺主要从阴部神经获得神经支配（Klopfenstein等，2006）。

14.3 营养摄入

14.3.1 乳房动静脉差异的测量

乳腺营养动静脉差异（AVD）和摄取的测量必须通过在动脉和排出乳腺实质的主要静脉系统放置插管。在这些体内研究中，通过腹静脉插管和颈动脉插管收集动、静脉血（Trottier等，1995a；图14.1）。乳腺系统前部的头端是用于代表性测量由位于胸部和腹部区域的腺体产生的营养物质浓度的优选插管部位。此外，乳腺收缩似乎开始于小窝母猪的腹股沟区域，而腹部和胸乳腺通常保持功能，直到哺乳期结束(Kim 等,2001)。

14.3.2 葡萄糖吸收

血糖是合成乳糖的主要前体。葡萄糖占母猪乳腺组织吸收总碳量的 40%～60%(Dourmad 等,2000；Linzell 等,1969b；Renaudeau 等,2003；Spincer 等,1969)。乳腺的葡萄糖摄取量占总体葡萄糖需求量的很大比例。一旦在乳腺中,葡萄糖被用作乳糖、甘油和脂肪酸合成的主要底物(Linzell 等,1969b),并且还可以为与维持乳腺相关的代谢过程提供能量。Linzell 等(1969a)报道了乳腺葡萄糖的提取率约为 26%。这与 Spincer 等(1969)估计的 31% 和 Renaudeau 等(2003)获得的26%一致,然而 Dourmad 等(2000)和 Trottier 等(1995b)报告了较低的值分别为20%和20.7%。使用标记碳葡萄糖,Linzell 等(1969b)估计,将 53%的葡萄糖用于乳糖合成,34%被氧化成 CO,其余 13%用于脂肪酸或氨基酸合成。例如,Dourmad 等(2000)显示,在葡萄糖 AVD 峰值期间,游离脂肪酸 AVD 下降到负值,表明利用葡萄糖合成脂肪酸的效率在吸收状态下增加。作者还报道了与饲喂前相比乳腺 AVD 增加,表明在吸收状态下葡萄糖氧化增加。乳腺似乎通过改变 AVD 来响应日粮葡萄糖的可利用性；在 Dourmad 等(2000)的研究中,葡萄糖的 AVD 在饲喂后 16 h 后由 24.2 mg/dL 显著下降至 12.2 mg/dL,与动物体内葡萄糖浓度从 126.8 mg/dL 下降至 62.4 mg/dL 直接相关。尽管仅基于一项研究,鉴于母猪乳腺的葡萄糖提取率在相对较宽的动脉血糖浓度范围内保持不变(即 62.4～126.8 mg/dL),乳腺的葡萄糖摄取可能是由高容量、低亲和力的葡萄糖转运蛋白所介导。

在妊娠结束时、整个哺乳期和断奶后,在乳腺组织中测量分别编码非胰岛素依赖性葡萄糖转运蛋白 GLUT1,胰岛素依赖性葡萄糖转运蛋白 GLUT4 和胰岛素受体蛋白的基因 SLC2A1,SLC2A4 和 INSR 的转录丰度(Manjarín 等,2012a)。所有 3 个基因的相对丰度在所有阶段相对较高,GLUT1 是 GLUT4 和胰岛素受体的两倍,在早期和高峰期之间表达没有变化。GLUT1 和 GLUT4 均是高容量、低亲和力和高 Km 转运的蛋白,因此,预计乳腺单糖摄取将随着动脉血糖浓度的增加而增加。鉴于葡萄糖作为母乳中乳糖合成的底物的重要性,这并不奇怪。Renaudeau 等(2003)估计需要 1300 g 葡萄糖来支持 11kg 的产奶量。根据 Guan 等(2004b)的工作,据估计,猪乳房每天消除约 2000 g 葡萄糖以支持 11.4 kg 的日产奶量(Farmer 等,2008a)。研究中乳糖产量与葡萄糖摄取的比例范围为 0.35～0.68(Dourmad 等,2000；Linzell 等,1969b；Spincer 等,1969)。

14.3.3 其他含能前体营养物质吸收

三项经典研究,即 Spincer 等(1969),Linzell 等(1969b)和 Spincer 和 Rook (1971)报道了母猪中甘油三酸酯、磷脂、醋酸酯、丙酸酯和乳酸的摄取量

(表 14.2)。除葡萄糖外，非氨基酸碳吸收的主要来源是甘油三酯和乳酸盐（表 14.2）。Linzell 等（1969b）使用五头母猪，从 12～61 天的哺乳期估计乳腺 AVD。乳酸和甘油三酯 AVD 是相当可观的，而磷脂、游离脂肪酸、挥发性脂肪酸和 β-羟基丁酸酯的 AVD 较小或不稳定。Spincer 等（1969）报道，血浆甘油三酸酯摄入总量中，血浆甘油三酯的比例为 11%，而乙酸酯仅占 2%。在血浆甘油三酸酯中，油酸酯（23%）、亚油酸酯（21%）、棕榈酸酯（19%）和硬脂酸酯（16%）是最突出的，当放射性甘油三酯静脉内灌注时，乳脂中超过 60% 的棕榈酸和 70% 的硬脂酸是由血浆甘油三酯摄取的（Spincer 和 Rook，1971）。Dourmad 等（2000）最近的研究结果证实了这些以往关于母猪乳腺中葡萄糖、甘油三酯和氨基酸相对摄取量的报告。

表 14.2　哺乳母猪血浆碳水化合物和脂质的动静脉血差异（mg/100 mL，挥发性脂肪酸和游离脂肪酸为 Meq/L）（Farmer 等，2008b）。

	Linzell 等，1969b	Spincer 等，1969	Spincer and Rook，1971
葡萄糖	19.8[1]	37.5	28.7
甘油三酯	7.6	7.2	10.2
乳酸盐	7.4	1.4	ND[2]
磷脂	0.3	4.1	ND
β-羟基丁酸酯	0.2	0.2	ND
挥发性脂肪酸	0.2[1,3]	0.6[4]	ND
游离脂肪酸	0.1[3]	ND	ND
柠檬酸盐	ND	0	ND

[1] 采用平均血细胞比容值为 31.7% 从全血校正到血浆。
[2] 未确定。
[3] 正值和负值均观察到。
[4] 该值仅用于醋酸盐。

最近有一些关于哺乳母猪乳腺中摄取矿物质和维生素的信息。Nielsen 等（2002a）和 Dourmad 等（2000）显示钙的正乳腺 AVD。Dourmad 等（2000）也注意到磷的正摄取，其饲喂后乳腺组织的 AVD 增加，而钙摄取在餐后保持不变。来自动脉血的钙和磷的吸收率（分别为 4.1% 和 3.1%）通常较其他主要营养素（20%～35%）低。此外，母猪乳腺几乎没有摄取（Dourmad 等，2000）核黄素（3.4 pmol/mL）、维生素 B（0 pg/mL）或叶酸（−1.2 ng/mL）。

14.3.4　能量摄取和奶合成能量效率

可以通过摄取葡萄糖、氨基酸和其他乳前体及其能量含量来计算乳腺能量摄取。根据 Linzell 等(1969b)和 Renaudeau 等(2003)的研究,可以计算 9.1 kJ/L 和 8.8 kJ/L 等离子体的能量 AVD(图 14.2)。在饲喂后动物体内测量能量的 AVD 值为 8.9 kJ/L 的相似值(Dourmad 等,2000),而禁食 16 h 后,能量的 AVD 趋于降低(7.6 kJ/L)。在饲喂的母猪中,葡萄糖和乳酸对乳腺总能量摄取的贡献平均为 45%,而在禁食母猪中,该值降至 28%(Dourmad 等,2000)。游离脂肪酸对能量摄入的贡献高度依赖于母猪的营养状况。在禁食母猪中,游离脂肪酸贡献了大约 35% 的能量摄入,而在饲喂后没有检测到吸收。限制性饲喂母猪的情况在饲喂母猪和禁食母猪之间(Renaudeau 等,2003),其中游离脂肪酸占总乳房能量摄取量的 15%。

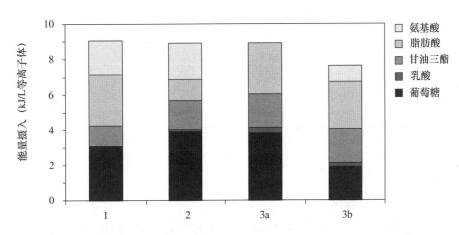

图 14.2 不同研究中乳腺能量吸收的划分(Farmer 等,2008b)。1,Linzell 等(1969b);2,Renaudeau 等(2003);3,Dourmad 等(2000 年)。3a,餐后;3b,禁食。

可以使用 Renaudeau 等(2003)测量的乳腺 AVD 计算血浆前体奶合成的能量效率(表 14.3),假设血浆流量为 550 L/L 牛奶。每天产奶 10 kg,前体和产品中的能量分别为 48.8 和 43.6 MJ/天,相当于能效为 0.89。这个值大于通过能源代谢研究得出的产奶 ME 效率,该值的通常估计为 0.72,但与报道的用身体储存计算的能量效率(0.88,Noblet 和 Etienne,1987)较为接近。另一方面,从能量代谢研究获得的效率估算与乳腺平衡研究之间的这种差异可能部分地与在乳腺外发生的其他代谢过程有关。然而,它也可能源自对营养物或血液流动的摄取低估。实际上,通过使用描述乳腺营养转化的化学计量的生物化学模型,Van Milgen 等(2003)表明,来自 Renaudeau 等(2003)血流量或 AVD 的数据必须增加约 10%,以确保观察和化学计量平衡之间的对应关系。

表 14.3 从前体估计牛奶合成的能量效率(Farmer 等,2008b)

	g/天	kJ/天
前体(摄取奶 10 L/天)[1]		
葡萄糖	1359	21328
乳酸盐	26	388
甘油	246	9363
甘油三酯	172	6714
脂肪酸	8	149
氨基酸	561	10883
总计	2373	48825
产品(摄取奶 10 L/天)[2]		
乳糖	560	8960
脂肪	580	22040
蛋白	530	12561
总计	1670	43561
效率(%)		89.2

[1] 前体由 Renaudeau 等(2002)测定的 AVD 计算假设需要相当于 550 L 血浆的血流来产生 1 kg 的牛奶(Trottier 等,1997)。

[2] 产品估计来自 Renaudeau 等(2002)。

14.3.5 氨基酸摄取

与其他营养素相反,母猪乳腺的氨基酸的 AVD 和净摄取量已经根据日粮(Guan 等,2002,2004a,b;Manjarín 等,2012b)、哺乳阶段和奶水需求(Guan 等,2004a,b;Nielsen 等,2002b)进行了广泛的测量。Guan 等(2004a,b)研究显示,乳腺对氨基酸的净摄取量随着哺乳期(大约 6 天)升至奶需求高峰期(即 14～18 天)而增加,随后减少。摄入量的增加与 AVD 的增加有关。相比之下,随着窝产仔数的增加观察到的净氨基酸摄取量增加与血流量增加相关,而不是与 AVD 的增加相关(Nielsen 等,2002b)。与哺乳期相关的 AVD 变化引起了关于这种变化是否与细胞转运蛋白丰度或氨基酸竞争性转运抑制相关的问题。所有脊椎动物细胞的氨基酸吸收受到氨基酸传入细胞时细胞膜中的蛋白质载体的协调活性控制(Broër 等,2008;Palacín 等,1998;Shennan 等,2000)。关于氨基酸转运蛋白的现有知识及其在乳腺(包括母猪)中的调控的综述超出了本书当前章的范围。简言之,已经在母猪乳腺组织中检测了编码属于两个钠依赖性和两个不依赖钠的系统下的氨基酸转运蛋白的基因的转录物。在钠依赖系统中,$B^{0,+}$ 和 y^+L,编码 $ATB^{0,+}$、$y^+LAT1/4F2hc$ 和 $y^+LAT2/4F2hc$ 的基因的转录本分别由 Pérez-Laspiur 等(2009)($ATB^{0,+}$)和 Manjarín 等(2011,2012b)($ATB^{0,+}$,$y^+LAT1/4F2hc$ 和 $y^+LAT2/$

4F2hc)所报道。对于钠独立系统，y^+ 和 $b^{0,+}$，为编码 CAT-1、CAT-2b 和 $b^{0,+}$ AT1/rBAT 基因的转录体分别由 Pérez-Laspiur 等（2009）（CAT-1、CAT-2b）和 Manjarín 等（2011，2012b）（CAT-1、CAT-2b 和 $b^{0,+}$ AT1）所报道。

这些基因中的几个基因的转录丰度显示增加以响应奶需求（Manjarín 等，2011）和日粮氨基酸可用性（Pérez-Laspiur 等，2009）。相比之下，在 Manjarín 等（2012b）的工作中，在这些相同基因的转录物丰度中没有发现依赖于日粮粗蛋白水平和氨基酸可用性的变化。Manjarín 等（2012b）认为，氨基酸 AVD 对日粮氨基酸可用性的变化可能是血液乳腺细胞层面竞争性抑制过程的结果，而不是基因表达的变化。氨基酸转运体系的持续特点及其各自的蛋白质，特别是对于日粮和生理挑战的反应，将能够更好地了解母猪如何调控产奶。例如，在 Manjarín 等（2011）的研究中，妊娠 112 天和哺乳期 17 天之间 CAT-1、$ATB^{0,+}$ 和 y^+LAT2 的转录本丰度增加。这三种基因的表达与编码 β-酪蛋白和 α-乳清蛋白的基因表达呈正相关，β-乳白蛋白是两种乳腺合成蛋白。这些作者（Manjarín 等，2011）提出，CAT-1、$ATB^{0,+}$ 和 y^+LAT2 可能成为改善母乳产量的分子靶点，因为这些蛋白质在哺乳母猪饲料中负责典型的限制性氨基酸赖氨酸的运输。

14.3.6　激素和乳腺摄取营养素

许多激素在调节奶水生成过程中发挥重要作用，乳腺组织中存在特定激素受体直接证明了这些激素参与奶水生成。生长激素受体（Manjarín 等，2012a）、IGF-I（Lee 等，1993；Manjarín 等，2012a；Theil 等，2006）、瘦蛋白（Palin 等，2004）、催乳素（Manjarín 等，2012a；Palin 等，2004；Plaut 等，1989；Theil 等，2005；Trott 等，2009）、胰岛素（Manjarín 等，2012a）、糖皮质激素（Manjarín 等，2012a）和催产素（Lundin-Schiller 等，1996）存在于猪乳腺组织中。这些受体处于复杂的调节之下，因为它不在本章的范围之内，所以不予讨论。

关于催乳素在母猪哺乳期如何调节乳腺营养摄取的信息很少。Pettigrew 等（1993）证明，循环胰岛素浓度与奶中蛋白质（$r=0.42$）、脂肪（$r=0.39$）和乳糖（$r=0.36$）含量相关，从而表明胰岛素可能参与哺乳期间乳腺吸收血浆底物。编码胰岛素受体蛋白的基因转录体在猪乳腺组织中相对丰富，并且在妊娠结束之后、在整个哺乳期和断奶后保持不变（Manjarín 等，2012a）。

一项体外研究表明，外源胰岛素与催乳素和皮质醇联合可以导致猪乳腺外植体脂质合成和葡萄糖氧化速率的显著增加。此外，报告显示乳腺组织中催乳素浓度和葡萄糖代谢速率之间的密切关系（Jerry 等，1989），并将催乳素与其母猪乳腺组织受体结合作为乳腺代谢率的主要效应因子（Plaut 等，1989）。鉴于其对母猪哺乳期起始和维持起到重要作用，催乳素这样参与乳腺组织的营养代谢并不奇怪

(Farmer 等,1998)。

体内研究表明,当给母猪饲喂不同水平的粗蛋白时,动脉胰岛素和催乳素浓度与乳腺氨基酸 AVD 密切相关,从而证明了胰岛素与催乳素浓度和乳腺氨基酸摄取之间存在联系。因此,可以通过血液中的胰岛素和结合在猪乳腺细胞上的催乳素来调节母猪乳腺的氨基酸利用(Farmer 等,2008)。然而,母猪乳腺组织中催乳素受体编码基因的 mRNA 丰度在第 17 天比哺乳期的第 5 天高,但与编码乳蛋白 α-乳白蛋白和 β-酪蛋白的基因表达无关(Manjarín 等,2012a)。最近的一项研究还表明,妊娠晚期催乳素的血液浓度增加导致仔猪生长速度增加,并且增加了编码 β-酪蛋白、α-乳清蛋白和葡萄糖转运蛋白 1 的基因 mRNA 表达的产后增加(Van Klompenberg 等,2013)。Farmer 等(2008)的研究却显示葡萄糖 AVD 与催乳素的循环浓度相关性较小。在同一研究中,葡萄糖 AVD 与胰岛素的血液浓度也不相关。胰岛素依赖猪乳腺葡萄糖转运的程度是未知的。大部分葡萄糖转运可能通过胰岛素独立的葡萄糖转运蛋白介导。如前所述,编码 GLUT1(胰岛素依赖性葡萄糖转运蛋白)的基因转录物的丰度是 GLUT4(胰岛素依赖性葡萄糖转运蛋白)的两倍。此外,葡萄糖和氨基酸的运输本身似乎均不处于 IGF-I 调控之下(Farmer 等,2008)。转基因母猪过表达乳腺 IGF-I 也具有与对照母猪相似的乳成分(Monaco 等,2005),进一步表明 IGF-I 对乳腺组织的营养吸收几乎没有作用。

最后,尚无关于糖皮质激素在母猪中作用的信息。在一项研究中,在产前至产后断奶期间编码糖皮质激素受体的基因表达与 α-乳白蛋白和 β-酪蛋白的表达呈正相关(Manjarín 等,2012a),表明可能其在氨基酸吸收起作用。

14.4 结论

哺乳母猪的产奶量最终取决于乳腺营养物的可用性(Boyd 等,1995)。这种营养物的可用性反过来又受到乳腺血流量和营养吸收的影响。乳腺血流量的估计是非常复杂的,需要血管插管。大多数报道的血浆流量值为 1.9~4.9 L/min,涉及调节乳腺血流的一个主要因素是放乳。基于乳腺 AVD 的研究表明,母猪乳腺主要吸收葡萄糖和氨基酸,而在较小程度上,也吸收其他能量前体,如甘油三酯、磷脂、乙酸盐、丙酸盐和乳酸盐。母猪乳房的营养摄入主要受日粮调节,也受哺乳期阶段和哺乳强度等因素影响。关于葡萄糖和氨基酸,存在位于乳腺上的转运蛋白,其吸收需要,但是这些营养和分子调节尚未阐明。显然,控制母猪产奶过程所涉及的机制尚未完全清楚。这种知识对于开发最佳适应管理策略至关重要,这将优化母猪产奶量。

参考文献

Barone, R., 1996. Anatomie comparée des mammifères domestiques. Tome cinquième angiologie, Ed. Vigot, Paris, France, pp. 372-373.

Boyd, R.D., Kensinger, R.S., Harrell, R.J. and Bauman, D.E., 1995. Nutrient uptake and endocrine regulation of milk synthesis by mammary tissue of lactating sows. Journal of Animal Science 73(2): 36-56.

Broër, S., 2008. Amino acid transport across mammalian intestinal and renal epithelia. Physiological Reviews 88: 249-286.

Busk, H., Sorensen, M.T., Mikkelsen, E.O., Nielsen, M.O. and Jakobsen, K., 1999. Responses to potential vasoactive substances of isolated mammary blood vessels from lactating sows. Comparative Biochemistry and Physiology Part C: Toxicology & Pharmacology 124: 57-64.

Dourmad, J.Y., Matte, J.J., Lebreton, Y. and Fontin, M.L., 2000. Influence du repas sur l'utilisation des nutriments et des vitamines par la mamelle chez la truie en lactation. Journées de Recherche Porcine en France 32: 265-273.

Eriksson, M., Lind, B., Uvnas-Moberg, K. and Hokfelt, T., 1996. Distribution and origin of peptide-containing nerve fibres in the rat and human mammary gland. Neuroscience 70: 227-245.

Farmer, C., Guan, X. and Trottier, N.L., 2008a. Mammary arteriovenous differences of glucose, insulin, prolactin and IGF-I in lactating sows under different protein intake levels. Domestic Animal Endocrinology 34: 54-62.

Farmer, C. and Hurley, W.L., 2015. Mammary development. Chapter 4. In: Farmer, C. (ed.) The gestating and lactating sow. Wageningen Academic Publishers, Wageningen, the Netherlands, pp. 73-94.

Farmer, C., Robert, S and Rushen, J., 1998. Bromocriptine given orally to periparturient sows inhibits milk production. Journal of Animal Science 76: 750-757.

Farmer, C., Trottier, N.L. and Dourmad, J.Y., 2008b. Review: current knowledge on mammary blood flow, mammary uptake of energetic precursors and their effects on sow milk yield. Canadian Journal of Animal Science 88: 195-204.

Fick, A., 1870. Ueber die Messung des Blutquantums in den Herzventrikeln. Physikalisch-Medizinische Gesellschaft zu Wiirzburg 2: 16-28.

Franke-Radowiecka, A. and Wasowicz, K., 2002. Adrenergic and cholinergic innervation of the mammary gland in the pig. Anatomy Histology and Embryology 31: 3-7.

Ghoshal, N.G., 1975. Porcine heart and arteries. In: Sisson, S., Grossman, J.D. and Getty, R. (eds.) The anatomy of the Domestic Animals, volume 2. W.B. Saunders Company, Philadelphia, PA, USA, pp. 1306-1342.

Guan, X., Bequette, B.J., Calder, G., Ku, P.K., Ames, K.N. and Trottier, N.L., 2002. Amino acid availability affects amino acid transport and protein metabolism in the porcine mammary gland. Journal of Nutrition 132: 1224-1234.

Guan, X., Bequette, B.J., Ku, P.K., Tempelman, R.J. and Trottier, N.L., 2004a. The amino acid need for milk synthesis is defined by the maximal uptake of plasma amino acids by porcine mammary glands. Journal of Nutrition 134: 2182-2190.

Guan, X., Pettigrew, J.E., Ku, P.K., Ames, N.K., Bequette, B.J. and Trottier, N.L., 2004b. Dietary protein concentration affects plasma arterio-venous difference of amino acids across the porcine mammary gland. Journal of Animal Science 82: 2953-2963.

Jerry, J., Stover, R.K. and Kensinger, R.S., 1989. Quantitation of prolactin-dependent responses in porcine mammary explants. Journal of Animal Science 67: 1013-1019.

Kent, J.C., Kennaugh, L.M. and Hartmann, P., 2003. Intramammary pressure in the lactating sow in response to oxytocin and during natural milk ejections throughout lactation. Journal of Dairy Research 70: 131-138.

Kety, S.S. and Schmidt, C.F., 1945. The determination of cerebral blood flow in man by the use of nitrous oxide in low concentrations. American Journal of Physiology 143: 53-66.

Kim, S.W., Easter, R.A. and Hurley, W.L., 2001. The regression of unsuckled mammary glands during lactation in sows: the influence of lactation stage, dietary nutrients, and litter size. Journal of Animal Science 79: 2659-2668.

Klindt, J., 2003. Influence of litter size and creep feeding on preweaning gain and influence of preweaning growth on growth to slaughter in barrows. Journal of Animal Science 81: 2434-2439.

Klopfenstein, C., Farmer, C. and Martineau, G.P., 2006. Diseases of the mammary glands. In: Straw, B.E., Zimmerman, J.J., D'Allaire, S. and Taylor, D.J. (eds.) Diseases of swine, 9th edition. Blackwell Publishing, Oxford, UK, pp. 57-85.

Lacasse, P., Farr, V.C., Davis, S.R. and Prosser, C.G., 1996. Local secretion of nitric oxide and the control of mammary blood flow. Journal of Dairy Science 79: 1369-1374.

Lapierre, H., Doepel, L., Milne, E. and Lobley, G.E., 2009. Responses in mammary and splanchnic metabolism to altered lysine supply in dairy cows. Animal 3: 360-371.

Lee, C.Y., Bazer, F.W. and Simmen, F.A., 1993. Expression of components of the insulin-like growth factor system in pig mammary glands and serum during pregnancy and pseudopregnancy: effects of oestrogen. Journal of Endocrinology 137: 473-483.

Lignereux, Y., Rossel, R. and Jouglard, J.Y., 1996. Note sur la vascularisation veineuse des mamelles chez la truie. Revue de Médecine Vétérinaire 3: 191-194.

Linzell, J.L., Mephan, T.B., Annison, E.F. and West, C.E., 1969a. Mammary metabolism in lactating sows: arteriovenous differences of milk precursors of milk. In: Larson, B.L. and Smith, V.R. (eds.) Lactation, volume 1. Academic Press, New York, NY, USA, 143.

Linzell, J.L., Mephan, T.B., Annison, E.F. and West, C.E., 1969b. Mammary metabolism in lactating sows: arteriovenous differences of milk precursors and the mammary metabolism of [^{14}C]glucose and [^{14}C]acetate. British Journal of Nutrition 23: 319-332.

Lundin-Schiller, S., Kreider, D.L., Rorie, R,W., Haresty, D., Mitchell, M.D. and Koike, T.I., 1996. Characterization of porcine endometrial, myometrial, and mammary oxytocin binding sites during gestation and labor. Biology of Reproduction 55: 575-581.

Manjarín, R., Steibel, J.P., Kirkwood, R.N., Taylor, N.P. and Trottier, N.L. 2012a. Transcript abundance of hormone receptors, mammalian target of rapamycin pathway-related kinases, insulin-like growth factor I, and milk proteins in porcine mammary tissue. Journal of Animal Science 90: 221-230.

Manjarín, R., Steibel, J.P., Zamora, V., Am-in, N., Kirkwood, R.N., Ernst, C.W., Weber, P.S., Taylor, N.P. and Trottier, N.L., 2011. Transcript abundance of amino acid transporters, β-casein, and α-lactalbumin in mammary tissue of periparturient, lactating, and postweaned sows. Journal of Dairy Science 94: 3467-3476.

Manjarín, R., Zamora, V., Wu, G., Steibel, J.P., Kirkwood, R.N., Taylor, N.P., Wils-Plotz, E., Trifilo, K. and Trottier, N.L., 2012b. Effect of amino acids supply in reduced crude protein diets on performance, efficiency of mammary uptake, and transporter gene expression in lactating sows. Journal of Animal Science 90: 3088-3100.

Monaco, M.H., Gronlund, D.E., Bleck, G.T., Hurley, W.L., Wheeler, M.B. and Donovan, S.M., 2005. Mammary specific transgenic over-expression of insulin-like growth factor-I (IGF-I) increases pig milk IGF-I and IGF binding proteins, with no effect on milk composition or yield. Transgenic Research 14: 761-773.

National Research Council (NRC), 2012. Nutrient requirements of swine, 11th revised edition. National Research Council of the National Academies. Washington, DC, USA.

Nielsen, T.T., Pierzynowski, S.G., Borsting C.F., Nielsen, M.O. and Jakobsen, K., 2002a. Catheterization of arteria epigastrica cranialis, measurement of nutrient arteriovenous differences and evaluation of daily plasma flow across the mammary gland of lactating sows. Acta Agricultura Scandinavia Section A: Animal Science 42: 113-120.

Nielsen, T.T., Trottier, N.L., Stein, H.H., Bellavers, C. and Easter, R.A., 2002b. The effect of litter size and day of lactation on amino acid uptake by the porcine mammary glands. Journal of Animal Science 80: 2402-2411.

Noblet, J. and Etienne, M., 1987. Metabolic utilisation of energy and maintenance requirements of lactating sows. Journal of Animal Science 64: 774-781.

Olsson, K., Malmgren, C., Olsson, K.K., Hansson, K. and Häggström, J., 2003. Vasopressin increases milk flow and milk fat concentration in the goat. Acta Physiologica Scandinavia 177: 177-184.

Palacin, M., Estevez, R., Bertran, J. and Zorzano, A., 1998. Molecular biology of mammalian amino acid transporters. Physiological Reviews 78: 969-1054.

Palin, M.F., Beaudry, D. and Farmer, C., 2004. Gene expression of leptin, leptin receptor, prolactin receptor and whey acidic protein in mammary glands of late-pregnant gilts from two breeds. Canadian Journal of Animal Science 84: 621-629.

Pérez Laspiur, J., Burton, J.L., Weber, P.S.D., Moore, J., Kirkwood, R.N. and Trottier, N.L., 2009. Dietary protein intake and stage of lactation differentially modulate amino acid transporter mRNA abundance in porcine mammary tissue. Journal of Nutrition 139: 1677-1684.

Pettigrew, J.E., McNamara, J.O., Tokach, M.D., King, R.H. and Crooker, B.A., 1993. Metabolic connections between nutrient intake and lactational performance in the sow. Livestock Production Science 35: 137-152.

Plaut, K.I., Kensinger, R.S., Griel, Jr., L.C. and Kavanaugh, J.F., 1989. Relationships among prolactin binding, prolactin concentrations in plasma and metabolic activity of the porcine mammary gland. Journal of Animal Science 67: 1509-1519.

Renaudeau, D., Lebreton, Y., Noblet, J. and Dourmad, J.Y., 2002. Measurement of blood flow through the mammary gland in lactating sows: methodological aspects. Journal of Animal Science 80: 196-201.

Renaudeau, D., Noblet, J. and Dourmad, J.Y., 2003. Effect of ambient temperature on mammary gland metabolism in lactating sows. Journal of Animal Science 81: 217-231.

Schummer, A., Wilkens, H., Vollmerhaus, B. and Habermehl, K.H., 1981. Skin and cutaneous organs. In: Nickel, R., Schummer, A. and Seiferle, E. (eds.) The anatomy of the domestic animals,. volume 3. The circulatory system, the skin, and the cutaneous organs of the domestic mammals. Springer-Verlag, New York, NY, USA, pp. 473-557.

Shennan, D.B. and Peaker, M., 2000. Transport of milk constituents by the mammary gland. Physiological Reviews 80: 925-951.

Spincer, J. and Rook, J.A.F., 1971. The metabolism of [U-^{14}C]glucose, [1-^{14}C]palmitic acid and [1-^{14}C]stearic acid by the lactating mammary gland of the sow. Journal of Dairy Research 38: 315-322.

Spincer, J., Rook, J.A.F. and Towers, K.G., 1969. The uptake of plasma constituents by the mammary gland of the sow. Biochemistry Journal 111: 727-732.

Theil, P.K., 2015. Transition feeding of sows. Chapter 7. In: Farmer, C. (ed.) The gestating and lactating sow. Wageningen Academic Publishers, Wageningen, the Netherlands, pp. 147-172.

Theil, P.K., Labouriau, R., Sejrsen, K., Thomsen, B. and Sorensen, M.T., 2005. Expression of genes involved in regulation of cell turnover during milk stasis and lactation rescue in sow mammary glands. Journal of Animal Science 83: 2349-2356.

Theil, P.K., Sejrsen, K., Hurley, W.L., Labouriau, R., Thomsen, B. and Sorensen, M.T., 2006. Role of suckling in regulating cell turnover and onset and maintenance of lactation in individual mammary glands of sows. Journal of Animal Science 84: 1691-1698.

Trott, J.F., Horigan, K.C., Gloviczki, J.M., Costa, K.M., Freking, B.A., Farmer, C., Hayashi, K., Spencer, T., Morabito, J.E. and Hovey, R.C., 2009. Tissue-specific regulation of porcine prolactin receptor expression by estrogen, progesterone and prolactin. Journal of Endocrinology 202: 153-166.

Trottier, N.L., Shipley, C.F. and Easter, R.A., 1995a. A technique for the venous cannulation of the mammary gland in the lactating sow. Journal of Animal Science 73: 1390-1395.

Trottier, N.L., Shipley, C.F. and Easter, R.A., 1995b. Arteriovenous differences for amino acids, urea nitrogen, ammonia, and glucose across the mammary gland of the lactating sow. Journal of Animal Science 73(2): 57-58.

Trottier, N.L., Shipley, C.F. and Easter, R.A., 1997. Plasma amino acid uptake by the mammary gland of the lactating sow. Journal of Animal Science 75: 1266-1278.

Turner, C.W., 1952. The anatomy of the mammary gland of swine. In: The mammary gland. I. The anatomy of the udder of cattle and domestic animals. Lucas Brothers, Columbia, MO, USA, pp. 279-314.

Van Milgen, J., Gondret, F. and Renaudeau, D., 2003. The use of nutritional models as a tool in basic research. In: Souffrant, W.B. and Metges, C.C. (eds.) Progress in research on energy and protein metabolism. European Federation of Animal Science (EAAP) publication no. 109, pp. 259-263.

VanKlompenberg, M.K., Manjarín, R., Trot, J.F., McMicking, H.F. and Hovey, R.C., 2013. Late-gestational hyperprolactinemia accelerates mammary epithelial cell differentiation that leads to increased milk yield. Journal of Animal Science 91: 1102-1111.

15. 仔猪肠道和肠道相关淋巴组织发育：母体环境的作用

I. Le Huërou-Luron * *and S. Ferret-Bernard*
INRA, UR1341 ADNC, 35590 Saint-Gilles, France; isabelle. luron@rennes. inra.fr

摘要：肠道发育是一种在胎儿早期开始、并且在产后的头几个月内持续的现象。肠道在产前发育期间获得的在结构与功能上的能力，能够为胎儿在子宫外的生活和适应营养和微生物挑战做好准备。黏膜免疫系统的发育在出生后的早期发生。被动、先天和适应性免疫系统能够提供针对有害病原体的适当保护，并伴随着对于饲料中普遍存在的抗原和微生物群的耐受。在妊娠期和哺乳期，许多与母体环境相关的因素可能影响幼龄仔猪肠道功能的发育模式。母猪-胎儿关系的受损是子宫内生长受限的主要原因，能够诱发因肠道发育不成熟引起的消化系统疾病。此外，出生后肠道微生物群的建立在仔猪肠道及其免疫系统的发育中起重要作用。能够诱导母体微生物群变化的母体环境因素（如饲料组成、抗生素治疗等）会对后代肠道生理学产生巨大影响。最后，有证据表明母猪饲料组成可能影响初乳和乳成分，从而导致仔猪肠道功能的变化。本文的结论部分对仔猪早期肠道生理学改变的长期影响的前景进行了讨论。

关键词：消化发育，肠道生理，黏膜免疫系统，上皮屏障，微生物群

15.1 引言

有相当多的证据表明，仔猪的肠道及其免疫系统在功能上与成年猪不同。肠道发育是一种在胎儿形成的早期开始，并且在产后的前几个月内持续的现象。在产前发育过程中，肠道在结构和功能上获得一定的能力，为胎儿应对在子宫外生活和新的挑战做好准备。在引发肠道特征发生改变的因素中，出生和断奶阶段饲料摄入的改变以及细菌在肠道定殖为主要的促进因素。在出生时，肠道已准备好处理第一份饲料。然而，在哺乳期间其表现出微妙的灵活性，以适应日粮组成在此阶段的变化，使其在断奶前逐渐获得处理断奶日粮的能力。另外，由于偶蹄类动物的胎盘屏障阻止了外部抗原的侵入，这使得新生仔猪的黏膜免疫系统，特别是获得性

免疫系统在出时生严重发育不良。但在产后,在一般环境中其发育迅速发生(Bailey 等,2001)。先天和适应性免疫系统之间的协作能够提供对有害病原体的适当保护,并同时伴随着对于饲料中普遍存在的抗原和微生物群的耐受性(Bailey 等,2006,2009)。针对"无害"抗原的过度发育或不适当的免疫反应可导致短暂的过敏性肠损伤。大量非致病性细菌的过度生长引起的过度调节反应,将会导致肠道微生物介导的损伤,而这种损伤经常与断奶后腹泻相关。

本章节的主要目的是讨论产后肠道发育与肠道微生物群落建立的关系,以及母猪的饲料组成如何调节其发育模式。首先,我们总结了正常新生仔猪的肠道及其相关淋巴组织的一般特征,以及母体环境如何影响这些特征。我们强调了通过对妊娠和哺乳母猪的抗生素治疗或在母猪日粮中补充益生元的方式,来调节细菌在肠道的定殖。同时,我们还关注了子宫内生长迟缓的仔猪其肠道成熟度的损伤。最后,讨论了母猪饲料补充 n-3 多不饱和脂肪酸(n-3 polyunsaturated fatty acids)的影响。

15.2 肠道和肠相关淋巴组织的发育

在胎儿期,肠道必须获得消化食物和吸收营养物质的能力,建立针对致病菌的防御机制,消除外源毒素和耐受共生微生物群和食物抗原。的确,肠道是一个复杂的组织,其主要功能为选择性吸收和消化。作为机体的第一道上皮屏障,肠上皮层必须能够控制外源物质的通过,这些物质能够参与到肠局部免疫系统的发育成熟(Calder 等,2006)。几种成分参与肠道的先天防御系统,如黏膜分泌物(黏液层、溶菌酶、抗菌肽等),物理上皮屏障(由细胞间紧密连接形成),吞噬细胞(巨噬细胞、嗜中性粒细胞),肥大细胞,表达于上皮细胞模拟细菌受体的糖复合物以及肠蠕动。肠相关淋巴组织(gut-associated lymphoid,GALT)的适应性防御系统(或特异性免疫)由与肠相关的淋巴组织[派伊尔淋巴集结(Peyer's patches,PP)和肠系膜淋巴结(mesenteric lymph nodes,MLN),其作为黏膜免疫应答的诱导剂]和更扩散的区域组成,固有层是局部免疫应答的效应因子。PP、MLN 和固有层含有抗原呈递细胞、B 和 T 淋巴细胞,它们都参与这种局部免疫活动。

15.2.1 肠相关淋巴组织的构成

肠道免疫系统(或 GALT)被认为是机体第一位的免疫器官,因为它含有体内 70%~85%的免疫细胞。GALT 通常分为诱导性(PP 和 MLN)和效应器(固有层,指 PP 外部的免疫细胞和上皮内淋巴细胞)位点(Burkey 等,2009)。肠上皮细胞单层提供了高度特异的物理性和功能性屏障,其能够促进腔内抗原选择性转移到用于维持局部组织稳态的底层 GALT(图 15.1)。

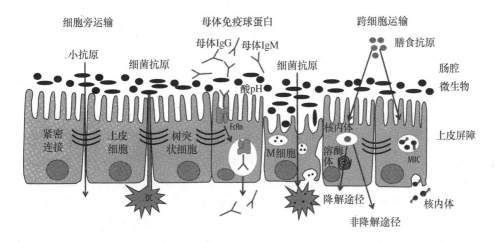

图15.1 通过肠上皮的分子通道。

细菌抗原的跨细胞转运通常归因于淋巴集结（Peyer's patches，PP）的滤泡相关上皮上面覆盖的微胶囊细胞（Corr 等，2008）。此外，微生物抗原会被树突状细胞直接取样，树突状细胞可以在肠细胞之间打开紧密连接，将树突延伸到管腔（Rescigno 等，2008）。存在于近侧小肠的日粮抗原可通过肠上皮细胞的内吞作用在顶膜被输送并朝向固有层胞转（跨细胞渗透性）（Menard 等，2010）。在肠细胞基底外侧，这些抗原以氨基酸或分解产物的形式释放（图 15.1）。饲料蛋白质也可以被肠细胞加工，作为可以扩散到基底膜中并与局部免疫细胞相互作用的上皮外泌体形式释放。同时，肠细胞之间也允许小分子通过（细胞通透性）（图 15.1）。最后，抗原呈递细胞〔猪肠上皮细胞上不存在主要组织相容性复合体（major histocompatibility complex，MHC）Ⅱ类分子〕到达排空的 MLN 以产生效应 T 细胞和效应 B 细胞。在羊和其他哺乳动物中，激活的淋巴细胞通过高内皮小静脉离开节点，但不通过传出淋巴管（因为猪淋巴结倒置）（Binns 等，1990），然后传回血流并返回到固有层来执行其特异性的获得性效应器功能。

猪的 PP 由位于小肠（空肠和上回肠）上部的离散的、有组织的淋巴组织，还有沿回肠大的连续斑块组成，两者均负责吸收抗原和诱导黏膜免疫应答。PP 由生殖中心的巨 B 细胞卵泡组成，邻近被滤泡树突状细胞包围的 T 细胞区域。PP 被卵泡相关上皮细胞所覆盖，这些上皮细胞包含有与上皮细胞内交错的微细胞。这种上皮表达有助于配体特异性转胞吞作用的 Toll 样受体（Toll-like receptors，TLR）。在猪中，与空肠 PP 相比，回肠 PP 的淋巴细胞产生率、总生殖卵泡数和平均卵泡直径均较大（Pabst 等，1988）。此外，回肠 PP 可以作为参与 B 细胞产生的初级淋巴器官（Butler 等，2009）。

免疫诱导后，固有层（由平滑肌细胞、成纤维细胞、血液和淋巴管组成）作为肠道免疫应答的调节器。效应器功能存在于巨噬细胞、树突细胞、嗜中性粒细胞、肥大细胞和淋巴细胞中(Bailey 等，2001)。与啮齿动物和人类相比，猪的肠黏膜具有更高的组织程度。事实上，浆细胞优先定位于肠隐窝中，CD4+ T 细胞定位于肠绒毛的核心，CD8+ 淋巴细胞组成多数上皮内淋巴细胞。

15.2.2 肠交换表面

肠通过增加长度、周长和绒毛尺寸来改变其表面积。这些变化具有空间和时间的模式(图 15.2)。在产后生命的前几周，从第一餐开始，小肠的重量便开始增加(Butler 等，2009)。这是由肠黏膜重量增加和蛋白质积累引起的(Xu 等，1992)。在隐窝区域产生的细胞在沿着隐窝迁移到绒毛轴线时分化和成熟。因此，出生后观察到的增加的隐窝细胞增生允许直至断奶时绒毛高度变化连续增加。通过小肠长度、黏膜密度、干物质含量和绒毛大小的综合升高来估计，这些变化导致从出生到哺乳期(即在产后的前 3 周)小肠的空肠与回肠部分交换表面增加 14 倍以上。相比之下，在哺乳期，结肠只有中等程度的出生后生长，然而在断奶后，当猪饲喂更复杂和富含谷物的饲料时，其生长开始加速(Montagne 等，2007)。值得注意的是，结肠中的隐窝大小比小肠高 3 倍以上。这可能反映了在无绒毛的结肠中保持大的交换面的必要性，以便保持其水和电解质吸收的主要功能并提供隐窝相关细菌储藏位置(Lee 等，2013)。

初乳在出生后的前两天刺激小肠显著生长。Zhang 等(1997)报道，初乳摄入 6 h 足以引起肠结构发生显著变化。这种变化主要涉及小肠的重量，其在哺乳的前两天增加了 80%，而结肠的重量仅增加 30%(图 15.2)。相应地，空肠绒毛和隐窝区分别增加了 120% 和 40%。初乳的营养属性取决于生长因子和免疫球蛋白(Ig)的存在(见第 9 章关于初乳和牛奶的组成；Hurley，2015)。这些观察结果支持了初乳生长因子的主要营养作用，即在饲喂不含初乳生长因子的配方饲料仔猪中(Godlewski 等，2005)或在剥夺了初乳生长因子，例如，表皮生长因子(Clark 等，2005)、胰岛素样生长因子(Xu 等，1994a)、瘦素(Wolinski 等，2001)的仔猪中观察到较高的细胞凋亡率。由于对初乳中 Ig 的内吞，黏膜生长也与肠细胞的暂时性肥大有关。在猪中，两种机制可以解释大分子以完整形式从肠腔进入体循环的早期转移，两种机制分别为依赖或不依赖于与 IgG 的 Fc 区段结合的一种肠受体(新生 Fc 受体，neonatal Fc receptor，FcRn)(Cervenak 等，2009，Stirling 等，2005)。在仔猪产后 48 h，通过一般被称为"肠闭合"(上皮闭合)的机制，IgG 的传递逐渐停止。最近的研究表明，FcRn 也可能参与防止肠道 Ig 降解的保护机制，因而在整个生命过程中在肠道中发挥免疫传感器的作用(Baker 等，2009)。

15.2.3 营养的消化和吸收

对于新生仔猪来说，对营养物质的消化与吸收及其重要，因为这些仔猪在产后第一周需要大量的营养来支持其快速生长与高代谢率。在出生时，仔猪便具有支

15.仔猪肠道和肠道相关淋巴组织发育:母体环境的作用

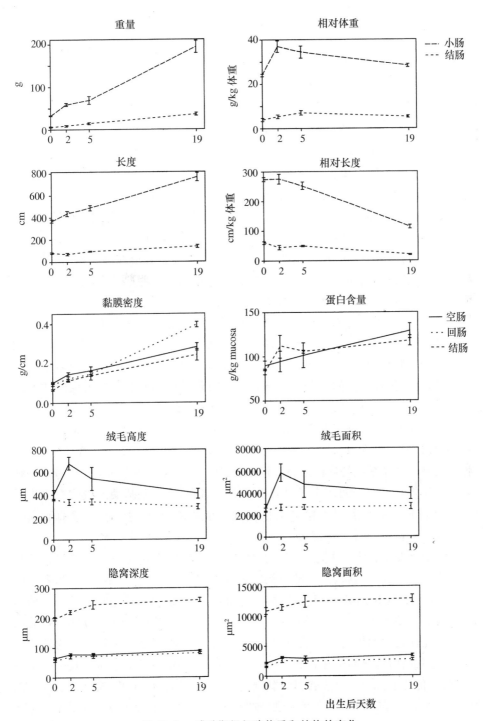

图15.2 哺乳期间仔猪体重和结构的变化。

持其在子宫外生活所有消化器官以及全部的转运器官,但这些器官在哺乳期间会发生巨大的改变(Boudry 等,2010,Le Huërou-Luron,2002,Le Huërou-Luron 等,2010)。的确,初乳的摄取会加快其消化酶样成熟(Jensen 等,2001)。

初生仔猪在对母乳进行乳糖酶性消化后,吸收其中的葡萄糖和半乳糖。肠乳糖酶活性在出生时达到最高峰并在随后降低,而麦芽糖酶和蔗糖酶活性在产后随年龄增加而逐步升高。葡萄糖结合转运蛋白(sodium-glucose linked transporter,SGLT1)将葡萄糖和半乳糖转运到顶膜上,同时葡萄糖转运蛋白2(glucose transporter 2,GLUT2)作为基底外侧膜。肠 SGLT1 和 GLUT2 活性在出生时和哺乳期间都很高,从而大大促进了新生儿的葡萄糖体内平衡(Gabler 等,2009,Yang 等,2011)。蛋白质消化在出生时也是高效的。肽酶活性,如氨基肽酶 N(aminopeptidase N)和二肽基肽酶 IV(dipeptidyl peptidase IV)在出生时已经发育良好。此外,出生时肠肽和氨基酸转运系统的高表达确保了蛋白质的最高吸收,而此后其逐步下降直至断奶。然而,由于肠的长度和重量的增加,肠吸收氨基酸的总量随着年龄增加而增加。二肽和三肽通过肽转运蛋白1(peptide transporter 1,PepT1)在顶膜被吸收,并且最终被肠细胞胞质中的肽酶水解。基底外侧膜中的运输系统介导游离氨基酸从细胞质运出到门静脉循环。从数量上来说,通过 PepT1 的肽吸收是摄取蛋白质消化产物的主要模式。抗细胞溶质肽酶水解的小肽确实可以进入血液,但它们只占相对较小的部分。在围产期,小肠中 PepT1 水平在出生时达到高峰,然后逐步下降直至断奶(Boudry and Le Huërou-Luron, personal communication)。新生猪结肠中 PepT1 的存在表明,仔猪出生后立即进行肽转运的能力较强(Boudry and Le Huërou-Luron, personal communication)。在刷状缘膜中,转运二肽和三肽的主要系统是广泛的特异性系统:$B^{o,+}$ 针对中性氨基酸,$B^{o,+}$(或 y^+)针对阳离子氨基酸,X^{2-} 针对阴离子氨基酸。在仔猪中,各种氨基酸的吸收在出生后 24 h 内急剧下降。然后,其在第7天恢复到出生的值,其后进一步下降直到断奶(Buddington 等,1996,Zhang 等,1997)。

脂肪消化发生在小肠中,脂肪酸的吸收几乎在回肠末端完成。脂肪消化的整个过程依赖于外分泌胰腺,其能够分泌三种酶分别为脂肪酶(lipase)、羧基酯水解酶(carboxyl ester hydrolase)和磷脂酶 A(phospholipase A),还包括一种协同消化日粮脂肪的辅助因子——肠杆菌肽(colipase)。尽管胰腺脂肪酶作为最重要的脂肪分解酶在仔猪中的表达量很低,但是乳猪可以有效地消化母猪乳汁中的营养成分,有报道其脂肪消化率能达到 96%(Jensen 等,1997)。肠道脂质吸收是一个多步骤过程,传统上分为三个步骤:顶端吸收入肠细胞,细胞内加工,随后释放入淋巴和门静脉循环。在吸收后,游离脂肪酸和 sn-2 单酰基甘油(sn-2 monoacylglycerols)与肠脂肪酸结合蛋白 FABP(intestinal fatty acid binding protein FABP,I-FABP)和肝脏 FABP(liver FABP,L FABP)结合,以防止其转运回肠腔,并促进其运输到内质网，其在此处用于甘油三酯和磷脂的从头合成。关于转运蛋白的发育调控知之甚少，但有证据表明细胞内脂质加工在出生时便可以运作。在哺乳期间，肠可以通过增加脂质的吸收来适应高脂摄入。

15.2.4　上皮屏障的通透性

肠道是营养和腔内成分的第一道屏障，在决定产后机体防御方面起核心作用。肠道通透性会在产后降低，初乳摄入加速了这一现象。当动物无法摄取初乳时，大分子（白蛋白、牛血清白蛋白）的转运则发生较晚（Westrom等，1984）。超过3周龄后，机体会停止转运大于1000 Da的大分子，而小分子的转运则以比出生后低10倍的速率持续（Westrom等，1989）。然而，尤斯灌流室（Ussing chambers）研究肠道渗透性的体外实验显示出重要的位点依赖性发育模式。空肠的细胞外通透性在产后前2周较高，然后逐渐下降直到哺乳期结束，然而在此期间其在回肠中则一直增加（De Quelen等，2011，Le Huërou-Luron等，2010）。食物和细菌抗原在早期通过上皮屏障有助于肠道免疫系统的成长，并促进耐受性的获得（Menard等，2010）。事实上，消化道腔中蛋白质的存在在肠道免疫系统的成熟中起关键作用（Menezes等，2003）。

15.2.5　小肠中肠相关淋巴组织的发育和细菌定殖

尽管全身免疫组织发育良好，但新生仔猪中黏膜免疫系统基本上不存在（Inman等，2005）。许多研究表明，肠道免疫系统的发展由微生物群落接触驱动（Pabst等，1988）。空肠PP的囊泡随着年龄增长，其宽度和长度在无特定病原体和常规猪中是无菌动物的两倍，从而表明肠腔微生物群的影响（Barman等，1997）。这些卵囊泡在出生后第1天和第10天之间的大小增加了100倍。这种发育在无菌动物中基本上不存在（Rothkotter等，1989）。此外，在无菌猪中，B细胞优先表达表面IgM而不是IgA。

黏膜免疫细胞以高度程序化的序列结构分布在小猪的肠道内（Vega-Lopez等，1995）。出生时，PP由少量T细胞包围的极小的未成熟卵泡和固有层中的非常少的抗原呈递细胞组成。在第1~2周龄，固有层内大量出现II类MHC＋树突状细胞，出现表达CD2表面标志物而不是CD4或CD8表面标志物的异常T细胞。然后PP开始组织，在10~15日龄的时候达到类似成年个体的结构。在2~4周，绒毛的核心可以发现成熟记忆CD4＋ T细胞，同时少量B细胞出现，主要表达IgM。5周后，CD8＋ T细胞进入上皮，隐窝区出现许多IgA＋ B细胞。到7周时，尽管细胞的绝对数量随着仔猪生长而不断增加，还是可以观察到类似于成年个体的的免疫学结构（Rothkotter，2009）。

15.2.6　被动免疫

仔猪在出生时免疫系统严重缺陷，其高度依赖于母体初乳和乳汁中存在的特异性和非特异性免疫因子的供应，用于仔猪的免疫保护、发育和存活（Salmon 等，2009）。妊娠母猪上皮胎盘呈现出相对不渗透，意味着新生仔猪是低蛋白或丙种球蛋白缺乏（hypo- or agammaglobulinemic）。与血浆相比，初乳富含 IgM 和 IgG，其原因是在妊娠结束时这两种 Ig 的选择性渗出。来自肠腔的大分子，例如存在于初乳中的 Ig（Salmon，2000），在出生后立即以非选择性方式被吸收（Westrom 等，1984）。全身免疫（IgG 和 IgM）由此从母猪转移到其仔猪体内。肠道对大分子的吸收的快速停止发生于产后 24～36 h。在出生 24 h 内，哺乳仔猪的血清 IgG 浓度通常与母体类似。尽管 IgG 和 IgM 可以到达仔猪血液循环，大多数 IgA 保留在肠腔中，以便通过覆盖这些细胞上的表面受体来抑制病原体细菌向肠细胞的第一黏附期（Salmon 等，2009）。迁移到乳腺的淋巴细胞（间接来源于母体 GALT）产生释放到母猪乳中的分泌性 IgA，用于维持后代的体液免疫（enteromammary axis，肠乳腺轴），其解释了母乳和肠 IgA 之间相同的特异性（Salmon，2000）。小猪从第 2 周开始可以合成其自身的分泌型 IgA（Allen 等，1973）。

15.2.7 天然免疫

肠道的天然免疫能够阻止病原菌从内腔进入固有层，进而阻止其到达仔猪的血液中。黏膜屏障发展出几种外部机制，包括肠蠕动、蛋白水解、酸性 pH 以及胃肠排空。上皮细胞自身也可产生抗菌的多肽（导管素和防御素）以及酶（肠碱性磷酸酶，intestinal alkaline phosphatase），来抑制细菌的跨上皮性转运和促炎性脂多糖（LPS）的解毒。具有分泌性 IgA 的上皮黏液层的存在是肠道保护性免疫的重要方面，其不仅防止病原微生物的定殖和侵袭（高亲和力 IgA），而且通过"免疫排斥"（低亲和力 IgA）的方法将共生细菌限制在肠腔。其他机制涉及由肠细胞或下游树突状细胞表达的先天感受受体的存在。这些受体为模式识别受体，包括 TLR 受体和识别细菌成分的细胞内核苷酸结合低聚结构域样受体（nucleotide-binding oligomerization domain，NOD），以及作为病毒感受器的视黄酸诱导型基因 I 样解旋酶（retinoic-acid-inducible gene I，RIG-I）。这些受体启动先天免疫反应，并调节对感染或组织损伤的适应性反应。细菌或细菌组分与模式识别受体（位于细胞表面或细胞内）的结合从而激活 NF-κB 通路，并最终激活下游促炎性细胞因子的转录。无论如何，出生后早期，其他机制使得共生细菌能够被耐受，以便严格控制肠道中过度的活性。在出生后，上皮细胞顶端表面的模式识别受体变得不易接触，并且最初由共生细菌激活的 TLR 的作用可被其由肠上皮细胞产生的表达产物负调节剂抑制（Abreu，2010）。

15.2.8 获得性免疫

年轻幼仔 Th1 和 Th2 反应之间的平衡更偏向于 Th2 分布，导致对细胞内病原体感染高度易感（Adkins 等，2004）。已经证明，新生仔猪不能对大量病原体给予有效免疫应答，因为 Th1 应答在多个步骤受损，包括不能由 CD4＋T 细胞和单核吞噬细胞产生 Th1 型细胞因了（Marodi，2001）。Th1/Th2 平衡随着的年龄的成熟导致细胞因子分泌的变化,例如促炎 IFNγ 分泌的增加,这被认为是出生后黏膜免疫发育成熟的标记。已经有研究显示几个能够促进 IFNγ 产生的基因其表达受肠细菌的影响。猪 IL-17 的 mRNA 在黏膜组织中高表达(Katoh 等,2004),再次,共生细菌特别是分裂的丝状细菌能够刺激肠黏膜中 Th17 细胞的发育和累积(Niess 等,2008)。

15.3　母体环境对调节肠和肠相关淋巴组织发育的重要性

在妊娠和哺乳期间,许多与母猪环境有关的因素可能会影响新生仔猪肠道及其黏膜免疫系统的出生后发育。子宫内环境不仅对胎儿发育和生存重要,而且还影响新生儿的出生后发育和健康(Morise 等,2008)。母猪-胎儿关系的受损是子宫内生长限制(intra-uterine growth restriction,IUGR)的主要原因,其能够导致由肠道发育不成熟引起的消化系统疾病。此外,出生后肠道微生物群落的建立对新生仔猪肠道及其免疫系统的发育起着重要作用。在产后的前几天,新生仔猪似乎在很大程度上分享了母亲的微生物群落(Bauer 等,2006)。因此,引起母体微生物群变化的母体环境因素(日粮组成、抗生素治疗等)对后代肠道生理情况有巨大的影响。最后,有证据表明,母体饲料组成可能影响初乳和乳汁的组成,导致仔猪肠道功能的变化。

15.3.1　日粮诱导的母体微生物群变化:对后代肠道发育的影响

毫无疑问,新生仔猪与环境和定殖的细菌早期接触对于健康的肠道和免疫成熟至关重要。通过对无菌仔猪的研究发现,细菌对消化道生长发育至关重要(Chowdhury 等,2007)。通过对比无菌与常规仔猪肠细胞中基因表达谱,显示了微生物群对肠屏障功能的影响(Chowdhury 等,2007)。更具体地来说,已经证明细菌定殖诱导了有利于肠上皮细胞更新和黏液生物合成的基因之表达。肠道微生物群还影响黏膜免疫系统的几个组分(如抗菌"潘氏细胞"的活性、IgA 的产生、上皮淋巴细胞的发育)的成熟以及功能,以预防会影响屏障功能的炎症反应(Chowdhury 等,2007,Hooper,2004)。微生物群在新生仔猪肠道发育中的主要作用,是在细菌定殖被早期改变的条件下得到确认的。在啮齿动物中,在给予新生幼仔抗生素会下调与先天宿主防御和抗原呈递相关基因的表达(Schumann 等,2005)。相反,新生仔猪通过饲喂补充有益生元的配方饲料会大大增加结肠中的细菌总量使其生长增加(Aufreiter 等,2011)。共生微生物群对降解其他难消化的复合碳水化合物来说也是至关重要的,特别是首次在日粮中引入固体饲料时。在小鼠中,肠道细菌可以直接调节食物的利用,这意味着食物的真正"热值"是随微生物群组成及其为宿主动员潜在能量的能力而变化(Turnbaugh 等,2006)。

健康种群中微生物群的组成是相当独特的,且在初次定殖后似乎相当稳定。

然而，微生物定居过程的早期干扰，如由抗生素治疗或母猪饲养环境的卫生状况引起的早期干扰会对后代肠道的发育程序产生了巨大影响。这种问题与猪生产能力有关，因为抗生素经常用于控制围产期和产后母猪的生殖泌尿道和乳腺的感染。Fåk 等（2008）证明，在年轻大鼠上，分娩前的母体抗生素治疗引起了细菌定殖的变化，增加了蔗糖酶活性，降低肠屏障特性。最新一项研究报告了母体抗生素治疗对母猪微生物群落及其新生仔猪后代具有有害影响（Boudry 等，2012；Ferret-Bernard 等，2013；Lallès 等，2013）。其对仔猪肠功能的影响也是显著的。根据具体时间框架，作为对 LPS 刺激的促炎反应，回肠和结肠通透性、上皮转录组和黏膜炎症特征会被改变。

日粮中包括特定的底物，以刺激假定的益生细菌的生长，并防止潜在病原体的生长和定殖的特定物质，被称为益生元，它们包括日粮纤维、抗性淀粉和寡糖。在饲料中引入的优势寡糖是低聚果糖（fructo-oligosaccharides，FOS）、甘露聚糖寡糖（mannan-oligosaccharides，MOS）和菊糖（inulin）。与人类乳汁相比，母猪和母牛乳汁中的半乳寡聚糖（galacto-oligosaccharides，GOS）浓度要低得多，这解释了为什么它们的性质在农场动物物种还没有被广泛地研究。幼龄仔猪或成年个体消耗 FOS 的量与其双歧杆菌（*Bifidobacteria*）等有益细菌的种群数量增加以及短链脂肪酸（short-chain fatty acids，SCFA）的高产相关（Howard 等，1995，Tsukahara 等，2003），从而改善肠道对病原体的抵抗，有利于相关的局部免疫系统的发育。在 SCFA 中，FOS 显著升高管状丁酸钠的浓度。由于丁酸盐是大肠上皮细胞的主要能量源，其在肠腔内的增加可以解释 FOS 诱导的上皮和黏蛋白细胞的增殖，从而有助于其有益效果。通过使用动物模型，一些研究证明了母猪饲料中补充益生素对泌乳免疫的有益作用，例如初乳或成熟乳汁中 Ig 含量的增加（Adogony 等，2007；Czech 等，2010；Gourbeyre 等，2012）。最近在养猪生产中证实了这种效果，其中在妊娠最后 1/3 和整个哺乳期间在母猪日粮中补充 scFOS（约 10 g/天的 scFOS）可以使初乳中 IgA 和 TGFβ1 浓度升高。母体饲喂 scFOS 也与后代肠道免疫的加速发展有关（Le Bourgot 等，2013），其为 PP 的 Th1 反应的极化提供了良好条件。此外，scFOS 通过刺激回肠 PP 分泌型 IgA 的产生来提高肠道的保护作用。这些结果强调了母体日粮在支持新生仔猪黏膜免疫发育中的关键作用。然而，在哺乳期间并没有观察到仔猪生长的改善（Le Bourgot 等，2013），但补充 scFOS 的母猪往往会增加仔猪断奶后的生长性能。在晚期妊娠和哺乳期间对母猪饲喂高纤维日粮可以观察到对仔猪的早期生长（Oliviero 等，2009），低出生体重仔猪的初乳摄入量和产仔性能（Loisel 等，2013）的有益作用。本书第 5 章中有更多的细节（Meunier-Salaün and Bolhuis，2015）。

15.3.2　胚胎子宫内生长迟缓和肠道成熟

在过去 10 年中，通过遗传选择增加了母猪的平均窝仔数量。但也伴随着平均仔猪出生体重的减少，每窝仔猪出生体重的均匀度变差，导致在产仔数多的窝里体重小的仔猪（出生体重小于 1 kg）的比例上升（Le Dividich，1999；Milligan 等，2002；Quiniou 等，2002）。与同窝出生的体重正常或较重的仔猪相比，具有 IUGR（子宫内生长迟缓）的猪在屠宰时表现出持续的较低的出生后生长率和较低的屠宰率（Bee，2004；Gondret 等，2006；Morise 等，2008；Poore 等，2004；Rehfeldt 等，2006）。

IUGR 的小猪通常消耗的初乳比正常出生体重的仔猪少。具有 IUGR 的新生仔猪初次爬到哺乳乳腺能力较低，反映出其出生时的活力较低（Quesnel 等，2012）。然而，IUGR 仔猪的生长衰竭也与肠尺寸减小有关，例如重量减轻（与体重成正比）、长度、壁厚、绒毛高度和隐窝深度和较低的产后早期由肠内食物出现引起的肠内营养性反应（D'Inca 等，2010，2011）。在出生后的第一周，通过估算回肠重量/长度比和绒毛尺寸的综合减少，IUGR 的仔猪具有更薄的肠道，并且其交换表面积至少低 40%。IUGR 仔猪肠黏膜密度和绒毛长度的减少可能由细胞死亡的增加引起。尽管肠酶的活性没有受到 IUGR 的显著影响，但是可以观察到远端肠道中 PepT 1 表达的降低，表明肠道发育迟缓。在哺乳期间正常出生体重的仔猪会减少空肠和增加回肠对大分子的渗透性，通过阻止这种改变，IUGR 仔猪上皮屏障功能的发育特征也发生了改变（Boudry 等，2011）。这一点似乎至关重要，因为肠道在将食物分子加工成机体的可用营养物质以及调节参与 GALT 成熟的抗原物质的通量方面发挥主要作用。此外，与同龄的正常出生体重仔猪相比，2 日龄的 IUGR 仔猪空肠 PP 中的卵泡表面积减少（S. Ferret-Bernard，personal communication）。IUGR 仔猪中参与应激反应的几种蛋白质的表达也同样上调（Wang 等，2008）。

肠黏膜与肠腔细菌的结合能力似乎在 IUGR 仔猪中受到影响。实际上，已经证明在 2 日龄 IUGR 仔猪中肠腔细菌的黏附性有所增加（D'Inca 等，2010）。因此，为了应对这一早期挑战，IUGR 仔猪的肠道建立了防御机制，通过增强黏蛋白合成和 IgA 分泌来维持肠道屏障保护（D'Inca 等，2010，2011）。同时，微生物群落的性质变化亦有报道。与正常出生体重的仔猪相比，在 IUGR 的断奶仔猪中其盲肠的发酵活性有明显的差异，乙酸和丙酸盐浓度更高（Michiels 等，2013）。IUGR 的大鼠提示，这些差异可能归因于在底物可用性方面的差异或肠道微生物群组成方面的改变（Fanca-Berthon 等，2010）。

据报道，在围产期时，断奶早期的 IUGR 猪（Michiels 等，2013）和成年个体（Alvarenga 等，2013）仍然存在很多差异（D'Inca 等，2010；Wang 等，2005；Xu 等，1994b）。除了消化缺陷因素，如代谢活动和激素失衡的变化，可能与生长潜力的降低有关（Wang 等，2008）。生长因子中最一致的抑制表达是血清中胰岛素样生长因子 1（insulin-like growth factor 1，IGF-1）水平和 IGF-1 基因在组织中（例如肌肉、肝脏、肾脏和脂肪组织）的表达（Chen 等，2011；Gondret 等，2013）。

总之,胎儿营养的缺陷对后代肠道生理具有巨大的影响和使肠道成熟延后。肠屏障和免疫系统发育的这种损伤导致 IUGR 存活仔猪的出生后死亡率更高,生长速度降低(Morise 等,2008;Quiniou 等,2002)。

15.3.3 母体 n-3 多不饱和脂肪酸(n-3 polyunsaturated fatty acids)和后代肠道成熟

日粮多不饱和脂肪酸(Dietary polyunsaturated fatty acids,PUFA),如 n-3 家族的亚油酸(linoleic acid,LA,18:2n-6)、α-亚麻酸(α-linolenic acid,ALA,18:3n-3)及其长链衍生物,包括二十碳五烯酸(eicosapentaenoic acid,EPA,20:5n-3)、二十二碳五烯酸(docosapentaenoic acid,DPA,22:5n-3)和二十二碳六烯酸(docosahexaenoic acid,DHA,22:6n-3)和 n-6 家族的花生四烯酸(arachidonic acid,ARA,20:4n-6)是所有组织细胞的重要组成部分,其有助于新生仔猪生理功能的正常生长和发育。仔猪组织中的脂肪酸组成与妊娠期和哺乳期中母猪日粮的组成有关。在妊娠母猪日粮中提供亚麻子油(富含 ALA)能够升高母猪的血浆和胎盘中的 PUFA,同时升高仔猪出生时和哺乳期间血浆和组织中的 PUFA(De Quelen 等,2013)。饲料中亚麻籽油也增加了母猪乳汁中 ALA、EPA 和 DPA 的含量(De Quelen 等,2013;Farmer 等,2009)。蔬菜来源的如亚麻子油、核桃油或菜籽油(压榨的)和海洋来源的如鲱鱼鱼油和鲨鱼肝油可以用来供应母猪饲料中的 n-3 PUFA(Fritsche 等,1993;Rooke 等,2001a,b)。在本书第 16 章中更详细地阐述了饲喂其他脂肪源(如 CLA 和 omega 3)对母猪免疫系统的影响(Bontempo and Jiang,2015)。

妊娠期和哺乳期母体补充 n-3 PUFA 对母猪和整窝的初生仔猪表现显著的影响(de Quelen 等,2010;Quiniou 等 2010)。但是,妊娠和哺乳期间在母猪饲料中添加榨取过的亚麻籽,能够增加接近平均窝仔出生体重的仔猪断奶时的存活率,但不能增加较小出生体重(<1 kg)的仔猪的存活率(Quiniou 等 2010)。对妊娠后期和哺乳期母猪饲喂亚麻籽亦能够对仔猪断奶后生长产生积极影响,其原因可能是通过提高其免疫抵抗力(Farmer 等,2010)。已知 n-3 PUFA 具有抗炎特性及其对肠道炎症性疾病具有益处。令人惊讶的是,母猪饲料中补充鱼油(EPA 和 DHA)会增加哺乳期大鼠的结肠通透性,导致其生命后期(3 个月龄)对结肠炎的易感性增加(Innis 等,2010)。最近证实了母猪饲料中添加亚麻子油(ALA)能使仔猪上皮屏障功能产生类似变化(De Quelen 等,2011)。肠道通透性增加,这可能与食物诱导的控制肠功能的肠神经系统的神经性变化有关。没有观察到对肠黏膜炎症状态的影响,但随后在生命中观察到 MHC II 类+抗原呈递细胞的增加(Boudry,Ferret-Bernard and Le Huërou-Luron,unpublished results)。

总之,母猪饲料中补充 n-3 PUFA 会改变肠道功能发育和肠道免疫系统成熟,在哺乳期其影响似乎不明显,因为初乳免疫力仍然存在,但对于断奶后的生长和健康来说可能更为重要。新生猪肠道免疫系统成熟的这些改变对于面临严重情况（营养和卫生挑战）的老龄猪的影响值得进一步研究。

15.4 结论

这部分的综述阐述了母体环境如何改变早期生活中肠道微生物群的建立和肠道免疫系统的形成。出生时，包括饲料在内的环境因素在确定微生物群组成中比遗传因素更为重要。饲料组成和营养平衡在维持出生后的上皮完整性、增殖和屏障功能中起重要作用。有证据表明，与妊娠晚期的母体应激相关的产前应激对猪后代的免疫系统有明显的破坏性影响（Couret 等，2009；Merlot 等，2013）。有趣的是，最近的研究提供了有力的证据，即在不改变肠道的屏障功能基础生理学特性及细胞因子模式的情况下，断奶前短时间的日粮变化或早期肠道细菌定殖的改变可以长期影响炎症反应的严重性（Boudry 等，2013；Chatelais 等，2011）。肠道健康方面的研究应该集中在饲喂如何改变肠道微生物群、GALT 和上皮屏障之间的关系，从而潜在地影响在生命后期对应激或营养挑战的敏感性，例如在断奶后遇到的圈养条件或环境卫生条件的降低。更好地了解这些变化和其机制（表观遗传）的长期影响，以及动物遗传选择的小心使用，都可以改善动物生产，减轻饲料限制和使用抗生素等管理过程中的负面影响。

参考文献

Abreu, M.T., 2010. Toll-Like Receptor Signalling in the Intestinal epithelium: how bacterial recognition shapes intestinal function. Nature Reviews Immunology 10: 131-144.

Adkins, B., Leclerc, C. and Marshall-Clarke, S., 2004. Neonatal adaptive immunity comes of age. Nature Reviews Immunology 4: 553-564.

Adogony, V., Respondek, F., Biourge, V., Rudeaux, F., Delaval, J., Bind, J.L. and Salmon, H., 2007. Effects of dietary Scfos on immunoglobulins in colostrums and milk of bitches. Journal of Animal Physiology and Animal Nutrition 91: 169-174.

Allen, W.D. and Porter, P., 1973. The Relative Distribution of Igm and Iga cells in intestinal mucosa and lymphoid tissues of the young unweaned pig and their significance in ontogenesis of secretory immunity. Immunology 24: 493-501.

Alvarenga, A.L.N., Chiarini-Garcia, H., Cardeal, P.C., Moreira, L.P., Foxcroft, G.R., Fontes, D.O. and Almeida, F.R.C.L., 2013. Intra-uterine growth retardation affects birthweight and postnatal development in pigs, impairing muscle accretion, duodenal mucosa morphology and carcass traits. Reproduction, Fertility and Development 25: 387-395.

Aufreiter, S., Kim, J.H. and O'Connor, D.L., 2011. Dietary oligosaccharides increase colonic weight and the amount but not concentration of bacterially synthesized folate in the colon of piglets. Journal of Nutrition 141: 366-372.

Bailey, M., Plunkett, F.J., Rothkotter, H.J., Vega-Lopez, M.A., Haverson, K. and Stokes, C.R., 2001. Regulation of mucosal immune responses in effector sites. Proceedings of the Nutrition Society 60: 427-435.

Bailey, M., 2009. The mucosal immune system: recent developments and future directions in the pig. Developmental and Comparative Immunology 33: 375-383.

Bailey, M. and Haverson, K., 2006. The postnatal development of the mucosal immune system and mucosal tolerance in domestic animals. Veterinary Research 37: 443-453.

Baker, K., Qiao, S.W., Kuo, T., Kobayashi, K., Yoshida, M., Lencer, W.I. and Blumberg, R.S., 2009. Immune and non-immune functions of the (not so) neonatal Fc receptor, Fcrn. Seminars in Immunopathology 31: 223-236.

Barman, N.N., Bianchi, A.T., Zwart, R.J., Pabst, R. and Rothkotter, H.J., 1997. Jejunal and ileal Peyer's patches in pigs differ in their postnatal development. Anatomy and Embryology 195: 41-50.

Bauer, E., Williams, B.A., Smidt, H., Mosenthin, R. and Verstegen, M.W., 2006. Influence of dietary components on development of the microbiota in single-stomached species. Nutrition Research Reviews 19: 63-78.

Bee, G., 2004. Effect of early gestation feeding, birth weight, and gender of progeny on muscle fiber characteristics of pigs at slaughter. Journal of Animal Science 82: 826-836.

Binns, R.M. and Licence, S.T., 1990. Exit of recirculating lymphocytes from lymph nodes is directed by specific exit signals. European Journal of Immunology 20: 449-452.

Bontempo, V. and Jiang, X.R., 2015. Feeding various fat sources to sows: effects on immune status and performance of sows and piglets. Chapter 16. In: Farmer, C. (ed.) The gestating and lactating sow. Wageningen Academic Publishers, Wageningen, the Netherlands, pp. 357-375.

Boudry, G., David, E.S., Douard, V., Monteiro, I.M., Le Huërou-Luron, I. and Ferraris, R.P., 2010. Role of intestinal transporters in neonatal nutrition: carbohydrates, proteins, lipids, minerals, and vitamins. Journal of Pediatric Gastroenterology and Nutrition 51: 380-401.

Boudry, G., Morise, A., Seve, B. and Le Huërou-Luron, I., 2011. Effect of milk formula protein content on intestinal barrier function in a porcine model of Lbw neonates. Pediatric Research 69: 4-9.

Boudry, G., Ferret-Bernard, S., Savary, G., Le Normand, L., Perrier, C., Perez-Gutteriez, O., Zhang, J., Smidt, H. and Le Huërou-Luron, I., 2012. The maternal microbiota influences the post-natal development of gut barrier function and immune system in piglets and their gut adaptation to a high fat diet later in life. Digestive Disease Week. May 2012, San Diego, U.S.A., Gastroenterology 142: supplement 1: S14

Boudry, G., Jamin, A., Chatelais, L., Gras-Le Guen, C., Michel, C. and Le Huërou-Luron, I., 2013. Dietary protein excess during neonatal life alters colonic microbiota and mucosal response to inflammatory mediators later in life in female pigs. Jounal of Nutrition 143: 1225-1232.

Buddington, R.K. and Malo, C., 1996. Intestinal brush-border membrane enzyme activities and transport functions during prenatal development of pigs. Journal of Pediatric Gastroenterology and Nutrition 23: 51-64.

Burkey, T.E., Skjolaas, K.A. and Minton, J.E., 2009. Board-invited review: porcine mucosal immunity of the gastrointestinal tract. Journal of Animal Science 87: 1493-1501.

Butler, J.E., Lager, K.M., Splichal, I., Francis, D., Kacskovics, I., Sinkora, M., Wertz, N., Sun, J., Zhao, Y., Brown, W.R., DeWald, R., Dierks, S., Muyldermans, S., Lunney, J.K., McCray, P.B., Rogers, C.S., Welsh, M.J., Navarro, P., Klobasa, F., Habe, F. and Ramsoondar, J., 2009. The piglet as a model for B cell and immune system development. Veterinary Immunology and

Immunopathology 128: 147-170.

Calder, P.C., Krauss-Etschmann, S., de Jong, E.C., Dupont, C., Frick, J.S., Frokiaer, H., Heinrich, J., Garn, H., Koletzko, S., Lack, G., Mattelio, G., Renz, H., Sangild, P.T., Schrezenmeir, J., Stulnig, T.M., Thymann, T., Wold, A.E. and Koletzko, B., 2006. Early nutrition and immunity – progress and perspectives. British Jounal of Nutrition 96: 774-790.

Cervenak, J. and Kacskovics, I., 2009. The neonatal Fc receptor plays a crucial role in the metabolism of Igg in livestock animals. Veterinary Immunology and Immunopathology 128: 171-177.

Chatelais, L., Jamin, A., Gras-Le Guen, C., Lalles, J.P., Le Huërou-Luron, I. and Boudry, G., 2011. The level of protein in milk formula modifies ileal sensitivity to Lps later in life in a piglet model. PLoS One 6: e19594.

Chen, R., Yin, Y., Pan, J., Gao, Y. and Li, T., 2011. Expression profiling of Igfs and Igf receptors in piglets with intrauterine growth restriction. Livestock Science 136: 72-75.

Chowdhury, S.R., King, D.E., Willing, B.P., Band, M.R., Beever, J.E., Lane, A.B., Loor, J.J., Marini, J.C., Rund, L.A., Schook, L.B., Van Kessel, A.G. and Gaskins, H.R., 2007. Transcriptome profiling of the small intestinal epithelium in germfree versus conventional piglets. BMC Genomics 8: 215.

Clark, J.A., Lane, R.H., Maclennan, N.K., Holubec, H., Dvorakova, K., Halpern, M.D., Williams, C.S., Payne, C.M. and Dvorak, B., 2005. Epidermal growth factor reduces intestinal apoptosis in an experimental model of necrotizing enterocolitis. American Journal of Physiology – Gastrointestinal and Liver Physiology 288: 4.

Corr, S.C., Gahan, C.C. and Hill, C., 2008. M-Cells: origin, morphology and role in mucosal immunity and microbial pathogenesis. FEMS Immunology and Medical Microbiology 52: 2-12.

Couret, D., Jamin, A., Kuntz-Simon, G., Prunier, A. and Merlot, E., 2009. Maternal stress during late gestation has moderate but long-lasting effects on the immune system of the piglets. Veterinary Immunology and Immunopathology 131: 17-24.

Czech, A., Grela, E.R., Mokrzycka, A. and Pejsak, Z., 2010. Efficacy of mannanoligosaccharides additive to sows diets on colostrum, blood immunoglobulin content and production parameters of piglets. Polish Journal of Veterinary Sciences 13: 525-531.

D'Inca, R., Kloareg, M., Gras-Le Guen, C. and Le Huërou-Luron, I., 2010. Intrauterine growth restriction modifies the developmental pattern of intestinal structure, transcriptomic profile, and bacterial colonization in neonatal pigs. Journal of Nutrition 140: 925-931.

D'Inca, R., Gras-Le Guen, C., Che, L., Sangild, P.T. and Le Huërou-Luron, I., 2011. intrauterine growth restriction delays feeding-induced gut adaptation in term newborn pigs. Neonatology 99: 208-216.

De Quelen, F., Boudry, G. and Mourot, J., 2010. Linseed oil in the maternal diet increases long chain-pufa status of the foetus and the newborn during the suckling period in pigs. British Jounal of Nutrition 104: 533-543.

De Quelen, F., Chevalier, J., Rolli-Derkinderen, M., Mourot, J., Neunlist, M. and Boudry, G., 2011. N-3 polyunsaturated fatty acids in the maternal diet modify the postnatal development of nervous regulation of intestinal permeability in piglets. The Journal of Physiology 589: 4341-4352.

De Quelen, F., Boudry, G. and Mourot, J., 2013. Effect of different contents of extruded linseed in the sow diet on piglet fatty acid composition and hepatic desaturase expression during the post-natal period. Animal 7: 1671-1680.

Fak, F., Ahrne, S., Molin, G., Jeppsson, B. and Westrom, B., 2008. Microbial manipulation of the rat dam changes bacterial colonization and alters properties of the gut in her offspring. American Journal of Physiology – Gastrointestinal and Liver Physiology 294: G148-G154.

Fanca-Berthon, P., Hoebler, C., Mouzet, E., David, A. and Michel, C., 2010. Intrauterine growth restriction not only modifies the cecocolonic microbiota in neonatal rats but also affects its activity in young adult rats. Journal of Pediatric Gastroenterology and Nutrition 51: 402-413.

Farmer, C., Giguere, A. and Lessard, M., 2010. Dietary supplementation with different forms of flax in late gestation and lactation: effects on sow and litter performances, endocrinology, and immune response. Journal of Animal Science 88: 225-237.

Farmer, C. and Petit, H.V., 2009. Effects of dietary supplementation with different forms of flax in late-gestation and lactation on fatty acid profiles in sows and their piglets. Journal of Animal Science 87: 2600-2613.

Ferret-Bernard, S., Boudry, G., Le Normand, L., Romé, V., Savary, G., Perrier,C., Lallès, J.P., Le Huërou-Luron, I., 2013. Perinatal antibiotic treatment of sows affects intestinal barrier and immune system in offspring. In: Proceedings of the 64th Annual Meeting of the European Federation of Animal Science. August 26-30, 2013. Nantes, France.

Fritsche, K.L., Huang, S.C. and Cassity, N.A., 1993. Enrichment of omega-3 fatty acids in suckling pigs by maternal dietary fish oil supplementation. Journal of Animal Science 71: 1841-1847.

Gabler, N.K., Radcliffe, J.S., Spencer, J.D., Webel, D.M. and Spurlock, M.E., 2009. Feeding long-chain N-3 polyunsaturated fatty acids during gestation increases intestinal glucose absorption potentially via the acute activation of ampk. Journal of Nutritional Biochemistry 20: 17-25.

Godlewski, M.M., Slupecka, M., Wolinski, J., Skrzypek, T., Skrzypek, H., Motyl, T. and Zabielski, R., 2005. Into the unknown – the death pathways in the neonatal gut epithelium. Journal of Physiology and Pharmacology 3: 7-24.

Gondret, F., Lefaucheur, L., Juin, H., Louveau, I. and Lebret, B., 2006. Low birth weight is associated with enlarged muscle fiber area and impaired meat tenderness of the longissimus muscle in pigs. Journal of Animal Science 84: 93-103.

Gondret, F., Pere, M.C., Tacher, S., Dare, S., Trefeu, C., Le Huërou-Luron, I. and Louveau, I., 2013. Spontaneous intra-uterine growth restriction modulates the endocrine status and the developmental expression of genes in porcine fetal and neonatal adipose tissue. General and Comparative Endocrinology 194: 208-216.

Gourbeyre, P., Desbuards, N., Gremy, G., Le Gall, S., Champ, M., Denery-Papini, S. and Bodinier, M., 2012. Exposure to a galactooligosaccharides/inulin prebiotic mix at different developmental time points differentially modulates immune responses in mice. Journal of Agricultural and Food Chemistry 60: 11942-11951.

Hooper, L.V., 2004. Bacterial contributions to mammalian gut development. Trends in Microbiology 12: 129-134.

Howard, M.D., Gordon, D.T., Pace, L.W., Garleb, K.A. and Kerley, M.S., 1995. Effects of dietary supplementation with fructooligosaccharides on colonic microbiota populations and epithelial cell proliferation in neonatal pigs. Journal of Pediatric Gastroenterology and Nutrition 21: 297-303.

Hurley, W.L., 2015. Composition of sow colostrum and milk. Chapter 9. In: Farmer, C. (ed.) The gestating and lactating sow. Wageningen Academic Publishers, Wageningen, the Netherlands, pp. 193-229.

Inman, C.F., Jones, P., Harris, C., Haverson, K., Miller, B., Stokes, C.R. and Bailey, M., 2005. The mucosal immune system of the neonatal piglet. The Pig Journal 55, 211-222.

Innis, S.M., Dai, C., Wu, X., Buchan, A.M. and Jacobson, K., 2010. Perinatal lipid nutrition alters early intestinal development and programs the response to experimental colitis in young adult rats. American Journal of Physiology – Gastrointestinal and Liver Physiology 299, G1376-1385.

Jensen, A.R., Elnif, J., Burrin, D.G. and Sangild, P.T., 2001. Development of intestinal immunoglobulin absorption and enzyme activities in neonatal pigs is diet dependent. Journal of Nutrition 131: 3259-3265.

Jensen, M.S., Jensen, S.K. and Jakobsen, K., 1997. Development of digestive enzymes in pigs with emphasis on lipolytic activity in the stomach and pancreas. Journal of Animal Science 75: 437-445.

Katoh, S., Kitazawa, H., Shimosato, T., Tohno, M., Kawai, Y. and Saito, T., 2004. Cloning and characterization of swine interleukin-17, preferentially expressed in the intestines. Journal of Interferon and Cytokine Research 24: 553-559.

Lallès, J.P., Arnal, M.E., Boudry, G., Ferret-Bernard, S., Le Huërou-Luron, I., Zhang, J., Smith, H., 2013. Effects of sow antibiotic treatment and offspring diet on microbiota and gut barrier throughout life. In: Proceedings of the 64th Annual Meeting of the European Federation of Animal Science. August 26-30, 2013. Nantes, France.

Le Bourgot, C., Ferret-Bernard, S., Blat, S., Apper-Bossard, E., Le Normand, L., Respondek, F., Le Huërou-Luron, I., 2013. Peri-partum scFOS supplementation modulates development and activity of the immune system of suckling piglets. In: Proceedings of the 64th Annual Meeting of the European Federation of Animal Science. August 26-30, 2013. Nantes, France.

Le Dividich, J., 1999. A review. Neonatal and weaner pig: management to reduce variation. In: Cranwell, P.D. (ed.) Manipulating Pig Production VII. Australian Pig Science Association, pp. 135-155.

Le Huërou-Luron, I., 2002. Production and gene expression of brush border disaccharidases and peptidases during development in pigs and calves. In: Zabielski, R., Gregory, P.C. and Weström, B. (eds.) Biology of the Intestine in Growing Animals. Elsevier Science, Amsterdam, the Netherlands, pp. 491-513.

Le Huërou-Luron, I., Blat, S. and Boudry, G., 2010. Breast- v. formula-feeding: impacts on the digestive tract and immediate and long-term health effects. Nutrition Research Reviews 23: 23-36.

Lee, S.M., Donaldson, G.P., Mikulski, Z., Boyajian, S., Ley, K. and Mazmanian, S.K., 2013. Bacterial colonization factors control specificity and stability of the gut microbiota. Nature 501: 426-429.

Loisel, F., Farmer, C., Ramaekers, P. and Quesnel, H., 2013. Effects of high fiber intake during late pregnancy on sow physiology, colostrum production, and piglet performance. Journal of Animal Science 91: 5269-5279.

Marodi, L., 2001. Il-12 and Ifn-gamma deficiencies in human neonates. Pediatric Research 49: 316.

Menard, S., Cerf-Bensussan, N. and Heyman, M., 2010. Multiple facets of intestinal permeability and epithelial handling of dietary antigens. Mucosal Immunology 3: 247-259.

Menezes, J.S., Mucida, D.S., Cara, D.C., Alvarez-Leite, J.I., Russo, M., Vaz, N.M. and de Faria, A.M., 2003. Stimulation by food proteins plays a critical role in the maturation of the immune system. International Immunology 15: 447-455.

Merlot, E., Quesnel, H. and Prunier, A., 2013. Prenatal stress, immunity and neonatal health in farm animal species. Animal 7: 2016-2025.

Meunier-Salaün, M.C. and Bolhuis, J.E., 2015. High-Fibre feeding in gestation. Chapter 5. In: Farmer, C. (ed.) The gestating and lactating sow. Wageningen Academic Publishers, Wageningen, the Netherlands, pp. 95-116.

Michiels, J., De Vos, M., Missotten, J., Ovyn, A., De Smet, S. and Van Ginneken, C., 2013. Maturation of digestive function is retarded and plasma antioxidant capacity lowered in fully weaned low birth weight piglets. British Journal of Nutrition 109: 65-75.

Milligan, B.N., Dewey, C.E. and de Grau, A.F., 2002. Neonatal-piglet weight variation and its relation to pre-weaning mortality and weight gain on commercial farms. Preventive Veterinary Medicine 56: 119-127.

Montagne, L., Boudry, G., Favier, C., Huerou-Luron, I., Lalles, J.P. and Seve, B., 2007. Main intestinal markers associated with the changes in gut architecture and function in piglets after weaning. British Journal of Nutrition 97: 45-57.

Morise, A., Louveau, I. and Le Huërou-Luron, I., 2008. Growth and development of adipose tissue and gut and related endocrine status during early growth in the pig: impact of low birth weight. Animal 2: 73-83.

Niess, J.H., Leithauser, F., Adler, G. and Reimann, J., 2008. Commensal gut flora drives the expansion of proinflammatory Cd4 T cells in the colonic lamina propria under normal and inflammatory conditions. Journal of Immunology 180: 559-568.

Oliviero, C., Kokkonen, T., Heinonen, M., Sankari, S. and Peltoniemi, O., 2009. Feeding sows with high fibre diet around farrowing and early lactation: impact on intestinal activity, energy balance related parameters and litter performance. Research in Veterinary Science 86: 314-319.

Pabst, R., Geist, M., Rothkotter, H.J. and Fritz, F.J., 1988. Postnatal development and lymphocyte production of jejunal and ileal peyer's patches in normal and gnotobiotic pigs. Immunology 64: 539-544.

Poore, K.R. and Fowden, A.L., 2004. The effects of birth weight and postnatal growth patterns on fat depth and plasma leptin concentrations in juvenile and adult pigs. The Journal of Physiology 558: 295-304.

Quesnel, H., Farmer, C. and Devillers, N., 2012. Colostrum intake: influence on piglet performance and factors of variation. Livestock Science 146: 105-114.

Quiniou, N., Dagorn, J. and Gaudré, D., 2002. Variation of piglets' birth weight and consequences on subsequent performance. Livestock Production Science 78: 63-70.

Quiniou, N., Goues, T., Mourot, J. and Etienne, M., 2010. Effect of the incorporation of 1.4% lipids from palm oil or extruded linseed in gestation and lactation diets on farrowing progress and piglets' survival before weaning. In: 42ème Journées de la Recherche Porcine. February 2-3, 2010. Paris, France, pp. 137-138.

Rehfeldt, C. and Kuhn, G., 2006. Consequences of birth weight for postnatal growth performance and carcass quality in pigs as related to myogenesis. Journal of Animal Science 84:Suppl: E113-123.

Rescigno, M., Lopatin, U. and Chieppa, M., 2008. Interactions among dendritic cells, macrophages, and epithelial cells in the gut: implications for immune tolerance. Current Opinion in Immunology 20: 669-675.

Rooke, J.A., Sinclair, A.G. and Edwards, S.A., 2001a. Feeding tuna oil to the sow at different times during pregnancy has different effects on piglet long-chain polyunsaturated fatty acid composition at birth and subsequent growth. British Journal of Nutrition 86: 21-30.

Rooke, J.A., Sinclair, A.G. and Ewen, M., 2001b. Changes in piglet tissue composition at birth in response to increasing maternal intake of long-chain N-3 polyunsaturated fatty acids are non-linear. British Journal of Nutrition 86: 461-470.

Rothkotter, H.J., 2009. Anatomical particularities of the porcine immune system – a physician's view. Developmental and Comparative Immunology 33: 267-272.

Rothkotter, H.J. and Pabst, R., 1989. Lymphocyte subsets in jejunal and ileal Peyer's patches of normal and gnotobiotic minipigs. Immunology 67: 103-108.

Salmon, H., 2000. Mammary gland immunology and neonate protection in pigs. homing of lymphocytes into the Mg. Advances in Experimental Medicine and Biology 480: 279-286.

Salmon, H., Berri, M., Gerdts, V. and Meurens, F., 2009. Humoral and cellular factors of maternal immunity in swine. Developmental and Comparative Immunology 33: 384-393.

Schumann, A., Nutten, S., Donnicola, D., Comelli, E.M., Mansourian, R., Cherbut, C., Corthesy-Theulaz, I. and Garcia-Rodenas, C., 2005. Neonatal antibiotic treatment alters gastrointestinal tract developmental gene expression and intestinal barrier transcriptome. Physiological Genomics 23: 235-245.

Stirling, C.M., Charleston, B., Takamatsu, H., Claypool, S., Lencer, W., Blumberg, R.S. and Wileman, T.E., 2005. Characterization of the porcine neonatal Fc receptor – potential use for trans-epithelial protein delivery. Immunology 114: 542-553.

Tsukahara, T., Iwasaki, Y., Nakayama, K. and Ushida, K., 2003. Stimulation of butyrate production in the large intestine of weaning piglets by dietary fructooligosaccharides and its influence on the histological variables of the large intestinal mucosa. Journal of Nutritional Science and Vitaminology (Tokyo) 49: 414-421.

Turnbaugh, P.J., Ley, R.E., Mahowald, M.A., Magrini, V., Mardis, E.R. and Gordon, J.I., 2006. An obesity-associated gut microbiome with increased capacity for energy harvest. Nature 444: 1027-1031.

Vega-Lopez, M.A., Bailey, M., Telemo, E. and Stokes, C.R., 1995. Effect of early weaning on the development of immune cells in the pig small intestine. Veterinary Immunology and Immunopathology 44: 319-327.

Wang, J., Chen, L., Li, D., Yin, Y., Wang, X., Li, P., Dangott, L.J., Hu, W. and Wu, G., 2008. Intrauterine growth restriction affects the proteomes of the small intestine, liver, and skeletal muscle in newborn pigs. Journal of Nutrition 138: 60-66.

Wang, T., Huo, Y.J., Shi, F., Xu, R.J. and Hutz, R.J., 2005. Effects of intrauterine growth retardation on development of the gastrointestinal tract in neonatal pigs. Biology of the Neonate 88: 66-72.

Westrom, B., Svendsen, J. and Tagesson, C., 1984. Intestinal permeability to polyethyleneglycol 600 in relation to macromolecular 'closure' in the neonatal pig. Gut 25: 520-525.

Westrom, B.R., Tagesson, C., Leandersson, P., Folkesson, H.G. and Svendsen, J., 1989. Decrease in intestinal permeability to polyethylene glycol 1000 during development in the pig. Journal of Developmental Physiology 11: 83-87.

Wolinski, J., Lesniewska, V., Biernat, M., Babelewska, M., Korczyniski, W. and Zabielski, R., 2001. Exogenous leptin influences gastrointestinal growth and *in vitro* small intestinal motility in neonatal piglets – preliminary results. Journal of Animal and Feed Sciences.

Xu, R.J., Mellor, D.J., Tungthanathanich, P., Birtles, M.J., Reynolds, G.W. and Simpson, H.V., 1992. Growth and morphological changes in the small and the large intestine in piglets during the first three days after birth. Journal of Developmental Physiology 18: 161-172.

Xu, R.J., Mellor, D.J., Birtles, M.J., Breier, B.H. and Gluckman, P.D., 1994a. Effects of oral Igf-I or Igf-Ii on digestive organ growth in newborn piglets. Biology of the Neonate 66: 280-287.

Xu, R.J., Mellor, D.J., Birtles, M.J., Reynolds, G.W. and Simpson, H.V., 1994b. Impact of intrauterine growth retardation on the gastrointestinal tract and the pancreas in newborn pigs. Journal of Pediatric Gastroenterology and Nutrition 18: 231-240.

Yang, C., Albin, D.M., Wang, Z., Stoll, B., Lackeyram, D., Swanson, K.C., Yin, Y., Tappenden, K.A., Mine, Y., Yada, R.Y., Burrin, D.G. and Fan, M.Z., 2011. Apical Na+-D-glucose cotransporter 1 (Sglt1) activity and protein abundance are expressed along the jejunal crypt-villus axis in the neonatal pig. American Journal of Physiology – Gastrointestinal and Liver Physiology 300: 28.

Zhang, H.Z., Malo, C. and Buddington, R.K., 1997. Suckling induces rapid intestinal growth and changes in brush border digestive functions of newborn pigs. Journal of Nutrition 127: 418-426.

16. 各种来源的脂肪饲喂母猪:对母猪和仔猪的免疫状态和性能的影响

V. Bontempo[*] *and X. R. Jiang*

Department of Health, Animal Science and Food Safety, Università degli Studi di Milano, via Celoria 10, 20133 Milan, Italy; valentino.bontempo@unimi.it

摘要:妊娠晚期胎儿能量需求大大增加,如果日粮能量供应不能满足要求,则会分解代谢母体储备。妊娠晚期和哺乳期饲料中通常会添加脂肪和油脂,以便作为高能量供应来提高新生仔猪存活率和母猪产奶量并降低动员体内储备。此外,使用来自脂质的代谢能的效率非常高,与其他营养物相比,它们具有最小的热量增加,从而减少夏季热应激。在妊娠时合成代谢阶段,增加母猪的能量摄入可能会增加妊娠晚期可用于动员的脂肪量,并可能导致母猪在哺乳期的表现增强。除了一般作为能源的重要性,脂肪和油也是必需脂肪酸的一种来源。最近的研究表明,几种脂肪酸可以改善机体功能,并且可能在生理过程如新幼仔免疫功能和母猪生育力中非常重要。本章主要关注在饲料中使用ω-脂肪酸(n-3 和 n-6 FA)和共轭亚油酸(conjugated linoleic acid)在提高母猪的繁殖性能和免疫力、胎儿和新幼仔的表现和免疫状态方面的最新进展。

关键词:共轭亚油酸,免疫状态,ω-脂肪酸,性能,仔猪,母猪

16.1 引言

母猪的繁殖性能对于决定分娩环节的获利能力方面至关重要,其取决于管理和母猪生物学因素之间复杂的相互作用,会影响母猪生育力、胚胎存活、胎儿发育及仔猪的分娩和存活(Tanghe 和 De Smet,2013)。妊娠后期母猪饲料中添加脂肪能够提高初乳产量并增加新生仔猪的存活率(Hansen 等,2012)。在过去 20 年中,许多研究已经评估了在妊娠期和哺乳期饲料中添加多不饱和脂肪酸(polyunsaturated fatty acids,PUFA)如二十二碳六烯酸(docosahexaenoic acid,DHA)、二十碳五烯酸(eicosapentaenoic acid,EPA)和共轭亚油酸(conjugated linoleic acid,CLA)的作用,通过提高胎儿生长、仔猪健康和母猪哺乳表现,从而有益于分解代谢条件下的妊娠母猪及泌乳母猪。

长链多不饱和脂肪酸(long-chain polyunsaturated fatty acids)是前列腺素(prostaglandins)和类二十烷酸(eicosanoids)的底物,因此影响细胞因子的表达(Alexander,1998)。n-3 和 n-6 脂肪酸(fatty acids,FA)是脑、中枢神经系统和血管系统中质膜的必需成分,使其成为组织快速形成(即妊娠和胎儿生长)期间的关键成分。这些 PUFA 产生不同类型的类二十烷酸(eicosanoids),其在调节炎症反应、血压和血小板聚集中起重要作用。饲料中 CLA 还影响初乳脂肪中的脂肪酸组成,但更有趣的是,它们对初乳中的免疫因子产生积极影响,比如 IgG 浓度的增加(Bontempo 等,2004)。

最近的研究表明,几种脂肪酸似乎可以改善机体功能,并且可能对一些生理过程极其重要。然而,这些发现往往是矛盾的,这可能是由于补充脂肪酸水平以及补充脂肪酸的持续时间不同。本综述总结了目前关于在母体日粮中补充 ω-FA 和 CLA 对母猪和仔猪的生产表现、脂肪酸分布和免疫状态的影响的相关知识。此外,还讨论了使用两种类型的脂肪酸的策略。

16.2　母猪饲料中脂肪和油的使用

乳汁中的脂肪和能量含量对仔猪生存起着重要作用,且发现通过在母猪饲料中添加脂肪可以提高整窝体重。妊娠晚期在母猪的的饲料中补充脂肪会增加初乳中的总脂质和乳糖,并且还会增加初乳 IGF-I 浓度(Farmer 和 Quesnel,2009)。为了减少哺乳期母猪所遇到的脂肪和能量损失,将动物脂肪或植物油添加到晚期妊娠或哺乳期母猪饲料中被广泛应用于商业农场。在妊娠晚期增加饲料脂肪可能会提高母猪的初乳产量,并增加新生仔猪的生存率(Hansen 等,2012)。然而,妊娠晚期营养水平的增加也可能对哺乳期的母猪产能产生负面影响。第 7 章更详细地介绍了过度饲喂的这个问题(Theil,2015)。在妊娠晚期过量的能量摄入被证明可以抑制乳腺分泌组织的发育,并因为采食量的减少而增加哺乳期的身体状况的损失(Laws 等,2009a)。还有人提出,饲料中脂肪的类型而不是总脂肪的含量对于决定幼崽的结局至关重要(Laws 等,2007a)。因此,补充饲料脂肪和油应该仔细并适当地进行。

16.3　ω-脂肪酸(n-3 与 n-6 FA)

n-3 和 n-6 FA 是多不饱和脂肪酸(polyunsaturated fatty acids,PUFA),并且可以通过甲基端的第一双键的位置彼此区分。因此,n-3 和 n-6 脂肪酸(FA)在代谢和功能上是不同的,不可互换,并且可能具有相反的生理功能(Calder,2003)。在 PUFA 中,α-亚麻酸(α-linolenic acid,ALA;C18:3n-3)和亚油酸(linoleic acid,

LA；C18：2n-6)被归类为营养必需脂肪酸(essential fatty acids,EFA),因为哺乳动物不能合成它们(Enser,1984)。作为营养和生理上都很重要的其他 PUFA 前体,ALA 和 LA 可以通过特异性酶的去饱和和延长而转化为更长链的 FA(图 16.1)。ALA 可以转化为 EPA(C20：5n-3)和 DHA(C22：6n-3),而 LA 可以转化为花生四烯酸(arachidonic acid,ARA；C20：4n-6)(Kurlak 和 Stephenson,1999)。因此,动物可以从饲料中直接获得 DHA 和 EPA,亦可从来源于饲料中的 ALA 从头合成。DHA 对于幼崽的正常生长发育至关重要(Calder,2003),ARA 是肠内稳态和损伤后修复的主要调节因子。胃肠道紊乱是新生幼仔发病和死亡的主要原因（Jacobi 等，2011)。

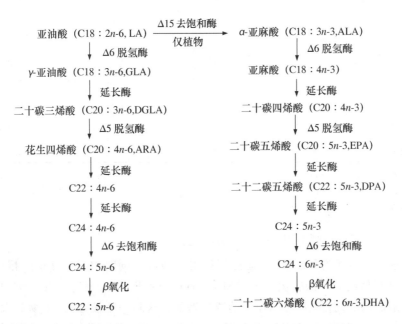

图 16.1 植物来源的 n-6 和 n-3 多不饱和脂肪酸转化为其长链形式(根据 Tanghe 和 De Smet,2013. 重新绘制)。

16.3.1 ω-脂肪酸的来源

不同植物油或鱼油的 FA 成分见表 16.1。LA(C18：2n-3)和 ALA(C18：3n-3)可以从谷物和植物油中获得,而 EPA(C20：5n-3)和 DHA(C22：6n-3)主要存在于许多类型的鱼油中,如鲑鱼、沙丁鱼、鲨鱼肝、金枪鱼和鲱鱼油。许多植物油,如大豆、玉米和向日葵油,富含 n-6 FA,主要为 LA,但亚麻籽是 ALA 的丰富来源(超过 50%)。核桃、油菜籽或大豆油也含有 7%～17% ALA 含量,这被认为是 n-3 FA 的丰富来源。

表 16.1 不同类型植物油或鱼油中 n-3 和 n-6 脂肪酸含量   %

项目	C18:2n-6 (LA)	C18:3n-3 (ALA)	C20:5n-3 (EPA)	C22:6n-3 (DHA)
植物油[1]				
亚麻油	15.18	54.24	—	—
菜籽油	20.12	8.37	—	—
豆油	56.02	7.15	—	—
玉米油	59.27	1.07	—	—
葵花籽油	71.17	0.45	—	—
核桃油[2]	51.60	17.82	—	—
鱼油[3]				
鲱鱼油	2.7	1.3	12.5	8.9
沙丁鱼油	2.5	1.3	9.6	8.5
三文鱼油	2.0	1.0	6.7	16.1
鲨鱼肝油	2.3	0.4	9.2	7.3
金枪鱼油	0.9	0.4	4.4	22.2

[1] 选自 Zambiazi 等(2007).
[2] 选自 Dogan 和 Akgul (2005).
[3] 选自 Gruger(1967).

16.3.2 多不饱和脂肪酸的生物放大作用

在子宫内生活的最后阶段,通过仔猪组织和母猪饲料中 FA 分布的相似性证明,FA 从母体到胎儿的转运逐渐增加(Farmert 和 Petit,2009)。在人类中,随着妊娠年龄的增加,脐带血中 PUFA 的浓度增加。这是一个被称为"生物放大"的过程,即 EFA 浓度的增加和其到胎儿循环、脐带、肝和脑的转移,其在妊娠的最后阶段达到最高水平(Uauy 和 Dangour,2006)。根据 Lauridsen 和 Jensen(2007)的报告,仔猪血浆和组织中 FA 分布在断奶后 3 周内都受到母体饲料成分的极大影响。多不饱和 n-3 FA 在许多生理活动中具有关键作用,例如中枢神经系统的发育、视网膜功能、细胞膜的流动性、前列腺素合成和免疫(Innis,2007;Kim 等,2007;Mateo 等,2009;Uauy 和 Dangour,2006)。特别地,DHA 是神经系统和细胞膜的主要成分,并且对膜稳定性和膜蛋白(多巴胺和神经递质)功能特性产生重要影响。

16.3.3 ω-脂肪酸对母猪和仔猪免疫水平的影响

据报道,n-3 和 n-6 FA 作为结构组分存储在细胞膜中,也作为生成二十烷酸如前列腺素(prostaglandins,PGE)的底物(Rossi 等,2010)。Omega-3(n-3)FA

可以减少炎症细胞因子产生,而 n-6 FA 则被证明促炎症(Calder,2003,2013)。如图 16.1 所示,LA 可以转换为 ARA,ALA 可以转换为 EPA。ARA 是 2 系列 PGE 和血栓素及 4 系列白三烯的前体,EPA 是 3 系列 PGE 和血栓素以及 5 系列白三烯的前体,也可转化为它们各自的二十烷酸,其在动脉粥样硬化、冠心病、支气管哮喘和其他炎性病症中起重要作用。2 系列的 PGE 调节促炎细胞因子的产生,而 3 系列的 PGE 是抗炎类二十烷酸。为了促进抗炎 PGE 的产生并阻止炎性 PGE 的产生,应在饲料中减少 ARA,并且采食适量的 n-3 FA(Rossi 等,2010)。

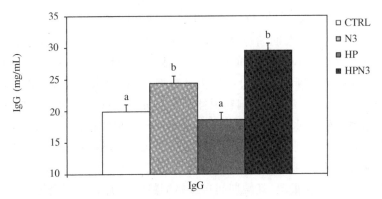

图 16.2 第一胎母猪单独饲喂 n-3 脂肪酸(N3)或高蛋白(high-protein,HP)饲料,或组合饲喂时(HPN3),在妊娠晚期和整个哺乳期([a,b] 不同,$P<0.01$)初乳中 IgG 浓度。选自 Mateo 等(2009)。

新生仔猪的生存依赖于足够的初乳摄取来提供体液免疫保护(Rooke 和 Bland,2002)。仔猪饲料中的长链 n-3 PUFA 可以供调节炎症免疫反应的强度和持续时间(Leonard 等,2011)。在被鱼油饲喂的母猪哺乳的小猪中证明了其抗炎症反应的效果;在妊娠晚期和哺乳期饲料中用鲱鱼鱼油(70% g/kg 饲料)代替猪油减少了哺乳仔猪免疫细胞体外二十烷酸的释放(Fritsche 等,1993)。来自鱼油的 Omega-3 FA 已被证明可以改善大脑的发育和新生仔猪的免疫力(Leskanich 和 Noble,1999)。Bazinet 等(2004)报道说,57 日龄仔猪从亚麻籽油(3.5%)中大量摄入 ALA,可以与佐剂一起调节其免疫功能和组织 FA 对抗原的反应。此外,n-3 FA 及其代谢物是过氧化物酶体增殖物受体激活剂 γ(peroxisome proliferator receptor activator gamma,PPARγ)的天然配体。通过其抑制炎性细胞因子表达的能力和引导免疫细胞分化成抗炎表型的功能,这种 PPARγ 在免疫应答中起着重要作用。据报道,n-3 PUFA 不仅减少了促炎细胞因子肿瘤坏死因子-α(TNF-α)和白细胞介素 6 和 8 的分泌,而且增强了 PPARγ 的表达(Calder,2013)。然而有人认为,摄取大量的 n-6 FA,特别是 ARA 可能有助于炎症过程,并且有利于或加重炎性疾病(Papadopoulos 等,2009)。Rooke 和 Bland(2002)报道,仔猪对 G 类免疫球蛋白(IgG)的合成和吸收与母体 IgG 的量呈正相关,从而增强从初乳摄取足够 IgG 的重要性。Mitre 等(2005)证明从妊娠 80 天到断奶期间鲨鱼肝油的补充增加了母猪的初乳 IgG 浓度,以及随后哺乳仔猪血清 IgG 浓度。Mateo 等(2009)观察到在母猪饲料中补充从鱼油获得的长链 n-3 FA 时,初乳 IgG 的浓度增加(图 16.2)。Farmer 等(2010)也观察到母猪在妊娠后期和哺乳期间饲喂亚麻籽饲料时仔猪的免疫反应增强,这可能导致仔猪存活率提高。然而,从妊娠 109 天到断奶期间母体鱼油的补充会上调肠促炎细胞因子表达,且不影响乳腺分泌物的免疫球蛋白组成(Leonard 等,2010b,2011)。

16.3.4 ω-脂肪酸对母猪和仔猪生产性能的影响

在养猪生产中,妊娠期间母体饲料补充鱼油对胎儿生长发育会有显著影响(Corson 等,2008;Laws 等,2009b)。研究表明,饲料中补充 n-3 FA 可能对仔猪活力和仔猪断奶前与断奶后生长产生积极影响,但对母猪繁殖性能没有明显影响(Tanghe 和 De Smet,2013)。饲料中补充长链 n-3 FA(包括 EPA 和 DHA)可以通过多种基因在转录水平的调控来影响许多分子和细胞信号通路,这被认为是 PUFA 影响繁殖的机制(Waters 等,2014)。在后备母猪和母猪的玉米-大豆饲料中添加鱼油能够显著提高出生时产仔数(Smits 等,2011;Spencer 等,2004;Webel 等,2004)。据报道,妊娠期间在母猪饲料中添加 1.75% 金枪鱼油可以提高仔猪生长速度,妊娠期不同时期饲喂金枪鱼油对仔猪的脑部 FA 水平有着不同的影响(Rooke 等,2001a)。Mateo 等(2009)报告说,无论在第几胎,在哺乳期饲料中仅添加 0.2% n-3 FA 均可提高保育仔猪的生长速度。此外,Laws 等(2007a)报道说,在妊娠前 60 天内饲喂 10% 鱼油的母猪,其在断奶前期显示出较高的新生仔猪预后及后续的子代生长性能,但在妊娠 60 日龄至分娩期间补充鱼油则对仔猪在出生时和断奶时的体重没有影响(Laws 等,2007b)。从妊娠 109 天至断奶(24 天)在母猪饲料中添加 100 g/天鱼油可以改善断奶后仔猪的小肠形态和生长性能(Leonard 等,2010a,2011)。然而,也有许多已发表的研究成果表明:无论在母猪妊娠早期、中期或晚期补充这些没有显著的反应(Estienne 等,2006;Lauridsen 和 Danielsen,2004;Perez-Rigau 等,1995;Rooke 等,1998,1999,2001a,b)。Rooke 等(2001c)观察到在妊娠和哺乳期间饲喂 1.65% 鲑鱼油的母猪其仔猪的断奶前死亡率降低,这主要是由于仔猪被母猪压死的可能性降低。研究亚麻籽或亚麻籽油中 n-3 FA 对母猪繁殖和仔猪性能的影响见表 16.2。在妊娠期和哺乳期的母猪饲料中添加亚麻籽可以增加其仔猪的出生体重和断奶重量(Baidoo 等,2003;Farmer 等,2010)。Boudry 等(2009)也观察到在妊娠和哺乳期的母猪饲料中添加 3.0% 亚麻油能够使其妊娠期变长。这与 Rooke 等(2001c)使用 1.65% 鲑鱼油的研究结果一致。在妊娠 108 天至哺乳 28 天这段时间内,将从动物脂肪、菜籽油、鱼油、椰子油、棕榈油或向日葵油中获得的脂肪(8%)补充至母猪饲料中能通过母猪中脂肪和能量的输出来提高哺乳仔猪的体重(Lauridsen 和 Danielsen,2004)。然而，妊娠期间在饲料中补充葵花籽油（提供 C18∶2n-6 LA）未能影响母猪的繁殖性能和仔猪在断奶前的生长（Corson 等，2008；Laws 等，2007a,b）。

16.3.5　ω-脂肪酸对于母猪和仔猪体内脂肪酸分布的影响

妊娠晚期和哺乳期母猪饲料中的 FA 组成对哺乳后代仔猪的 FA 组成有重大影响，且这种影响持续至断奶后。实际上，母体饲料中 n-6∶n-3 FA 的比例影响了断奶前仔猪和断奶后仔猪体脂肪这些 FA 的比例（Lauridsen 和 Jensen，2007）。当用亚麻籽饲喂母猪时，母乳组成的变化也表明不饱和和 n-3 FA 的增加以及饱和 FA（saturated FA，SFA）的降低（Baidoo 等，2003）。Gunnarsson 等（2009）观察到，在分娩前三周，在母猪饲料中补充 n-3 FA（来自亚麻籽油）会使仔猪大脑组织中 DHA 含量增加而使 n-6∶n-3 PUFA 的比例降低。许多研究表明，在妊娠晚期和哺乳期间用鱼油饲喂母猪可以增加母猪的乳汁、血清中和仔猪组织中长链的 n-3 FA 水平（Fritsche 等，1993；Perez Rigau 等，1995；Rooke 等，1998，1999，2000，2001a,b）。表 16.2 显示亚麻油或亚麻籽油中的 n-3 FA 对母猪和仔猪体内 FA 分布的影响。Boudry 等（2009）在妊娠和哺乳期间对母猪饲料补充 3.0％亚麻籽油时，观察到仔猪回肠的 FA 成分的改变。当母猪在妊娠晚期和哺乳期饲喂亚麻籽油时，可以观察到母猪乳汁中的 n-3 FA 和 n-3∶n-6 比例增加，同时仔猪组织中，特别是大脑中 n-3 FA 和 n-3∶n-6 比例的增加（Bazinet 等，2003；Farmer and Petit，2009；De Quelen 等，2010，2013）。Missotten 等（2009）观察到，从妊娠 45 日龄开始在母猪饲料中提供亚麻油可以增加仔猪肝脏、肌肉和脂肪组织中的 EPA 浓度（5.6±0.3 日龄），但不会增加 DHA 浓度。此外，Farmer 等（2007）报道，88～212 日龄在饲料中添加亚麻可增加 PUFA 的浓度，并降低了后备母猪血液和乳腺外体中的 n-6∶n-3 比例。

表 16.2　研究母猪日粮中添加植物油 n-3 脂肪酸对其脂肪酸分布和繁殖能力及仔猪生长影响的试验。

添加物	饲喂时间	母猪	仔猪	参考文献
2.0％亚麻籽油	分娩前 10 天到哺乳期 14 天	增加乳汁中的 n-3 FA 和 n-3∶n-6 比例	增加组织中的 n-3 FA 和 n-3∶n-6 比例	Bazinet 等，2003
10％亚麻籽，6.5％亚麻籽粉或 3.5％亚麻籽油	88～212 日龄	增加 PUFA，减少血浆和乳腺外体的 n-6∶n-3 比例	无观察	Farmer 等，2007
3.0％亚麻籽油＋2.0％猪油	妊娠 28 天至断奶	增加妊娠时长，增加乳汁和红细胞的 n-3 FA，降低 n-6∶n-3 比例	改变回肠中肠 FA 的组成	Boudry 等，2009

续表 16.2

添加物	饲喂时间	母猪	仔猪	参考文献
10%亚麻籽，6.5%亚麻籽粉或3.5%亚麻籽油	妊娠68天到哺乳期21天	增加乳汁和血浆中PUFA和n-3 FA，降低n-6∶n-3比例	增加大脑中n-3 FA，降低n-6∶n-3比例	Farmer and Petit，2009
妊娠期为1.5%亚麻油；哺乳期为5.5%亚麻籽	妊娠28天至断奶	增加血浆中n-3和n-6 FA，降低C18∶2n-6与C18∶3n-3比例，增加初乳与乳汁中n-3 FA	增加大脑中C22∶6n-3含量	de Quelen 等，2010
10%亚麻籽，6.5%亚麻籽粉或3.5%亚麻籽油	妊娠68天到哺乳期21天	增加乳汁中蛋白质含量	降低死亡率，改善断奶后生长	Farmer 等，2010
3.92%或5.25%的压榨亚麻油在妊娠或哺乳期	妊娠28天至断奶	在初乳和乳汁中增加n-3 FA	改变组织中长链n-3 PUFA水平	de Quelen 等，2013

16.3.6　ω-脂肪酸的饲喂策略

如前所述，一般认为妊娠和哺乳期母猪饲料中的 FA 组成会影响其乳汁中的 FA 组成，并进而影响仔猪组织中的 FA 组成。然而，在妊娠和哺乳期间补充 FA 的时间和量可能会不同地影响仔猪的生长和免疫状态。许多研究选择妊娠晚期和哺乳期作为补充的最佳时机（Bazinet 等，2003；Fritsche 等，1993；Leonard 等，2010a,b，2011），但许多研究者也选择从早期妊娠至断奶这一区间（Boudry 等，2009；De Quelen 等，2010，2013；Farmer 等，2010；Mitre 等，2005；Rooke 等，2001c）。亦有其他研究者仅在妊娠期测试了 ω-FA 的作用（Laws 等，2007a,b，2009a,b）。Leonard 等（2010b）发现，从妊娠期的 109 日龄开始饲喂鱼油时，对初乳中 IgG 浓度没有影响。这种作用的缺乏可能是由于补充时间太短，因为其他研究表明在妊娠早期开始补充则具有积极的作用（Mitre 等，2005）。需要更多的剂量反应方面的工作来确定 n-3 FA 酸补充的最佳用量。

将不同来源的 FA（例如鱼油、亚麻籽油）加入母猪的饲料中，并对其转移至后代和对后代的影响进行了调查。母猪补充鱼油（0.3%～10%）改善了出生后仔猪生长（Mateo 等，2009；Laws 等，2007a；Rooke 等，2001a；Smits 等，2011），以及母猪初乳和仔猪的免疫水平（Fritsche 等，1993；Mateo 等，2009；Rooke 和 Bland，2002）。在妊娠晚期向母猪饲喂亚麻籽油（3.5%～10%）改善了断奶后仔猪的生长(Baidoo 等,2003;Farmer 等,2010)。另一方面,高水平的鱼油补充可能不会改善仔猪生长。Danielsen 和 Lauridsen(2001)报道,在哺乳期母猪饲料中添加 10% 的鱼油时,断奶时的仔猪数量明显较小,同时产奶量下降。有证据表明,饲料中补充 n-3 FA 和 n-6 FA 能有效增加其在猪胎儿中的可用性。高 n-6:n-3 比值被认为是胰岛素抵抗和动脉粥样硬化的关键因素(Papadopoulos 等,2009)。哺乳母猪饲料中 n-6 和 n-3 PUFA 的比例变化可能会影响母猪的免疫因子组成,包括免疫球蛋白和细胞因子,以及仔猪的免疫状态。Yao 等(2012)报道,与 3:1 和 13:1 的比例相比,当母猪日粮中 n-6:n-3 比例为 9:1 时,可以增加初乳中的 IgG 浓度和乳汁中的 IgM 浓度。

16.4 共轭亚油酸

具有共轭双键的脂肪酸天然存在于来自反刍动物的食用脂肪中,例如在乳脂和牛油中(Pariza 等,2001)。共轭亚油酸是一组 LA 的几何和位置异构体的集合名称,其中双键被单个碳—碳键而不是亚甲基分隔。对其生物学效应研究最多的两种异构体是顺式-9、反式-11CLA(c9,t11)(反刍动物中产生的主要异构体)和反式-10、顺式-12CLA(t10,c12)。两种异构体已被用于实验性研究,代表了研究最广泛的 CLA 异构体。由 Pariza 等(2001)总结了 CLA 有许多有利的生物学效应。顺式-9、反式-11 异构体主要负责抗致癌作用,但反式-10、顺式-12 异构体可降低机体脂肪,并被称为影响血脂的最有效异构体(Rossi 等,2010)。哺乳期的 CLA 摄取可以增加母乳中 CLA 的浓度(Schmid 等,2008)。哺乳期的饲料中补充 CLA 还导致初乳和乳汁中 CLA 的富集,因此允许哺乳仔猪获得 CLA 异构体(Bee,2000a)。因此,似乎 CLA 可以通过乳汁从母猪转移到其后代。此外,当在妊娠期和哺乳期用 0.5% CLA 饲喂母猪时,在哺乳仔猪中证明了 CLA 刺激免疫功能的能力(Bontempo 等,2004;Corino 等,2009)。

16.4.1 共轭亚油酸对母猪和仔猪免疫状况的影响

根据先前的研究报道,CLA 在几个实验研究中都具有免疫调节作用(Bassaganya-Riera 等,2001;Corino 等,2002;Whigham 等,2001)。据报道,在猪饲料中添加 CLA 可以增加免疫球蛋白的产量(Corino 等,2009;Moraes 等,2012),CLA 的抗炎症特性可从免疫应激的反应过程中促炎细胞因子减少得到证明(Lai 等,2005)。还有报道 CLA 可以降低免疫刺激诱导的分解代谢反应,而不会对免疫功能产生不利影响。这种分解代谢反应由细胞因子介导,并由 PGE2 合成调节(Miller 等,1994)。因此,降低免疫应答可以为骨骼肌中的合成代谢过程提供更多的能量。Corino 等（2002）报道,日粮中 CLA 可以增加血清 IgG 和溶菌酶的产生,但对断奶仔猪饲喂 4 周的 CLA 对其血清 α-1-酰基糖蛋白（α-1-acylglycoprotein,AGPAGP）水平没有影响。这个发现与 Sugano（1998）等的研究是一致的,他们发现大鼠日粮中补充 0.5％和 1％CLA 其血清 IgG 会升高。此外,日粮中补充 CLA 可以帮助减轻 LPS 注射仔猪的炎症反应,这是通过提高促炎细胞因子（即 IL-6 和 TNF-α）的产生和表达及上调 PPARγ 表达来实现的（Lai 等，2005）。然而,CLA 对哺乳母猪和仔猪的潜在益处目前还不太清楚。有证据表明,成年免疫保护在产后发育,并受到 PUFA 摄入的影响（Field 等,2001）。也许在这个快速生长和发育期间暴露于 PUFA 以及更大的初乳免疫球蛋白摄入量,可能会增强免疫功能并降低各种疾病的风险。

被动免疫或仔猪对初乳免疫球蛋白的充分摄入在其产后生命的头几周起着重要的作用。不仅在产后极短时间内（Drew 和 Owen,1988）,且在断奶后（Varley 等,1986）,若母体转移给新生儿的免疫球蛋白不足会增加了感染的易感性。Rooke 和 Bland（2002）提出,仔猪的 IgG 合成吸收与母体 IgG 的量呈正相关,因此增强从初乳中摄取足够的 IgG 极为重要。Krakowski 等（2002）也报道,初乳中 IgG 含量的增加,不仅决定了在哺乳期、且决定了在断奶后仔猪的免疫力。表 16.3 显示了母猪日粮中补充 CLA 对母猪和其仔猪的初乳及血清免疫球蛋白影响的研究结果。Bontempo 等（2004）报道,在妊娠晚期和哺乳期间,在母猪的日粮中补充 0.5％ CLA,可以增加初乳中 IgG 浓度、母猪血清 IgG 和溶菌酶水平,以及哺乳期仔猪血清中 IgG 和溶菌酶浓度（表 16.4）,因此可能对母猪和哺乳仔猪的免疫反应有积极的影响。补充有 CLA 母猪断奶的小猪其免疫状态指标升高,如肠黏膜炎症的减少和血清 IgG 和 IgA 升高（Patterson 等,2008）。Corino 等（2009）观察到补充 CLA 可以增加母猪初乳中 IgG、IgA 和 IgM 浓度,证实了 CLA 异构体在猪免疫调节中的作用（表 16.5）,这与 Corino 等（2002）和 Bontempo 等（2004）所述相同。

表 16.3 母猪日粮中添加共轭亚油酸(CLA)对母猪初乳和
其仔猪血清中免疫球蛋白的影响的研究

CLA 添加剂	初乳中免疫球蛋白	血清中免疫球蛋白	参考文献
母猪			
0.5％	IgG,IgA 和 IgM 增加	—	Corino 等,2009
0.5％	IgG 增加	IgG 增加	Bontempo 等,2004
仔猪			
0.5％	—	IgG 增加	Bontempo 等,2004；Corino 等,2009
2％	—	IgA 和 IgG 增加	Patterson 等,2008

表 16.4 在哺乳期间母猪日粮补充共轭亚油酸(CLA)对初乳 IgG 水平和母猪及其仔猪血清中的免疫变量的影响(Bontempo 等,2004)。

	哺乳期天数			
	0	2	10	20
母猪				
初乳 IgG(g/L)				
对照	33.59[a]			
CLA	49.44[b]			
血清 IgG(g/L)				
对照		10.78[a]	14.98[a]	17.41[a]
CLA		20.85[b]	21.33[b]	25.06[b]
血清溶菌酶(mg/L)				
对照		0.996[a]	1.398[a]	1.462[a]
CLA		1.866[b]	2.013[b]	2.256[b]
仔猪				
血清 IgG(g/L)				
对照		22.92[a]	12.28[a]	9.28[a]
CLA		29.40[b]	20.86[b]	16.22[b]
血清溶菌酶(mg/L)				
对照		0.47	0.50	1.29[a]
CLA		0.65	0.71	2.19[b]

[a,b] 同一列平均数有不同上标则差异显著($P<0.05$)。

表 16.5 在分娩前 7 天至产后 7 天(T1),或分娩前 7 天至妊娠(T2)给予对照饲喂的或 0.5% 共轭亚油酸(CLA)的母猪其初乳中 IgG、IgA 和 IgM 浓度(Corino 等,2009)。　g/L

	对照	T1	T2
IgG	82.6[a]	108.0[b]	101.4[b]
IgA	11.3[a]	18.4[b]	21.0[b]
IgM	6.1[a]	7.9[b]	9.1[b]

[a,b] 同一行平均数有不同上标则差异显著($P<0.05$)。

16.4.2 共轭亚油酸对母猪和仔猪生产性能的影响

研究表明,哺乳期在日粮中补充 CLA 可以导致初乳和乳汁中 CLA 的富集,从而使 CLA 异构体进入哺乳仔猪(Bee,2000a)。Bee(2000b)观察到妊娠期和哺乳期饲喂 CLA 的母猪,其哺育的仔猪在断奶后生长得更快。在妊娠晚期和哺乳期在母猪饲料中补充CLA也可增加仔猪在断奶时体重（Corino等，2009）。然而另一些研究表明，在分娩至断奶前7或8天（Bontempo等，2004；Corino等，2009），或在妊娠40~75天至断奶时（Poulos等，2004），用CLA异构体饲喂母猪其生产性能（采食量、体重减轻或身体状况评分）没有差异。其他研究也显示，在妊娠和哺乳期间补充CLA的母猪所产的仔猪其生长速度没有改善（Bontempo等，2004；Patterson等，2008；Peng等，2010；Poulos等，2004）。Krogh等（2012）观察到，在饲料中添加1.3% CLA（顺式-9，反式-11和反式-10，顺式-12）之后初乳产量和仔猪性能降低，但是泌乳量增加。此外，据报道，饲喂1% CLA可以减少哺乳期母猪背膘厚度的损失，导致断奶时仔猪重量的增加（Cordero等，2011）。然而，这与Park等（2005），Harrell等（2002）的研究结果有所不同，Park等（2005）发现从配种前15天至断奶期间饲料中补充CLA倾向于降低母猪背膘厚度和仔猪在出生或断奶时重量；Harrell等（2002）发现仔猪重量和整窝重量、母猪体重和背膘厚度变化不受在哺乳期5~20天饲喂的CLA的影响。

16.4.3 共轭亚油酸对母猪和仔猪脂肪酸分布的影响

哺乳期的CLA摄入会改变乳汁的成分。在妊娠和哺乳期，母猪饲料中补充CLA可提高总饱和脂肪酸（saturated fatty acid，SFA）的浓度，但对乳汁中的单不饱和脂肪酸（monounsaturated fatty acids，MUFA）则没有影响（Bee，2000a；Peng等，2010）。Schmid等（2008）表明，富含CLA的高山黄油在不会导致乳脂肪抑制或改变总SFA，MUFA和PUFA的组成的情况下，对乳汁中的CLA含量有积极的影响。此外，脂质代谢中的酶活性、乳FA组成的变化以及乳脂的总含量和产量的显著降低等很多的直接效应一般归因于CLA（Bontempo等，2004；Cordero等，2011）。Harrel等（2002）发现，在分娩后5天至断奶时饲料中添加1%CLA会使乳脂含量降低36%。日粮中添加CLA可以改变FA数量，且这种变化在乳脂中的幅度比初乳脂质更高（Cordero等，2011）。可以通过调节母体CLA摄入来改变子代的身体组成。妊娠期和哺乳期母猪CLA的补充使新生仔猪和断奶仔猪背部SAF含量增加而总MUFA和LA浓度下降（Peng等，2010）。Peng等（2010）表明，母猪在妊娠期和哺乳期日粮中增加CLA会改变仔猪出生时血浆FA组成，且这种改变在出生时大于断奶时。母猪饲料中添加CLA会增加脐带血浆、初乳和牛奶中CLA异构体的浓度，并随之增加新生仔猪和断奶仔猪血浆、背膘厚度和长尾肌的CLA浓度（Bee，2000a，b；Peng等，2010；Poulos等，2004；Schmid等，2008）。有趣的是，在Peng（2010）等的研究结果中，可以在初乳和牛奶中检测到顺式-9，反式-11-18:2和反式-10，顺式-12-18:2，但在脐带血浆中检测到只有反式-10，顺式-12-18:2，表明脐带血和乳汁之间的CLA转移模式可能存在一些差异。

16.4.4 共轭亚油酸的饲喂策略

许多研究表明,在母猪饲料中补充 CLA 显著提高了母猪及其后代的免疫状况。然而,母猪饲料中补充 CLA 的持续时间可能会影响哺乳母猪及其仔猪的生长程度、免疫组分以及代谢和激素反应(Corini 等,2009)。Peng 等(2010)表明,在妊娠晚期和哺乳期间日粮中补充 CLA 会改脐带血和乳汁的 FA 分布,并导致从母猪到仔猪的复杂转移模式。因此,许多研究使用从妊娠晚期开始并持续到哺乳期的区间进行 CLA 补充(Bontempo 等,2004;Codero 等,2011;Krogh 等,2012;Peng 等,2010)。而在一些研究中,从妊娠中期直到断奶期间一直在母猪饲料中补充 CLA(Patterson 等,2008;Poulos 等,2004),在其他研究中,则在整个妊娠和哺乳期间饲喂 CLA (Bee,2000a,b)。Corino 等(2009)评估了在两个不同时期饲喂 0.5% CLA 的日粮,分别是从分娩前 7 天到产后 7 天,或分娩前的 7 天直至断奶。他们发现在仔猪断奶时的体重和免疫组分方面,从分娩前 7 天到产后 7 天补充更加有益。Park 等(2005)观察到,与从交配后 74 天至断奶时饲喂 CLA 的母猪相比,从配种前 15 天到断奶期间饲喂 CLA 的母猪其仔猪在 21 日龄时体重更低。然而,Poulos 等(2004)却没有发现当母猪在妊娠 40 天至断奶,或者妊娠 75 天至断奶期间饲喂 CLA 时其仔猪生长有任何差别。

当 CLA 的补充水平为 0.5% 时,其对哺乳母猪和仔猪免疫力(Bontempo 等,2004)以及仔猪生长发育(Corino 等,2009)有益。较高的补充水平,如 1%(Codero 等,2011)、1.3%(Krogh 等,2012)或 2%(Bee,2000a,b;Patterson 等,2008)时同样对母猪或其后代有影响。Peng 等(2010)报道,当 CLA 补充水平从 0.5% 增加到 1.0%时,新生仔猪(2 日龄)和断奶仔猪(26 日龄)脐带血、血浆、背膘厚度和背最长肌中 CLA 异构体的浓度呈线性增加。Park 等(2005)发现,当饲料中 CLA 水平较高(2%)和饲喂时间较长(从配种前 15 日至断奶)时仔猪出生体重反而较低。

16.5 结论

由于生物放大作用,可以通过在母猪日粮中补充 n-3 FA 的方式,使 n-3 长链 PUFA(EPA 和 DHA)富集于仔猪的组织(包括神经组织)中。ω-脂肪酸与 CLA (共轭亚油酸)可用于改善母猪的 FA 分布和其后代的生长性能。此外,当母猪在妊娠和哺乳期饲喂 n-3 FA 和 CLA 时,n-3 FA 和 CLA 的免疫调节活性可能会改善仔猪在初生期、断奶前和断奶后期的健康和福利。然而,需要做更多的工作才能建立 CLA 或 ω-FA 的最佳补充水平以及为获得最有利的结果所需的理想补充时间。

参考文献

Alexander, J.W., 1998. Immunonutrition: the role of n-3 fatty acids. Nutrition 14: 627-633.

Baidoo, S.K., Azunaya, G. and Fallad-Rad, A., 2003. Effects of feeding flaxseeds on the production traits of sows. Journal of Animal Science 81(1): 320.

Bassaganya-Riera, J., Hontecillas-Magarzo, R., Bregendahl, K., Wannemuehler, M.J. and Zimmerman, D.R., 2001. Effects of dietary conjugated linoleic acid in nursery pigs of dirty and clean environments on growth, empty body composition, and immune competence. Journal of Animal Science 79: 714-721.

Bazinet, R.P., McMillan, E.G. and Cunnane, S.C., 2003. Dietary alpha-linolenic acid increases the n-3 PUFA content of sow's milk and the tissues of the suckling piglet. Lipids 38: 1045-1049.

Bazinet, R.P., Douglas, H., McMillan, E.G., Wilkie, B.N. and Cunnane, S.C., 2004. Dietary 18:3omega3 influences immune function and the tissue fatty acid response to antigens and adjuvant. Immunology Letters 95: 85-90.

Bee, G., 2000a. Dietary conjugated linoleic acids alter adipose tissue and milk lipids of pregnant and lactating sows. The Journal of Nutrition 130: 2292-2298.

Bee, G., 2000b. Dietary conjugated linoleic acid consumption during pregnancy and lactation influences growth and tissue composition in weaned pigs. The Journal of Nutrition 130: 2981-2989.

Bontempo, V., Sciannimanico, D., Pastorelli, G., Rossi, R., Rosi, F. and Corino, C., 2004. Dietary conjugated linoleic acid positively affects immunologic variables in lactating sows and piglets. The Journal of Nutrition 134: 817-824.

Boudry, G., Douard, V., Mourot, J., Lallès, J.P. and Le Huërou-Luron, I., 2009. Linseed oil in the maternal diet during gestation and lactation modifies fatty acid composition, mucosal architecture, and mast cell regulation of the ileal barrier in piglets. The Journal of Nutrition 139: 1110-1117.

Calder, P.C., 2003. N-3 polyunsaturated fatty acids and inflammation: from molecular biology to the clinic. Lipids 38: 343-352.

Calder, P.C., 2013. Omega-3 polyunsaturated fatty acids and inflammatory processes: nutrition or pharmacology? British Journal of Clinical Pharmacology 75: 6456-6462.

Cordero, G., Isabel, B., Morales, J., Menoyo, D., Piñeiro, C., Daza, A. and Lopez-Bote, C.J., 2011. Conjugated linoleic acid (CLA) during last week of gestation and lactation alters colostrum and milk fat composition and performance of reproductive sows. Animal Feed Science and Technology 168: 232-240.

Corino, C., Bontempo, V. and Sciannimanico, D., 2002. Effects of dietary conjugated linoleic acid on some specific immune parameters and acute phase protein in weaned piglets. Canadian Journal of Animal Science 82: 115-117.

Corino, C., Pastorelli, G., Rosi, F., Bontempo, V. and Rossi, R., 2009. Effect of dietary conjugated linoleic acid supplementation in sows on performance and immunoglobulin concentration in piglets. Journal of Animal Science 87: 2299-2305.

Corson, A.M., Laws, J., Litten, J.C., Dodds, P.F., Lean, I.J. and Clarke, L., 2008. Effect of dietary supplementation of different oils during the first or second half of pregnancy on the glucose tolerance of the sow. Animal 2: 1045-1054.

Danielsen, V. and Lauridsen, C., 2001. Fodringens indflydelse påmælkemængde og-sammensætning. In: Jakobsen, K. and Danielsen, V. (eds.) Intern Rapport, Danmarks JordbrugsForskning 141: 29-36. In Danish.

De Quelen, F., Boudry, G. and Mourot, J., 2010. Linseed oil in the maternal diet increases long chain-PUFA status of the foetus and the newborn during the suckling period in pigs. The British Journal of Nutrition 104: 533-543.

De Quelen, F., Boudry, G. and Mourot, J., 2013. Effect of different contents of extruded linseed in the sow diet on piglet fatty acid composition and hepatic desaturase expression during the post-natal period. Animal 7: 1671-1680.

Dogan, M. and Akgul, A., 2005. Fatty acid composition of some walnut (*Juglans regia L.*) cultivars from east Anatolia. Grasas y Aceites 56: 328-331.

Drew, M.D. and Owen, B.D., 1988. The provision of passive immunity to colostrum-deprived piglets by bovine or porcine serum immunoglobulins. Canadian Journal of Animal Science 68: 1277-1284.

Enser, M.I., 1984. The chemistry, biochemistry and nutritional importance of animal fats. In: Wiseman, J. (ed.) Fats in Animal Nutrition. Butterworths, London, UK, pp. 23-51.

Estienne, M.J., Harper, A.F. and Estienne, C.E., 2006. Effects of dietary supplementation with omega-3 polyunsaturated fatty acids on some reproductive characteristics in gilts. Reproductive Biology 6: 231-241.

Farmer, C., Petit, H.V, Weiler, H. and Capuco, A.V., 2007. Effects of dietary supplementation with flax during prepuberty on fatty acid profile, mammogenesis, and bone resorption in gilts. Journal of Animal Science 85: 1675-1686.

Farmer, C., Giguère, A. and Lessard, M., 2010. Dietary supplementation with different forms of flax in late gestation and lactation: effects on sow and litter performances, endocrinology, and immune response. Journal of Animal Science 88: 225-237.

Farmer, C. and Petit, H.V., 2009. Effects of dietary supplementation with different forms of flax in late-gestation and lactation on fatty acid profiles in sows and their piglets. Journal of Animal Science 87: 2600-2613.

Farmer, C. and Quesnel, H., 2009. Nutritional, hormonal, and environmental effects on colostrum in sows. Journal of Animal Science 87: 56-64.

Field, C.J., Clandinin, M.T. and Van Aerde, J.E., 2001. Polyunsaturated fatty acids and T-cell function: implications for the neonate. Lipids 36: 1025-1032.

Fritsche, K.L., Huang, S.C. and Cassity, N.A., 1993. Enrichment of omega-3 fatty acids in suckling pigs by maternal dietary fish oil supplementation. Journal of Animal Science 71: 1841-1847.

Gruger, E.H., 1967. Fatty acid composition of fish oils. In: Stansby, M.E. (ed.) Fish oils, their chemistry, technology, stability, nutritional properties, and uses. Avi Publishing Company, Westport, CT, USA, 440 pp.

Gunnarsson, S., Pickova, J., Högberg, A., Neil, M., Wichman, A., Wigren, I., Uvnäs-Moberg, K. and Rydhmer, L., 2009. Influence of sow dietary fatty acid composition on the behaviour of the piglets. Livestock Science 123: 306-313.

Hansen, A.V., Lauridsen, C., Sørensen, M.T., Bach Knudsen, K.E. and Theil, P.K., 2012. Effects of nutrient supply, plasma metabolites and nutritional status of sows during transition on performance in the following lactation. Journal of Animal Science 90: 466-480.

Harrell, R.J., Phillips, O., Boyd, R.D., Dwyer, D.A. and Bauman, D.E., 2002. Effects of conjugated linoleic acid on milk composition and baby pig growth in lactating sows. North Carolina State University Annual Swine Report 2002.

Innis, S.M., 2007. Human milk: maternal dietary lipids and infant development. Proceedings of the Nutrition Society 66: 397-404.

Jacobi, S.K., Lin, X., Corl, B.A., Hess, H.A., Harrell, R.J. and Odle, J., 2011. Dietary arachidonate differentially alters desaturase-elongase pathway flux and gene expression in liver and intestine of suckling pigs. The Journal of Nutrition 141: 548-553.

Kim, S.W., Mateo, R.D., Yin, Y.L. and Wu, G., 2007. Functional amino acids and fatty acids for enhancing production performance of sows and piglets. Asian-Australasian Journal of Animal Science 20: 295-306.

Krakowski, L., Krzyzanowski, J., Wrona, Z., Kostro, K. and Siwicki, A.K., 2002. The influence of nonspecific immunostimulation of pregnant sows on the immunological value of colostrum. Veterinary Immunology and Immunopathology 87: 89-95.

Krogh, U., Flummer, C., Jensen, S.K. and Theil, P.K., 2012. Colostrum and milk production of sows is affected by dietary conjugated linoleic acid. Journal of Animal Science 90(4): 366-368.

Kurlak, L.O. and Stephenson, T.J., 1999. Plausible explanations for effects of long chain polyunsaturated fatty acids (LCPUFA) on neonates. Archives of Disease in Childhood – Fetal and Neonatal Edition 80: F148-F154.

Lai, C., Yin, J., Li, D., Zhao, L. and Chen, X., 2005. Effects of dietary conjugated linoleic acid supplementation on performance and immune function of weaned pigs. Archives of Animal Nutrition 59: 41-51.

Lauridsen, C. and Danielsen, V., 2004. Lactational dietary fat levels and sources influence milk composition and performance of sows and their progeny. Livestock Production Science 91: 95-105.

Lauridsen, C. and Jensen, S.K., 2007. Lipid composition of lactational diets influences the fatty acid profile of the progeny before and after suckling. Animal 1: 952-962.

Laws, J., Laws, A., Lean, I.J., Dodds, P.F. and Clarke, L., 2007a. Growth and development of offspring following supplementation of sow diets with oil during early to mid gestation. Animal 1: 1482-1489.

Laws, J., Laws, A., Lean, I.J., Dodds, P.F. and Clarke, L., 2007b. Growth and development of offspring following supplementation of sow diets with oil during mid to late gestation. Animal 1: 1490-1496.

Laws, J., Amusquivar, E., Laws, A., Herrera, E., Lean, I.J., Dodds, P.F. and Clarke, L., 2009a. Supplementation of sow diets with oil during gestation: sow body condition, milk yield and milk composition. Livestock Science 123: 88-96.

Laws, J., Litten, J.C., Laws, A., Lean, I.J., Dodds, P.F. and Clarke, L., 2009b. Effect of type and timing of oil supplements to sows during pregnancy on the growth performance and endocrine profile of low and normal birth weight offspring. The British Journal of Nutrition 101: 240-249.

Leonard, S.G., Sweeney, T., Pierce, K.M., Bahar, B., Lynch, B.P. and O'Doherty, J.V., 2010a. The effects of supplementing the diet of the sow with seaweed extracts and fish oil on aspects of gastrointestinal health and performance of the weaned piglet. Livestock Science 134: 135-138.

Leonard, S.G., Sweeney, T., Bahar, B., Lynch, B.P. and O'Doherty, J.V., 2010b. Effect of maternal fish oil and seaweed extract supplementation on colostrum and milk composition, humoral immune response, and performance of suckled piglets. Journal of Animal Science 88: 2988-2997.

Leonard, S.G., Sweeney, T., Bahar, B., Lynch, B.P. and O'Doherty, J.V., 2011. Effect of dietary seaweed extracts and fish oil supplementation in sows on performance, intestinal microflora, intestinal morphology, volatile fatty acid concentrations and immune status of weaned pigs. The British Journal of Nutrition 105: 549-560.

Leskanich, C.O. and Noble, R.C., 1999. The comparative roles of polyunsaturated fatty acids in pig neonatal development. The British Journal of Nutrition 81: 87-106.

Mateo, R.D., Carroll, J.A., Hyun, Y., Smith, S. and Kim, S.W., 2009. Effect of dietary supplementation of n-3 fatty acids and elevated concentrations of dietary protein on the performance of sows. Journal of Animal Science 87: 948-959.

Miller, C.C., Park, Y., Pariza, M.W. and Cook, M.E., 1994. Feeding conjugated linoleic acid to animals partially overcomes catabolic responses due to endotoxin injection. Biochemical and Biophysical Research Communications 198: 1107-1112.

Missotten, J., De Smet, S., Raes, K. and Doran, O., 2009. Effect of supplementation of the maternal diet with fish oil or linseed oil on fatty-acid composition and expression of Δ5- and Δ6-desaturase in tissues of female piglets. Animal 3: 1196-1204.

Mitre, R., Etienne, M., Martinais, S., Salmon, H., Allaume, P., Legrand, P. and Legrand, A.B., 2005. Humoral defence improvement and haematopoiesis stimulation in sows and offspring by oral supply of shark-liver oil to mothers during gestation and lactation. British Journal of Nutrition 94: 753-762.

Moraes, M.L., Ribeiro, A.M.L., Kessler, A.M., Ledur, V.S., Fischer, M.M., Bockor, L., Cibulski, S.P. and Gava, D., 2012. Effect of CLA on performance and immune response of weanling piglets. Journal of Animal Science 90: 2590-2598.

Park, J.C., Kim, Y.H., Jung, H.J., Moon, H.K., Kwon, O.S. and Lee, B.D., 2005. Effects of dietary supplementation of conjugated linoleic acid (CLA) on piglets' growth and reproductive performance in sows. Asian-Australasian Journal of Animal Sciences 18: 249-254.

Papadopoulos, G.A., Maes, D.G.D., Van Weyenberg, S., Van Kempen, T.A.T.G., Buyse, J. and Janssens, G.P.J., 2009. Peripartal feeding strategy with different n-6: n-3 ratios in sows: effects on sows' performance, inflammatory and periparturient metabolic parameters. The British Journal of Nutrition 101: 348-357.

Pariza, M.W., Park, Y. and Cook, M.E., 2001. The biologically active isomers of conjugated linoleic acid. Progress in Lipid Research 40: 283-298.

Patterson, R., Connor, M.L., Krause, D.O. and Nyachoti, C.M., 2008. Response of piglets weaned from sows fed diets supplemented with conjugated linoleic acid (CLA) to an *Escherichia coli* K88+ oral challenge. Animal 2: 1303-1311.

Peng, Y., Ren, F., Yin, J.D., Fang, Q., Li, F.N. and Li, D.F., 2010. Transfer of conjugated linoleic acid from sows to their offspring and its impact on the fatty acid profiles of plasma, muscle, and subcutaneous fat in piglets. Journal of Animal Science 88: 1741-1751.

Perez Rigau, A., Lindemann, M.D., Kornegay, E.T., Harper, A.F. and Watkins, B.A., 1995. Role of dietary lipids on fetal tissue fatty acid composition and fetal survival in swine at 42 days of gestation. Journal of Animal Science 73: 1372-1380.

Poulos, S.P., Azain, M.J. and Hausman, G.J., 2004. Conjugated linoleic acid (CLA) during gestation and lactation does not alter sow performance or body weight gain and adiposity in progeny. Animal Research 53: 275-288.

Rooke, J.A. and Bland, I., 2002. The acquisition of passive immunity in the new-born piglet. Livestock Production Science 78: 13-23.

Rooke, J.A., Bland, I.M. and Edwards, S.A., 1998. Effect of feeding tuna oil or soyabean oil as supplements to sows in late pregnancy on piglet tissue composition and viability. The British Journal of Nutrition 80: 273-280.

Rooke, J.A., Bland, I.M. and Edwards, S.A., 1999. Relationships between fatty acid status of sow plasma and that of umbilical cord, plasma and tissues of newborn piglets when sows were fed on diets containing tuna oil or soyabean oil in late pregnancy. The British Journal of Nutrition 82: 213-221.

Rooke, J.A., Shanks, M. and Edwards, S.A., 2000. Effect of offering maize, linseed or tuna oils throughout pregnancy and lactation on sow and piglet tissue composition and piglet performance. Animal Science 71: 289-299.

Rooke, J.A., Sinclair, A.G. and Edwards, S.A., 2001a. Feeding tuna oil to the sow at different times during pregnancy has different effects on piglet long-chain polyunsaturated fatty acid composition at birth and subsequent growth. The British Journal of Nutrition 86: 21-30.

Rooke, J.A., Sinclair, A.G. and Ewen, M., 2001b. Changes in piglet tissue composition at birth in response to increasing maternal intake of long-chain n-3 polyunsaturated fatty acids are non-linear. The British Journal of Nutrition 86: 461-470.

Rooke, J.A., Sinclair, A.G., Edwards, S.A., Cordoba, R., Pkiyach, S., Penny, P.C., Penny, P., Finch, A.M. and Horgan, G.W., 2001c. The effect of feeding salmon oil to sows throughout pregnancy on pre-weaning mortality of piglets. Animal Science 73; 489-500.

Rossi, R., Pastorelli, G., Cannata, S. and Corino, C., 2010. Recent advances in the use of fatty acids as supplements in pig diets: a review. Animal Feed Science and Technology 162: 1-11.

Schmid, A., Collomb, M., Bee, G., Bütikofer, U., Wechsler, D., Eberhard, P. and Sieber, R., 2008. Effect of dietary alpine butter rich in conjugated linoleic acid on milk fat composition of lactating sows. British Journal of Nutrition 100: 54-60.

Smits, R.J., Luxford, B.G., Mitchell, M. and Nottle, M.B., 2011. Sow litter size is increased in the subsequent parity when lactating sows are fed diets containing n-3 fatty acids from fish oil. Journal of Animal Science 89: 2731-2738.

Spencer, J.D., Wilson, L., Webel, S.K., Moser, R.L. and Webel, D.M., 2004. Effect of feeding protected n-3 polyunsaturated fatty acids (Fertilium™) on litter size in gilts. Journal of Animal Science 82(1): 211.

Sugano, M., Tsujita, A., Yamasaki, M., Noguchi, M. and Yamada, K., 1998. Conjugated linoleic acid modulates tissue levels of chemical mediators and immunoglobulins in rats. Lipids 33: 521-527.

Tanghe, S. and De Smet, S., 2013. Does sow reproduction and piglet performance benefit from the addition of n-3 polyunsaturated fatty acids to the maternal diet? Veterinary Journal 197: 560-569.

Theil, P.K., 2015. Transition feeding of sows. Chapter 7. In: Farmer, C. (ed.) The gestating and lactating sow. Wageningen Academic Publishers, Wageningen, the Netherlands, pp. 147-172.

Uauy, R. and Dangour, A.D., 2006. Nutrition in brain development and aging: Role of essential fatty acids. Nutrition Reviews 64: S24-S33

Varley, M.A., Maitland, A. and Towle, A., 1986. Artificial rearing of piglets: the administration of two sources of immunoglobulins after birth. Animal Production 43: 121-126.

Waters, S.M., Coyne, G.S., Kenny, D.A. and Morris, D.G., 2014. Effect of dietary n-3 polyunsaturated fatty acids on transcription factor regulation in the bovine endometrium. Molecular Biology Reports 41: 2745-2755.

Webel, S.K., Otto-Tice, E.R., Moster, R.L. and Orr, Jr., D.E., 2004. Effect of feeding duration of protected n-3 polyunsaturated fatty acid (Fertilium™) on litter size and embryo survival in sows. Journal of Animal Science 82(1): 212.

Whigham, L.D., Cook, E.B., Stahl, J.L., Saban, R., Bjorling, D.E., Pariza, M.W. and Cook, M.E., 2001. CLA reduces antigen-induced histamine and PGE(2) release from sensitized guinea pig tracheae. American Journal of Physiology – Regulatory, Integrative and Comparative Physiology 280: R908-R912.

Yao, W., Li, J., Wang, J.J., Zhou, W., Wang, Q., Zhu, R., Wang, F. and Thacker, P., 2012. Effects of dietary ratio of n-6 to n-3 polyunsaturated fatty acids on immunoglobulins, cytokines, fatty acid composition, and performance of lactating sows and suckling piglets. Journal of Animal Science and Biotechnology 3: 43.

Zambiazi, R.C., Przybylski, R., Zambiazi, M.W. and Mendonca, C.B., 2007. Fatty acid composition of vegetable oils and fats. B. CEPPA Curitiba 25: 111-120.

17. 优化哺乳期和断奶母猪繁殖生理及性能的最佳技术

N. M. Soede* and B. Kemp

Adaptation Physiology Group, Department of Animal Sciences, Wageningen University, P.O. Box 338, 6700 AH Wageningen, the Netherlands; nicoline.soede@wur.nl

摘要：为了优化断奶母猪的繁殖性能，需要在哺乳期、断奶到发情的间期、发情期和妊娠期间进行优化管理。基本上来说，管理应集中于在哺乳期和断奶后优化卵泡的发育，并且在断奶后至第一次发情期间使这些卵泡产生的卵母细胞受精最大化。在哺乳期，卵泡发育主要受泌乳性体重减轻和哺乳影响。因此，管理策略应旨在刺激采食量，如可能减少哺乳期结束时的哺乳数量。在断奶至发情期间隔的短时间内，母猪营养和与公猪的接触最为重要。为了使受精状态最佳，良好的发情检测和授精策略至关重要。

关键词：母猪，哺乳期，断奶，繁殖，卵泡

17.1 引言

在哺乳期间，仔猪的吸吮强度和母猪能量或蛋白质的负平衡通常抑制排卵期前卵泡的生长，因而防止泌乳的发生。促性腺激素（gonadotrophin）释放和卵泡发育可以被抑制到相当程度，这种程度下断奶后卵泡发育和随后的繁殖参数（断奶-发情间隔，排卵率，胚胎存活和胚胎发育）受到损害，最终影响分娩率、窝仔数（参见Prunier等，2003；Soede等，1995综述）和窝均一性（Steverink等，1997；Wientjes等，2013）。这在体重明显减轻的头胎母猪中是最明显的（Hoving等，2012），但在体重减轻较多的老龄经产母猪中亦可观察到（Thaker和Bilkei，2005）。断奶后的饲养和环境条件也可能影响后续繁殖能力，例如影响断奶后卵泡发育的条件，如慢性应激（Turner和Tilbrook，2006）；或影响发情检测质量和授精时间（如公猪接触）的条件（Kemp等，2005）。

本综述讨论了管理哺乳期和断奶母猪最优繁殖性能的最佳策略。要了解这些建议的生理背景，我们首先提供了相关繁殖生理学的概述和其影响因素。我们对

初产母猪给予了大量关注,因为在这些动物中能特别看到哺乳期对随后的繁殖性能的影响(例如所谓的第二胎综合征)。疾病对繁殖性能的影响将不予讨论。第一胎和第二胎母猪的最佳繁殖性能也受到后备母猪发育和管理的影响,包括到哺乳期前的过渡期,其同时也受妊娠期间各种条件的影响。但是,这些方面将在本书的其他章节中讨论。

17.2 繁殖生理

以下介绍的繁殖生理方面的信息主要来自 Soede 等(2011)。

17.2.1 泌乳性乏情期

在泌乳期间,由于吮吸诱导的促性腺释放激素(gonadotropin-releasing hormone,GnRH)冲动发生器的抑制,促黄体素(luteinising hormone,LH)的外周浓度和 LH 脉冲被抑制(De Rensis 等,1993)。LH 抑制程度也与母猪的能量负平衡有关;与饲喂水平高的母猪相比,饲喂水平较低的初产母猪其 LH 外周血浓度降低(Quesnel 和 Prunier,1998;Van den Brand 等,2000a)。哺乳对卵泡刺激素(follicle-stimulating hormone,FSH)的影响则不太一致,主要是由于抑制素(由卵泡达到 3mm 产生的,Noguchi 等 (2010))的反馈反应,而不是吸吮作用(由 Prunier 等,2003 综述)。在泌乳过程中,LH 脉冲通常恢复正常(Van den Brand 等,2000a),这可能与吸吮频率的降低和/或垂体 LH 对 GnRH 的反应性增加有关(Bevers 等,1981;Rojanasthien 等,1987)。伴随着但不仅仅是由于这种 LH 脉冲的增加,卵泡直径在哺乳期间增加(由 Britt 等,1985 综述)。因此,随着哺乳的进行,卵巢卵泡池中的窦状卵泡达到更大的直径,尽管直到断奶之后大多数母猪都不会产生直径 3~4 mm 的卵泡(Lucy 等,2001)。偶尔的,母猪会出现 LH 释放的早期恢复,在泌乳期间发育出排卵前直径大小的卵泡(约 8 mm)并排卵(Kemp 等,1998)。哺乳期的前 2 周不可能发生这种情况,因为在哺乳期间母猪具有达到足够量级的排卵前 LH 波动以诱发排卵发育的能力(Sesti 和 Britt,1993)。这种哺乳期排卵在低窝仔数(低哺乳强度)和/或低泌性体重减少(高采食量)的多胎母猪中发生频率更高。

17.2.2 断奶至发情间隔

断奶时的囊状卵泡池可以由大约 100 个卵泡组成,直径不定可达 6 mm(由 Knox,2005 综述)。从卵泡池中募集卵泡发生于当脉动性 GnRH/LH 释放从低频/高幅度模式转变到高频/低振幅模式时(Shaw 和 Foxcroft,1985)(图 17.1)。当母猪在哺乳期间其能量负平衡更加严重时,其在断奶之前和之后的 LH 脉冲的频率便会受损(Quesnel 等,1998;Van den Brand 等,2000c),且这情况会导致断奶-发情间隔的延长(Quesnel 和 Prunier,1998;Shaw 和 Foxcroft,1985;Van den

Brand 等，2000c)。因此，与经产母猪相比，初产母猪在断奶时通常具有较小的卵泡(<2~3 mm 对 3~4 mm)和更长的断奶－排卵间隔(Gerritsen，2008；Langendijk 等，2000c)。当卵泡生长到排卵大小时，卵泡中的雌激素产生增加，LH 脉动和 FSH 释放逐渐减少(Noguchi 等，2010；Prunier 等，1987)。在一定浓度阈值下，雌激素通过正反馈诱导 LH 波动，导致雌二醇生产立即下降，诱导排卵并引起卵泡壁的黄体生成，从而引发孕激素的产生。在猪中，LH 波动达到顶点之后平均(30±3) h 发生排卵(例如 Soede 等，1994)。

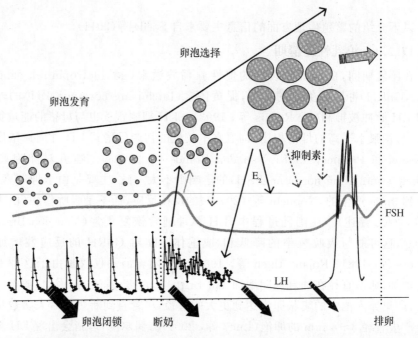

图 17.1　哺乳晚期和断奶期的卵泡发育的调控，显示最相关的生殖激素(图片由 George Foxcroft，Edmonton，Canada 提供并改编)。

17.2.3　发情和排卵

发情是排卵期间，当公猪出现时，母猪以表现出"静立反应"为特征的接受行为。母猪站立不动，竖起她的耳朵，旋转耳朵(Signoret，1970)，从而允许公猪进行配种。除了这些行为变化外，还看到外阴变化(发红、肿胀和黏液生成增加)。尽管这些发情现象是由排卵前卵泡产生的雌激素引起的，但在一般动物中雌激素的外周浓度与发情症状的强度和持续时间似乎没有直接的关系(Soede 等，1994)。来自公猪的嗅觉和触觉刺激对于唤起静立反应是最重要的(Signoret，1970)，公猪存在时，一般通过使用所谓的"压背测试"来模拟这些触觉刺激进行发情检测。有公猪存在的情况下，发情平均持续 40~60 h，但对于不同母猪具发生时间为 24~96 h (由 Soede 和 Kemp，1997 综述)。不同农场之间发情期时常也变化很大(Steverink 等，1999)，其受到发情检测时公猪接触强度、母猪的压力水平、胎次(通常在后备母猪中较短)以及断奶-发情间隔(断奶至发情间隔超过 4 天时较短)等因素的影响。导致这些影响的确切机制尚不清楚。由于排卵发生在发情期约 2/3 时(由 Soede 和 Kemp，1997 综述)，从发情到排卵发生的时间可能会有很大差异。

17.3 哺乳期：在断奶时优化卵泡发育

哺乳期抑制卵泡发育可能导致随后的繁殖力下降，为了克服这种下降，可以应用几种管理策略。这些策略的目的是改善母猪的能量平衡，降低哺乳后期的哺乳强度，其主要是通过减少每窝仔数(分别断奶)或限制哺乳期时长(间歇性哺乳)。因为哺乳期长度本身是断奶后卵巢活动的主要决定因素，本章还讨论了这一点。

17.3.1 体重损失的重要性

在哺乳期间，大多数母猪大量损失其身体蛋白质和脂肪储备。当体重和蛋白质损失超过 12% 时，可能会发生繁殖问题，如断奶-发情间隔延长，妊娠率较低和窝产仔数降低(Thaker 和 Bilkei，2005)。尤其是初产母猪，因为它们在发育至成熟期时具有显著的营养需求，且具有较低的采食量和可代谢脂肪及蛋白质储备(Hoving 等，2011)。因此，在评估体重损失对繁殖性能影响的方面的研究中，经常在对初产母猪进行饲料限制。

在 20 世纪 70 年代和 80 年代初期，哺乳期饲料限制被证明对断奶-发情间期具有相当显著的影响(增加约 10 天)，对排卵率和胚胎存活几乎没有影响。在最近的数据中，饲料限制对断奶-发情间期的影响相当小(不到一天)，而对排卵率和胚胎存活的影响更为明显(表 17.1)。采食限制可能使排卵率降低 2%～4%，胚胎存活率降低 10%～20%。母猪在哺乳期间对饲喂限制反应性的这种变化是通过遗传来选择缩短断奶至发情间期的结果。其结果是，多年来哺乳期低采食量对断奶至发情间期的影响不太明显，但对仔猪数量和妊娠率的影响变得更加明显。然而，最近的研究仍然报告了在哺乳期为 19 天的母猪中，采食量与随后的断奶-发情间隔(Yoder 等，2012；2013)之间的关系，在哺乳晚期当采食量下降发生时，断奶至发情间隔受到显著影响。

表 17.1 哺乳期的采食量(高对低[1])对断奶-发情间隔(WEI)、排卵率和胚胎存活率的影响

参考文献	断奶时间	断奶-发情间隔(天)		排卵率(%)[2]		胚胎存活率(%)[3]	
		高	低	高	低	高	低
King and Williams (1984)	d32	7.6a	19.9b	14.4	13.5	70	72
Kirkwood 等 (1987)	d35	4.3a	5.8b	18.1	18.6	83a	68b
Kirkwood 等 (1990)	d28	6.0a	8.9b	17.6	17.7	83a	72b

续表 17.1

参考文献	断奶时间	断奶-发情间隔(天) 高	低	排卵率[2] 高	低	胚胎存活率(%)[3] 高	低
Baidoo 等（1992）	d28	5.9	7.5	16.2	16.7	85a	64b
Zak 等（1997a）L:wk1-3[4]	d28	3.7	5.6	19.9a	15.4b	88a	87
Zak 等（1997a）L:wk4	d28		5.1		15.4b		64b
Zak 等（1998）	d28	4.2a	6.3b	14.4	15.6	83	72
Quesnel 和 Prunier（1998）	d24	5.7	5.9	19.2	20.7	—	
Van den Brand 等（2000c）	d22	5.1	5.7	18.1a	16.4b	68	68
Terletski 等，unpublished[5]	d21	6.6	6.7	18.6a	16.7b	64	69
Vinsky 等（2006）[6]	d21	5.3	5.4	18.3	18.2	79a	68b
Foxcroft 等，unpublished	d21	5.7	5.5	18.5	17.5	65a	78b
Patterson 等（2011）L:wk3	d20	5.0	5.3	19.7	20.2	71	70

[1] 高为自由采食量的 80%～90%，低为自由采食量的 40%～60%。
[2] 每只母猪排卵卵母细胞数。
[3] a 与 b 之间差异（$P<0.05$）。
[4] L:wk：采用采食量处理的哺乳周数。
[5] 高：控制饲喂 >190 kg 生产时体重；低：控制饲喂 <170 kg 生产时体重。
[6] 限制母猪的雌性胚胎数量下降。

这些哺乳期采食量对后续繁殖参数的影响源于其对 LH 释放和初期卵泡发育的影响。Quesnel 等（1998）显示，与自由采食的母猪相比，以自由采食摄入量的 50% 饲养的第一胎母猪在断奶前后具有较低的 LH 波动率，且断奶后其卵泡较小。断奶后第 2 天，大于 4 mm 的卵泡的平均数在限制饲喂的母猪为（6.8±2.7）个，而自由采食的母猪为（12.2±1.3）个。Zak 等（1997b）也证实，不仅卵泡大小，而且卵泡和卵母细胞质量受饲料限制的影响。在断奶后约 4 天，在限制饲喂的母猪中能够在体外（in vitro）成熟的卵母细胞较低，并且这些卵母细胞的卵泡液导致成熟卵母细胞数量减少。卵泡和卵母细胞成熟的这些差异可能有助于饲料限制对胚胎存活的影响。

Morrow 等（1992）发现，135 个美国农场中有 40% 的农场其第二胎的窝仔数比第一胎低或相同。在农场内，在作为后备母猪第一次受精至第一次断奶之间体重增加较轻的初产母猪似乎在第二胎时，其窝仔数减少的风险会增加，称为第二胎综合征（Hoving 等，2010）。这种综合征与孕期体重增加和哺乳期体重下降有关。此外，第二窝的窝仔低的母猪其随后的胎次中窝仔数量亦较低，且早一个胎次淘汰（Hoving 等，2011），这强调了第二胎综合征的经济重要性。

体重损失包括蛋白质和脂肪的损失。Clowes 等（2003）指出，与脂肪损失相比，第一次哺乳期间的蛋白质损失似乎与后续的繁殖性能更加相关，但 Schenkel

等(2010)发现在第二次哺乳期,背脂损失和蛋白质损失对产仔数的影响相似。Schenkel 等(2010)进一步发现,与分娩时体重较轻的母猪相比,分娩期间体重最高的 50%的母猪(207~245 kg)其脂肪丢失对随后的仔猪体重影响较大。因此,哺乳期蛋白质损失与繁殖性脂肪损失的相对重要性可能取决于许多因素,包括体重和基因型。

防止哺乳期繁殖问题的最合乎逻辑的方法是防止哺乳期过度体重损失。这可以通过优化后备母猪发育、优化妊娠和哺乳期的饲料或营养摄取量,或通过减少哺乳后期哺乳的策略来实现。这些方面将在下文讨论。

后备母猪的发育

后备母猪的发育,尤其是第一次配种的年龄和生长率,不仅会对第一胎次,且对第二胎次的繁殖力也有重大影响(由 Bortolozzo 等 (2009)综述;Hoving 等,2010)。关于后备母猪最佳发展的综述,请参见第 1 章(Rozeboom,2015)。

妊娠期间的采食量

在过去 25 年中,有几项研究表明,在分娩期间采食过量或分娩时体重过高会降低哺乳期间的自愿采食量(如 Dourmad,1991;Mullan 和 Williams,1989;Quesnel 等,2005a;Yang 等,1989),这可能与这些动物中较高的葡萄糖不耐受有关(Quesnel 等,2005b;Weldon 等,1994)。另一方面,yang 等(1989)已经表明,第一胎母猪妊娠期间采食量不足(导致母猪瘦小)不能通过在哺乳期间增加自愿采食量来补偿,且会导致断奶-发情期间隔的增大。因此,应根据维持、繁殖和生长的要求饲喂后备母猪,但不可过量。此外,妊娠期间饲喂富含纤维的日粮可以刺激哺乳期采食量(第 5 章;Meunier-Salaün 和 Bolhuis,2015)。

环境温度

哺乳母猪的温度要求与仔猪要求的温度大不相同。新生仔猪具有较低的临界温度,即低于该温度的仔猪需要额外能量以维持体内平衡的温度(为 32~35℃,其在 4 周龄时逐渐降至 24~27℃)。对于母猪来说,其上限临界温度没有详细研究,但可能低于 22~25℃(Makkink 和 Schrama,1998)。高于上限临界温度,母猪会降低饲料摄入量来防止过热。基于 9 个实验,Black 等(1993)计算出,当温度高于 16℃时,温度升高 1℃母猪的日常自愿摄食量则会降低 2.4 MJ 的可消化能(正常的泌乳母猪的饲料中提供约 13.8 MJ 的消化能量(DE)/kg)。Quiniou 和 Noblet (1999)报道在 25~27℃,自愿采食量以 214 g/℃下降。在同样的温度范围内,Silva 等(2009a)表明,当湿度较高[相对湿度(relative humidity,RH)94%]时,采食量下降剧烈 (492 g/(℃·天)。因此,高环境温度可以显著降低泌乳期间的采食量。皮肤润湿、滴水和喷雾降温可能会在高温环境下增加母猪的哺乳期摄食量(由 Makkink 和 Schrama,1998 综述)。Wagenberg 等(2006)和 Silva 等(2009b)研究显示,地板冷却使母猪的采食量增加了 12%--18%,仔猪断奶重增加,母猪体重损失降低。当乳汁产量和采食量高时,周围环境温度低对仔猪的影响在出生后的前几天是最严重的,高环境温度对母猪的负面影响是主要在哺乳中期和晚期。因此,在哺乳期的第一周应提供较高的环境温度,此后应提供较低的温度。根据Makkink和Schrama(1998)的报告,当有一个良好的小气候可用于仔猪时,哺乳晚期的室温甚至可低至16℃,低环境温度会增加哺乳期母猪的采食量,从而有助于预防繁殖问题。

进食模式

一些报告显示,哺乳母猪每天饲喂超过两次会提高采食量,而在使用自由采食系统时有时会发现更好的进食量。例如,Hoofs和Elst-Wahle(1993)发现,当使用自动喂料器而不是每天饲喂两次时,饲料采食量增加10%。一般来说,当采用自由采食法时,建议每天从槽中取出饲料,以防止饲料结块和变酸。Koketsu等(1996a,1996b)在鉴别了哺乳期间的饲料摄入模式,并研究了其对繁殖性能的影响。当母猪采食量增加时(无论是增加快速还是逐渐增加),如在哺乳期间没有下降则具有最佳的繁殖性能。为了防止哺乳期采食量下降,建议在哺乳期间使用逐步添加进行饲料补充。Everts等(1995)建议在分娩时给予2 kg饲料,并以每天0.5 kg的方式逐步增加进食量,直到达到推荐的进食量。最近,Thingnes等(2012)发现逐步增加在哺乳后期可能也很重要。这些作者都是第9天之前慢慢增加母猪的饲料补充,之后母猪仍然保持这个逐步添加时间表,或者随意进食饲料。在剩余的35天哺乳期间内,自由采食的猪随后发生采食量下降(定义为至少3天采食量下降超过3 kg)的数量是另一组的两倍(50%对25%),且采食量下降的母猪具有较低的总饲料摄取量和更高的哺乳期体重损失。然而,关于在哺乳期间使用的最佳采食模式的建议可能是基因型依赖性,与母猪自愿采食量的差异以及影响该采食量的因素有关。在问题农场,建议通过连续评估不降低采食量的较高的饲料补充,来优化饲料采食量。

哺乳期的饲料蛋白质和脂肪水平

饲料蛋白质水平可能影响母猪断奶后的表现。Yang等(2000b)比较了哺乳期间的5种日粮中赖氨酸(lysine)浓度,从总赖氨酸0.6%(低)到总赖氨酸1.6%(高)。尽管日粮中赖氨酸浓度的增加会使初产母猪的赖氨酸和蛋白质摄入量线性增加,但也导致采食量的线性下降,因此,使得随后的仔猪数量线性减少。因此,高蛋白质水平限制了采食量,从而对随后的繁殖性能产生不利影响。另一方面,日粮中蛋白质较少可能会增加机体蛋白质的损失(Mejia-Guadarrama等,2002),从而对繁殖表现产生负面影响,例如卵泡发育和卵母细胞成熟受损(Yang等,2000a)及较低的排卵率(Mejia-Guadarrama等,2002)。据此可以得出结论,随后繁殖所需的赖氨酸(蛋白质)需求不大于最佳产奶量所需(Yang等,2000b),因而,可以在对随后的繁殖没有不利影响的情况下对哺乳期母猪的每日日粮赖氨酸需求进行跟踪(见 NRC,2012)。另一种影响身体储存动员的方法是增加日粮中脂肪含量。虽然高脂日粮通常导致采食量下降,但 Drochner(1989)报道,与日粮中脂肪水平较低的组相比,在高龄母猪中总代谢能(metabolisable energy,ME)摄入量仍然平均增加了 12%(3%～32%)。使用日粮中脂肪作为能量来源增加了乳脂肪含量,在某些情况下,总产奶量也增加(Drochner,1989)。Van den Brand 等(2000b)发现,在 21 天的哺乳期间内,富含脂肪的饲料(13.5%对 3.4%)导致母乳脂肪含量增加(+1.5%)和初产母猪体重损失明显增加(+3.8 kg)。因此,高脂日粮对乳脂的控制效果可能不会妨碍身体状况的损失。另一方面,高脂日粮可能导致在炎热的环境中摄入更高的能量,因为当日粮中脂肪用于生产乳汁时,母猪的热量产生较低。因此,高能量饲料可能有益于繁殖表现,特别是如其可用这种饲料使能量摄入超过生产乳汁所需的额外能量。最近,Smits 等(2012)发现在使用两种能量水平(5 级,13.0～15.3 MJ/kg DE)和蛋白水平(5 级,218～259 g/kg)都上升的日粮时,每日可消化能量摄入量呈线性增加(61～72 MJ/天 DE)虽然 5 种日粮在哺乳期采食量相似,当日粮中能量达到 14.7 MJ/kg DE 和蛋白水平达到 248 g/kg 时,发现其断奶至发情间隔期线性下降(8.1～5.7 天),断奶母猪分娩第二窝比例有所增加(49%～62%)。因此,增加哺乳期日粮中的能量和蛋白质水平可以使胎次较小的母猪生产力提高。

最佳饮水量

由于猪舍温度高、且每天生产约 10 kg 的乳汁,因而母猪对水的需求量很大。Fraser 和 Phillips(1989)报道,在具有乳头饮水器的系统中,在分娩期和哺乳期早期为母猪输送 0.7 L/min 的水明显太少。母猪的饮水量与仔猪的生长和死亡率之间有关联。因此,建议在哺乳期间自由饮水,并定期检查乳头饮水器的输水量应为 2～4 L/min。

17.3.2 哺乳期长度

选择特定的哺乳期通常不是基于繁殖表现,而是基于有效管理、良好的仔猪表现或动物福利条例等因素。世界各地农场之间的哺乳期长短差异很大。在美国,大多数农场在哺乳期 18～21 天断奶(Knox 等,2013),而欧盟在 21 天内不允许断奶,例如芬兰在 28 天内不允许断奶。1982 年,Varley 总结了哺乳期时间长短对后续繁殖的影响。在他的研究中,哺乳期 3 周内断奶时,母猪断奶至发情期间隔期平均为 7 天,当母猪在哺乳期第 10 天断奶时则其变为 8～16 天。哺乳期长短不影响排卵率,但当时短于 21 天会增加着床时的胚胎死亡率,从而降低了产仔数。短的哺乳期时长对仔猪数的负面影响体现在断奶后子宫恢复受损。现在的问题是这些发现是否仍然可用于现代母猪,原因是遗传选择不仅改变了繁殖性能(减少断奶至发情间隔、增加排卵率和提高产仔数),也改变了机体成分(高瘦肌肉组织,低背脂肪水平)。

对不同长度哺乳期的母猪其生殖生理学知之甚少。Willis 等（2003）在比较了在哺乳期第 14 天与 24 天断奶的母猪，发现在第 14 天的母猪中，断奶之前和之后 LH 波动（LH-pulsatility）较低，与之相伴的是一条未完全恢复的下丘脑-垂体-卵巢轴。当 14 天断奶时，FSH 的浓度也较大，并且断奶后雌激素浓度的升高有延迟，同时卵泡发育受到抑制。一些研究分析了哺乳期天数与母猪繁殖表现关系的农场数据，这些研究表明，当哺乳期短于 21 天时，断奶至发情间隔增加（Belstra 等，2004；2001；Knox 和 Rodriguez Zas，2001；Tummaruk 等，2001）。当哺乳期较短在断奶后第 6 天没有出现发情的母猪，同样断奶后 3 天其卵泡较小（Knox 和 Rodriguez Zas，2001），说明卵泡发育受到抑制。此外，即使发情不被延迟，生殖过程仍可能受到影响。例如，Knox 和 Rodriguez Zas（2001）指出，在哺乳期较短（小于 16 天）的母猪中，母猪在断奶后 8 天内显示发情的比例有所下降（35% 对 94%～98%，母猪哺乳期超过 17 天），排卵的母猪其百分比也下降（78% 对 98%，母猪哺乳期为 25～31 天）。排卵失败的母猪（2%）或者显示出只有中小卵泡或卵巢囊肿的短暂/间歇性发情，或者具有较长的发情期，这个发情期具有在第一次发情期 5 天之内未排卵的正常大小的排卵前卵泡。在排卵母猪中，哺乳期长度不影响排卵时的卵泡大小。因此，尽管母猪在短期哺乳期的雌激素产量足以诱发发情行为，但是不成熟的正反馈系统可能会阻止/禁止 LH 波动的发生（Sesti 和 Britt，1993），因此导致这些母猪排卵失败。如果具有正常大小的排卵前卵泡的母猪发生 LH 波动失败，那么这些卵泡可能会变成囊性。最近没有报告概述哺乳期长度对囊性卵泡或囊性卵巢发育的影响。在早期报告中，Svajgr 等（1974）发现，哺乳期为 13 天的母猪其囊泡卵数是哺乳期为 24 天母猪的两倍（1.3 对 0.6）。与 Varley（1982）早期的综述一致，Willis 等（2003）没有发现哺乳期长度对排卵率的影响。

由 Varley（1982）综述，哺乳期长度也可能影响后续的分娩率和产仔数。在最近的文献中，发现哺乳期较短（少于 3 周）对随后的产仔数和分娩率有不利影响（Koketsu 等，1997；Le Cozler 等，1997）。有限的关于哺乳期较长的母猪繁殖性能方面的信息表明，在 4 周以上哺乳期对分娩率（+3%）和仔猪数（+0.6 头仔猪）均有正面影响（Gaustad-Aas 等，2004），虽然这种影响（Tummaruk 等，2001）并不总是被发现，且并不会导致每头母猪每年的产仔数增多（Xue 等，1993）。另一方面，在需要延长哺乳期的系统中（如有机农场系统），老龄经产母猪可能具有非常高的排卵量。Leenhouwers 等（2011）显示，有机农场系统的母猪比常规母猪分娩时窝仔数更大。Wientjes 等（2012）报道，有机农场母猪的产仔数为 17.0～18.8 头，平均哺乳期为 41 天。这样高的排卵量可能与这些母猪在 6 周的哺乳期内变成合成代谢的事实有关，这可能是恢复卵泡发育的原因。

总之，短期哺乳期（<3 周）对断奶后卵泡发育、随后的发情间隔、排卵反应，甚

至分娩率和窝仔数均有明显的负面影响。最近的文献中较短哺乳期对断奶至发情间隔的影响似乎不太明显,这可能与持续遗传选择断奶-发情间隔较短的母猪有关。较长的哺乳期(直到 5 周或 6 周)可能会增加随后的产仔数,因为母猪的身体状况在持续的哺乳期间得以恢复。较长哺乳期在繁殖方面可能的缺点是会使泌乳性排卵的母猪数量增加。这可能特别发生在来自特定高产品系的经产母猪中,但亦可发生于哺乳期间具有较低哺乳仔猪数量或高采食量的任意母猪中。哺乳期排卵往往没有被注意到,这样的母猪就会被标记为"延迟发情"。另一方面,刺激哺乳期排卵可被用作进行哺乳期授精的管理策略(见下文)。

17.3.3 仔猪数量-逐步断奶

逐步断奶是在完全断奶之前几天逐步移出一部分仔猪断奶的方法。它减少了哺乳晚期仔猪的哺乳刺激,从而改善了母猪的能量平衡。在不同研究中,断奶制度有很大变化。逐步断奶在全部断奶前持续 3~7 天,在持续 17~35 天的剩余哺乳期剩下 2~6 只仔猪。Matte 等(1992)对逐步断奶母猪的繁殖性能进行了总结,并得出结论,从哺乳期到发情期间隔的减少很大程度上受哺乳期最后几天母猪保留仔猪数量的影响。当最后 4.7 天母猪只有 3 头仔猪留下时,发现断奶至发情期间隔最大。有趣的是,Grant 表明(Varley 和 Foxcroft,1990),与单纯的应用逐步断奶相比,当前 6 只乳头被覆盖时逐步断奶(在哺乳期的最后 7 天留下 5 头仔猪给母猪)对 LH 浓度和卵泡发育的影响较大。由于仔猪在两种处理方式中生长速度相似,所以对母猪的代谢需求相似,因而对 LH 和卵泡发育的影响主要由哺乳性的神经刺激来介导。因此,只有少数实验评估了逐步断奶对生殖生理或母猪表现的影响,而且这些实验都集中在胎次较低的母猪上(Mahan,1993;Vesseur 等,1997;Zak 等,2008;Foxcroft 等,未出版)。这些研究同样一致地表明,当应用逐步断奶时断奶至发情/授精间隔减少。最近的两项研究已经评估了内分泌状态对这种影响的作用。Degenstein 等(2006)发现,在整个逐步断乳期间(哺乳期 16~19 天),催乳素(prolactin)浓度降低,FSH 浓度升高。Zak 等(2008)也发现在哺乳期第 18 天时,逐步断奶后 10 h 内 LH 浓度和脉冲频率急剧增加。在断奶时(3 天后),LH 平均浓度与对照组相似。断奶后的第二天,逐步断奶组的母猪有更多的卵泡直径大于 3 mm。因此,逐步断奶制度期间 LH 和 FSH 浓度增加,以较低的催乳素浓度,可能刺激了卵泡发育并缩短了断奶期与发情期的间期。

几乎没有关于逐步断奶的母猪繁殖能力方面的信息。然而,已经测量的生殖特征显示出排卵率略有增加(17.7 对 15.5 (Zak 等,2008)),以及第 28 天胚胎存活率升高(64% 对 60% (Zak 等,2008));后续窝仔猪数量增加(Vesseur 等,1997),并且第二胎母猪的分娩率也有所提高(97% 对 86% (Vesseur 等,1997))。因此,通过刺激滤泡发育,主要是减少哺乳刺激,在哺乳期最后 3~5 天移出的绝大多数仔猪通常会缩短断奶期至发情期间隔。然而,这种断奶-发情期间隔的改善仅在特定的母猪群中看到,这群猪除此之外断奶-发情时间间隔便会延长。此外,为数不多的研究调查了逐步断奶之外的断奶后之影响,总体上其结论并不确定,因而从整体来看,其对后续繁殖能力的影响尚不清楚。这些结果也意味着哺乳期仔猪数量的减少往往对随后的繁殖能力没有重大影响,部分原因是剩余的仔猪每个个体消耗的乳汁将更多(Vesseur 等,1997)。

17.3.4 间歇性哺乳和哺乳期排卵

减少仔猪的哺乳刺激的一种方法(因其刺激滤泡发育和随后的哺乳期排卵),可以在哺乳期的最后一段时间采取将母猪和其仔猪每日分离的方式。这个过程称为减少或限制哺乳,或中断哺乳或间歇的哺乳。Langendijk 等(2007)和 Gerritsen 等(2008)回顾了间歇性哺乳对随后繁殖能力的影响。他们表明,如果在哺乳期间歇性哺乳开始的不是太早(最好晚于第 18 天)并且每天持续至少 10 h,高达 90% 的母猪可能会显示泌乳期发情。在分开过程中,母猪饲养于在远离仔猪视线和听觉的位置,最好允许一些公猪接触。母猪的基因型和胎次是非常重要的,因为初产母猪对间歇性哺乳的发情反应较低(Soede 等,2012)。有趣的是,对于治疗有反应的猪,通过显示发情来完成这种反应,以一种在大约开始治疗后正常的间隔(4~5天)时发生的同步性方式完成(Gerritsen 等,2008)。因此,与 7 天相比,14 天的治疗期没有增加发情反应的发生,没有出现泌乳期发情的母猪在断奶后具有"正常"断奶-发情期间隔(Soede 等,2012)。因此,间歇性泌乳制度下,母猪的哺乳期发情反应似乎是一个"全有或无"现象,或者是"正常"的卵泡期持续时间,亦或是根本就没有反应。

在间歇性哺乳的早期研究中,大多数母猪在断奶后排卵。因此,从这些研究中,仍然不清楚间歇性哺乳和/或哺乳期授精是否以及如何影响妊娠的。最近的间歇性哺乳方面的研究(由 Gerritsen 等(2008)综述)表明,如果间歇性哺乳诱导的排卵发生在分娩后 19~21 天且如果间歇性哺乳持续至排卵后 20 天,受胎率、胚胎存活和胚胎发育将受到不利影响。后一种作用可能与这些母猪中较低的孕酮浓度有关。当母猪在分娩后 3 周以上进行授精和间歇性哺乳在排卵后 2 天或 9 天停止时,仔猪大小和分娩率都受到哺乳期授精的不利影响(Soede 等,2012)。同样,哺乳期进行的授精如果在分娩后超过 3 周进行时则不会对产仔数或分娩率产生负面影响(Gaustad-Aas 等,2004)。在哺乳期延长的系统中,在排卵期间对母猪授精可能具有经济上的吸引力。最近的研究因此评估了分群圈养(Hultén 等,2006)、每日公猪接触、将母猪带到检测/配种区域(van Wettere 等,2013)或每日围栏公猪的接触,与分别断奶(Terry 等,2013)结合起来,对刺激哺乳期排卵的影响。

总之,最近的研究表明,哺乳期可以诱导卵泡发育和排卵。然而在哺乳期显示发情的母猪数量仍有很大差异,其受到治疗开始的时间、基因型和母猪胎次的影响。大多数显示发情的母猪在有标准排卵率的正常断奶间隔之后都是这样做的。当母猪在间歇哺乳期制度区间排卵时,与断奶后授精的母猪相比,它们具有相似的分娩率和排卵量,只要分娩-授精间隔少于 24 天,授精-断奶间隔不超过 10 天。

17.3.5 结论

短的哺乳期(短于 3 周)对随后的繁殖性能(断奶-发情间隔、分娩率、窝产仔数)产生负面影响,因为基本的生殖相关过程(如下丘脑-垂体-卵巢轴、卵泡发育、子宫环境)需要在分娩后的前几周进行恢复。断奶后生产性能降低常见于哺乳期间体重大量丢失的母猪,因为在断奶时卵泡发育被抑制。因此,管理策略应旨在限制这种体重损失。这样的策略不仅包括优化的饲喂计划、哺乳期水的摄入量和环境温度,还包括在哺乳前对母猪的最佳营养管理。哺乳期采食和哺乳都是卵泡发育的主要动力。因此,管理技术,例如间歇性哺乳或逐步断奶等在哺乳期的最后时间内可以成功刺激卵泡发育,从而提高断奶后的生产表现。

17.4 断奶至发情间隔:支持卵泡发育

正如前面一段关于哺乳期的讨论,哺乳期繁殖问题的主要原因在于哺乳期的采食能力有限,导致体重明显减轻。然而,有可能进行哺乳后修复。解决方案旨在正常的断奶-发情间隔期间优化卵泡发育,或者在延长的断奶-发情期间修复卵泡发育。

17.4.1 在断奶至发情间隔期刺激卵泡发育

断奶后饲喂:水平和组成

一般来说,断奶后饲喂和管理对断奶-发情间期的影响仅在母猪断奶-发情间隔延长时才能看到。在断奶-发情间隔较短的母猪中,在断奶后直接募集卵泡,并在 5~6 天内发育成排卵前的大小。这表明,最佳的断奶后饲养对于我们现在用的断奶期-发情期间隔较短的现代化母猪而言,已经变得不那么重要。然而,对于第一胎母猪,优化的断奶后营养管理可能仍然是相关的。例如,在断奶时机体条件较差 (148 kg 和 13.4 mm 背膘厚度)的第一胎母猪中,与高脂日粮饲喂的母猪相比,断奶后高碳水化合物日粮可以将第一胎母猪在断奶后 9 天内发情的比例从 52% 提高到 67% (Van den Brand 等,2001)。最近的数据 (Van den Brand 等,2006;2009)表明,断奶至发情之间的葡萄糖补充并没有导致已经较短的断奶-发情间隔变得更短,而是导致了出生仔猪数量上的升高和更均一的出生体重,并增加小仔猪的断奶前存活率。此外,从断奶到配种这段时间内在日粮中包含可发酵的非淀粉多糖 (non-starch polysaccharides,NSP,例如甜菜渣)增加了总产仔数和产活仔猪数

(Sörensen,1994;Van der Peet-Schwering 等,2003)。添加甜菜糖浆和葡萄糖的效果可能是通过其胰岛素和 IGF-1 的刺激作用来实现的。实际上,有迹象表明胰岛素和 IGF-1 支持卵泡发育,使得断奶和发情之间的卵泡发育更均匀,从而减少发育中的胚胎和仔猪变化(Wientjes,2013)。

公猪刺激

断奶后的母猪与公猪的接触可能会刺激 LH 释放、断奶后卵泡发育和排卵时机,从而影响显示发情和排卵的母猪的百分比(由 Kemp 等,2005 综述)。在断奶身体状况不佳的第一胎母猪中(148 kg,13.4 mm 背膘厚度),断奶后强烈的公猪接触将母猪在断奶后 9 天内显示发情的百分比提高到 51%,而没有公猪刺激的母猪为 30%(Langendijk 等,2000b)。因此,断奶母猪与公猪的接触可能刺激卵泡发育,特别是对一些母猪有益,这些母猪要不然就会有延长的断奶-排卵间隔(如第一胎母猪)。刺激卵泡发育所需的公猪接触水平不是很清楚,可能在不同母猪之间有所不同。例如,Knox 等(2002)发现,从断奶 3 天开始,无论是每天与公猪接触一次、两次还是三次的母猪,其在断奶后 8 天内排卵的比例(95%)和断奶-排卵间隔(150 h)都很相似。虽然将母猪圈养在一起可能会对发情表现和检测产生负面影响(Knox 等,2004),不知道这是否也会影响卵泡发育和排卵时间。尽管有这些不确定因素,建议在断奶后的第二天便开始与公猪接触,每天两次鼻-鼻接触约 10 min,以确保充分的刺激。

猪舍条件

断奶母猪的舍饲条件可能有很大差异。母猪可能是单独或分组地饲养,空间容量也可能不同,并且环境条件(如环境温度)也可能不同。对于这些因素能否以及如何影响卵泡发育知之甚少。Langendijk 等(2000b)发现,相比单个圈养的母猪,以 4 个为一栏(自由进入栏位)饲养的母猪,母猪在断奶-发情时间间隔会有约 10 h 的延迟(105 h 对 94 h),断奶至排卵间隔期间亦延迟 10 h(138 h 对 127 h)。尽管两组的断奶-发情期间隔都较短,但是群体饲养的母猪的发情和排卵相对延迟可能与母猪的压力有关,这种压力来源于已经发情的母猪的爬跨行为。实际上,发情期间的刺激应激[可由反复注射促肾上腺皮质激素(adrenocorticotropic hormone,ACTH)来模拟]被发现可以阻碍卵泡发育(Lang 等,2004)、LH 波动以及排卵(Turner 和 Tilbrook,2006)。

外源激素的使用

在预期断奶时卵泡发育较低的母猪组中(例如第一胎母猪),可以使用外源激素来刺激卵泡发育。一个知名的产品达到这种效果,其通过结合主要起 FSH 作用的 400 IU 的马绒毛膜促性腺激素(equine chorionic gonadotropin,eCG,以前称为 pregnant mare serum gonadotropin,PMSG 或妊娠母马血清促性腺激素)和主要起

LH 作用的 200 IU 的人绒毛膜促性腺激素(hCG,human chorionic gonadotropin)。在许多研究中,在断奶时或断奶后的第二天给予这种激素组合,能够使断奶-发情期改善,但有时候导致较低的受胎率或较低的窝仔数（Kirkwood,1999）。Vargas 等(2006)不仅发现在断奶后 24 h 内治疗会有更多的母猪在断奶后的前 10 天发情(95%对 80%),而且还发现下一胎的窝仔数会更高（11.2 对 10.4）,3 胎后的淘汰率没有受到影响。不同研究之间在促性腺激素治疗后繁殖性能的差异可能与注射时卵泡发育的差异有关。因此,如果给予外源激素以刺激断奶后发情,建议检查这些处理对母猪产仔数和分娩率的影响。

17.4.2 断奶后卵泡发育

流行病学资料表明,断奶-发情间隔短是第二窝病综合征的风险因素。实际上,断奶-发情间隔较短的母猪产生第二窝仔数变少的可能性更低（Morrow 等,1992）。因此,通过给予母猪在哺乳后更长的恢复期,似乎可以改善卵泡发育,从而改善胚胎的存活。这个更长的恢复期可以通过不同的方式来实现,可以在第二次断奶后发情期间进行母猪授精,或在较短的断奶后期间向母猪饲喂孕酮类似物。

第二发情期配种(Skip-a-heat,HNS)

在断奶后的第二次发情期间而不是在第一次发情期间（Skip-a-heat,HNS)受精,能够使母猪从哺乳期恢复,这可能对繁殖性能产生重大影响。跳过初产母猪的第一次发情后则母猪的受胎率提高了 15%～17%,且随后的仔猪数量增加了 1.3～2.5 头(Clowes 等,1994；Vesseur,1997；Werlang 等,2011)。Clowes 等(1994)报道,第二次发情后的母猪在 LH 激增后 50 h 时具有更高浓度的 IGF-1 和孕激素,这可能表明卵泡和黄体的发育更加良好。当使用这种配种方式时,频繁的发情检查对于确保母猪显示第二次发情是重要的。是否应用这些技术要考虑经济可行性的问题。

使用孕酮类似物(progesterone analogues)

为了人为延长断奶-发情间隔并使得卵泡质量恢复,母猪可以给予口服孕酮类似物——烯丙孕素(altrenogest),其剂量为 15～20 mg/天。后续繁殖性能的结果取决于治疗时机。当仅仅在断奶后治疗 3～4 天时,受胎率和窝仔数量可能不受影响(van Leeuwen 等,2011)或增加(Boland,1983；Martinat-Botté 等,1995)。然而,6～8 天的治疗时间可能会降低随后的繁殖表现(van Leeuwen 等,2011；Werlang 等,2011),而长达 14 天的治疗则会同时在妊娠第 50 天将产仔提高 1.8 头(Patterson 等,2008；van Leeuwen 等,2011),在仔猪出生时提高 2.5 头（van Leeuwen 等,2011)。这些后果与烯丙孕素治疗期间卵泡的生长模式有关。初期卵泡生长在 6～8 天达到约 5 mm,其后卵泡大小轻微下降并在 4.0～4.5 mm 处稳定（van Leeuwen 等,2010)。卵泡发育的这种模式说明烯丙孕素治疗的第 6～8

天卵泡发生周转。因此,在这个时期结束烯丙孕素治疗可能导致老年卵泡的排卵,由此造成这些母猪生殖表现较差(van Leeuwen,2011)。断奶后的最佳阿仑膦酸盐治疗策略仍然需要进一步探讨。

17.4.3 小结

尽管大多数母猪的断奶-发情间隔通常较短(4～6 天),但仍然可以通过提供适当的公猪接触、饲养和猪舍条件来优化卵泡发育。在预期断奶后繁殖率较低的母猪中(例如过度体重减轻的第一胎母猪),可以通过使用能够导致 21 个非生产性天数(周期长度)的第二发情期配种的策略,或使用 10～14 天的孕酮类似物治疗来延长断奶-发情期/断奶-授精期间隔。

17.5 发情和授精:确保最大受精

为了达到最大受精,应该在排卵的适当时间进行授精,使用高质量的精液和足量的精子,使用卫生、非创伤和低应激授精的方法。本节讨论了授精环节中最大的挑战。

17.5.1 授精时间

在母猪中,受精结果非常依赖于授精相对于排卵的时间(由 Kemp 和 Soede,1997 综述)。在授精后,子宫收缩将精子细胞运送到子宫-输卵管交界处的精子库(Rodriguez-Martinez,2007)。精子细胞中仅有一小部分达到精子库(Rodriguez-Martinez,2007),大部分在子宫内被吞噬,或被子宫组织吸收或通过阴道排出(Steverink 等,1998;Woelders and Matthijs,2001)。当排卵接近时,内分泌和生理条件发生变化,精子细胞开始从储存地进入输卵管。当授精时间超过排卵期时,精子细胞可能会老化并失去其受精能力。另一方面,当排卵时间超过授精时间时,卵母细胞在精子细胞到达输卵管中部的受精部位时已经老化。在每 4 h 以超声波检查排卵时间的研究中,显示母猪在排卵前 8 h 受精可以获得最佳受精结果(图 17.2)。在这一时期,90%的母猪受精率达 100%。当授精进行得较早或者较晚时,受精结果会下降,但在排卵前 24 h 内的授精可使超过 90%的母猪中至少 80%的卵母细胞受精。由于母猪的排卵率很少低于 20,这使得在妊娠第 5 天至少有 16 只良好的胚胎。Nissen 等(1997)也表明排卵前 24 h 授精则产卵率最高,产仔数大。

在实践中,授精时间不能基于排卵时间,因为排卵时间既不能预测也不可看到。因此,授精时间必须基于发情的迹象,因为母猪在发情期间排卵(见本章的"生殖生理学"部分)。不幸的是,发情期间在不同农场之间(平均 40～60 h)和同一农场内差异很大(从小于 24 h 到超过 96 h)(由 Soede 和 Kemp,1997 综述)。因此,在标准发情时母猪应每天进行授精,以便在最佳间隔期间至少有一次授精。评估

17. 优化哺乳期和断奶母猪繁殖生理及性能的最佳技术

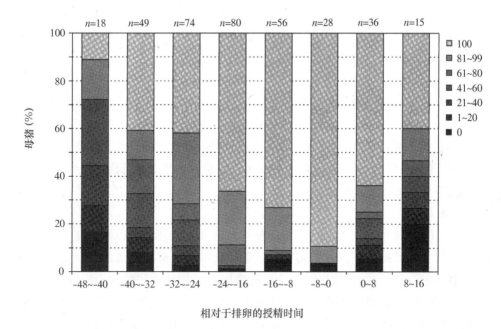

图 17.2 经产母猪的受精分布结果取决于授精相对于排卵的时间。受精率定义为冲洗的卵母细胞中的胚胎百分比和排卵后第 5 天时的胚胎,以 8 h 授精间隔中母猪达到 0,1%～20%,21%～40%,41%～60%,61%～80%,81%～99%和100%受精的百分比呈现(Kemp 和 Soede,2005)。

发情时 12 h 或 24 h 区间内受精的繁殖力方面的研究显示其有类似的表现(Castagna 等,2003)。然而,可能需要特别注意后备母猪,因为它们可能具有较短的最佳间隔(Waberski 等,1994)。只要在发情期间进行授精,相对于排卵来说,母猪过度授精的风险并不大。然而,授精母猪在排卵后可能会引起亚临床感染的风险(Rozeboom 等,1997),从而导致分娩率和窝仔数量下降。这种影响被证明与能降低子宫清除感染的孕酮浓度上升有关(De Winter 等,1996)。另一方面,在超过 70%的母猪接受至少一次排卵后授精的农场中,分娩率和产仔数不受影响(Castagna 等,2003)。

17.5.2 影响发情持续时间的因素

早期的文献(Soede 和 Kemp,1997)列出了影响发情持续时间的已知因素。下面提到一些主要因素。

公猪接触

用公猪更多的刺激能增加发情的表现,从而增加可监测发情的持续时间。在公猪刺激的强度/数量方面(例如,将母猪送到检测/配种区域将增加发情持续时间(Jongman 等,1996))以及发情检测的频率方面(Knox 等,2002)都是真实的。另一方面,公猪持续存在可能会使母猪习惯于公猪刺激并减少可见发情的持续时间(例如,Knox 等,2004)。因此建议将公猪饲养于母猪之外和/或使用另一只公猪进行发情检测,根据母猪的猪舍状况,可以将母猪带到公猪处进行发情检测,或者将公猪带到母猪处。在任何一种情况下,在检查期间母猪需要与公猪有良好的鼻一鼻接触,以便有适当的外激素和嗅觉刺激(Signoret,1970)。因此,重要的是使用聚焦于母猪的成熟公猪(具有适当的野猪气味)。最好是每天进行两次发情检测,从而对爬跨发情进行评估,对于最佳的授精时间是重要的。使用相同的发情检测方法是非常重要的,因为母猪会针对这种发情检测方案对其行为作出调整(Langendijk 等,2000a)。

应激

在 1970 年,Liptrap 已经表明注射 ACTH 和皮质类固醇可以减少猪的发情期持续时间。从那时起,许多研究证实了应激状况可能会抑制母猪的发情表现。许多不同的因素可以引起母猪应激。主要应激因素是母猪之间的社会应激,这可能会抑制发情表现,特别是在等级低的母猪中,也可能是对人类的恐惧,这可能会降低母猪在人类存在时的发情表现(Pedersen,2007)。因此,在评估母猪的发情反应时,认识到应激状况是很重要的。

断奶至发情间隔

在大多数农场,但不是所有的农场,相比于断奶-发情间隔较长的母猪,断奶-发情间隔较短(少于 5 天)的母猪发情期更长,因而断奶-发情间隔较长的母猪应在发情后立即受精。

17.5.3 发情检测和授精策略的评估

在发情表现不佳或多变的农场,分娩率和/或排卵量小于预期,评估发情期持续时间和发情时受精的时间可能是值得的。在每头母猪身上建立发情期的开始和结束都至少需要一周,以便考虑到断奶-发情间期的变化。发情开始定义为:当母猪没有对公猪显示出站立反应的最后一次发情检查时,以及当母猪显示出站立反应时的第一次发情检查之间的中点,这也适用于发情的结束。发情持续时间的变化是发情表现和发情检测质量的量度。排卵时间估计发生在发情持续时间的 2/3。这个时间点可能与授精的时间有关,并用于评估授精策略,因为授精应在排卵前 0~24 h 进行。图 17.3 显示了一组评估断奶母猪授精策略的示例。

17.5.4 授精方面的其他因素

除了适当的授精时间,与授精有关的其他方面也可能影响受精。精液用量应该是质量好的,这意味着在收集、稀释和储存方面都要质量好,以及适当的质量检查。在本综述中,合适精液品质是前提条件。为了最佳的子宫收缩,建议在授精期

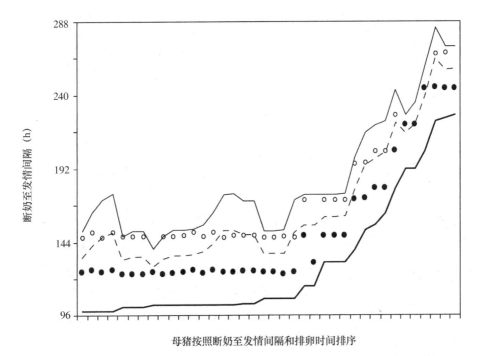

图 17.3 一批断奶母猪授精时机评估实例。发情期检测每天进行两次,以确定发情开始和发情结束(直线),排卵时间估计发生在发情的 2/3 处(虚线)。在这组母猪中,发情较晚的话发情时间会明显下降。大多数母猪的第一次授精(闭合圆)在完美的确定在授精前的 0~24 h。然而,绝大多数母猪在排卵后接受了第二次授精(开放圆)。第二次授精是多余的,可能是有害的(从 P. Langendijk 得到的数据)。

间出现公猪。公猪接触引起催产素释放,随后刺激子宫收缩至少 1 h(Langendijk 等,2005)。当仅仅应用人工刺激时,催产素释放较低,甚至可能不存在(Langendijk 等,2005;Madej 等,2005)。为了从公猪接触中获益最多,对精子运输产生最大的益处,建议避免母猪在人工授精前与公猪接触。此外,公猪接触刺激母猪的静立反应,这可能会限制精液回流,从而增加良好授精的机会。发情和授精过程中的应激也可能通过不同的机制干扰受精概率。例如,在一系列研究中,在发情前期重复注射 ACTH 会降低发情期持续时间和干扰排卵,并且在发情期间注射 ACTH 会给管腔内环境和最终的胚胎恢复带来负面影响[由 Einarsson 等(2008)综述]。

17.5.5 人工授精技术的发展

人工授精技术的发展主要目的是使用较少精子细胞进行高产的人工授精,减少对发情和授精过程中的劳动投入。提高优秀公猪利用效率,这最终有利于遗传发展,并可能最终使得有效利用冷冻保存或性别分类的精液(由 Roca 等,2011

综述)。另外,低剂量人工授精对授精的时间和质量提出了更高的要求。

授精技术和剂量

直到2000年左右,所有母猪的人工授精都是子宫颈输精。将输精管固定在进入子宫颈一半的地方,使用的精子剂量为80~100 mL,含有高达$(2～3)×10^9$个精子细胞。目前,世界上绝大多数人工授精仍然使用子宫颈输精,但也已开发用于宫内或宫颈后授精的输精管,并且其使用越来越多。将输精管固定在子宫颈中后,一根较细的管子从输精管中心延伸约10 cm,直至伸到子宫内子宫角的分叉处,然后将40～60 mL的精液注入[含$(1～2)×10^9$精子细胞]。子宫角授精减少了所需的精子细胞数(Hernández-Caravaca等,2012;Watson和Behan,2002),其原因是由于精子细胞的回流减少,宫颈褶皱中精液减少(Hernández-Caravaca等,2012)。当前不建议在后备母猪中进行子宫角输精,它们具有较低的导管通过率,较高的宫颈损伤概率和与之相关的低繁殖力。同时子宫角输精相关的管理措施也是不同的。例如,与宫颈输精相反,子宫角输精期间的公猪接触可能阻碍插管的通过。选择子宫颈还是子宫角人工授精在不同农场之间可能不同,其取决于公猪的遗传价值、精液、输精管的成本、劳动力成本和使用这两种技术的繁殖性能。

定时人工授精

由于不能充分预测排卵时间,母猪标准发情期间必须每天授精,以便在排卵前的最佳24 h内进行至少一次授精。因此,大多数母猪在发情期间被授精2～3次。如果排卵时间在母猪中能够被成功诱导,则可以进行定时人工授精。这减少了与发情检测和授精相关的劳动,并且使种公猪的利用更有效。最近,Driancourt(2013)回顾了固定时间人工授精情况,主要内容如下。可以使用LH激动剂(例如人绒毛膜促性腺激素-hCG或pLH)来模拟排卵前内源性LH波动,或使用GnRH激动剂(例如布舍瑞林-buserelin、戈舍瑞林-goserelin、内皮素-depherelin和曲普瑞林- triptorelin)可诱导排卵前的内源性LH波动。这些产品施用的时间通常为:在用孕酮类似物治疗后的固定时间(后备母猪)、断奶后的固定时间或发情开始后(母猪)。如果所选时机适合,排卵将在35～44 h后进行。由于母猪可能会显示出不同的断奶后卵泡发育和发情开始,其取决于胎次、哺乳期体重损失、季节、遗传和哺乳期长短等因素,制定适用于在这些情况下的母猪策略是一个真正的挑战。制药公司目前正在制定固定时间的人工授精策略并进行评估。

17.6 结论

哺乳期的管理很大程度上影响后续的繁殖表现。应避免哺乳期体重明显减

17.优化哺乳期和断奶母猪繁殖生理及性能的最佳技术

轻,特别是在第一胎母猪中,哺乳期长度应大于 21 天以允许足够的卵泡发育。哺乳期管理策略如间歇性哺乳和逐步断奶可能刺激乳腺卵泡发育,从而使断奶后发情开始。这是通过减少母乳强度(最重要的)和改善母猪的能量平衡的共同作用实现的。然而,这种策略的好处似乎微不足道,因为现代母猪已被选择具有较短的断奶-发情间隔,且并不是所有的母猪都对这种治疗做出反应。断奶后,适当的营养、公猪接触和避免应激条件可以进一步刺激卵泡发育。这应该引起母猪具有较短且同步的断奶－排卵间隔。然而,由于断奶后的滤泡期通常很短,其至最佳的断奶后状况可能不足以优化哺乳后的繁殖能力。另一种可能性是允许初产母猪通过推迟断奶后发情和授精(例如通过短期使用孕酮类似物)从哺乳期恢复。随后,优质卵泡的最佳受精率可以通过适当时间的优质授精来实现。图 17.4 中总结了本章所涉及的各个方面。重要的是要注意,生殖成功的一个关键方面取决于存在有高技能的工作人员,他们认识到自己的行为对母猪生殖生理的重要性。

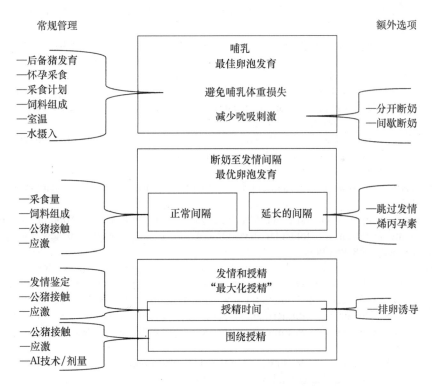

图 17.4 影响断奶母猪繁殖性能的主要管理方法概要。

参考文献

Baidoo, S.K., Aherne, F.X., Kirkwood, R.N. and Foxcroft, G.R., 1992. Effect of feed intake during lactation and after weaning on sow reproductive performance. Canadian Journal of Animal Science 72: 911-917.

Belstra, B.A., Flowers, W.L. and See, M.T., 2004. Factors affecting temporal relationships between estrus and ovulation in commercial sow farms. Animal Reproduction Science 84: 377-394.

Bevers, M.M., Willemse, A.H., Kruip, T.A.M. and Van de Wielm, D.F.M., 1981. Prolactin levels and the LH-response to synthetic LH-RH in the lactating sow. Animal Reproduction Science 4: 155-163.

Black, J.L., Mullan, B.P., Lorschy, M.L. and Giles, L. R., 1993. Lactation in the sow during heat stress. Livestock Production Science 35: 153-170.

Boland, M.P. 1983. Control of oestrus in primiparous sows. Theriogenology 19: 377-384.

Bortolozzo, F.P., Bernardi, M.L., Kummer, R. and Wentz, I., 2009. Growth, body state and breeding performance in gilts and primiparous sows. In: Rodriguez-Martinez, H., Vallet, J.L. and Ziecik, A.J. (eds.) Control of Pig Reproduction VIII. Nottingham University Press, Nottingham, UK, pp. 281-291.

Britt, J.H., Armstrong, J.D., Cox, N.M. and Esbenshade, K.L., 1985. Control of follicular development during and after lactation in sows. Journal of Reproduction and Fertility, Supplement 33: 37-54.

Castagna, C.D., Peixoto, C.H., Bortolozzo, F.P., Wentz, I., Ruschel, F., Neto, G.B. and De Suinos, S., 2003. The effect of post-ovulatory artificial insemination on sow reproductive performance. Reproduction in Domestic Animals 38: 373-376.

Clowes, E. J., Aherne, F.X. and Foxcroft, G.R., 1994. Effect of delayed breeding on the endocrinology and fecundity of sows. Journal of Animal Science 72: 283-291.

Clowes, E. J., Aherne, F.X., Foxcroft, G.R. and Baracos, V.E., 2003. Selective protein loss in lactating sows is associated with reduced litter growth and ovarian function. Journal of Animal Science 81: 753-764.

De Rensis, F., Hunter, M.G. and Foxcroft, G.R., 1993. Suckling-induced inhibition of luteinizing hormone secretion and follicular development in the early postpartum sow. Biology of Reproduction 48: 964-969.

De Winter, P.J.J., Verdonck, M., De Kruif, A., Coryn, M., Deluyker, H.A., Devriese, L.A. and Haesebrouck, F., 1996. The relationship between the blood progesterone concentration at early metoestrus and uterine infection in the sow. Animal Reproduction Science 41: 51-59.

Degenstein, K., Wellen, A., Zimmerman, P., Shostak, S., Patterson, J., Dyck, M. and Foxcroft, G.R., 2006. Effect of split weaning on hormone release in lactating sows. Advances in Pork Production 17: 17.

Dourmad, J. Y., 1991. Effect of feeding level in the gilt during pregnancy on voluntary feed intake during lactation and changes in body composition during gestation and lactation. Livestock Production Science 27: 309-319.

Driancourt, M. A., 2013. Fixed-time artificial insemination in gilts and sows: tools, schedules and efficacy. In: H. Rodriguez-Martinez, Soede, N.M. and W. L. Flowers (eds.) Control of Pig Reproduction IX No. IX. Context, Leicestershire, UK. pp. 89-99.

Drochner, W. 1989. Einflusse von Fetzulagen an Sauen auf Aufzuchtleistung und Fruchtbarkeit. Ubersicht Tierernährung 17: 99-139.

Einarsson, S., Brandt, Y., Rodriguez-Martinez, H. and Madej, A., 2008. Conference Lecture: Influence of stress on estrus, gametes and early embryo development in the sow. Theriogenology 70: 1197-1201.

Everts, H., Blok, M.C., Kemp, B., Van der Peet-Schwering, C.M.C. and Smits, C.H.M., 1995. Requirements of lactating sows, Lelystad, the Netherlands.

Fraser, D. and Phillips, P.A., 1989. Lethargy and low water intake by sows during early lactation: A cause of low piglet weight gains and survival? Applied Animal Behaviour Science 24: 13-22.

Gaustad-Aas, A.H., Hofmo, P.O. and Karlberg, K, 2004. The importance of farrowing to service interval in sows served during lactation or after shorter lactation than 28 days. Animal Reproduction Science 81: 287-293.

Gerritsen, R., 2008. Lactational oestrus in sows: follicle growth, hormone profiles and early pregnancy in sows subjected to intermittent suckling, Phd thesis, Wageningen University, Wageningen, the Netherlands.

Gerritsen, R., Soede, N.M., Langendijk, P., Hazeleger, W. and Kemp, B, 2008. The intermitted suckling regimen in pigs; Consequences for reproductive performance of sows. Reproduction in Domestic Animals, Supplement 5 43: 29-35.

Hernández-Caravaca, I., Izquierdo-Rico, M.J., Matas, C., Carvajal, J.A., Vieria, L., Abril, D., Soriano-Ubeda, C. and Garcia-Vazquez, F.A., 2012. Reproductive performance and backflow study in cervical and post-cervical artificial insemination in sows. Animal Reproduction Science 136: 14-22.

Hoofs, A.I.J. and Elst-Ter Wahle, L., 1993. Self feeding mangers for lactating sows. Report Praktijkonderzoek, Rosmalen, the Netherlands.

Hoving, L.L., Soede, N.M., Feitsma, H.. and Kemp, B., 2012. Lactation weight loss in primiparous sows: consequences for embryo survival and progesterone and relations with metabolic profiles. Reproduction in Domestic Animals 47: 1009-1016.

Hoving, L.L., Soede, N.M., Graat, E.A.M., Feitsma, H. and Kemp, B., 2010. Effect of live weight development and reproduction in first parity on reproductive performance of second parity sows. Animal Reproduction Science 122: 82-89.

Hoving, L.L., Soede, N.M., Graat, E.A.M., Feitsma, H. and Kemp, B., 2011. Reproductive performance of second parity sows: Relations with subsequent reproduction. Livestock Science 140: 124-130.

Hultén, F., Wallenbeck, A. and Rydhmer, L., 2006. Ovarian activity and oestrous signs among group-housed, lactating sows: Influence of behaviour, environment and production. Reproduction in Domestic Animals 41: 448-454.

Jongman, E.C., Hemsworth, P.H. and Galloway, D.B., 1996. The influence of conditions at the time of mating on the sexual behaviour of male and female pigs. Applied Animal Behaviour Science 48: 143-150.

Kemp, B. and Soede, N.M., 1997. Consequences of variation in interval from insemination to ovulation on fertilization in pigs. Journal of Reproduction and Fertility. Supplement 52: 79-89.

Kemp, B., Soede, N.M. and Hazeleger, W., 1998. Control of Ovulation. In: Progress of Pig Science: 285-302.

Kemp, B., Soede, N.M. and Langendijk, P., 2005. Effects of boar contact and housing conditions on estrus expression in sows. Theriogenology 63: 643-656.

King, R.H. and Williams, I.H., 1984. The effect of nutrition on the reproductive performance of first-litter sows 1. feeding level during lactation, and between weaning and mating. Animal Production 38. 241-247.

Kirkwood, R., Baidoo, S.K., Aherne, F.X. and Sather, A.P., 1987. The influence of feeding level during lactation on the occurence and endocrinology of the postweaning estrus in sows. Canadian Journal of Animal Science 67: 405-415.

Kirkwood, R.N, 1999. Pharmacological intervention in swine reproduction. Journal of Swine Health and Production 7: 29-35.

Kirkwood, R.N., Baidoo, S.K. and Aherne, F.X., 1990. The influence of feeding level during lactation and gestation on the endicrine status and reproductive performance of second parity sows. Canadian Journal of Animal Science 70: 1119-1126.

Knox, R.V., 2005. Recruitment and selection of ovarian follicles for determination of ovulation rate in the pig. Domestic Animal Endocrinology 29: 385-397.

Knox, R.V., Breen, S.M., Willenburg, K.L., Roth, S., Miller, G.M., Ruggiero, K.M. and Rodriguez-Zas, S.L., 2004. Effect of housing system and boar exposure on estrus expression in weaned sows. Journal of Animal Science 82: 3088-3093.

Knox, R.V., Miller, G.M., Willenburg, K.L. and Rodriguez-Zas, S.L., 2002. Effect of frequency of boar exposure and adjusted mating times on measures of reproductive performance in weaned sows. Journal of Animal Science 80: 892-899.

Knox, R.V. and Rodriguez- Zas, S.L., 2001. Factors influencing estrus and ovulation in weaned sows as determined by transrectal ultrasound. Journal of Animal Science 79: 2957-2963.

Knox, R.V., Rodriguez Zas, S.L., Sloter, N.L., McNamara, K.A., Gall, T.J., Levis, D.G., Safranski, T.J. and Singleton, W.L., 2013. An analysis of survey data by size of the breeding herd for the reproductive management practices of North American sow farms. Journal of Animal Science 91: 433-445.

Koketsu, Y., Dial, G.D. and King, V.L., 1997. Influence of various factors on farrowing rate on farms using early weaning. Journal of Animal Science 75: 2580-2587.

Koketsu, Y., Dial, G.D., Pettigrew, J.E. and King, V.L., 1996a. Feed intake pattern during lactation and subsequent reproductive performance of sows. Journal of Animal Science 74: 2875-2884.

Koketsu, Y., Dial, G.D., Pettigrew, J.E., Marsh, M.E. and King, V.L., 1996b. Characterization of feed intake patterns during lactation in commercial swine herds. Journal of Animal Science 74: 1202-1210.

Lang, A., Kaeoket, K., Kindahl, H., Madej, A. and Einarsson, S., 2004. Influence of CRH and ACTH administration on endocrine profile and ovulation in sows. Reproduction in Domestic Animals 39: 181-189.

Langendijk, P., Berkeveld, M., Gerritsen, R., Soede, N.M. and Kemp, B., 2007. Intermittent suckling: tackling lactational anoestrus and alleviating weaning risks for piglets. In: Wiseman, J., Varley, M.A., McOrist, S. and Kemp, B. (eds.) Paradigms in pig science. Nottingham University Press, Nottingham, UK. pp. 359-384.

Langendijk, P., Soede, N.M., Bouwman, E.G. and Kemp, B., 2000a. Responsiveness to boar stimuli and change in vulvar reddening in relation to ovulation in weaned sows. Journal of Animal Science 78: 3019-3026.

Langendijk, P., Soede, N.M. and Kemp, B., 2000b. Effects of boar contact and housing conditions on estrus expression in weaned sows. Journal of Animal Science 78: 871-878.

Langendijk, P., Soede, N.M. and Kemp, B., 2005. Uterine activity, sperm transport, and the role of boar stimuli around insemination in sows. Theriogenology 63: 500-513.

Langendijk, P., Van Den Brand, H., Soede, N.M. and Kemp, B., 2000c. Effect of boar contact on follicular development and on estrus expression after weaning in primiparous sows. Theriogenology 54: 1295-1303.

Le Cozler, Y., Dagorn, J., Dourmad, J.Y., Johansen, S. and Aumaître, A., 1997. Effect of weaning-to-conception interval and lactation length on subsequent litter size in sows. Livestock Production Science 51: 1-11.

Leenhouwers, J.I., Ten Napel, J., Hanenberg, E.H.A.T. and Merks, J.W.M., 2011. Breeding replacement gilts for organic pig herds. Animal 5: 615-621.

Liptrap, R.M., 1970. Effect of corticotrophin and corticosteroids on oestrus, ovulation and oestrogen excretion in the sow. Journal of Endocrinology 47: 197-205.

Lucy, M.C., Liu, L., Boyd, C.K. and Bracken, C.J., 2001. Ovarian follicular growth in sows. Reproduction Supplement 58: 31-45.

Madej, A., Lang, A., Brandt, Y., Kindahl, H., Madsen, M.T. and Einarsson, S., 2005. Factors regulating ovarian function in pigs. Domestic Animal Endocrinology 29: 347-361.

Mahan, D. C., 1993. Effect of weight, split-weaning, and nursery feeding programs on performance responses of pigs to 105 kilograms body weight and subsequent effects on sow rebreeding interval. Journal of Animal Science 71: 1991-1995.

Makkink, C.A. and Schrama, J.W., 1998. Thermal requirements of the lactating sow. In: Verstegen, M.W.A., Moughan, P.J. and Schrama, J.W. (eds.) The lactating sow. Wageningen Pers, Wageningen, the Netherlands, pp. 271-283.

Martinat-Botté, F., Bariteau, F., Forgerit, Y., Macar, C., Poirier, P. and Terqui, M., 1995. Synchronization of oestrus in gilts with altrenogest: effects on ovulation rate and foetal survival. Animal Reproduction Science 39: 267-274.

Matte, J. J., Pomar, C. and Close, W.H., 1992. The effect of interrupted suckling and splitweaning on reproductive performance of sows: a review. Livestock Production Science 30: 195-212.

Mejia-Guadarrama, C.A., Pasquier, A., Dourmad, J.Y., Prunier, A. and Quesnel, H., 2002. Protein (lysine) restriction in primiparous lactating sows: effects on metabolic state, somatotropic axis, and reproductive performance after weaning. Journal of Animal Science 80: 3286-3300.

Meunier-Salaün, M.C. and Bolhuis, J.E., 2015. High-Fibre feeding in gestation. Chapter 5. In: Farmer, C. (ed.) The gestating and lactating sow. Wageningen Academic Publishers, Wageningen, the Netherlands, pp. 95-116.

Morrow, W.E.M., Leman, A.D., Williamson, N.B., Morrison, R.B. and Robinson, R.A., 1992. An epidemiological investigation of reduced second-litter size in swine Preventive Veterinary Medicine 12: 15-26.

Mullan, B.P. and Williams, I.H., 1989. The effect of body reserves at farrowing on the reproductive performance of first-litter sows. Animal Production 48: 449-457.

Nissen, A.K., Soede, N.M., Hyttel, P., Schmidt, M. and D'Hoore, L., 1997. The influence of time of insemination relative to time of ovulation on farrowing frequency and litter size in sows, as investigated by ultrasonography. Theriogenology 47: 1571-1582.

Noguchi, M., Yoshioka, K, Itoh, S., Suzuki, C., Arai, S., Wadda, Y., Hasegawa, Y. and Kaneko, H., 2010. Peripheral concentrations of inhibin A, ovarian steroids, and gonadotropins associated with follicular development throughout the estrous cycle of the sow. Reproduction 139: 153-161.

NRC. 2012. Nutrient requirements of swine. Eleventh revised edition ed. National Academic Press, Washington, D.C., USA.

Patterson, J., Wellen, A., Hahn, M., Pasternak, A., Lowe, J., DeHaas, S., Krraus, D., Williams, N. and Foxcroft, G.R., 2008. Responses to delayed estrus after weaning in sows using oral progestagen treatment. Journal of Animal Science 86: 1996-2004.

Patterson, J.L., Smit, M.N., Novak, S., Wellen, A.P. and Foxcroft, G.R., 2011. Restricted feed intake in lactating primiparous sows. I. Effects on sow metabolic state and subsequent reproductive performance. Reproduction, Fertility and Development 23: 889-898.

Pedersen, L. J. 2007. Sexual behaviour in female pigs. Hormones and Behavior 52: 64-69.

Prunier, A., Martinat-Botté, F., Ravault, J.P. and Camous, S., 1987. Perioestrous patterns of circulating LH, FSH, prolactin and oestradiol-17[beta] in the gilt. Animal Reproduction Science 14: 205-218.

Prunier, A., Soede, N.M., Quesnel, H. and Kemp, B., 2003. Productivity and longevity of weaned sows. In: Pluske, J.R., Le Dividich, J. and Verstegen, M.W.A. (eds.) Weaning the pig. Wageningen Academic Publishers, Wageningen, the Netherlands, pp. 385-419.

Quesnel, H., Mejia-Guadarrama, C.A., Dourmad, J.Y., Farmer, C. and Prunier, A., 2005a. Dietary protein restriction during lactation in primiparous sows with different live weights at farrowing: I. Consequences on sow metabolic status and litter growth. Reproduction Nutrition Development 45: 39-56.

Quesnel, H., Mejia-Guadarrama, C.A., Pasquier, A., Dourmad, J.Y. and Prunier, A., 2005b. Dietary protein restriction during lactation in primiparous sows with different live weights at farrowing: II. Consequences on reproductive performance and interactions with metabolic status. Reproduction Nutrition Development 45: 57-68.

Quesnel, H., Pasquier, A., Mounier, A.M. and Prunier, A., 1998. Influence of feed restriction during lactation on gonadotropic hormones and ovarian development in primiparous sows. Journal of Animal Science 76: 856-863.

Quesnel, H. and Prunier, A., 1998. Effect of insulin administration before weaning on reproductive performance in feed-restricted primiparous sows. Animal Reproduction Science 51: 119-129.

Quiniou, N. and Noblet, J., 1999. Influence of high ambient temperatures on performance of multiparous lactating sows. Journal of Animal Science 77: 2124-2134.

Roca, J., Parrilla, I., Rodriguez-Martinez, H., Gil, M.A., Cuello, C., Vazquez, J.M. and Martinez, E.A., 2011. Approaches towards efficient use of boar semen in the pig industry. Reproduction in Domestic Animals 46: 79-83.

Rodriguez-Martinez, H., 2007. Role of the oviduct in sperm capacitation. Theriogenology 68, Supplement 1: S138-S146.

Rojanasthien, S., Madej, A., Lundeheim, N. and Einarsson, S., 1987. Luteinizing hormone response to different doses of synthetic gonadotropin-releasing hormone during early and late lactation in primiparous sows. Animal Reproduction Science 13: 299-307.

Rozeboom, D.W., 2015. Conditioning of the gilt for optimal reproductive performance. Chapter 1. In: Farmer, C. (ed.) The gestating and lactating sow. Wageningen Academic Publishers, Wageningen, the Netherlands, pp. 13-26.

Rozeboom, K.J., Troedsson, M.H.T., Shurson, G.C., Hawton, J.D. and Crabo, B.G., 1997. Late Estrus or Metestrus Insemination after Estrual Inseminations Decreases Farrowing Rate and Litter Size in Swine. Journal of Animal Science 75: 2323-2327.

Schenkel, A.C., Bernardi, M.L., Bortolozzo, F.P. and Wentz, I., 2010. Body reserve mobilization during lactation in first parity sows and its effect on second litter size. Livestock Science 132: 165-172.

Sesti, L.A. and Britt, J.H., 1993. Influence of stage of lactation, exogenous luteinizing hormone-releasing hormone, and suckling on estrus, positive feedback of luteinizing hormone, and ovulation in sows treated with estrogen. Journal of Animal Science 71: 989-998.

Shaw, H.J. and Foxcroft, G.R., 1985. Relationships between LH, FSH and prolactin secretion and reproductive activity in the weaned sow. Journal of Reproduction and Fertility 75: 17-28.

Signoret, J.P., 1970. Reproductive behaviour of pigs. Journal of Reproduction and Fertility Supplement 11: 105-117.

Silva, B.A.N., Noblet, J., Donzele, J.L., Oliveira, R.F.M., Primot, Y., Gourdine, J.L. and Renaudeau, D., 2009a. Effects of dietary protein level and amino acid supplementation on performance of mixed-parity lactating sows in a tropical humid climate. Journal of Animal Science 87: 4003-4012.

Silva, B.A.N. Noblet, J., Oliveira, R.F.M., Donzele, J.L., Fernandez, H.C., Lima, A.L., Renaudeau, D. and Noblet, J., 2009b. Effect of floor cooling and dietary amino acids content on performance and behaviour of lactating primiparous sows during summer. Livestock Science 120: 25-34.

Smits, R.J., Henman, D.J. and King, R.H., 2012. Increasing the energy content of lactation diets fed to first-litter sows reduces weight loss and improves productivity over two parities. Animal Production Science 53: 23-29.

Soede, N.M., Helmond, F.A. and Kemp, B., 1994. Periovulatory profiles of oestradiol, LH and progesterone in relation to oestrus and embryo mortality in multiparous sows using transrectal ultrasonography to detect ovulation. Journal of Reproduction and Fertility 101: 633-641.

Soede, N.M. and Kemp, B., 1997. Expression of oestrus and timing of ovulation in pigs. Journal of Reproduction and Fertility Supplement 52: 91-103.

Soede, N.M., Langendijk, P. and Kemp, B., 2011. Reproductive cycles in pigs. Animal Reproduction Science 124: 251-258.

Soede, N.M., Laruenssen, B., Abrahamsse-Berkeveld, M., Gerritsen, R., Dirx-Kuijken, N., Langendijk, P. and Kemp, B., 2012. Timing of lactational oestrus in intermittent suckling regimes: consequences for sow fertility. Animal Reproduction Science 130: 74-81.

Soede, N.M., Wetzels, C.C.H., Zondag, W., Hazeleger, W. and Kemp, B., 1995. Effects of a second insemination after ovulation on fertilization rate and accessory sperm count in sows. Journal of Reproduction and Fertility 105: 135-140.

Sörensen, G., 1994. Pulpetter til soer, Danske Slagterier, Copenhagen, Denmark.

Stevenson, J.S. and Davis, D.L., 1984. Influence of reduced litter size and daily litter separation on fertility of sows at 2 to 5 weeks postpartum. Journal of Animal Science 59: 284-293.

Steverink, D.W., Soede, N.M., Groenland, G.J., Van Schie, F.W., Noordhuizen, J.P.T.M. and Kemp, B., 1999. Duration of estrus in relation to reproduction results in pigs on commercial farms. Journal of Animal Science 77: 801-809.

Steverink, D.W.B., Soede, N.M., Bouwman, E.G. and Kemp, B., 1997. Influence of insemination-ovulation interval and sperm cell dose on fertilization in sows. Journal of Reproduction and Fertility 111: 165-171.

Steverink, D.W.B., Soede, N.M., Bouwman, E.G. and Kemp, B., 1998. Semen backflow after insemination and its effect on fertilisation results in sows. Animal Reproduction Science 54: 109-119.

Svajgr, A.J., Hays, V.W., Cromwell, G.L. and Dutt, R.H., 1974. Effect of Lactation Duration on Reproductive Performance of Sows. Journal of Animal Science 38: 100-105.

Terry, R., Kind, K.L., Hughes, P.E., Kennaway, D.J., Herde, P.J. and Van Wettere, W.H.E.J., 2013. Split weaning increases the incidence of lactation oestrus in boar-exposed sows. Animal Reproduction Science 142: 48-55.

Thaker, M.Y.C. and Bilkei, G., 2005. Lactation weight loss influences subsequent reproductive performance of sows. Animal Reproduction Science 88: 309-318.

Thingnes, S.L., Ekker, A.S., Gaustad, A.H. and Framstad, T., 2012. *Ad libitum* versus step-up feeding during late lactation: the effect on feed consumption, body composition and production performance in dry fed loose housed sows. Livestock Science 149: 250-259.

Tummaruk, P., Lundeheim, N., Einarsson, S. and Dalin, A.M., 2001. Reproductive performance of purebred Hampshire sows in Sweden. Livestock Production Science 68: 67-77.

Turner, A.I. and Tilbrook, A.J., 2006. Stress, cortisol and reproduction in female pigs. Society of Reproduction and Fertility Supplement 62: 191-203.

Van den Brand, H., Dieleman, S.J., Soede, N.M. and Kemp, B., 2000a. Dietary energy source at two feeding levels during lactation of primiparous sows: I. Effects on glucose, insulin, and luteinizing hormone and on follicle development, weaning-to-estrus interval, and ovulation rate. Journal of Animal Science 78: 396-404.

Van den Brand, H., Heetkamp, M.J.W., Soede, N.M., Schrama, J.W. and Kemp, B., 2000b. Energy balance of lactating primiparous sows as affected by feeding level and dietary energy source. Journal of Animal Science 78: 1520-1528.

Van den Brand, H., Langendijk, P., Soede, N.M. and Kemp, B., 2001. Effects of postweaning dietary energy source on reproductive traits in primiparous sows. Journal of Animal Science 79: 420-426.

Van den Brand, H., Soede, N.M. and Kemp, B., 2000c. Dietary energy source at two feeding levels during lactation of primiparous sows: I. Effects on glucose, insulin, and luteinizing hormone and on follicle development, weaning-to-oestrus interval, and ovulation Journal of Animal Science 78: 396-404.

Van den Brand, H., Soede, N.M. and Kemp, B., 2006. Supplementation of dextrose to the diet during the weaning to estrus interval affects subsequent variation in within-litter piglet birth weight. Animal Reproduction Science 91: 353-358.

Van den Brand, H., Van Enckevort, L.C.M., Van der Hoeven, E.M. and Kemp, B., 2009. Effects of dextrose plus lactose in the sows diet on subsequent reproductive performance and within litter birth weight variation. Reproduction in Domestic Animals 44: 884-888.

Van Der Peet-Schwering, C.M.C., Kemp, B., Binnendijk, G.P., Den Hartog, L.A., Spoolder, H.A.M. and Verstegen, M.W.A., 2003. Performance of sows fed high levels of nonstarch polysaccharides during gestation and lactation over three parities. Journal of Animal Science 81: 2247-2258.

Van Leeuwen, J.J., Williams, S.I., Martens, M.R., Jourquin, J., Driancourt, M.A., Kemp, B. and Soede, N.M., 2011. The effect of different postweaning altrenogest treatments of primiparous sows on follicular development, pregnancy rates, and litter sizes. Journal of Animal Science 89: 397-403.

Van Leeuwen, J.J.J., 2011. Post weaning altrenogest use in sows, Wageningen University, Wageningen, the Netherlands.

Van Leeuwen, J.J.J., Williams, S.I., Kemp, B. and Soede, N.M., 2010. Post-weaning Altrenogest treatment in primiparous sows; the effect of duration and dosage on follicular development and consequences for early pregnancy. Animal Reproduction Science 119: 258-264.

Van Wettere, W.H.E.J., Kaisler-Smith, C.R., Terry, R., Weaver, A.C., Herde, P.J., Kennaway, D.J., Hughes, P.E. and Kind, K.L., 2013. Boar contact is an effective stimulant of ovulation during early lactation. Livestock Science 155: 454-458.

Vargas, A.J., Bernardi, M.L., Wentz, I., Neto, G.B. and Bortolozzo, F.P., 2006. Time of ovulation and reproductive performance over three parities after treatment of primiparous sows with PG600. Theriogenology 66: 2017 2023.

Varley, M.A., 1982. The time of weaning and its effects on reproductive function. In: Cole, D.J.A. and Foxcroft, G.R. (eds.) Control of Pig Reproduction. Butterworth, London, pp. 459-478.

Varley, M.A. and Foxcroft, G.R., 1990. Endocrinology of the lactating and weaned sow. Journal of Reproduction and Fertility, Supplement 140: 47-61.

Vesseur, P.C., 1997. Causes and consequences of variation in weaning to oestrus interval in the sow, Wageningen University, Wageningen, the Netherlands.

Vesseur, P.C., Kemp, B., den Hartog, L.A. and Noordhuizen, J.P.T.M., 1997. Effect of split-weaning in first and second parity sows on sow and piglet performance. Livestock Production Science 49: 277-285.

Vinsky, M.D., Novak, S., Dixon, W.T., Dyck, M.K. and Foxcroft, G.R., 2006. Nutritional restriction in lactating primiparous sows selectively affects female embryo survival and overall litter development. Reproduction, Fertility, and Development 18: 347-355.

Waberski, D., Weitze, K.F., Gleumes, T., Schwarz, M., Willmen, T. and Petzoldt, R., 1994. Effect of time of insemination relative to ovulation on fertility with liquid and frozen boar semen. Theriogenology 42: 831-840.

Wagenberg, V.A.V., Van der Peet-Schwering, C.M.C., Binnendijk, G.P. and Claessens, P.J.P.W., 2006. Effect of floor cooling on farrowing sows and litter performance: field experiment under Dutch conditions. Transactions of the American Society of Agricultural and Biological Engineers 49: 1521-1527.

Watson, P. F. and Behan, J.R., 2002. Intrauterine insemination of sows with reduced sperm numbers: Results of a commercially based field trial. Theriogenology 57: 1683-1693.

Weldon, W.C., Lewis, A.J., Louis, G.F., Kovar, J.L. and Miller, P.S., 1994. Postpartum hypophagia in primiparous sows: II. Effects of feeding level during gestation and exogenous insulin on lactation feed intake, glucose tolerance, and epinephrine-stimulated release of nonesterified fatty acids and glucose. Journal of Animal Science 72: 395-403.

Werlang, R. F., Argenti, L.E., Fries, H.C.C., Bernardi, M.I., Wentz, I. and Bortolozzo, F.P., 2011. Effects of breeding at the second oestrus or after post-weaning hormonal treatment with altrenogest on subsequent reproductive performance of primiparous sows. Reproduction in Domestic Animals 46: 818-823.

Wientjes, J.G.M., 2013. Piglet birth weight and litter uniformity: importance of pre-mating nutritional and metabolic conditions, Wageningen University, Wageningen, the Netherlands.

Wientjes, J.G.M., Soede, N.M., Van der Peet-Schwering,.C.M.C., Van den Brand, H. and Kemp, B., 2012. Piglet uniformity and mortality in large organic litters: effects of parity and pre-mating diet composition. Livestock Science 114: 218-229.

Wientjes, J.G.M., Soede, N.M., Knol, E.F., Van den Brand, H. and Kemp, B., 2013. Piglet birth weight and litter uniformity: effects of weaning-to-pregnancy interval and body condition changes in sows of different parities and crossbred lines. Journal of Animal Science 91: 2099-2107.

Willis, H.J., Zak, L.J. and Foxcroft, G.R., 2003. Duration of lactation, endocrine and metabolic state, and fertility of primiparous sows. Journal of Animal Science 81: 2088-2102.

Woelders, H. and Matthijs, A., 2001. Phagocytosis of boar spermatozoa *in vitro* and *in vivo*. Reproduction Supplement 58: 113-127.

Xue, J.L., Dial, G.D., Marsh, W.E., Davies, P.R. and Momont, H.W., 1993. Influence of lactation length on sow productivity. Livestock Production Science 34: 253-265.

Yang, H., Eastham P.R., Phillips, P. and Whittemore, C.T., 1989. Reproductive performance, body weight and body condition of breeding sows with differing body fatness at parturition, differing nutrition during lactation and differing litter size. Animal Production 48: 181-201.

Yang, H., Foxcroft, G.R., Pettigrew, J.E., Johnston, L.J., Shurson, G.C., Costa, A.N. and Zak, L.J., 2000a. Impact of dietary lysine intake during lactation on follicular development and oocyte maturation after weaning in primiparous sows. Journal of Animal Science 78: 993-1000.

Yang, H., Pettigrew, J.E., Johnston, L.J., Shurson, G.C. and Walker, R.D., 2000b. Lactational and subsequent reproductive responses of lactating sows to dietary lysine (protein) concentration. Journal of Animal Science 78: 348-357.

Yoder, C.L., Schwab, C.R., Fix, J.S., Duttlinger, V.M. and Baas, T.J. 2012. Lactation feed intake in purebred and F1 sows and its relationship with reproductive performance. Livestock Science 150: 187-199.

Yoder, C.L., Schwab, C.R., Fix, J.S., Stalder, K.J., Dixon, P.M., Duttlinger, V.M. and Baas, T.J., 2013. Estimation of deviations from predicted lactation feed intake and the effect on reproductive performance. Livestock Science 154: 184-192.

Zak, L.J., Foxcroft, G.R., Aherne, F.X. and Kirkwood, R.N., 2008. Role of luteinizing hormone in primiparous sow responses to split weaning. Reproduction in Domestic Animals 43: 445-450.

Zak, L.J., Cosgrove, J.R., Aherne, F.X. and Foxcroft, G.R., 1997a. Pattern of feed intake and associated metabolic and endocrine changes differentially affect postweaning fertility in primiparous lactating sows. Journal of Animal Science 75: 208-216.

Zak, L.J., Williams, I.H., Foxcroft, G.R., Pluske, J.R., Cegielski, A.C., Clowes, E.J. and Aherne, F.X., 1998. Feeding lactating primiparous sows to establish three divergent metabolic states: i. Associated endocrine changes and postweaning reproductive performance. Journal of Animal Science 76: 1145-1153.

Zak, L.J., Xu, X., Hardin, R.T. and Foxcroft, G.R., 1997b. Impact of different patterns of feed intake during lactation in the primiparous sow on follicular development and oocyte maturation. Journal of Reproduction and Fertility 110: 99-106.

18. 母猪健康

R. M. Friendship* and T. L. O'Sullivan

Department of Population Medicine, University of Guelph, Guelph, ON, N1E 2W1, Canada; rfriends@uoguelph.ca

摘要：在母猪群体中控制疾病有助于防止疾病在其他生产阶段蔓延，并且对于母猪健康的关注是创建群体免疫力，也很有可能消除限制性疾病的关键。母猪为哺乳仔猪提供被动免疫力。理想情况下，强壮的健康仔猪将会被断奶，而不是可能会将疾病传染给产房内同圈同伴的弱势的感染小猪。由于疾病和受伤引起的母猪高死亡率和过早淘汰导致畜群生产力降低，因此母猪在达到最大生产力之前被淘汰，并被生产力较低的后备母猪代替。分娩期间母猪容易受到诸如子宫感染和乳腺炎等健康问题的影响，更容易发生系统性感染（如猪丹毒）。母猪猝死的主要原因包括胃溃疡和胃肠道扭转引起的出血，以及热应激、肾炎和心力衰竭。由于之前环境条件的适应，母猪通常对农场存在的大多数地方性疾病是免疫的，但是当新引入时，一些病原体会引起畜群暴发疾病。例如，猪繁殖与呼吸综合征病毒（porcine reproductive and respiratory syndrome virus）和猪流感病毒（swine influenza virus）可能导致广泛的母猪疾病，并可能导致母猪死亡。很多情况下，母猪感染猪病原体导致胎儿或哺乳仔猪发生最严重的损失或死亡。由于细小病毒（parvovirus）感染而发生的繁殖损失会对胚胎和胎儿造成重大损失，但通常在繁育母猪中没有疾病的迹象。在哺乳期间对母猪造成损伤的疾病可能导致仔猪死亡率升高，因为患病母猪可能无法提供足够的乳汁来防止其仔猪的饥饿。为了使畜群的生产表现达到最高，母猪的健康必须进行优化。

关键词：疾病，群体免疫力，母猪死亡率，乳腺炎，跛行

18.1 引言

母猪健康不仅对繁育种群的生产性能产生影响，而且对生产周期的各个阶段都有影响。本章重点讨论了哺育阶段和生长育肥阶段的母猪群体健康状况对预防或最小化疾病传播的重要性，以及减少母猪群中疾病问题的管理程序，例如常规疫苗接种和寄生虫防治计划。最后，将介绍一些重要的母猪疾病。

18.2 母猪群体的疾病管理影响整个猪群

许多猪病的控制取决于母猪群给予仔猪针对农场中所有重要病原体的免疫力,同时不将这些病原体传播给仔猪。换句话说,猪群中的每只母猪不仅为其仔猪提供营养,而且通过富含免疫球蛋白的初乳和乳汁向仔猪提供被动免疫,同时给予仔猪健康的微生物菌群。随着群体数量的增加,建立强大的整个群体免疫力变得困难。从来没有感染过当地病原体的母猪群可以容易地在大的群体中存活。如果母猪在分娩中没有接触过病原体,她的仔猪不会受到针对这种病原体的被动保护,从而使其比其他来自于免疫母猪的仔猪更容易感染。没有获得被动保护的仔猪可能在产后的头几周内遇到疾病,并被感染所吞噬。这些感染的猪可能会变得非常脆弱甚至会死亡,但也可能向其他的猪群和环境散布病原体。在这种情况下,同一圈内分娩的所有仔猪患病概率都会增加,周围环境的污染也可能会对后续在这个圈里出生的仔猪造成威胁。此外,生病的仔猪可能会将疾病携带到哺育舍,甚至是生长育肥舍。在妊娠最后期首次感染病原体的母猪可能是病原体的来源并感染其仔猪。因此,要尽力确保整个母猪群体对畜群中流行的病原体具有免疫力。这对于进入正在繁殖群体、暴露于各种没有免疫力的病原体的年轻替代后备母猪尤为重要。

一般来说,母猪通过与其他母猪混合过程中接触到潜在病原体获得免疫力,也可以从与环境微生物的接触中获得。当农场较小而进行粗放饲养时,比较容易保证母猪都暴露于同样的微生物且没有天然的动物亚群。然而,现代化的密集限制设施使得将年轻种畜暴露于农场中所有可能的疾病变得困难。实现母猪免疫的策略包括有目的地接触(特别适用于隔离设施中的替代后备母猪)和疫苗接种。此外,还应强调防止新进种群带来新的疾病(生物安全)。

一种新的疾病进入母猪群最大风险之一发生在引进后备猪时。原畜群的健康状况必须尽可能与接受群体相类似。单一来源的后备猪引进总是优于多种来源。在进入主要畜群之前,新进的后备母猪应安置在隔离设施中。使用隔离舍的主要优点是,如果后备母猪是一种疾病的亚临床携带者其可能会被迅速诊断,这些后备母猪可以运往市场,而不是融入主要群体而使整个群体受到感染。但是隔离舍也提供了一个机会,使得备母猪暴露于存在于接受群体而不是源群体中的潜在致病因子。后备母猪可能会经历一种温和的疾病形式,在此期间它们排出了大量的病原体,然后发展出强大的免疫力。有时接受农场中的疾病对新进的后备母猪造成严重的健康威胁。隔离期间是对后备母猪接种的一个良好机会,可以对给新进后备母猪健康构成威胁的疾病进行免疫接种,使其在与常住繁殖种群混合时不会发生严重疾病。

青年母猪的免疫力可能比老年母猪要小,使得后备母猪接种疫苗成为优先。

从免疫的角度来看，最好是有一个低的母猪周转率，其中尽可能少的后备母猪被引入繁殖群体。将新引进动物封闭几个月，是控制或消除一些猪疾病的战略的重要组成部分，例如传染性胃肠炎（transmissible gastroenteritis，TGE）或猪繁殖与呼吸综合征（porcine reproductive and respiratory syndrome，PRRS）。这两种疾病都是由病毒引起的，且在这两种情况下疫苗都不是很有效。普遍认为在突然暴发这些疾病的情况下，应尽可能快地将所有母猪暴露于病原体以发展群体免疫。这种广泛接触病原体理想地产生均匀免疫的母猪群。免疫母猪将能够保护仔猪免受疾病的侵害，并防止病毒从哺乳猪进入保育猪。在有目的地暴露和传播疾病给所有母猪这一时期，繁殖群体对所有新进的后备种猪都是封闭的，这被认为是消除猪群疾病的重要的第一步。只有证明病毒不在母猪群中传播，并且它不再存在于环境中，开始将新引进后备种猪引入群体才是安全的。

疾病进入畜群还有许多其他途径，管理这种风险在现代猪业生产中变得非常重要。已经开发了许多基于风险的方法来帮助设计猪场的生物安全计划（Neumann，2012）。主要风险因素除引进后备种猪种群外，还包括：运送牲畜和饲料的运输车、鸟类和啮齿动物以及进出圈舍的人类。

18.3 母猪健康管理

一般来说，相对于畜群中其余部分而言，个体母猪的健康常常被忽略。许多现代猪场，由于疾病和伤害造成的死亡率和淘汰率可能超过 10%（Abiven 等，1998）。健康计划往往旨在减少哺乳期和断奶仔猪的损失，而较少关注种畜群中个体健康最大化。许多给予母猪的疫苗其目的是最大限度地提高能够从母猪转移到哺乳仔猪中的被动免疫和泌乳免疫的质量，例如预防在新生仔猪中由大肠杆菌引起的腹泻（colibacillosis，大肠杆菌病）的疫苗。其他母猪预防接种计划主要旨在减少与胎儿死亡相关的损失。在猪场中使用的疫苗接种或治疗方案很少，其主要目的是改善母猪的健康。

在一个特殊的时期，母猪感染传染病的风险很高。可能最大的风险发生于当替换后备母猪到达农场时。当它们处于运输和混群的应激下以及可能发情期开始时，后备母猪可能暴露于新的病原体。从具有非常高的健康条件的遗传繁殖群体购买替换后备母猪是很常见的。如果新进后备母猪在进入繁殖群之前没有暴露于普通病原体如细胞内劳森氏菌（*Lawsonia intracellularis*），那么它们可能会出现急性出血性腹泻导致高死亡率，除非使用预防性抗生素或后备母猪在引进前免疫接种。同样地，在没有嗜血杆菌的农场饲养的后备母猪在被引入具有更常规健康状况的繁殖群后，不久之后就会发展出急性多发性肌炎（格拉热氏病）。这两种细菌通常可在大多数农场中被发现且在繁殖动物中不会引起临床疾病，在自繁自养的农场这些病原体为地方性流行。

母猪第二次面临较大健康风险大约在分娩时。分娩的过程对机体要求很高，因为这是母猪易于疲惫、心力衰竭和身体受伤的时候，这也是母猪免疫力下降的时候。系统性感染如猪丹毒（erysipelas，以高热和皮肤损伤为特征的细菌性疾病）在分娩期间比在其他任何时间更易发生。此外，分娩和哺乳期早期疾病微生物有更多机会进入机体，例如通过产道进入从而导致子宫感染或子宫炎，以及通过造成乳腺炎的乳头损伤进入。分娩母猪需要仔细监测，不仅可以作为防止死胎和新生儿死亡的手段，而且可以在疾病早期诊断母猪以便及时进行干预和治疗。通常当母猪在分娩过程中发生疾病时，用抗生素治疗可以挽救母猪，但是母猪可能不会为其幼仔产生足够的乳汁。尽管早期开始对病猪进行治疗，但将仔猪寄养至其他母猪通常是必要的。在分娩时减少母猪疾病的一些步骤包括：在畜群内保持良好的胎次分配（减少畜群中老龄母猪的数量），如果在分娩期间需要援助则应使用良好的卫生程序，接种疫苗以预防疾病如猪丹毒等，适当的营养和饲养管理以预防妊娠期间的肥胖，以及控制环境以防止热应激。

18.4 母猪疾病概述

在表18.1中总结了母猪疾病。

突然死亡是母猪群体中相对较常见的事件。据报道，心力衰竭是母猪死亡的主要原因之一（D'Allaire等，1991；Sanz等，2007），但这是一个难以确定的诊断。母猪心脏重量与身体重量之比是家畜中最小的一种，使母猪容易引起循环系统的不可逆过载，导致急性心力衰竭。母猪死亡率通常在炎热的夏季达到高峰。在限制饲养系统中的母猪特别容易产生热应激。母猪没有机会沉浸在泥里，其使自己凉快只限于喘气，除非与喷雾或喷淋以增加热量流失。

突然死亡也可由胃溃疡导致的突然失血以及胃或肠扭转相关的休克引起。母猪胃的食管部溃疡是常见的病症，但严重的失血发生很少。病变通常自愈，然后可能多次复发。瘢痕组织可以导致狭窄形成并食道至胃的开口变窄。慢性胃溃疡的临床症状包括呕吐、黑色焦油粪便、苍白、身体状况不佳。因胃脾扭转或其他类型的肠道事故死亡的母猪同样显得非常苍白，但通常它们的腹部有突起的外观。在尸检时，易于看到扭曲的坏死性肠道。扭转通常与饲喂期间的刺激有关且可以通过使用自动进料系统大大降低，从而使饲料到达的等待时间和预期时间最小化。

造成大量母猪死亡的灾难性事故可能发生在现代的限制性建筑物中。两个最常见的原因是：电力故障导致机械通气系统停止，导致温度突然升高和由于热应激导致的死亡损失；其次是由于粪便气体从粪便坑中上升而引起的意外窒息，通常由于在没有足够的通风条件下搅动粪便引起。

表 18.1 基于机体系统受影响的母猪疾病或主要临床症状

生殖系统	运动/肌肉骨骼	消化系统
钩端螺旋体病	关节炎（各种细菌）	传染性胃肠炎
细小病毒	骨软骨病	猪流行性腹泻
猪丹毒	骨软化症	增生性出血性肠病
伪狂犬病	蹄损伤	猪痢疾
膀胱炎/肾盂肾炎	脊柱脓肿	猪霍乱
乳腺炎	猪水疱病	炭疽病
子宫炎	口蹄疫	
难产	捷申病毒	
猝死	**皮肤**	**呼吸系统**
增生性出血性肠病	疥癣	PRRS
心内膜炎	虱子	猪流感
猪应激综合征	猪丹毒	伪狂犬病
胃溃疡	癣	胸膜肺炎
器官或肠的扭转	脓肿	巴氏杆菌病
心脏病	晒伤/冻伤	萎缩性鼻炎
电击死	肩疮	后圆线虫病
难产	伤口	蛔虫迁移
粪便气体吸入	水疱病	猪霍乱
		非洲猪瘟

18.5 影响母猪健康的传染病

以下对一些影响母猪健康的常见和重要疾病的概述，根据机体系统排列。

18.5.1 繁殖系统

在大多数管理良好现代养猪场，母猪的受孕率高达 90％ 左右，其生产了大量仔猪。近年来，引进超高产的繁殖品系使得每次分娩平均出生 15 头或更多的猪变得平常。疾病可影响受孕率以及胚胎和胎儿的生存。妊娠引起的死亡率结果包括规则和不规则的返情、流产或每窝出生的猪数量少的方面。正常发情被定义为母猪分娩后大约 21 天内开始发情，而不正常发情是指母猪在分娩后＞24 天显示发情迹象。在活仔数量较小的窝中存在木乃伊胎或死胎仔猪，表明妊娠期间子宫内的疾病。特定的疾病往往会留下其独特的生殖障碍模式。例如，细小病毒（parvovirus）和肠道病毒（enterovirus）感染导致有时被称之为 SMEDI(stillbirths, mummified fetuses, embryonic death and infertility)的模式，即死胎、木乃伊胎、胚胎死亡和不育症的简写。胎儿在受孕后约 70 天可以使抗细小病毒免疫达到高峰,因此在妊娠期间感染细小病毒的母猪只有小木乃伊仔猪(<17 cm 长度)出生。胎儿骨骼在受孕后 35 天开始发育,在骨骼形成之前胚胎会被再吸收且在出生时不以木乃伊样出现。

流产在母猪中往往不常见。其可以由非感染性原因引起,例如限制性饲喂与突然的冷应激、晒伤以及继发的高烧(发热症)。传染性疾病(如 PRRS 和钩端螺旋体病)穿过胎盘,杀死胎儿并引起流产,通常影响妊娠最后三个月的母猪。与晚期流产相关的疾病通常以死产或弱仔猪为特征。

普通的猪生殖疾病可以通过在配种前 4 周和 2 周免疫后备母猪、在断奶时免疫母猪来预防。常规使用的商业母猪疫苗含有钩端螺旋体(Leptospira sp.)、丹毒丝菌(Erysipelothrix rhusiopathiae)和细小病毒。在某些情况下,也可以在配种前使用 PRRS 病毒疫苗,尽管目前可用的 PRRS 疫苗疗效不佳。

分娩后不久,产道可能会流出阴道分泌物,它是正常仔猪生产过程的一部分,这种正常的分泌称为恶露。分娩是母猪比较脆弱的时期,阴道或子宫的感染可能在此时发生,特别是出生有困难时。未产出的仔猪会导致毒血症和死亡。阴道分泌物也可能在配种时或者 21 天后母猪可能会再次发情时被注意到。子宫感染的母猪会产生大量黏液脓性阴道分泌物。阴道分泌物的发生可能发生在畜群中,并且与随后降低的受孕率有关,因而需要解决。一般建议在配种区改善卫生条件、配种时注意卫生以及淘汰受影响的动物。

膀胱炎和肾盂肾炎常与阴道炎和子宫炎有关。通常,正常排空尿液会阻止感染性微生物从尿道上行并进入膀胱,但是如果饮水量受限制和排尿减少,则可能会发生上行感染。跛行、老年和肥胖也是膀胱炎的危险因素。受影响的母猪可能表现出微小或微妙的临床症状,或者可能频繁排尿同时体重减轻和厌食,有时可以在尿液中检测到血液或脓液,在严重急性病例中出现临床症状之前动物已经死亡。确保水容易获得,注意卫生以防止蔓延是控制膀胱炎和肾盂肾炎的关键。如果发生疾病,对动物要尽早使用抗生素治疗。

乳腺炎可能涉及一个或仅仅几个乳腺,但有时整个乳房变硬,使得整窝仔猪挨饿。一般来说,乳腺炎在分娩后的第一天或第二天被发现,并且经常伴有局部炎症征象,包括皮肤充血和发红,以及一些全身性症状,如发烧和厌食。最常见的感染是由革兰氏阴性大肠菌群细菌引起的(Gerjets 和 Kemper,2009)。母猪一般对抗生素治疗反应良好且能康复,但是仔猪则需要被寄养至未感染母猪,否则可能导致饥饿。慢性乳腺炎也可能发生,通常由于伤口导致细菌进入。其可能形成肉芽肿和脓肿。致病因子通常是隐秘杆菌(Arcanobacterium sp.)、葡萄球菌(Staphylococcus sp.)或者链球菌属(Streptococcus sp.)。较老的母猪更可能受到乳腺炎的影响,虽然母猪可能没有发现全身性疾病的迹象,但受影响的腺体失去产奶能力。

18.5.2 运动/肌肉骨骼系统

跛行一般是母猪过早淘汰的第二个最常见的原因，第一个是繁殖障碍。第19章(Calderón Díaz 等，2015)中有更多专门针对母猪利用年限的细节。最有可能的是由于跛行难以诊断并且可能导致其他问题，如哺乳期表现差和繁殖损失，跛行的普遍性和影响力被低估。动物福利的一个重要组成部分与跛足有关。

跛行可能是由脚部骨骼或关节的各种病理或损伤(由外伤引起，例如断腿)、撕裂的蹄子，骨软骨病(osteochondrosis，OC)、关节病、传染病(包括关节炎)和脚腐病引起的，跛行与各种重要的风险因素有关，如遗传学、体型结构、饲喂管理、猪舍(尤其是地板条件)和运动水平。

OC被认为是后备母猪和年轻母猪中腿部软弱和跛蹄的非常常见的原因(Jorgensen，2000)。由 OC 引起的病变可以完全愈合，进入一种不可逆转的衰弱障碍，在成年动物中称之为关节的剥脱性骨软骨病(osteochondrosis dissecans，OCD)或骨关节病(osteoarthrosis，OA)。OC/OA 是非传染性疾病，是由于关节软骨和生长板的软骨内骨化失败引起的。OC 过程的特点是，骨骺脱落的碎片或被侵蚀的关节软骨导致已矿化软骨的暴露。在 OA 中关节表面变硬(硬化)。OCD 和 OA 之间的主要区别涉及 OA 关节周围组织的矿化和 OA 情况下骨骼的过度矿化。在某些情况下，软骨可能会重新附着并愈合以形成起皱的关节表面。滑膜钙化，周围组织被钙化过度物生长(骨赘)侵入。

各种研究表明多因素病因参与这些病变的发展，包括各种环境和遗传因素。OC 倾向于与体重快速增加有关，其对关节施加压力。创伤也是最广泛提出的病因之一，然而，其作用更可能是对已损坏的骨骺软骨的最终伤害，而不是作为早期病变发展的起始因素。地板太滑和运动不足也被认为是促成因素。

与这些结构性骨骼状况相关的具体临床体征是前肢弯膝、前腿和后肢翻出(turned-out fore 和 hind legs)、蹄腕直立(upright pasterns)、僵硬运动和易于滑倒。行走困难的范围可以从缩短的步幅到不负重的跛行，或从四肢的僵硬到不能或不愿意站立。然而，跛行偏移是常见的，因为多个关节通常受到影响。

炎症性关节炎是另一种情况，其伴随关节囊的增厚产生关节组织的增生，并增加可能含有血液或纤维蛋白的滑液积聚。炎症性关节炎可以由多种病原体引起，包括肠球菌属(*E. rhusiopathiae*)，链球菌属(*Streptococcus* sp.)，副猪嗜血菌属(*H. parasuis*)和支原体(*Mycoplasma hyosynoviae*)。如果早期发现这些病症可以用抗生素治疗，但难以在关节中维持药物的最小抑制浓度，则可能需要延长治疗才能取得成功。

骨质疏松症是猪中另一种常见的骨骼疾病，涉及矿化骨组织的损失或变薄。骨质疏松症可导致猪跛行，特别是在哺乳期结束时的母猪中。矿化骨组织的缺乏导致骨折，其是繁殖种群淘汰的主要原因。骨质疏松症主要是由矿物质摄入不足或 Ca:P 比例异常引起的。

蹄子的健康也是一个关于舒适、利用年限和生产力的问题。蹄子病变在猪中很常见,可能是跛行的重要潜在原因。据报道,5%~20%的跛行病例与蹄部病变相关(Pluym 等,2013)。蹄子腐烂和感染可能起源于粗糙地板的损伤,病变允许细菌进入从而导致感染、严重的疼痛和临床中跛行。一般来说,最普遍的蹄损伤在蹄跟上,其比较柔软且构成承重表面的最大部分。当外爪比内爪大得多时,外爪的蹄跟倾向于变得发育过度,并且更容易受到机械伤害,特别是在硬行走表面上。其他导致蹄跟区域过度生长和破裂的主要因素包括较差的卫生条件和粗糙的地板(Anil 等,2007)。蹄壁上也会发生裂缝,水平和垂直裂缝都是常见的,并且可能与跛行有关。垂直蹄裂可由外伤引起,有些猪在躺卧或试图站起时,会不断地摩擦后蹄外侧爪与地面接触的墙壁。水平裂缝通常是由于由于真皮中血管损坏而导致的角质生长中断引起的,也可能由胃肠道内产生的内毒素引起,或与由革兰氏阴性细菌引起的感染有关,而不是由于蹄子受到的特定创伤引起。不是所有的蹄裂都表示跛行。最令人担忧的病变是那些穿透蹄部角质层,达到真皮层并引起炎症反应的病变。

18.5.3 胃肠道系统

有许多严重的非传染性状态涉及消化系统,可以影响成年母猪,包括口腔病变、胃溃疡、胃肠扭转和直肠脱垂。大多数常见的地方性胃肠道病原体很少引起繁育群体的疾病,因为在动物达到繁育年龄时已经获得了免疫力。在某些情况下,后备母猪在进入繁育群时会暴露于新病原体,例如猪增生性肠炎菌(*L. intracellularis*,增生性出血性肠病的原因)和猪痢疾短螺旋体(*Brachyspira hyodysenteriae*,猪痢疾原因)。母猪在这种情况下可出现腹泻和变得相当虚弱,如果不及时治疗可能死亡。

大多数情况下,只有在新的胃肠道病毒病已经进入畜群,或者摄入了有毒物质,成年母猪才会出现腹泻或呕吐症状。能够引起影响所有年龄组(包括母猪)的水样腹泻的群体暴发的两种最常见的病毒是 TGEV 和猪流行性腹泻病毒(porcine epidemic diarrhea virus,PEDV)。这些都是冠状病毒,尽管没有相关性,这两种疾病的临床症状是相似的。两种病毒都攻击小肠上皮细胞,导致肠道吸收水分和营养物质的能力急剧下降,从而导致腹泻。这些疾病通过粪便、口途径非常迅速地从猪传播到猪,从农场到农场。母猪的疾病一般是轻度的,只持续几天且母猪死亡率很低,但仔猪的死亡率可能接近 100%。患有 TGE 或 PED 的母猪通常会不再进食,而且生产的乳汁很少。没有治疗手段可用,只有包括电解质和液体的支持治疗以防止脱水。预防措施包括防止向农场引入病毒的生物安全措施,其特别强调来源于运输车和污染物(如靴子)的粪便污染。当发生疾病时,通常的做法是确保将其快速免疫于所有动物,特别是妊娠母猪,以便它们能够在两三周内产生免疫力,然后能够通过初乳和乳汁保护它们的幼仔。针对这些疾病的疫苗往往疗效很差,除非疾病在畜群中变得流行,否则不广泛使用。将母猪暴露于含有活病毒的粪便,创造病原免疫,开展卫生措施和精细转群措施,可以从感染猪群中消除这些疾病。

在限制饲养中母猪往往不会有肠道寄生虫的问题,尽管通常建议所有群体应在分娩之前进行常规的驱虫以控制线虫($Ascaris\ suum$,蛔虫)的传播。牧场上的母猪易发生各种寄生虫,包括红腹蠕虫($Hyostrongylus\ rubidus$)和结节虫($Oesophagostomum$ sp.)。户外饲养母猪的慢性寄生可能导致体重减轻和衰弱,建议对这些畜群进行驱虫计划。

18.5.4 皮肤/体表

外部寄生虫病如疥癣(由螨虫引起,$Sarcoptes\ scabiei$)和虱病(由虱子引起,$H.\ suis$)曾经是母猪群中的主要问题,因为引入非常有效治疗手段,这些病原体已被大多数现代养猪场消灭。有一些非常有效的杀螨剂可以采取战略的方式进行使用,以便从畜群中消除这些疾病。虱子和螨虫都会使母猪不安,并使它们不断摩擦。这些情况的存在引起哺乳期性能降低,导致更轻的断奶重量和更大的断奶仔猪死亡率(Davies,1995)。

由于缺少毛发覆盖,母猪皮肤的病变与其他动物物种相比非常明显。皮肤的变化可能是系统性疾病的第一个症状。对猪丹毒的诊断通常是在升高的、红色的、菱形的皮肤病变的基础上进行的。对PRRS诊断的怀疑是建立在在母猪耳朵充血(蓝耳病)基础上。

对皮肤的伤口通常很容易看到,正是由于如此,从福利角度来看肩胛疮的问题已经引起了人们的关注。哺乳母猪花费大量时间侧躺下哺乳,易于在肩胛骨棘突上发生疮。被选为很瘦且高产奶量的母猪倾向于具有非常少的保护性措施。分层分娩的板条箱地板导致对母猪肩膀上特定点的集中压力增加,从而增加发生疮的可能性。瘦母猪和老母猪的肩胛疮风险更大。在哺乳期间应注意饲喂以确保充分的身体状况,并改善分娩条形箱设计,使用垫料可能减少皮肤伤口的流行。

18.5.5 呼吸道或全身系统

与肠道疾病一样,对已经在农场中存在的大多数地方流行性疾病,成年猪暴露其中已经产生了免疫力。因此,呼吸系统疾病的临床症状如咳嗽和打喷嚏通常不是繁殖群体的常见特征。然而,新进的替代后备母猪可能没有保护性免疫的情况下抵达农场,而这些原来的动物中呼吸系统疾病并不罕见。例如,后备母猪可能在进入繁殖群体时首次暴露于猪肺炎支原体($Mycoplasma\ hyopneumoniae$),导致后备母猪不再进食,并显示一种非生产性干嗽。受感染的动物更容易发生继发性细菌感染，例如巴斯德氏菌病(pasteurellosis)，并可能病重。用猪肺炎支原体菌苗免疫接种新进后备母猪可以大大减少这些健康问题。

有一些重要的呼吸道病毒疾病在不同农场之间蔓延，并导致母猪群群体性疾病暴发。猪流感可能是这些疾病中最常见和重要的之一。当流感病毒进入敏感群体时，所有年龄的猪都可能受到影响，但老年动物往往表现出最严重的临床症状，包括高烧、咳嗽、呼吸困难、鼻出血和厌食。其死亡率一般较低，但患有流感的母猪可能不能很好地保育仔猪，从而使得乳猪的死亡率增加。此外，患有流感的猪其免疫功能会降低，继发或并发感染可能会导致更严重的疾病。如果需要，治疗是减少继发性细菌感染的直接手段。

如前所述，PRRS 病毒也可以从农场到农场传播，并导致所有年龄组的暴发。通常，畜群中的第一个临床体征是在妊娠舍通过一波厌食症和呼吸道疾病，随后是流产症状和和哺乳仔猪损失。PRRS 病毒可以快速突变，有许多不同的毒株具有广泛变化的能力来引起疾病。PRRS 病毒有一些高致病性毒株导致极高的母猪死亡率。历史上，除野毒株之外，疫苗一直是改造在猪群中传播的活病毒。不幸的是，针对突变病毒的保护，需要先暴露于与新的毒株非常相似的病毒，因此通过接种商业产品进行保护其价值往往有限。养猪业通常依赖于暴露于野外毒株来产生群体免疫力。

有一些非常重要的有报告的疾病导致母猪的全身性疾病，包括经典猪瘟(hog cholera，猪霍乱)和非洲猪瘟。这些疾病的讨论超出了本章的范围，但是当母猪群发生疾病的突然暴发时，尤其是死亡率显著时，需要始终考虑这些疾病。

历史上，寄生虫一直是母猪呼吸道疾病的来源，但是隔离饲养和现代驱虫药使这些疾病在今天关联不大，除了使用牧场饲养的畜群。在这些情况下最令人担心的两种寄生虫是猪线虫($A.\ suum$)和猪肺虫($Metastrongylus$ sp.)。线虫的幼虫从肠穿过肝脏移动到肺部，并被咳出和吞咽，重新进入胃肠道，以便在成年个体肠道成熟，并完成其生命周期。猪肺虫在肺泡支气管和细支气管中长至成体，在这里它们缠结并被黏液覆盖，阻塞气道。虫卵被咳出并被吞下，然后通过母猪胃肠道。蚯蚓是一个中间宿主，因此限制饲养有效地打断了感染的循环。

18.6　结论

在两个时间点后备母猪/母猪最容易患病：一是在作为后备猪进入繁殖群时，二是分娩期和哺乳期的前几天。最佳母猪健康取决于：首先是良好的生物安全性以防止将新的病原体引入畜群，与此同时通过特意将所有年轻繁殖群暴露或者常规免疫，可以最大限度地发挥母猪对地方流行疾病的免疫力。母猪个体的健康和福利取决于良好的饲养管理技巧,特别是仔细观察早期的疾病征兆并迅速做出反应。通常来说哺乳期母猪可以成功地治愈,但是为防止哺乳仔猪的损失,可能需要将仔猪寄养到健康的母猪下。饲养、环境和圈舍因素通常有助于临床疾病,需要进行优化以使母猪健康最大化。

参考文献

Abiven, N., Seegers, H., Beaudeau, F., Laval, A. and Fourichon, C., 1998. Risk factors for high sow mortality in French swine herds. Preventive Veterinary Medicine 33: 109-119.

Anil, S.S., Anil, L., Deen, J., Baidoo, S.K. and Walker, R.D., 2007. Factors associated with claw lesions in gestating sows. Journal of Swine Health and Production 15: 78-83.

Calderón Díaz, J.A., Nikkilä, M.T. and Stalder, K., 2015. Sow longevity. Chapter 19. In: Farmer, C. (ed.) The gestating and lactating sow. Wageningen Academic Publishers, Wageningen, the Netherlands, pp. 423-452.

D'Allaire, S., Drolet, R. and Chagnon M., 1991. The cause of sow mortality: a retrospective study. Canadian Veterinary Journal 32: 241-243.

Davies, P.R., 1995. Sarcoptic mange and production performance of swine: a review of the literature and studies of associations between mite infestation, growth rate and measures of mange severity in growing pigs. Veterinary Parasitology 60: 249-264.

Gerjets, I. and Kemper, N., 2009. Coliform mastitis in sows: a review. Journal of Swine Health and Production 17: 97-105.

Jorgensen, B., 2000. Osteochondrosis/osteoarthrosis and claw disorders in sows, associated with leg weakness. Acta Veterinaria Scandinavica 41: 123-138.

Neumann, E.J., 2012. Disease transmission and biosecurity. In: Zimmerman, J.J., Karriker, L.A., Ramirez, A., Schwartz, K.J. and Stevenson, G.W. (eds.) Diseases of swine. John Wiley and Sons Inc., Oxford, UK, pp. 141-164.

Pluym, L.M., Van Nuffel, A., Van Weyenberg, S. and Maes, D., 2013. Prevalence of lameness and claw lesions during different stages in the reproductive cycle of sows and the impact on reproduction results. Animal 7: 1174-1181.

Sanz, M., Roberts, J.D., Perfumo, C.J., Alvarez, R.M., Donovan, T. and Almond, G.W., 2007. Assessment of sow mortality in a large herd. Journal of Swine Health and Production 15: 30-36.

19. 母猪利用年限

J. A. Calderón Díaz[1], M. T. Nikkilä[2] and K. Stalder[1]
[1] Iowa State University, Animal Science, 109 Kildee Hall, Ames, IA 50011-3150, USA; stalder@iastate.edu
[2] Figen Ltd., Urheilutie 6 D, 01370 Vantaa, Finland

摘要：商品化繁殖种群中的母猪利用年限较短可以导致经济效率低和动物福利问题。繁殖母猪是畜群中最有价值的动物，据估计40%~50%的母猪在其第三或第四胎之前被淘汰，但在这时初始替换成本尚未达到。淘汰母猪的原因和淘汰率可能受到猪舍、基因型、管理、疾病、营养和市场趋势等许多因素的影响。不管原因如何，利用年限较短导致较高的淘汰率，因此，更多的后备母猪储备意味着更高的生产成本。后备母猪的生产力比经产母猪要低；它们的后代死亡率更高，生长率更低，保育时饲喂效率和育成阶段的生产力更低。此外，当将新后备猪引入繁殖群体时，存在疾病风险。利用年限的改善可能导致替换成本和生产成本的降低，以及已经达到最大生产力的畜群中成熟母猪的比例更大。

关键词：淘汰，利用年限，胎次分布，畜群生产力

19.1 引言

母猪利用年限是高效和盈利的养猪业的重要组成部分；然而，母猪淘汰率稳步上升表明近年来繁育群母猪利用年限已经下降。商品化繁育群中很大比例的母猪在仅仅生产一到两窝之后就被替换，母猪在生产一两窝猪之后被更换，这时母猪还没有达到其最大生产力，而且替换成本还没有得到收回（Carroll，2011；Stalder等，2003）。高淘汰率是猪生产者关注的问题，因为它影响农场的利润和生产效率。此外，高清除率可能会引起动物福利问题，因为一些淘汰和死亡原因可能表明这些动物正在经历较差的动物福利（Barnett等，2001）。此外，由于高淘汰率和死亡率导致的较短的利用年限可能会影响消费者对养猪业的看法及对现代猪肉生产实践和猪肉产品的可接受性。

大约70%的母猪淘汰是由于诸如生殖失败和腿部问题而引起的过早和计划外淘汰的结果（Engblom等，2011）。在美国，每年人约54%的育种母猪被淘汰（即

淘汰率加上死亡率);平均淘汰胎次在 3.3~3.8(D'Allaire 等，1987；Lucia 等，2000b)。在欧洲，平均每年淘汰率为 43%~52%,平均淘汰胎次为 4.3~4.6(Boyle 等，1998；Dijkhuizen 等，1989；Engblom 等，2007)。这种早期淘汰对生产者的盈利能力是不利的，因而育种母畜应该保留在猪群中，直到至少与其更换相关的初始投资成本被收回。Stalder 等(2003)和 Carroll(2011)认为，母猪在繁育群中应至少保留至其达到第三胎时,以便支付其自身的开销;不过,他们建议,收回初始投资成本所需的胎次因农场和生产系统而异。不管生产系统如何,母猪在畜群中停留的时间越长,更多的窝数和猪,其初始更换费用可以分摊。Sehested(1996)报道,将母猪利用年限提高一胎以上其经济影响与瘦肉含量提高 0.5% 一样,不过,他也报告说这一影响在五胎之外没有作用。

高淘汰率增加了非生产性天数，增加了猪群中的后备母猪数量，导致平均窝仔数下降和每头母猪每年断奶的猪数量减少(Hughes 和 Varley,2003)。另外,与多胎母猪所产的猪相比,初胎母猪所生仔猪在出生和断奶时体重较轻,有较高的死亡率风险和疾病易感性(Smits,2011)。此外，在商业育成场中母猪群占到饲料成本的约 20%,较低的淘汰率可能对育种猪饲料消耗变化引起的猪群饲料转化有较小的直接影响,因为替换后备母猪在第一次配种前一直要饲养 6~12 周(Smits,2011)。因此，鉴别和了解影响母猪利用年限的不同因素或做法至关重要，以增加母猪在繁育群中的时间,提高农场的利润率。

19.2 用于评估母猪利用年限的常用指标

科学文献中对母猪利用年限的定义没有明确的共识,它通常指的是生产性利用年限的长短,而不是母猪的自然利用年限。然而,母猪利用年限的定义根据研究的目的或评估利用年限的目的不同而有所不同。D'Allaire 等(1992)提出,有几种评估母猪利用年限的方法,包括生命的长短、畜群利用年限、生产利用年限和停留时间(表明母猪是否存活到预定的胎数或时间点的分类特征)。母猪利用年限的其他指标还包括：移除率、淘汰率、替代率、畜群中的后备母猪的百分比、存栏母猪平均胎次、淘汰时平均胎次。此外,也可以使用经济指标来测量母猪利用年限,例如利用年限中每天断奶仔猪数、每头猪断奶时畜群天数、达到净现值(母猪经济利润)时的胎次(Culbertson 和 Mabry,1995；Stalder 等,2003)。在许多情况下,商业运作所采用的管理做法使得很难使用一些母猪利用年限的定义。例如,生命长度,定义为出生日期和移除日期之间的天数,当后备母猪是购买的且不提供其出生日期时则只能估算 (Stalder 等,2007a)。

畜群生命定义为进入育种群体的日期与淘汰日期之间的天数。Lucia 等(2000b)报告说,后备母猪和母猪平均畜群生命为 583 天；然而,当将后备母猪从数据中排除时,它增加到 691 天。Koketsu(2003)报道了类似的为 682 天畜群生命,Babot 等(2003)报道西班牙商品母猪的畜群生命为 1105 天。生产利用年限是第一次受孕或第一次分娩和淘汰日期之间的天数。Rodríguez-Zas 等(2003)发现美国商品母猪第一次服务到淘汰的平均生产利用年限为 467 天,而 Engblom 等(2007)报告说,瑞典商品母猪从首次受孕到淘汰的平均生产利用年限为 735 天。芬兰纯种长白猪(Landrace)和大白(Large White)种群和杂交母猪从初次分娩开始,分别在畜群中保留 439 天(Serenius 和 Stalder,2004)和 536 天(Serenius 和 Stalder,2007)。在奥地利,大白母猪从第一次分娩开始平均在畜群中保留 531 天,而长白母猪平均生产利用年限为 615 天(Mészáros 等,2010)。在瑞士核心区,大白母猪的平均生产利用年限从第一次分娩开始为 602 天(Tarrés 等,2006a)。瑞典约克夏母猪和长白母猪初次分娩之后,分别在畜群中平均存在 353 天(Yazdi 等,2000a)和 618 天(Yazdi 等,2000b)。

淘汰率是由于健康或性能问题或管理决定(不包括死亡和安乐死)而从畜群中淘汰的母猪的数量。在不同国家所报道的商业畜群平均每年淘汰率为 43%~50%(Boyle 等,1998;Dijkhuizen 等,1989;Engblom 等,2007;Koketsu,2007;PigCHAMP,2012)。Lucia 等(2000b)和 Knauer 等(2011)报告说,以后备母猪形式被淘汰的母畜比例分别为 19% 和 28%。此外,Lucia 等(2000b)报告说,46% 的母猪在第三胎之前被移除,对猪生产者造成不利的经济影响。去除率包括淘汰、自然死亡和安乐死。其定义为一年内从畜群中除去的动物数量,除以平均存栏量再乘以 100。存栏可能仅指母猪,或者母猪和后备母猪(当后备母猪在其生产周期的不同时间被引入时)。为了更好地标准化术语,有人建议在计算去除率时只考虑配种过的母猪。然而,在一些农场淘汰的母猪很多(这些母猪引进繁殖群体但尚未配种),因而可能需要更多的调查研究。在不同的研究中报道了母猪年均去除率为 35%~55%(Boyle 等,1998;Dagorn 和 Aumaître 1979;D'Allaire 等,1987;Dijkhuizen 等,1989;Engblom 等,2007;Friendship 等,1986;Knauer 等,2007)。建议的目标为 39%~40%,其中包括淘汰率为 35%~36%,死亡率为 3%~5%(Dial 等,1992;Muirhead,1976)。这样的目标值应该针对个体农场进行调整,因为去除率受到几个农场特定因素的影响,如畜群大小、所需遗传周转量、品种、平均存栏的定义等。市场趋势和经济状况也影响生产者的淘汰决策和时间(Stalder 等,2004)。

母猪的利用年限也可以通过替代率进行评估,其定义为进入畜群的动物总数除以平均存栏再乘以 100。在稳定的群体中,如果存栏保持不变则移除率和替代率应该相似。然而,在正在扩张的畜群中,替代率可能高于去除率。相反,当畜群存栏大小减小时,替换率可能低于去除率。因此,在分析这些比率时,应考虑群体动态。

去除时母猪的平均胎次表示母猪在群体中的平均时间长度。几项研究表明，去除时的平均胎次在 2～5.6，但可高至 8(D'Allaire 等，1987；Koketsu 等，1999；Lucia 等，2000b；Pedersen 1996；Stein 等，1990)。在北美，报告的平均去除胎次为 3.1～4.1 (D'Allaire 等，1987；Koketsu，2003；Lucia 等，2000b；Rodríguez-Zas 等，2003)。在荷兰、爱尔兰、瑞典和日本的商业畜群中，当母猪去除时可完成 4.3～4.6 胎次(Boyle 等，1998；Dijkhuizen 等，1989；Engblom 等，2007；Koketsu，2007)。

至少来说，种母猪应该保存在猪群中，直到与其更换相关的投资成本得到恢复。根据 Lucia 等(2000a)报道，与盈利能力相关的最大去除胎次在 5～8。Rodríguez-Zas 等(2006)报告说，为了最大化育种至断奶群体的获利能力，淘汰 4 或 5 胎次的母猪是最佳选择。此外，Abell 等(2010)确定，在畜群中为额外的胎次而保留母猪的遗传损失其经济价值相对较低，因此美国商业畜群的最佳去除胎次大于 7。

此外，Hurnik 和 Lehman(1985)提出，利用年限较长是动物福利和动物需求是否得到满足的指标。相反，高清除率是一个福利问题，因为淘汰和死亡的一些原因可能涉及动物福利破坏的指示器(Barnett 等，2001)。较短的利用年限通常与涉及一定程度痛苦的死亡率的增加有关，同时也涉及由一些痛苦的病理学(例如腿部问题)导致的淘汰原因(Deen，2003)。此外，从动物福利的角度来看，将猪肉生产建立在母猪(这些母猪很多胎后已经不能应付仔猪生产的生理压力)基础上在道德上是不能接受的(Serenius 和 Stalder，2006)。

19.3　母猪利用年限的经济重要性：对母猪群的影响

19.3.1　胎次分布

繁殖群体的胎次分布表明，当其影响到母猪群体的生物和经济表现时，母猪更有可能被去除。Pinilla 和 Lecznieski(2014)将"理想"胎次分布定义为母猪去除率、后备母猪可用性、猪市场价和饲料成本的数学函数。据 D'Allaire 等(2012)报道，育种群体的"理想"胎次模式是，在自动淘汰发生之前母猪的最大产胎数量和年度淘汰率之间的函数，一旦这两个值被确定，胎次较早时被淘汰母猪的分布或百分比是一个线性函数。

一种"理想"胎次分布对于保持一致的生产性能和避免替换后备母猪其数量的严重波动至关重要。然而，很难建立一个"理想的"或"标准的"胎次分布(Dial 等，1992)，原因是其在不同农场之间变化很大，因其取决于特定的农场特征，例如品种、替代成本、设施种类和养殖技能(D'Allaire 等，2012)。表 19.1 显示了由几位研究人员报告的母猪群的理想胎次分布。为了达到这些目标，无论去除或淘汰原因是什么,有必要限制从畜群中淘汰年轻母猪。

表 19.1　来源于已发表文章推荐的产次分布(以母猪/产次的百分比表示)

	胎次								
	0	1	2	3	4	5	6	7	8
Straw (1984)	20	18	17	16	15	14	—	—	—
Parson 等 (1990)	30	23	19	14	10	5	2	1	—
Muirhead and Alexander (1997)	17	15	14	13	12	11	10	5	3
Carroll (1999)	—	17	16	15	14	13	11	10	4
Morrison 等 (2002)	19.1	16.5	16.9	14.1	10.2	8.2	5.1	4.9	4.9
Pinilla and Lecznieski (2014)	20	17	16	15	14	13	5	—	—

19.3.2　生产力

能繁母畜是群体中最有价值的动物(Schenck 等, 2008),据估计,40%~50%的母猪在其第三胎之前被淘汰(Boyle 等,1998;D'Allaire 等,1987;Jørgensen,2000a),此时它们没有达到初始更换成本(Stalder 等,2003)。母猪替换决定会影响母猪的预期生命、年替代率和生产成本。例如,严重跛蹄的母猪必须被安乐死,这意味着屠宰收入的丧失和安乐死或销毁尸体的额外费用(Pluym 等,2011)。在像丹麦这样的国家,将跛蹄母猪送去屠宰是一种刑事犯罪,这会导致因安乐死引起的农场母猪死亡率上升和兽医费用的增加(Boyle 等,2012)。很可能将来,欧盟其他国家也将实施类似的立法,这可能对猪生产者产生巨大的经济影响。

此外,还将导致生产者进一步产生开支,这些开支与开发和适应替代后备母猪相关(Stalder 等,2000)。Faust 等(1992,1993)开发了分层猪繁殖系统的生物经济模型,以确定各种性状(包括淘汰)对盈利能力的影响。结果表明,替代率最低的养猪系统利润最好。他们还发现,在低替代率下,后备母猪价值高达市值的450%;然而,在较高的替代率下,替代后备母猪的价值不超过市场价值的175%。Sehested(1996)报道说,将母猪胎次提高一次与将瘦肉含量提高0.5%具有类似的经济影响。Parson 等(1990)提出,稳定猪群(大小方面稳定)其生产力和其经济状况不受胎次分布及其淘汰策略的影响;然而,作者是在第一胎母猪的百分比约为30%的畜群中进行比较,与商业畜群中目前约为50%的替代率相比,这一百分比被认为低得多。

母猪利用年限在仔猪生产中起重要作用,原因有几个。首先,生产利用年限的长短与母猪生产生命期内生产的仔猪数量直接相关。例如,第一胎动物每窝生产的仔猪数量少于老年母猪生产的猪数量,早期从畜群中清除母猪导致不育率较高,每头母猪的窝数变低,平均窝产仔数降低,每头母猪每年断奶的仔猪数量降低,非生产天数较多(Anil 等，2009；D'Allaire 和 Drolet，1999；D'Allaire 等，1987；Dourmad 等，1994；Engblom 等，2007；Friendship 等，1986)。Grandjot(2007)报告说，由于跛蹄而淘汰的母猪在整个生产利用年限中比非跛蹄母猪少生产 1.5 窝。King 等(1998 年)报告说，当存栏中的后备母猪的比例上升 1%，平均非生产天数增加了 2.6 天。然而 Stein 等(1990)报道，利用年限较短可导致每窝出生时活着和断奶的仔猪数量增加，因为引进了具有较高平均繁殖力的遗传优势母猪。

因此，增加母猪利用年限的好处应包括：从较多胎次的母猪获得更大的窝产仔数且幼仔更重，较少的非生产性天数，较高的母猪预计价值，较低的替换成本，总收入增加(因为老年母猪有更多的仔猪产出，这反过来可能导致更多的猪用于出售)，并促进动物福利关切(D'Allaire 等，1987；Stalder 等，2000，2003)。

19.3.3 免疫

将后备母猪引入繁殖群体可能会给现有母猪带来健康风险，因其免疫系统不成熟(Sanz 等，2002)。后备母猪比例较高的农场经常会更多地面临猪链球菌(*Streptococcus suis*)、猪痢疾杆菌(*Actinobacillus suis*)、副嗜血杆菌(*Haemophilus parasuis*)、猪肺炎支原体(*Mycoplasma hyopneumoniae*)、痢疾葡萄球菌(*Staphylococcus hyicus*)和巴斯德氏菌(*Pasteurella organisms*)的感染(Sanz 等，2002)。本场选留后备母猪因暴露于畜群中地方流行的病原体，并对其产生免疫力。然而，在现代的猪生产系统中，许多后备母猪来源多个地方，并且很可能它们在被带到母猪群之前不会接触特定病原体。Loula(2000)指出，这可能有利于生产，但不利于免疫力的发展，因为后备母猪第一次接触某些生物体是在其进入母猪群时。

无论后备母猪的来源，一个时期的隔离和适应对繁殖种群的后备母猪的长期繁殖力至关重要。新进的后备母猪可能是健康的，但可能是带有感染或作为病原体的携带者。这些动物在装载、混群和运输中相关的应激增加了易感后备母猪疾病传播的可能性(Loula，2000；Wrathall 等，2003)。

19.4 为什么母猪通常离开繁殖群？

19.4.1 主动和被动淘汰

主动和被动的原因都可以使得母猪被替换。主动淘汰的发生是当管理层决定从育种群体中去除一头母猪时。例如，当母猪达到预先设定的最大胎次时会被淘汰，或者当母猪的生产水平变得不可接受时会被淘汰。被动淘汰的发生在当生产者必须从繁殖群体中移去母猪时，其原因可能是多种原因，如繁殖失败(不发情或不能受精)、哺乳期失败(乳汁产量低或断奶仔猪数量不足)和运动问题(Dagorn 和 Aumaître，1979；D'Allaire 等，1987)导致的母猪没有生产力，在这种情况下，由于母猪的状态，淘汰决定基本上是自动进行的。不幸的是，胎次较小的母猪经常由于被动的淘汰被去除（Pinilla 和 Lecznieski，2014）。

老龄是一个相对的术语，其定义可以在畜群之间有很大的不同（D'Allaire 等，1987）。但是，这种去除原因在第5~6胎之前并不经常使用。据报道，9%~31%的母猪去除是归因于年老（Boyle 等，1998；D'Allaire 等，1987；Dijkhuizen 等，1989；Friendship 等，1986；Lucia 等，2000b；Engblom 等，2007；Hughes 等，2010）。

繁殖失败（包括繁殖周期失败、受孕失败和分娩失败）是最常见的被动淘汰原因，占所有去除量的20%~43%（Boyle 等，1998；D'Allaire 等，1987；Dijkhuizen 等，1989；Engblom 等，2007；Hughes 等，2010；Lucia 等，2000b；Tarrés 等，2006b）。由于生殖问题而移除的母猪中有34%~43%是胎次较低的母猪（D'Allaire 等，1987；Hughes 等，2010；Lucia 等，2000b）。此外，因生殖障碍而移除的母猪每年出生的仔猪数量较少，且与其他原因淘汰的母猪相比其一生中非生产性天数更多（Sasaki 和 Koketsu，2011）。这可能有一部分原因是遗传因素，但是需要改进生殖管理实践来减少被淘汰的母猪数量。

产仔性能差，包括分娩或断奶窝仔数较小，仔猪出生体重较轻，个体的断奶体重低，乳房问题和母性差，平均占淘汰母猪的20%~30%，文献中报道其占到淘汰母猪的11%~56%（Boyle 等，1998；D'Allaire 等，1987；Dijkhuizen 等，1989；Engblom 等，2007；Hughes 等，2010；Lucia 等，2000b；Tarrés 等，2006b）。一般来说，在因产仔性能差而被淘汰之前，青年母猪被允许表现出其生殖能力。因此，对于成熟母猪（即3胎母猪及以上），产仔性能和年龄作为去除原因其重要性开始增加，而生殖失败导致的淘汰其重要性逐渐降低（Boyle 等，1998；D'Allaire 等，1987；Lucia 等，2000b）。Anil 等（2008）分析了美国中西部地区商业种群的2066头母猪中影响母猪利用年限的围产期危险因素，包括健康相关问题。277头母猪在分娩后35天内被淘汰。Brandt 等（1999）报告说，在每个胎次中，母猪在断奶5天时处于较高的淘汰风险。同样，Tarrés 等（2006a）观察到，淘汰集中在每次分娩或断奶后的第一天。

跛蹄也是使母猪从繁育种群中过早淘汰的最重要原因。跛行这个词广泛地包括诸如蹄和腿部问题、形态缺陷、跛行、不健全、四肢受伤和脓肿（Boyle 等，1998；Deen，2009；Hughes 和 Varley，2003）。据报道，在母猪繁殖群体中，6%~40%（平均为10%）的清除是由于运动问题（Anil 等，2009；D'Allaire 等，1987；Dargon 和 Aumaître 1979；Engblom 等，2007；Friendship 等，1986；Hill，1992）。Kirk 等（2005）对丹麦10个不同群体的265头淘汰母猪（93头自发死亡母猪和172头被处死母猪）进行尸检。在超过70%的母猪中，蹄和腿部疾病是安乐死的主要原因。因跛行而被淘汰的母猪一般在年龄较小时即被去除，且比因其他原因而去除的母猪产仔数更少（D'Allaire 等，1987；Lucia 等，2000a），这使得猪群中母猪的平均利用年限较短（Pluym 等，2011）。在商业农场，Mote 等（2008）发现 7%的跛蹄后备母猪未能生产一窝，其中 13%的青年母猪在第二个窝之前被清除，只有约 50%的母猪达到了第 4 胎（D'Allaire 等，1987；Boyle 等，1998）。D'Allaire 等（1987）估计，由于跛行而被淘汰的跛蹄母猪平均生产 2.93 窝。

跛行似乎直接和间接影响母猪利用年限（Anil 等，2009）。严重跛蹄的母猪可能立即从猪群中除去；然而，不太严重的跛行形式可能会影响性能并间接导致母猪去除（Oldham，1985）。许多因其他原因而被移除的母猪也可能是跛蹄的。Anil 等（2009）在分娩时和 350 天后，对 674 头母猪跛行程度进行了打分。他们发现，跛蹄母猪在检查出跛行的 350 天内，从猪群移除的风险比非跛蹄的母猪高 1.7 倍。同时作者还报道，跛蹄母猪的生存时间的中位值为跛行评估后第一次分娩 140 天之后，而非跛蹄母猪的生存时间中位值为 302 天。Anil 等（2008）报道说，在分娩后 35 天内淘汰的 217 头母猪中，93 头母猪以跛行为主要去除原因，28 只母猪跛行为次要原因。与跛蹄母猪相比，非跛蹄母猪在下一次分娩之前从畜群去除的可能性降低了 37%。

19.4.2　死亡率

死亡率，包括安乐死或被发现死亡的动物，其比例为 7%～15%（D'Allaire 等，1987；Engblom 等，2007；Lucia 等，2000b；Tarrés 等，2006b）；但其最高可达 20%（Abiven 等，1998）。在美国，平均死亡率在 5%～7.4%（USDA，2001）。根据美国农业部（2001）的报告，限饲母猪的死亡率不应超过 3%。尽管死亡率的风险因素尚未得到广泛研究，但较大的群体数量、胎次、围产期周期（即预测分娩前 3 天至分娩后 3 天），哺乳期时间过短和断奶被确定为危险因素（Chagnon 等，1991；D'Allaire，1992；Koketsu，2000；Sanz 等，2007；Sasaki 和 Koketsu，2008；Tiranti 等，2003；USDA，2001）。

死亡率随猪群大小的增加而增加（Abiven 等，1998；Koketsu，2000）。美国农业部（2001）报告说，死亡率在少于 250 头母猪的农场为 2.5%，其在超过 500 头母猪的农场的中增加至 3.7%。这可能与每个人需要处理的母猪数量增加及其导致的个体关注缺乏有关（Koketsu，2000）。

死亡时平均胎次在 3.4～4.2（Chagnon 等，1991；D'Allaire 等，1987；Madec，1984）。此外，某些母猪死亡发生率似乎与年龄相关。突然死亡和一些运动问题在第一胎和第二胎的母猪中更频繁地发生（Dewey 等，1993；Sanz 等，2007），而老年母猪的膀胱炎-肾盂肾炎发病频率更高（Madec，1984）。

Engblom 等（2007）报道，商业繁殖种群中母猪最高死亡率出现在刚刚分娩后，而分娩后的 1 周和断奶后 4 周时安乐死母猪的比例较大。据 Anil 等（2006）报道，围产期周期是生殖周期中风险最大的时期,42%的母猪死亡在这个短暂的时间内发生(Chagnon 等,1991)。

在文献中报道的死亡原因有几个且在不同研究之间有所不同,如畜群大小和数量、品种、营养、地理区域等都可能造成这些差异。Chagnon 等(1991)报道,加拿大种群中母猪死亡的主要原因是心力衰竭(31%)、腹部器官的扭转和事故(15%)和膀胱炎-肾盂肾炎(8%)。在 13%的病例中,死亡是由子宫内膜炎或子宫脱垂引起的。在法国母猪群中,尿道感染、子宫炎或跛行的高发率被确定为高死亡率的危险因素(Abiven 等,1998)。Kirk 等(2005)调查了丹麦母猪群中母猪死亡率的原因,并指出死亡主要是因为胃肠系统(45%)和生殖系统(24%)病变。在瑞典母猪群中,循环系统或心脏衰竭(24%)和骨折或内伤创伤相关的损伤(24%)通常在被发现死亡的母猪中观察到(Engblom 等,2008)。Kirk 等(2005)报道,关节病自发死亡(93%)或安乐死(88%)的母猪中几乎都是一个常规的二次诊断。在 70%以上的死亡病例中,运动系统紊乱是安乐死的主要原因;验尸发现关节炎是 24%~44%,骨折是 13%~16%(Engblom 等,2008;Kirk 等,2005)。同样,Sanz 等(2007)报道,在美国猪群被安乐死的母猪中,关节炎是最常见的验尸发现,频率为 37%。

降低死亡率的原因包括福利问题、员工的士气和经济损失(Sanz 等,2007)。由于死亡造成的经济损失包括母猪的价值、失去猪生产、替代早期死亡的额外成本以及由于淘汰能力降低导致的母猪群的质量降低(Sanz 等,2002)。Deen 和 Xue(1999)估计由于死亡(包括替换和机会成本)造成的经济损失为 400~500 美元。

19.5 提高母猪利用年限

19.5.1 蹄和腿完整性的重要性

蹄和腿的问题被认为是青年母猪的第二大清除原因(Boyle 等,1998;Engblom 等,2007;Lucia 等,2000b),因此是母猪淘汰的主要贡献因素。蹄部和腿部不健全会导致养猪业的经济损失,这些损失可能包括被动淘汰、临床监管和治疗所需的额外劳动力、兽医治疗费用、繁殖性能下降以及部分或全部胴体不能销售方面。蹄和腿部缺陷也可能损害动物福利,因为具有严重形态问题的母猪可能获得食物和水的机会有限,在站立或移动时经历不适,且由于缺乏运动而被迫改变正常行为的表达(Fernández de Sevilla 等,2008)。因此,在育种母猪中改善蹄部和腿部健壮性在道德和经济方面都至关重要。

多个个体形态特征与提高利用年限和生存性有关。Brandt 等(1999)报告说,较大骨架的动物在其第四胎或第五胎时淘汰风险增加。Jørgensen(2000a)观察到,丹麦约克夏猪和杂交猪中,膝盖弯曲、后肢软弱和后半身摇摆与利用年限降低相关。Grøndalen(1974)指出,与具有直线蹄腕的猪相比,具有柔软蹄腕的猪具有更好的步态得分。Fernández de Sevilla 等(2008)记录了西班牙大白、长白和杜洛克母猪中存在或不存在强直和极软(弱)蹄腕。他们报告说,与同一品种中具有正常或软(弱)的蹄腕的母猪相比,有强直蹄腕条件的西班牙大白猪母猪其淘汰风险更高。此外,与具有更正常定位或至少不那么严重偏软蹄腕的母猪相比,具有极软蹄腕的西班牙长白、大白和杜洛克母猪过早淘汰风险更大。此外,后腿张开会增加西班牙杜洛克母猪的去除风险。Tarrés 等(2006a)报道说,向外转动后腿和后腿内小内趾曾加了瑞士大白母猪淘汰的风险。Fernández de Sevilla 等(2009a)对从西班牙杜罗克品种收集的数据进行了竞争风险分析。从总体上来说腿部构象与母猪死亡率无关,但是与结构良好的母猪相比,具有次于最佳蹄和腿部形态的母猪其生殖性能很差的风险较高。此外,据报道,腿部形态随年龄恶化(Fernández de Sevilla 等,2009b)。

选择具有最佳形态特征的后备母猪是防止跛行的重要因素。Tiranti 和 Morrison(2006)对961头后备母猪监测了一年,并记录了前腿和后腿的形态分数。他们报告说,当后腿形态分数变差时,跛行而导致的去除风险增加。Tiranti 和 Morrison(2006)的研究结果与 Grindflek 和 Sehested(1996)的研究一致,后肢在站立时位置太远的母猪其利用年限较短。此外,Tiranti 和 Morrison(2006)的研究结果支持了 Jørgensen(2000b)报道的结果,这个特征与6月龄母猪的跛行有关。在选择时具有不良肢体形态评分的后备母猪,因为跛蹄问题而被从畜群中移除并淘汰或安乐死的风险较高。由于母猪利用年限延长以及随之而来的生产力增加,因此肢体结构的选择可能导致猪群其性能随时间提高(Tiranti 和 Morrison, 2006)。

在母猪中,趾(也称为爪)的病变非常普遍,繁殖群中多达100%的母猪有至少一个爪损伤(Anil 等,2007;Calderón Díaz 等,2013;Dewey 等,1993;Pluym 等,2011)。承重表面上穿透敏感组织的损伤是最严重的(Brooks 等,1977),因为它们会允许能影响关节的感染性病原体进入(Anil 等,2007;Zoric 等,2004)。在母猪中发现的一些最常见的蹄趾损伤包括:外壁和白线上的裂缝、蹄跟和趾的过度生长及蹄趾侵蚀(Anil 等,2007;KilBride 等,2010;Kirk 等,2005)。蹄趾病变可能会减少母猪利用年限和繁殖性能。据报道,西班牙长白和杜洛克母猪的蹄趾的非正常增长会增加淘汰风险(Fernández de Sevilla 等,2008)。Fitzgerald 等(2012)观察到,与没有相应病变的母猪相比,爪壁上有裂缝的母猪每窝断奶仔猪较少,爪子过度生长的母猪断奶时仔猪较轻。蹄趾病变被认为是跛行的原因(Anil 等,2007;Dewey 等,1993;Pluym 等,2011)。尽管轻度病变的母猪可能没有明显的疼痛迹象,但严重的蹄趾病变可能会疼痛并引起跛行(Deen,2009)。据估计,5%～20%的跛行是由蹄趾病变引起的(Dewey 等,1993),但是该领域研究较少。趾部病变是否引起跛行取决于病变的位置(Anil 等,2007)和严重程度(Gjein 和 Larssen,1995;Heinonen 等,2006)。

19.5.2 后备母猪发育的重要性

配种时的机体组成

配种时后备母猪的身体组成(背膘厚度、腰部深度和体重)可以影响母猪利用年限。后备母猪需要建立身体储备,使它们有长期的生产性种群生命。后备母猪发展的关键是减缓蛋白质沉积并建立脂肪、矿物质和其他营养储备,当哺乳期饲料摄入量不足以满足需要时,后备母猪可加以利用(Stalder 等,2007b)。为了最大限度地发挥其生命中出生的活仔猪的数量,后备母猪需要最低水平的背膘厚度。Brisbane 和 Chenais(1996)报告说,与非常瘦的后备母猪(即<10 mm 的背膘厚度)相比,背膘厚度>18 mm 的后备母猪有多出 10%的可能性留在畜群中直到至少第四胎。Yang 等(1989)建议,后备母猪在配种时至少要有 125 kg 体重,背膘厚度至少要有 13 mm。另一方面,Tarrés 等(2006b)报道,如在生长期结束后背膘厚度小于 16 mm,将会增加了第三次分娩后的淘汰风险,主要是由于生产力低下。同一作者报道说,在第一次分娩时背膘厚度大于 19 mm 的母猪有显著增加的淘汰风险。此外,由于较大的跛蹄问题,在第一次分娩时背膘厚度小于 15 mm、腰部深度小于 40 mm 的母猪,与较高的淘汰风险相关。Geiger 等(1999)发现,母猪在分娩期间背膘厚度小于 18 mm 会使得母猪死亡率升高。Challinor 等(1996)报道说,与背膘厚度为 14～16 mm 的后备母猪相比,背膘厚度为 18～22 mm 体重平均为 150 kg 后备母猪,在五胎之后平均多生产 7.2 个仔猪。然而,Rozeboom 等(1996)的结果表明,与在育种时具有更多背膘(即 48 mm)的后备母猪相比,青年后备母猪在繁殖时(发情期或第二次发情)具有较少的背膘(即 12.5 mm)对利用年限没有负面影响。

发情期以及第一次分娩的年龄

养猪生产者有一个共识,即如果后备母猪在较早的时候达到发情期,那么母猪利用年限或繁殖表现将会得到改善。Chapman 等(1978)发现,选择较早到达发情期和受孕的后备母猪改善了繁殖表现。这些结果似乎得到 Young 和 King(1981)的支持,初次发情与第三次发情时配种带来的差异,对出生猪总数、出生活猪数量、断奶猪数量、第一或第二窝断奶之后的断奶-发情间期没有显著影响,但是要保持后备母猪直到第三次发情额外需要大约 105 kg 的饲料,从而增加生产成本。同样,Holder 等(1995)报道,当其提前 11 天达到发情期时,能够生产 5 个胎次的后备母猪的比例较高(58.8%对 39.4%),尽管没有统计学意义。这些结果与 Brooks 和 Smith(1980)报道的结果相反,当发情期在 160 天或 200 天被诱导时,在完成 5 个胎次的母猪数量上或从这两组母猪的第一个五胎出生的猪的数量或重量都没有发现差异。MacLean 等(2001)也报道，在 155～174 天、174～195 天或＞199 天达到发情期的后备母猪之间，总出生仔猪没有差异。此外，Patterson 等(2010)没有发现发情期年龄和将杂交后备母猪保留在畜群中直至第三次分娩之间的关系，而 MacPherson 等(1977)报告说，无论后备母猪在第一、第二或第三次发情时配种，其第三胎后所生猪的总数没有差异。

初期受孕或初次分娩的年龄较小与预期利用年限较长相关(Le Cozler 等，1998；Saito 等，2011)，但也似乎与第一胎产仔数较小相关(Le Cozler 等，1998；Saito 等，2011；Schukken 等，1994；Serenius 等，2008；Tummaruk 等，2001)。不成熟的后备母猪不应该配种，而且根据 Schukken 等(1994)，首次受孕的最佳日龄为 200～220 天。同样地，Serenius 和 Stalder(2007)建议在 200～210 日龄对后备母猪配种。然而，Babot 等(2003 年)报告说，与配种时年龄偏小或者偏大的后备母猪相比，在 221～240 日龄初次配种的后备母猪其利用年限和终身繁殖更高。据 Schukken 等(1994)报道，后备母猪妊娠时超过 220 天以上与预期繁殖群体利用年限明显缩短有关，但是当将窝仔数和群体利用年限的影响综合在一起时，每头母猪的收入并没有受到初次受孕时年龄的显著影响。

Hoge 和 Bates(2011)研究了北美约克郡母猪的长寿和生命繁殖力的几项措施，并得出结论，无论怎么定义利用年限或生命繁殖力，初次分娩时年龄较小会显著降低淘汰的风险。一直以来，很多研究都报道说，在第一次分娩时年龄较小是提高生存率的因素 (Fernández de Sevilla 等，2008，2009a；Holder 等，1995；Serenius 和 Stalder，2004，2007；Yazdi 等，2000a,b)。根据 Knauer 等(2010)，在发情期和初次分娩时年龄较小的商品化母猪(即分别为 208 和 353 日龄)有更大的可能性到达第四胎。另外，Rozeboom 等(1996)没有发现第一次配种的年龄和完成三个胎次的能力之间的关联，或在第一、第二、第三或所有的胎次中出生或断奶时的窝仔数。然而，同一研究报告说，第一次配种时年龄的增加与在第一、第二、第三或所有的胎次中仔猪的出生体重和断奶时体重的增加相关。在对奥地利大白和长白种群进行的一项研究中，在 43 周龄或 60 周龄之前拥有第一窝的母猪遇到更大的淘汰风险(Mészáros 等，2010)。

利用年限的营养需求

影响利用年限的营养学理论可能是通过一些机制，例如减少氨基酸摄入量来降低瘦肉脂肪比例和增加在骨骼和其他组织发育中至关重要的日粮微量矿物质和维生素(Kitt，2010)。矿物质补充对于母猪蹄趾软或者硬组织结构的发育至关重要。骨骼正常生长、发育和维护中都需要钙以提供强度与结构。磷是骨形成的必需元素，它也是骨骼适当矿化必需的(Crenshaw，2001)。镁对骨骼的力量和完整性方面有作用，并且在神经脉冲传递中也是重要的(Patience 和 Zijlstra，2001)。

微量矿物质如锌、铜和锰对于所有动物的各种生理过程至关重要。当在饲料中添加到药理学浓度（125～250 mg/kg；Barber 等，1957）时，铜可作为生长兴奋剂。锰对于动物的生长和生育是至关重要的（Underwood 和 Suttle，1999），并且在骨的有机基质的形成中起关键作用。锌是生长、发育、繁殖和代谢活动所必需的（Hill 和 Spears，2001）。锌的缺乏可导致免疫功能降低，抗体滴度降低和其他缺陷（Richards 等，2010）。日粮矿物质和维生素缺乏症可能对骨骼、关节软骨质量和角的质量有不利影响（Van Riet 等，2013）。缺乏钙和磷饲料饲喂的动物显示出跛行、步态缓慢、麻痹和自发性骨折（Crenshaw，2001）。锰缺乏症的迹象包括不愿站立、蹄腕虚弱和失去平衡（McDowell，1992）。锰缺乏导致骨骼异常，包括较短和较厚的前腿，前腿腿内弧和扩大的蹄（Plumlee 等，1956）。另外，饲料中的钙和磷水平已被报道会影响繁殖失败以及因此的母猪利用年限（Arthur 等，1983；Koketsu 等，1996；Kornegay 等，1984）。其他矿物质也表现出对母猪利用年限及其组成特征的有益影响。Hagen 等（2000）显示补充铬对母猪死亡率的有益影响。然而，Crenshaw（2003）指出，通过饲料措施使骨矿化来减少跛行造成的母猪死亡率并不成功。Mahan 和 Newton（1995）发现，与妊娠的后备母猪相比，已经完成了3个胎次的母猪的矿物质组成有所下降。这项研究的结果表明，完成3个胎次的母猪中最显著的矿物质下降是钙、磷、镁、钾、钠和锌。因此，与育肥猪的日粮相比，通常建议对后备母猪应提供较高的钙、磷、铜和锌日粮浓度（NRC，2012），以便预防在哺乳期间由于矿物质储备过度减少引起的腿部问题（Whitney 和 Masker，2010）。

19.6 猪舍对利用年限的影响

19.6.1 后备母猪和母猪的空间要求

对种母猪的空间要求的建议很少，其中许多是基于消费者的看法而不是科学知识。像欧盟这样的国家认为，集约化饲养的后备母猪和经产母猪的占地面积至少分别为 1.64 m^2 和 2.25 m^2，对于小于 6 头的动物群体的空间容量增加 10%，40 头或以上动物群体同样减少 10%。这些增加是为了确保母猪的福利。

从养猪业的角度来看，空间要求对繁殖变量的影响是至关重要的。Ford 和 Teague（1978）报道，与对照相比，并没有发现 50% 或 70% 空间限制的后备母猪在发情期平均年龄上有差异。将对照后备母猪放置在 0.37 m^2 的空间中，之后每增加 13.6 kg 体重空间会增加 0.09 m^2。Cronin 等（1983）观察到，当大群饲养的后备母猪每头空间容量小于 0.9 m^2 时，发情期后检测到发情的后备母猪增长了 4.2%。

Young 等(2008)发现,在饲养期间饲喂于 1.13 m² 或 0.77 m² 猪舍的后备母猪,对三胎总产仔数以及去除率没有影响;虽然较大空间饲养大的母猪更有可能在年龄较小时达到发情期。在 Kuhlers 等(1985)的研究中,相较于只提供 0.62 m² 的后备母猪,在饲养期间提供了 1.25 m² 的占地面积的后备母猪可以分娩较多的仔猪且更多活仔猪。然而,这些研究结果需要谨慎解释,因为空间限制是通过组群大小的变化实现的。

使用生理措施作为动物应对环境能力的指标,以评估空间对后备母猪和母猪福利的影响。Hemsworth 等(1986)在空间容量减少(1、2 或 3 m²)的后备母猪上观察到慢性应激反应,Barnett 等(1992)指出,与在 2 m² 圈养的后备母猪相比,以 1 m² 的空间容量重新分群的后备母猪在分群 21 天后,其具有升高的皮质醇浓度和受损的免疫系统。

19.6.2　妊娠猪舍和利用年限

猪舍系统必须满足动物和生产者的要求。生产力、管理、福利、健康和经济是设计妊娠母猪舍系统时一些要考虑的主要标准(Den Hartog 等,1993)。在本书第 3 章(Spoolder 和 Vermeer,2015)中详细讨论了妊娠猪舍(Spoolder 和 Vermeer,2015)。猪舍系统可以分为独立的(限位栏、妊娠栏)和群组系统。不同群体猪舍系统的分类是基于所使用的饲养系统,包括静态或动态群体。通过对群体和单独饲养的母猪之间的比较,可以影响母猪利用年限的不同特征包括:创伤或损伤(跛行、肩疮、外阴撕咬等),一般管理(工作条件、母猪观察),健康(交叉感染、新感染、运动等),营养、福利和其他特征(Stalder 等,2007a)。

Bates 等(2003)没有发现母猪妊娠期间,饲养在妊娠栏或者配有电子母猪喂料机的组群饲养中的母猪,其返情比例有所差别。然而,群养的母猪其在断奶后 7 天恢复发情的百分比为 72%,而饲养在妊娠栏中的母猪只有 68.4%。与妊娠栏饲养的母猪(89.4%)相比,群养的母猪的分娩率(94.3%)也较高。Backus 等(1997)报道说,与配有电子喂料机的群养相比,限位栏饲养的妊娠母猪断奶后回到发情所需的时间较少;然而两组中出生的活仔猪数相似(饲养于单独饲喂栏中的母猪为 10.9 头活仔猪,而群养的母猪为 10.9 头活仔猪)。Barbari(2000)报道说,群养的农场的分娩率比单独妊娠栏饲喂有所下降。母猪不恢复发情则往往被淘汰,从而缩短其生产利用年限。

比较在妊娠期单独饲养和拴系饲养的母猪之间的死亡率的结果是矛盾的。Abiven 等(1998)指出,与在妊娠期间拴系饲养的母猪相比,母猪饲养在单独饲喂栏的农场死亡率的风险更大。另一方面,D'Allaire 等(1992)报道说,与拴系的母猪相比,在妊娠期间安置单独饲喂栏中的母猪不太可能死亡。Stone(1981)和 Friendship 等(1986)都报告说,在妊娠期间母猪利用年限的提高,似乎不与母猪妊娠期间在单独饲喂栏还是集体圈养有关。群组饲养提供更多的行动自由和空间以及一致的运动可能性,因而增加肌肉重量可能会使母猪敏捷性提高(Marchant 和 Broom,1996)。然而,群组饲养呈现出其他类型的挑战,例如母猪之间更高层次的相互作用和侵略,还有更多的蹄和腿部疾病(Backus 等 1997;Kroneman 等,1993)。此外,当母猪以群体饲喂时,它们对运动系统有很高的要求(Kroneman 等,1993),因为它们必须为饲料和栖身处等资源竞争(Muirhead,1983)。有研究比较了母猪饲养在限位栏中或在群体中对其运动能力的影响,其报道说,与保持在限位栏中的母猪相比,妊娠期间饲养在没有垫料的群体中的母猪运动问题明显更多(Anil 等,2005;Backus 等,1997;Harris 等,2006;Kroneman 等,1993;Pluym 等,2011)。研究表明,超过 96% 的妊娠期群体饲养母猪和 80% 的妊娠期限位饲养母猪至少有一个趾部病变(Anil 等,2007;Enokida 等 2011;Gjein 和 Larssen,1995;Grégoire 等,2013;Pluym 等,2011)。Olsson 和 Svendsen(2002)报道,在不同组群中,粪便区域或是用混凝土漏缝地板,或用塑料漏缝地板,在群养的母猪之间趾部损伤没有差异。Olsson 和 Svendsen(1971)报道说,妊娠期间在部分漏缝地板限位栏饲养或者群养的母猪会过度淘汰。D'Allaire 等(1989)发现,相比于不漏缝饲养的母猪的淘汰率,在妊娠期具有全部或部分漏缝地板的畜群中年淘汰率更高。MAFF(1981)报道,当母猪发生蹄部和腿部受伤时,在漏缝或非漏缝地板上饲养的母猪之间没有区别。

  地板的质量与蹄腿部受伤和由跛行导致的剔除率有关(Barnett 等,2001)。在自然条件下,地面通常是软和潮湿的(Heinonen 等,2006)。然而,大多数用于猪生产的地板类型是硬表面,很少或没有垫层。地板不应变形、变质或需要精心维护(Baxter 和 Mitchell,1977;Peet,1983)。此外,母猪猪舍设施使用的地板应尽量减少动物的不适和受伤,并为干燥和潮湿条件下的安全运动提供机会(Cowin,1978)。此外,应该给猪提供使用正常步态行走的可能性,而不是强迫它们改变步态来减少滑倒和受伤的风险(Von Wachenfelt 等,2008)。低摩擦会导致滑动和跌倒损坏关节(McKee 和 Dumelow,1995)。硬地板可能会导致母猪躺在地板上时有瘀伤和肿胀的腿。过度磨砂的地板上可能会导致母猪蹄趾过度磨损,而磨砂太少可能导致蹄趾过度生长(McKee 和 Dumelow,1995)。

  水泥地板与跛行和趾部磨损的增加有关(Webb 和 Nilsson,1983),因为地板的粗糙度和磨砂性有助于从趾部去除角质层(Zoric 等,2008)。此外,长期站立在裸露的混凝土上可能会增加施加于蹄趾的压力(Hinterhofer 等,2006),并使得真皮发炎。与饲养在坚实的混凝土地板材料上的母猪相比,当母猪饲养在漏缝的混凝土上时其跛行风险增加(Heinonen 等,2006;Nakano 等,1981)。漏缝式地板对动物造成一些挑战,例如不平坦的行走表面,承重表面减少,缺乏垫层和锋利板条边缘(KilBride 等,2009)。Heinonen 等(2006)在 21 个芬兰畜群中对 846 头母猪和后备母猪的跛行性进行了打分。报告说,与养在水泥非漏缝地板上的动物相比,位于漏缝地板上的动物的跛蹄可能性是其两倍,严重跛蹄的概率是其3倍。

在可能发生攻击行为的情况下应提供垫层。虽然垫层不会减少攻击行为,但可以降低与之相关的腿部问题的风险(Spoolder等,2009)。Heinonen等(2006)发现,与非漏缝混凝土地板和部分漏缝式地板上的母猪相比,秸秆或深层枯枝落叶上饲养的母猪其的蹄趾损伤较少。Andersen和Bøe(1999)报道说,与混凝土地板上饲养的的群体相比,饲养在秸秆垫层上的畜群的运动问题得分较低。Christensen等(1995)也报道,秸秆垫层似乎提供了针对胃肠道疾病的保护。然而,尽管坚实的地板可以为母猪提供积极方面,例如提供垫层,但它也存在一些缺点。这些缺点包括更高的生产成本、水分的保持可能会导致蹄趾角质层的软化、板条地板和液体肥料处理系统之间的不相容性(Tuyttens,2005)。垫层的另一个缺点是一些垫料如秸秆可能会对栏里面的卫生条件产生负面影响,并促进疾病的扩散。例如,在栏里面的休息区域中使用秸秆垫层可以增加感染沙门氏菌的可能性,因为动物与粪便的接触更多(Alsop,2005;Davies等,1997)。

19.7 疾病对母猪利用年限的影响

评估疾病对母猪利用年限的直接影响的研究在科学文献中很少。在Sanz等(2002)的研究中,报道说18.5%的母猪由于增殖性肠病死亡,16.6%因泌尿道感染(膀胱炎和/或肾盂肾炎)死亡,9.2%因支气管肺炎或胸膜肺炎而死亡。然而,作者没有报告某些胎次群体是否因这些疾病而死亡的风险较高。Pijoan(1986)发现成年母猪中因猪肺炎导致的死亡并不常见,其占死亡数不会超过5%。

像肠病(enteropathy)、猪繁殖与呼吸综合征(porcine reproductive 和 respiratory syndrome,PRRS)、伪狂犬病(pseudorabies,PRS)、猪肠道病毒(porcine enterovirus PEV)、猪流感(swine influenza,H3N2)、慢性猪丹毒(chronic erysipelas)、急性出血性回肠炎(acute haemorrhagic ileitis)等疾病只在少数母猪群中普遍(Loula,2000;Stalder等,2004)。这些疾病可能会对母猪生产利用年限的长短产生负面影响,因为一些微生物经常导致母猪流产,生产者面临着需要决定是否淘汰这些母猪,还是留这些母猪并对它们是否能够重新繁殖冒一次险。通常,生产者选择淘汰这些母猪,使得畜群生命缩短。本书第18章提供了对母猪重要疾病的更为详细的描述(Friendship和O'Sullivan,2015)。

19.8 季节对母猪利用年限的影响

许多猪肉生产者观察到他们母猪群的繁殖性能随着季节而变化。Peltoniemi

等(1999)报道,在配种率、分娩率和第一次配种的年龄均受季节影响,不论猪舍系统如何。然而,Hurtgen 和 Leman(1980)观察到,与群养系统中的母猪相比,在妊娠期间饲养在限位栏中的母猪受到季节性不育的影响较小。Love 等(1993)发现在妊娠期间单独饲养的母猪其分娩率没有季节性影响。相反,Thacker(2002)报道,与限位栏中饲养的母猪相比,群养的母猪的季节性不孕症更高。这种季节性变化可能导致利用年限问题,或者在一年中某些时候增加淘汰和/或死亡率(Stalder 等,2007b)。

高温环境导致哺乳期母猪食欲、泌乳量和身体储备动员下降(Prunier 等, 1997)。同样地,Koketsu 和 Dial(1997)报道说,与其他季节分娩的母猪相比,在夏季分娩的母猪具有较低的窝仔体重,断奶至发情间隔更长。所有这些因素可能使发情延迟、母猪淘汰升高(通常在高温条件下断奶后发生)。

多个研究表明,在气温升高的月份,死亡率更为频繁(Chagnon 等,1991; Deen 和 Xue,1999;Engblom 等,2008;Koketsu,2000)。在 Chagnon 等(1991)的研究中 1/3 的母猪死亡发生在 7 月和 8 月,Sanz 等(2007)也报道说温度大于 32℃提高了母猪死亡率。Jones(1967,1968)观察到,超过 55%的母猪在冬季死亡。因此,通过各种冷却或加热装置管理环境温度对于降低母猪死亡率至关重要。此外,生产者不应忽视人员或工人对母猪利用年限的影响(Thacker,2002)。夏季是工作人员休假的时间,因而缺乏经验的人员在母猪场处理更多的工作。这反过来可以解释与这几个月相关的一部分问题。

19.9 人员/管理或饲养管理对母猪利用年限的影响

管理实践和在养猪场内工作的人员的饲养管理技巧会影响一些导致母猪淘汰或死亡率升高的因素。Castro 和 Piva(1999)建议,管理和人力资源在现代猪业的所有生产力方面起着重要的作用。此外,English(2002)指出了良好的饲养管理技巧对猪的福利和生产力的重要性。劳动力,特别是熟练、有知识、经验丰富的劳动力的短缺,可能导致对动物的照料和管理不足,导致母猪死亡率和淘汰率提高(Loula,2000;Young 和 Aherne,2005)。Loula(2000)指出,观察母猪具有疾病或损伤的迹象是降低母猪死亡率的重要关键,至少在淘汰或安乐死的情况下,工作人员的主观决定决定了母猪是否将被淘汰(Engblom 等,2008)。因此,针对员工的适当培训计划至关重要,特别是对于没有畜牧经验的人员,以便发展他们在成功的养猪操作中需要的观察技能和对细节的注意(Stalder 等,2007a)。

许多研究报道,母猪死亡率与平均母猪存栏显著相关(Ahiven 等,1998;Anil 等,2003;Christensen 等,1995;Koketsu,2000)。工作人员在一对一观察上面的时间较少,因而可能无法尽早识别出生病或受伤或体重减轻的动物。大型农场的另一个担忧是,由于其比小农场更频繁地购买后备母猪,所以更有可能难以消除或控制疾病(Koketsu,2000)。此外,强调产出可能导致存栏过剩和尽力产出最大化(Loula,2000)。这些情况可能导致动物过度拥挤,受伤或生病的动物的空间不足(Young和Aherne,2005)。

19.10 结论

母猪利用年限较短可能对经济效益和动物福利产生不利影响。母猪利用年限是一个复杂的特征,其中许多因素有助于在商业种群中的母猪具有较长且高产的利用年限。因为要减少每年从繁殖猪群中移除的母猪的数量,所以提高母猪利用年限的挑战是巨大的。关于母猪利用年限经济学以及去除原因、影响母猪利用年限的营养和管理实践,在科学文献中有广泛的信息。在管理后备母猪来增加它们留在繁殖群体中的可能性至少到其生产成本被覆盖时,应该特别强调后备母猪发育和选择具有最佳蹄和腿部形态的后备母猪。此外,猪生产者应注意培养和提高这些每天与猪工作的工人的良好饲养管理技巧。动物护理和管理方面的不足可能导致母猪死亡率和淘汰率的上升。

参考文献

Abell, C.E., Jones, G.F., Stalder, K.J. and Johnson. A.K., 2010. Using the genetic lag value to determine the optimal maximum parity for culling in commercial swine breeding herds. The Professional Animal Scientist 26: 404-411.

Abiven, N., Seegers, H., Beaudeau, F., Laval, A. and Fourichon, C., 1998. Risk factors for high sow mortality in French swine herds. Preventative Veterinary Medicine 33: 109-119.

Alsop, J.E., 2005. An outbreak of salmonellosis in a swine finishing barn. Journal of Swine Health and Production 3(5): 265-268.

Andersen, I.L. and Bøe, K.E., 1999. Straw bedding or concrete floor for loose-housed pregnant sows: consequences for aggression, production and physical health. Acta Agriculturae Scandinavica, Section A – Animal Science 49: 190-195.

Anil, L., Anil, S.S., Deen, J., Baidoo, S.K. and Wheaton, J.E., 2005. Evaluation of well-being, productivity, and longevity of pregnant sows housed in groups in pens with and electronic sow feeder or separately in gestation stalls. American Journal of Veterinary Research 66: 1630-1638.

Anil, S.S., Anil, L. and Deen, J., 2008. Analysis of periparturient risk factors affecting sow longevity in breeding herds. Canadian Journal of Animal Science 88: 381-389.

Anil, S.S., Anil, L. and Deen, J., 2009. Effect of lameness on sow longevity. Journal of the American Veterinary Medical Association 235 (6): 734-738.

Anil, S.S., Anil, L., Deen, J., Baidoo, S.K. and Walker, R.D., 2006. Association of inadequate feed intake during lactation with removal of sows from the breeding herd. Journal of Swine Health and Production 14(6): 296-301.

Anil, S.S., Anil, L., Deen, J., Baidoo, S.K. and Walker, R.D., 2007. Factors associated with claw lesions in gestating sows. Journal of Swine Health and Production 15: 78-83.

Anil, S.S., Deen, J. and Anil, L., 2003. Herd-level analysis of sow longevity. In: Proceedings of the Allen D. Leman Swine Conference. September 13, 2003. St. Paul, MN, USA, pp. 199-202.

Arthur, S.R., Kornegay, E.T., Thomas, H.R., Veit, H.P., Notter, D.R., Webb, Jr., K.E. and Baker, J.L., 1983. Restricted energy intake and elavated calcium and phosphorus intake for gilts during growth. IV. Characterization of metacarpal, metatarsal, femur, humerus and turbinate bones of sows during three parities. Journal of Animal Science 57: 1200-1214.

Babot, D., Chavez, E.R. and Noguera, J.L., 2003. The effect of age at the first mating and herd size on the lifetime productivity of sows. Animal Research 52: 49-64.

Backus, G.B.C., Vermeer, H.M., Roelofs, P.F.M.M., Vesseur, P.C., Adams, J.A.H.N., Binnendijk, G.P., Smeets, J.J.J., Van der Peet-Schwering, C.M.C. and Van der Wilt, F.J., 1997. Comparative study of four housing systems for nonlactating sows. In: Proceedings of The 5th International Livestock Environment Symposium. May 29-31, 1997. Bloomington, MN, USA, pp. 273-279.

Barbari, M., 2000. Analysis of reproductive performances of sows in relation to housing systems. American Society of Agricultural Engineers' In: Proceedings of the 1st International Conference on Swine Housing. October 9-11, 2000. Des Moines, IA, USA, pp. 188-196.

Barber, R.S., Braude, R., Mitchell, D.G., Rock, J.A. and Rowell, J.G., 1957. Further studies on antibiotic and copper supplements for fattening pigs. British Journal of Nutrition 11: 70-79.

Barnett, J.L., Hemsworth, P.H., Cronin, G.M., Jongman, E.C. and Hutson, G.D., 2001. A review of the welfare issues for sows and piglets in relation to housing. Australian Journal of Agricultural Research 52: 1-28.

Barnett, J.L., Hemsworth, P.H., Cronin, G.M., Newman, E.A., McCallum, T.H. and Chilton, D., 1992. Effects of pen size, partial stalls and method of feeding on welfare related behavioural and physiological responses of group-housed pigs. Applied Animal Behaviour Science 34: 207-220.

Bates, R.O., Edwards, D.B. and Korthals, R.L., 2003. Sow performance when housed either in groups with electronic sow feeders or stalls. Livestock Production Science 79: 29-35.

Baxter, S.H. and Mitchell, C.D., 1977. Developments in floor construction in animal production. Veterinary Annual 17: 286.

Boyle, L. Carroll, C., McCutcheon, M., Clarke, S., McKeon, M., Lawlor, P., Ryan, T., Fitzgerald, T., Quinn, A., Calderón Díaz, J. and Lemos Teixeira, D., 2012. Towards January 2013. Updates, implications and options for group housing pregnant sows. Teagasc, Pig Development Department. Moorepark, Fermoy, Co. Cork, Ireland, 75 pp.

Boyle, L., Leonard, F.C., Lynch, B. and Brophy, P., 1998. Sow culling patterns on sow welfare. Irish Veterinary Journal 51: 354-357.

Brandt, H., Von Brevern, N. and Glodek, P., 1999. Factors affecting survival rate of crossbred sows in weaner production. Livestock Production Science 57: 127-135.

Brisbane, J.R. and Chenais, J.P., 1996. Relationship between backfat and sow longevity in Canadian Yorkshire and Landrace pigs. In: Proceedings of the 21st National Swine Improvement Conference and Annual Meeting Ottawa, Ontario, Canada. Available at: www.nsif.com/Conferences/1996/brisbane.htm.

Brooks, P.H. and Smith, D.A., 1980. The effects of mating on the reproductive performance, food utilisation and liveweight of the female pig. Livestock Production Science 7: 67-78.

Brooks, P.H., Smith, D.A. and Irwin, V.C.R., 1977. Biotin-supplementation of diets: the incidence of foot lesions and the reproductive performance of sows. Veterinary Record 101: 6-50.

Calderón Díaz, J.A., Fahey, A.G., Kilbride, A.L., Green, L.E. and Boyle, L.A., 2013. Longitudinal study of the effect of rubber slat mats on locomotory ability, body, limb and claw lesions, and dirtiness of group housed sows. Journal of Animal Science 91: 3940-3954.

Carroll, C., 1999. Sow culling and parity profiles. In: Proceedings of the National Pig Farmers' Conference. October 18-20, 1999. Ireland. pp. 35-41.

Carroll, C., 2011. The economics of early culling. In: Proceedings of the National Pig Farmers' Conference. October 18-19, 2011. Ireland, pp. 7-12.

Castro, G. and Piva, J., 1999. Comparing and contrasting sow management in the US, Brazil and Chile. In: Proceedings of the Allen D. Leman Swine Conference. September 17, 1999. St. Paul, MN, USA, pp. 123-128.

Chagnon, M., D'Allaire, S. and Drolet, R., 1991. A prospective study of sow mortality in breding herds. Canadian Veterinary Journal 55: 180-184.

Challinor, C.M., Dams, G., Edwards, B. and Close, W.H., 1996. The effect of body composition of gilts at first mating on long-term sow productivity. Animal Science 62: 60.

Chapman, J.D., Thompson, L.H., Gaskins, C.T. and Tribble, L.F., 1978. Relationship of age at first farrowing and size of first litter to subsequent reproductive performance in sows. Journal of Animal Science 47: 780-787.

Christensen, G., Vraa-Andersen, L. and Mousing, J., 1995. Causes of mortality among sows in Danish pig herds. Veterinary Record 137: 395-399.

Cowin, A., 1978. Floors and feet. Farm Buildings Digest 13(4): 24.

Crenshaw, T.D., 2001. Calcium, phosphorus, vitamin D, and vitamin K in swine nutrition. In: Lewis, A.J. and Southern, L.L. (eds.) Swine Nutrition, 2nd edition. CRC press, Boca Raton, FL, USA, pp. 187-212.

Crenshaw, T.D., 2003. Nutritional manipulation of bone mineralization in developing gilts. In: Proceedings of the Allen D. Leman Swine Conference. September 13, 2003. St. Paul, MN, USA, pp. 183-189.

Cronin, G.M., Hemsworth, P.H., Winfield, C.G., Muller B. and Chamley, W.A., 1983. The incidence of and factors associated with, failure to mate by 245 days of age in the gilt. Animal Reproduction Science 5: 199-205.

Culbertson, M.S. and Mabry, J.W., 1995. Effect of age at first service on first parity and lifetime sow performance. Journal of Animal Science 73(1): 21.

D'Allaire, S. and Drolet, R., 1999. Culling and mortality in breeding animals. In: Leman, A.D., Straw, B.E., Mengeling, W.L., D'Allaire, S. and Taylor, D.J. (eds.) Diseases of Swine, 7th edition. Wolfe Publishing Ltd., Ames, IA, USA, pp. 1003-1016.

D'Allaire, S., Drolet, R. and Stalder, K., 2012. Longevity in breeding animals. In: Straw, B.E., Zimmerman, J.J., Karriker, L.A., Ramirez, A., Schwartz, K.J. and Stevenson, G.W. (eds.) Diseases of swine, 10th edition. John Wiley & Sons, Inc., Ames, IA, USA, pp. 50-59.

D'Allaire, S., Leman, A.D. and Drolet, R., 1992. Optimizing longevity in sows and boars. Veterinary Clinics of North America. Food Animal Practice 8: 545-557.

D'Allaire, S., Morris, R.S., Martin, F.B., Robinson, R.A. and Leman, A.D., 1989. Management and environmental factors associated with annual sow culling rate: a path analysis. Preventative Veterinary Medicine. 7: 255-265.

D'Allaire, S., Stein, T.E. and Leman, A.D., 1987. Culling patterns in selected Minnesota swine breeding herds. Canadian Journal of Veterinary Research 51: 506-512.

Dargon, J. and Aumaître, A., 1979. Sow culling: reasons for and effect in productivity. Livestock Production Science 6: 167-177.

Davies, P.R., Morrow, W.M., Jones, F.T., Deen, J., Fedorka-Cray, P.J. and Harris, I.T., 1997. Prevalence of Salmonella in finishing swine raised in different production systems in North Carolina, USA. Epidemiology and Infection 119(2): 237-244.

Deen, J., 2003. Sow longevity measurement. In: Proceedings of the Allen D. Leman Swine Conference. September 13, 2003. St. Paul, MN, USA, pp. 192-193.

Deen, J., 2009. Lameness as a welfare and productivity concern. Proceeding of the 28th centralia swine research update. January 28, 2009. Kirkton, Ontario, Canada, pp.11-12.

Deen, J. and Xue, J., 1999. Sow mortality in the US: an industry-wide perspective. In: Proceedings of Allen D. Leman Conference. September 17, 1999. St. Paul, MN, USA, pp. 91-94.

Den Hartog, L.A., Backus, G.B.C. and Vermeer, H.M., 1993. Evaluation of housing systems for sows. Journal of Animal Science 17: 1339-1344.

Dewey, C.E., Friendship, R.M. and Wilson, M.R., 1993. Clinical and postmorten examination of sows culled for lameness. Canadian Journal of Animal Science 34(9): 555-556.

Dial, G.D., Marsh, W.E., Polson, D.D. and Vaillancourt, J.P., 1992. Reproductive failure: differential diagnosis. In: Leman, A.D., Straw, B.E., Mengeling, W.L., D'Allaire, S. and Taylor, D.J. (eds.) Diseases of swine, 7th edition. Wolfe Publishing Ltd., Ames, IA, USA, pp. 88-137.

Dijkhuizen, A.A., Krabenborg, R.M.M. and Huirne, R.B.M., 1989. Sow replacement: a comparison of farmers' actual decisions and model recommendations. Livestock Production Science 23: 207-218.

Dourmad, J.Y., Etienne, M., Prunier, A. and Noblet, J., 1994. The effect of energy and protein intake of sows on their longevity: a review. Livestock Production Science 40: 87-97.

Engblom, L., Eliasson-Selling, L., Lundeheim, N., Belák, K., Andersson, K. and Dalin, A.M., 2008. Post mortem findings in sows and gilts euthanised or found dead in a large Swedish herd. Acta Veterinaria Scandinavica. 50: 25.

Engblom, L., Lundeheim, N., Dalin, A.M. and Andersson, K., 2007. Sow removal in Swedish commercial herds. Livestock Science 106: 76-86.

Engblom, L., Stalder, K. and Lundeheim, N., 2011. Premature removal and mortality of commercial sows. In: Book of abstracts of the 62nd annual meeting of the European Federation of Animal Science. Wageningen Academic Publishers, Wageningen, the Netherlands, p. 364.

English, P., 2002. Overview of the evaluation of stockmanship. In: Proceedings of the Symposium on Swine Housing and Well-Being. Des Moines, IA, USA, pp. 19-31.

Enokida, M., Sasaki, Y., Hoshino, Y., Saito, H. and Koketsu, Y., 2011. Claw lesions in lactating sows on commercial farms were associated with postural behaviour but not with suboptimal reproductive performance or culling risk. Livestock Science 136: 256-261.

Faust, M.A., Robison, O.W. and Tess, M.W., 1993. Genetic and economic analyses of sow replacement rates in the commercial tier of a hierarchical swine breeding structure. Journal of Animal Science 71: 1400-1406.

Faust, M.A., Tess, M.W. and Robison, O.W., 1992. A bioeconomic simulation for a hierarchical swine breeding structure. Journal of Animal Science 70: 1760-1774.

Fernández de Sevilla, X., Fábrega, E., Tibau, J. and Casellas, J., 2008. Effect of leg conformation on survivability of Duroc, Landrace, and Large White sows. Journal of Animal Science 86: 2392-2400.

Fernández de Sevilla, X., Fábrega, E., Tibau, J. and Casellas, J., 2009a. Competing risk analyses of longevity in Duroc sows with a special emphasis on leg conformation. Animal 3(3): 446-453.

Fernández de Sevilla, X., Fábrega, E., Tibau, J. and Casellas, J., 2009b. Genetic background and phenotypic characterization over two farrowings of leg conformation defects in Landrace and Large White sows. Journal of Animal Science 87: 1606-1612.

Fitzgerald, R.F., Stalder, K.J., Karriker, L.A., Sadler, L.J., Hill, H.T., Kaisand, J. and Johnson, A.K., 2012. The effect of hoof abnormalities on sow behavior and performance. Livestock Science 145: 230-238.

Ford, J.J. and Teague, H.S., 1978. Effect of floor space restriction on age at puberty in gilts and on performance of barrows and gilts. Journal of Animal Science 47: 828-832.

Friendship, R.M. and O'Sullivan, T.L., 2015. Sow health. Chapter 18. In: Farmer, C. (ed.) The gestating and lactating sow. Wageningen Academic Publishers, Wageningen, the Netherlands, pp. 409-421.

Friendship, R.M., Wilson, M.R., Almond, G.W., McMillian, R.R., Hacker, R.R., Pieper, R. and Swaminathan, S.S., 1986. Sow wastage: reasons for and effect on productivity. Canadian Journal of Veterinary Research 50: 205-208.

Geiger, J.O., Irwin, C. and Pretzer, S., 1999. Assessing sow mortality. In: Proceeding of the Allen D. Leman Swine Conference. September 17, 1999. St. Paul, MN, USA, pp. 84-87.

Gjein, H. and Larssen, R.B., 1995. The effect of claw lesions and claw infections on lameness in loose housing of pregnant sows. Acta Veterinaria Scandinavica 36: 451-459.

Grandjot, G., 2007. Claw problems cost money. SUS-Schweinezucht und Schweinemast. Landwirtschaftsverlag GmbH 5: 28-31.

Grégoire, J., Bergeron, R., D'Allaire, S., Meunier-Salaün, M.C. and Devilles, N., 2013. Assesment of lameness in sows using gait, footprints, postural behaviour and foot lesion analysis. Animal 8: 1-11.

Grindflek, E. and Sehested. E., 1996. Conformation and longevity in Norwegian pigs. In: Proceedings of the Nordiska Jordbruksforskares Forening Seminar 265-longevity of Sows. Research Centre Foulum, Tjele, Denmark.

Grøndalen, T., 1974. Leg weakness in pigs. I. Incidence and relationship to skeletal lesions, feeding level, protein, and mineral supply, exercise and exterioir conformation. Acta Veterinaria Scandinavica 15: 555-573.

Hagen, C.D., Lindemann, M.D. and Purser, K.W., 2000. Effect of dietary chromium tripicolinate on productivity of sows under commercial conditions. Journal of Swine Health and Production 8: 56-63.

Harris, M.J., Pajor, E.A., Sorrells, A.D., Eicher, S.D., Richert, B.T. and Marchant-Forde, J.N., 2006. Effect of stall or small group gestation housing on the production, health and behavior of gilts. Livestock Science 102: 171-179.

Heinonen, M., Oravainen, J., Orro, T., Seppa-Lassila, L., Ala-Kurikka, E., Virolainen, J., Tast, A. and Peltoniemi, O.A.T., 2006. Lameness and fertility of sows and gilts in randomly selected loose-housed herds in Finland. Veterinary Record 159: 383-387.

Hemsworth, P.H., Barnett, J.L., Hansen, C. and Winfield, C.G., 1986. Effects of social environment on welfare status and sexual behaviour of female pigs: II. Effects of space allowance. Applied Animal Behaviour Science 16: 259-267.

Hill, G.M. and Spears, J.W., 2001. Trace and ultratrace elements in swine nutrition. In: Lewis, A.J. and Southern, L.L. (eds.) Swine nutrition, 2nd edition. CRC press, Boca Raton, FL, USA, pp. 238-239.

Hill, M.A., 1992. Skeletal system and feet. In: Leman, A.D., Straw, B.E., Mengeling, W.L., D'Allaire, S. and Taylor, D.J. (eds.) Diseases of swine, 7th edition. Wolfe Publishing Ltd., Ames, IA, USA, pp. 163-195.

Hinterhofer, C., Ferguson, C., Apprich, V., Haider, H. and Stanek, C., 2006. Slatted floors and solid floors: stress and strain on the bovine hoof capsule analyzed in finite elements analysis. Journal of Dairy Science 89: 155-162.

Hoge, M.D. and Bates, R. O., 2011. Developmental factors that influence sow longevity. Journal of Animal Science 89: 1238-1245.

Holder, R.B., Lamberson, W.R., Bates, R.O. and Safranski, T.J., 1995. Lifetime productivity in gilts previously selected for decreased age at puberty. Animal Science 61: 115-121.

Hughes, P.E., Smits, R.J., Xie, Y. and Kirkwood, R.N., 2010. Relationships among gilt and sow live weight, P2 backfat depth, and culling rates. Journal of Swine Health and Production 18: 301-305.

Hughes, P.E. and Varley, M.A., 2003. Lifetime performance of the sow. In: Wiseman, J., Varley, M.A. and Kemp, B. (eds.) Perspectives in pig science. Nottingham University Press, Nottingham, UK, pp. 333-355.

Hurnik, J.F. and Lehman, H., 1985. A contribution to the assessment of animal well-being. In: Proceedings of the 2nd European Symposium on Poultry Welfare. June 1985. Celle, Germany, pp. 67-76.

Hurtgen, J.P. and Leman, A.D., 1980. Seasonal influence on the fertility of sows and gilts. Journal of the American Veterinary Medical Association 177: 631-635.

Jones, J.E.T., 1967. An investigation of the causes of mortality and morbidity in sows in a commercial herd. British Veterinary Journal 123: 327-339.

Jones, J.E.T., 1968. The cause of death in sows: a one year survey of 106 herds in Essex. British Veterinary Journal 124: 45-54.

Jørgensen, B., 2000a. Osteochondrosis/osteoarthrosis and claw disorders in sows, associated with leg weakness. Acta Veterinaria Scandinavica 41: 123-138.

Jørgensen, B., 2000b. Longevity of breeding sows in relation to leg weakness symptoms at six months of age. Acta Veterinaria Scandinavica 41: 105-121.

KilBride, A.L., Gillman, C.E. and Green, L.E., 2010. A cross-sectional study of prevalence and risk factors for foot lesions and abnormal posture in lactating sows on commercial farms in England. Animal Welfare 19: 473-480.

KilBride, A.L., Gillman, C.E., Ossent, P. and Green, L.E., 2009. Impact of flooring on the health and welfare of pigs. In Practice 31: 390-395.

King, V.L., Koketsu, Y., Reeves, D. Xue, J. and Dial, G.D., 1998. Management factors associated with swine breeding-herd productivity in the United States. Preventive Veterinary Medicine 35: 255-264.

Kirk, R.K., Svensmark, B., Ellegaard, L.P. and Jensen, H.E., 2005. Locomotive disorders associated with sow mortality in Danish pig herds. Journal of Veterinary Medicine Series A 52: 423-428.

Kitt, S.J., 2010. Feeding the high performing sow herd. In: Proceeding of the Allen D. Leman Swine Conference. September 18, 2010. St. Paul, MN, USA, pp. 141-144.

Knauer, M., Stalder, K.J., Serenius, T., Baas, T.J., Berger, P.J., Karriker, L., Goodwin, R.N., Johnson, R.K., Mabry, J.W., Miller, R.K., Robison, O.W. and Tokach. M.D., 2010. Factors associated with sow stayability in 6 genotypes. Journal of Animal Science 88: 3486-3492.

Knauer, M.T., Cassady, J.P., Newcom, D.W. and See, M.T., 2011. Phenotypic and genetic correlations between gilt estrus, puberty, growth, composition, and structural conformation traits with first-litter reproductive measures. Journal of Animal Science 89: 935-942.

Koketsu, Y, Takahashi, H. and Akachi, K., 1999. Longevity, lifetime pig production and productivity, and age at first conception in a cohort of gilts observed over six years on commercial farms. Journal of Veterinary Medical Science 61: 1001-1005.

Koketsu, Y., 2000. Retrospective analysis of trends and production factors associated with sow mortality on swine-breeding farms in USA. Preventive Veterinary Medicine 46: 249-256.

Koketsu, Y., 2003. Re-serviced females on commercial swine breeding farms. Journal of Veterinary Medical Science 65: 1287-1291.

Koketsu, Y., 2007. Longevity and efficiency associated with age structures of female pigs and herd management in commercial breeding herds. Journal of Animal Science 85: 1086-1091.

Koketsu, Y. and Dial, G.D., 1997. Factors influencing the postweaning reproductice performance of sows on commercial farms. Theriogenology 47: 1445-1461.

Koketsu, Y., Dial, G.D., Pettigrew, J.E. and King, V.L., 1996. The influence of nutrient intake on biological measures of breeding herd productivity. Journal of Swine Health and Production 4: 85-94.

Kornegay, E.T., Diggs, B.G., Hale, O.M., Handlin, D.L., Hitchcok, J.P. and Barezwski, R.A., 1984. Reproductive performance of sows fed elevated calcium and phosphorus levels during growth and development. A cooperative study. Report S-145 of the committee on nutritional systems for swine to increase reproductive efficiency. Journal of Animal Science 59(1): 253.

Kroneman, A., Vellenga, L., Van der Wilt, F.J. and Vermeer, H.M., 1993. Review of health problems in group-housed sows, with special emphasis on lameness. Veterinary Quarterly 15: 26-29.

Kuhlers, D.L., Jungst, S.B., Marple, D.N. and Rahe, C.H., 1985. The effect of pen density during rearing on subsequent reproductive performance in gilts. Journal of Animal Science 61: 1066-1069.

Le Cozler, Y., Dargon, J., Lindberg, J.E., Aumaître, A. and Dourmad, J.Y., 1998. Effect of age at first farrowing and herd management on long-term productivity of sows. Livestock Production Science 53: 135-142.

Loula, T.J., 2000. Increasing sow longevity: the role of people and management. In: Proceedings of the Allen D. Leman Swine Conference. August 11, 2000. St. Paul, MN, USA, pp. 139-142.

Love, R.J., Evans, G. and Klupiec, C., 1993. Seasonal effects on fertility in gilts and sows. Journal of Reproduction and Fertility 48: 191-206.

Lucia, T., Dial, G.D. and Marsh, W.E., 2000a. Lifetime reproductive and financial performance of female swine. Journal of the American Veterinary Medical Association 216: 1802-1809.

Lucia, T., Dial, G.D. and Marsh, W.E., 2000b. Lifetime reproductive performance in female pigs having distinct reasons for removal. Livestock Production Science 63: 213-222.

Luther, H., Schwörer, D. and Hofer, A., 2007. Heritabilities of osteochondral lesions and genetic correlations with production and exterior traits in station-tested pigs. Animal 1: 1105-1111.

MacLean, M., Willis, H., Monaghan, R. and Foxcroft, G., 2001. The effect of gilt age at puberty on lifetime reproductive performance. Advances in Pork Production 12: 30.

MacPherson, R.M., Hovell, F.D. and Jones, A.S., 1977. Performance of sows first mated at puberty, or second, or third oestrous, and carcass assesement of once bred gilts. Animal Production 24: 333-342.

Madec, F., 1984 Analyse des causes de mortalite des truies en cours de periode d'elevage. Recueil de Medecine Veterinaire 160: 329-335.

Mahan, D.C. and Newton, E.A., 1995. Effect of initial breeding weight on macro- and micromineral composition over a three-parity period using a high-producing sow genotype. Journal of Animal Science 73: 151-158.

Marchant, J.N. and Broom, D.M., 1996. Effects of dry sow housing conditions on muscle weight and bone strength. Animal Welfare 62: 105-113.

McDowell, L.R., 1992. Minerals in animal and human nutrition. Academic Press, San Diego, CA, USA, 524 pp.

McKee, C.I. and Dumelow, J., 1995. A review of the factors involved in developing effective non-slip floors for pigs. Journal of Agricultural Engineering Research 60(1): 35-42.

Mészáros, G., Pálos, J., Ducrocq, V. and Sölkner, J., 2010. Heritability of longevity in Large White and Landrace sows using continuous time and grouped data models. Genetics Selection Evolution 42: 13.

Ministry of Agriculture, Food and Fisheries (MAFF), 1981. Injuries caused by flooring: a survey in pig health scheme herds. In: Proceedings of the Pig Veterinary Society. Pig Journal 8: 119-125.

Morrison, B., Larriestra, A., Yan, J. and Deen, J., 2002. Determining optimal parity distribution with a push model of gilt supply. In: Proceedings of the Allen D. Leman Swine Conference. September 14, 2002. St. Paul, MN, USA, pp. 173-177.

Mote, B.E., Stalder, K.J. and Rothschild, M.F., 2008. Reproduction, culling and mortality levels on current commercial sow farms. Iowa State University Animal Industry Report 2008; ASL R2360.

Muirhead, M.R., 1976. Veterinary problems of intensive pig husbandry. Veterinary Record 99(15): 288-292.

Muirhead, M.R., 1983. Pig housing and environment. Veterinary Record 113: 587.

Muirhead, M.R. and Alexander, T.J.L., 1997. Managing pig health and the treatment of disease. 5M Enterprises Ltd., Sheffield, UK, 610 pp.

Nakano, T., Aherne, F.X. and Thompson, J.R. 1981. Leg weakness and osteochondrosis in pigs. Pig News and Information 2: 29-34.

National Research Council (NRC), 2012. Nutrient requirements of swine: eleventh revised edition. The National Academies Press, Washington, DC, USA.

Oldham, J.G., 1985. Clinical measurement of pain, distress and discomfort in pigs. In: Proceedings of the British Veterinary Association Animal Welfare Foundation 2nd Symposium. April 16, 1985. pp. 89-91.

Olsson, A.C. and Svendsen, J., 2002. Claw lesions in gestating sows: comparison of different flooring. In: Proceedings of the 17th International Pig Veterinary Society. June 2-5, 2002. Ames, IA, USA, pp. 343.

Parson, T.D., Johnstone, C. and Dial, G.D., 1990. On the economic significance of parity distribution in swine herds. In: Proceedings of the 11th International Pig Veterinary Society. July 1-5, 1990. Lausanne, Switzerland 11: 380.

Patiente, J.F. and Zijlstra, R.T., 2001. Sodium, potassium, chloride, magnesium, and sulfur in swine nutrition. In: Lewis, A.J. and Southern, L.L. (eds.) Swine Nutrition, 2nd edition. CRC Press, Boca Raton, FL, USA, pp. 213-227.

Patterson, J.L., Beltranena, E. and Foxcroft. G.R., 2010. The effect of gilt age at first estrus and breeding on third estrus on sow body weight changes and long-term reproductive Performance. Journal of Animal Science 88: 2500-2513.

Pedersen, P.N., 1996. Longevity and culling rates in the Danish sow production and the consequences of a different strategy of culling. In: Proceedings of the Nordiska Jordbruksforskares Forening Seminar 265-longevity of Sows. Research Centre Foulum, Tjele, Denmark, pp. 28-33.

Peet, B., 1983. Pig flooring facts. Farm Building Digest 18(1): 16-17.

Peltoniemi, O.A.T., Love, R.J., Heinonen, M., Tuovinen, V. and Saloniemi, H., 1999. Seasonal and management effects on fertility of the sow: a descriptive study. Animal Reproduction Science 55: 47-61.

PigCHAMP, 2012. Summary of the 2012 benchmarking data. PigCHAMP Inc., Ames, IA, USA. Available at: http://tinyurl.com/q7ft382.

Pijoan, C., 1986. Respiratory system. In: Leman, A.D., Straw, B., Glock, R.D., Mengeling, W.L., Penny, R.H.C. and Scholl, E. (eds.) Diseases of Swine, 6th edition. Iowa State University Press, Ames, IA, USA, pp. 152-162.

Pinilla, J.C. and Lecznieski, L., 2014. Sow removal and parity distribution management. Available at: http://tinyurl.com/ofgyv3w.

Plumlee, M.P., Thrasher, D.M., Beeson, W.M., Andrews, F.N. and Parker, H.E., 1956. The effects of a manganese deficiency upon the growth, development and reproduction of swine. Journal of Animal Science 15: 352.

Pluym, L., Van Nuffel, A., Dewulf, J., Cools, A., Vangroenweghe, F., Van Hoorebeke, S. and Maes, D., 2011. Prevalence and risk factors of claw lesions and lameness in pregnant sows in two types of group housing. Veterinary Medicine – Czech 56(3): 101-109.

Prunier, A., Messias de Braganca, M. and Le Dividich, J., 1997. Influence of high ambient temperature on performance of reproductive sows. Livestock Production Science 52: 123-133.

Richards, J.D., Zhao, J., Harrell, R.J., Atwell, C.A. and Dibner, J.J., 2010. Trace mineral nutrition in poultry and swine. Asian-Aust Journal of Animal Science 23(11): 1527-1534.

Rodríguez-Zas, S.L., Davis, C.B., Ellinger, P.N., Schnitkey, G.D., Romine, N.M., Connor, J.F., Knox, R.V. and Southey, B.R., 2006. Impact of biological and economic variables on optimal parity for replacement in swine breed-to-wean herds. Journal of Animal Science 84: 2555-2565.

Rodríguez-Zas, S.L., Southey, B.R., Knox, R.V., Connor, J.F., Lowe, J.F. and Roskamp, B.J., 2003. Bioeconomic evaluation of sow longevity and profitability. Journal of Animal Science 81: 2915-2922.

Rozeboom, D.W., Pettigrew, J.E., Moser, R.L., Cornelius, S.G. and El Kandelgy, S.M., 1996. Influence of gilt age and body composition at first breeding on sow reproductive performance and longevity. Journal of Animal Science 74: 138-150.

Saito, H., Sasaki, Y. and Koketsu, Y., 2011. Associations between age of gilts at first mating and lifetime performance or culling risk in commercial herds. Journal of Veterinary Medical Science 73: 555-559.

Sanz, M., Roberts, J., Almond, G., Alvarez, R., Donovan, T. and Perfumo, C., 2002. What we see with sow mortality. In: Proceedings of the Allen D. Leman Swine Conference. September 14, 2002. St. Paul, MN, USA, pp. 181-184.

Sanz, M., Roberts, J.D., Perfumo, C.J., Alvarez, R.M., Donovan, T. and Almond, G.W., 2007. Assessment of sow mortality in a large herd. Journal of Swine Health and Production 15(1): 30-36.

Sasaki, Y. and Koketsu, Y., 2008. Mortality, death interval, survivals, and herd factors for death in gilts and sows in commercial breeding herds. Journal of Animal Science 86: 3159-3165.

Sasaki, Y. and Koketsu, Y., 2011. Reproductive profile and lifetime efficiency of female pigs by culling reason in high-performing commercial breeding herds. Journal of Swine Health and Production 19: 284-291.

Schenck, E.L., McMunn, K.A., Rosenstein, D.S., Stroshine, R.L., Nielsen, B.D., Richert, B.T., Marchant-Forde, J.N. and Lay, Jr., D.C., 2008. Exercising stall-housed gestating gilts: effects on lameness, the musculo-skeletal system, production and behavior. Journal of Animal Science 86: 3166-3180.

Schukken, Y.H., Buurman, J., Huirne, R.B.M., Willemse, A.H., Vernooy, J.C.M., Van den Broek, J. and Verheijden, J.H.M., 1994. Evaluation of optimal age at first conception in gilts from data collected in commercial swine herds. Journal of Animal Science 72: 1387-1392.

Sehested, E., 1996. Economy of sow longevity. In: Proceedings of the Nordiska Jordbruksforskares Forening Seminar 265-Longevity of sows. Research Centre Foulum, Tjele, Denmark, pp. 101-108.

Serenius, T. and Stalder, K.J., 2004. Genetics of length of productive life and lifetime prolificacy in the Finnish Landrace and Large White pig populations. Journal of Animal Science 82: 3111-3117.

Serenius, T. and Stalder, K.J., 2006. Selection for sow longevity. Journal of Animal Science 84: E166-E171.

Serenius, T. and Stalder, K.J., 2007. Length of productive life of crossbred sows is affected by farm management, leg conformation, sow's own prolificacy, sow's origin parity and genetics. Animal 1: 745-750.

Serenius, T., Stalder, K.J. and Fernando, R.L., 2008. Genetic associations of sow longevity with age at first farrowing, number of piglets weaned, and wean to insemination interval in the Finnish Landrace swine population. Journal of Animal Science 86: 3324-3329.

Smith, W.J. and Robertson, A.M., 1971. Observations on injuries to sows confined in part slatted stalls. Veterinary Record 89: 531-533.

Smits, R.J., 2011. Impact of the sow on progeny productivity and herd feed efficiency. Recent Advances in Animal Nutrition-Australia 18: 61-67.

Spoolder, H.A.M., Geudeke, M.J., Van der Peet-Schwering, C.M.C. and Soede, N.M., 2009. Group housing of sows in early pregnancy: a review of success and risk factors. Livestock Science 125: 1-14.

Spoolder, H.A.M. and Vermeer, H.M., 2015. Gestation group housing of sows. Chapter 3. In: Farmer, C. (ed.) The gestating and lactating sow. Wageningen Academic Publishers, Wageningen, the Netherlands, pp. 47-71.

Stalder, K., Morrison, B. and Baas, T., 2007b. Non-genetic factors influencing sow longevity. Available at: http://tinyurl.com/pjo2ju3.

Stalder, K.J., Karriker, L.V. and Johnson, A.K., 2007a. The impact of gestation housing on sow longevity. In: Proceedings of the Sow Housing Forum. June 6, 2007. Des Moines, IA, USA.

Stalder, K.J., Knauer, M, Baas, T.J., Rothschild, M.F. and Mabry, J.W., 2004. Sow longevity. Pig News and Information 25: 53N-74N.

Stalder, K.J., Lacy, R.C., Cross, T.L. and Conatser, G.E., 2003. Financial impact of average parity of culled females in a breed-to-wean swine operation using replacement gilt net present value analysis. Journal of Swine Health and Production 11: 69-74.

Stalder, K.J., Lacy, R.C., Cross, T.L., Conaster, G.E. and Darroch, C.S., 2000. Net present value of sow longevity and the economic sensitivity of net present value to changes in production, market price, feed cost, and replacement gilt costs in a farrow-to-finish operation. The Professional Animal Scientist 16: 33-40.

Stein, T.E., Dijkhuizen, A.A., D'Allaire, S. and Morris, R.S., 1990. Sow culling and mortality in commercial swine breeding herds. Preventive Veterinary Medicine 9: 85-94.

Stone, M.W., 1981. Sow culling survey in Alberta. Canadian Veterinary Journal 22: 363.

Straw, B., 1984. Causes and control of sow losses. Modern Veterinary Practice 65: 349-353.

Tarrés, J., Bidanel, J.P., Hofer, A. and Ducrocq, V., 2006a. Analysis of longevity and exterior traits on Large White sows in Switzerland. Journal of Animal Science 84: 2914-2924.

Tarrés, J., Tibau, J., Piedrafita, J., Fàbrega, E. and Reixach, J., 2006b. Factors affecting longevity in maternal Duroc swine lines. Livestock Science 100: 121-131.

Thacker, B.J., 2002. Seasonal infertility: management practices – causes and fixes. In: Proceedings of the Swine Disease Conference for Swine Practitioners. Iowa State University, Ames, IA, USA, pp. 123-130.

Tiranti, K., Hanson, J., Deen, J. and Morrison, B., 2003. Description of removal patterns in a selected sample of sow herds. In: Proceedings of the Allen D. Leman Swine Conference. September 13, 2003. St. Paul, MN, USA, pp. 194-198.

Tiranti, K.I. and Morrison, R.B., 2006. Association between limb conformation and retention of sows through the second parity. American Journal of Veterinary Research 67(3): 505-509.

Tummaruk, P., Lundeheim, N., Einarsson, S. and Dalin, A. M., 2001. Effect of birth litter size, birth parity number, growth rate, backfat thickness and age at first mating of gilts on their reproductive performance as sows. Animal Reproduction Science 66: 225-237.

Tuyttens, F.A.M., 2005. The importance of straw for pig and cattle welfare: a review. Applied Animal Behaviour Science 92: 261-282.

Underwood, E.J. and Suttle, N.F., 1999. The mineral nutrition of livestock, 3rd edition. CABI Publishing, New York, NY, USA, 614 pp.

United States Department of Agriculture (USDA), 2001. Part I: reference of swine health and management in the United States, 2000. National Animal Health Monitoring System. USDA:APHIS, Fort Collins, CO, USA.

Van Riet, M.M.J., Millet, S., Aluwé, M. and Janssens, G.P.J., 2013. Impact of nutrition on lameness and claw health in sows. Livestock Science 156: 24-35.

Von Wachenfelt, H., Pinzke, S., Nilsson, C., Olsson, O. and Ehlorsson, C.J., 2008. Gait analysis of unprovoked pig gait on clean and fouled concrete surfaces. Biosystems Engineering 101: 376-382.

Webb, N.G. and Nilsson, C., 1983. Flooring and injury -an overview. In: Baxter, S.H., Baxter, M.R. and MacCormack, J.A.C. (eds.) Farm animal housing and welfare. Martinus Nijhoff Publishers, Boston, MA, USA, pp. 226-259.

Whitney, M.H. and Masker, C., 2010. Replacement gilt and boar nutrient recommendations and feed management. In: Meisinger, D. (ed.) National swine nutrition guide. U.S. Pork Center of Excellence, Des Moines, IA, USA, pp. 97-107.

Wrathall, A.E., Simmons, H.A., Bowles, D.J. and Jones, S., 2003. Biosecurity strategies for conserving valuable livestock genetic resources. Reproduction Fertility and Development 16(2): 103-112.

Yang, H., Eastham, P.R., Phillips, P. and Whittemore, C.T., 1989. Reproductive performance, body weight and body condition of breeding sows with differing body fatness at parturition, differing nutrition during lactation and differing litter size. Animal Production 48: 181-201.

Yazdi, M.H., Lundeheim, N., Rydhmer, L., Ringmar-Cederberg, E. and Johansson, K., 2000a. Survival of Swedish Landrace and Yorkshire sows in relation to osteochondrosis: a genetic study. Animal Science 71: 1-9.

Yazdi, M.H., Rydhmer, L., Ringmar-Cederberg, E., Lundeheim, N. and Johansson, K., 2000b. Genetic study of longevity in Swedish Landrace sows. Livestock Production Science 63: 255-264.

Young, L.G. and King, G.J. 1981. Reproductive performance of gilts bred on first versus third estrus. Journal of Animal Science 53: 19-25.

Young, M. and Aherne, F., 2005. Monitoring and maintaining sow condition. Advances in Pork Production 16: 299-313.

Young, M.G., Tokach, M.D., Aherne, F.X., Dritz, S.S., Goodband, R.D. Nelssen, J.L. and Loughin, T.M., 2008. Effect of space allowance during rearing and selection criteria of performance of gilts over three parities in a commercial swine production system. Journal of Animal Science 86: 3181-3193.

Zoric, M., Nilsson, E., Mattsson, S., Lundeheim, N. and Wallgren, P., 2008. Abrasions and lameness in piglets born in different farrowing systems with different types of floor. Acta Veterinaria Scandinavica 50: 37.

Zoric, M., Sjölund, M., Persson, M., Nilsson, E., Lundeheim, N. and Wallgren, P., 2004. Lameness in piglets. Abrasions in nursing piglets and transfer of protection towards infections with Streptococci from sow to offspring. Journal of Veterinary Medicine B, Infectious Diseases and Veterinary Public Health 51: 278-284.